FUNDAMENTALS OF SYSTEMS ANALYSIS
with Application Design

Programming books from boyd & fraser

Structuring Programs in Microsoft BASIC
BASIC Fundamentals and Style
Applesoft BASIC Fundamentals and Style
Complete BASIC: For the Short Course
Fundamentals of Structured COBOL
Advanced Structured COBOL: Batch and Interactive
Comprehensive Structured COBOL
Pascal
WATFIV-S Fundamentals and Style
VAX Fortran
Fortran 77 Fundamentals and Style
Learning Computer Programming: Structured Logic, Algorithms, and Flowcharting

Also available from boyd & fraser

Database Systems: Management and Design
Using Pascal: An Introduction to Computer Science I
Using Modula-2: An Introduction to Computer Science I
Data Abstraction and Structures: An Introduction to Computer Science II
Fundamentals of Systems Analysis with Application Design
Data Communications for Business
Data Communications Software Design
Microcomputer Applications: Using Small Systems Software
The Art of Using Computers
Using Microcomputers: A Hands-On Introduction

Shelly, Cashman and Forsythe books from boyd & fraser

Computer Fundamentals with Application Software
Workbook and Study Guide to accompany Computer Fundamentals with Application Software
Learning to Use SUPERCALC®3, dBase III®, and WORDSTAR® 3.3: An Introduction
Learning to Use SUPERCALC®3: An Introduction
Learning to Use dBASE III®: An Introduction
Learning to Use WORDSTAR® 3.3: An Introduction
BASIC Programming for the IBM Personal Computer
Structured COBOL — Flowchart Edition
Structured COBOL — Pseudocode Edition
Introduction to Turbo Pascal Programming

FUNDAMENTALS OF SYSTEMS ANALYSIS
with Application Design

PAUL S. LICKER
University of Calgary

BOYD & FRASER PUBLISHING COMPANY
BOSTON

CREDITS:

Editor: Tom Walker
Director of Production: Becky Herrington
Ancillaries Editor: Donna Villanucci
Manufacturing Director: Erek Smith

Art Design and Execution: Ken Russo
Cover Design: Becky Herrington
Cover Art: M.C. Escher
        Courtesy of the Vorpal Gallery: San Francisco & New York City

Manufactured in the United States of America

Library of Congress Cataloging-in-Publication Data

Licker, Paul S. (Paul Steven), 1944-
   Fundamentals of systems analysis with application design.

   Includes index.
   1. System design. 2. System analysis. 3. Electronic data processing--Structured techniques. I. Title.
QA76.9.S88L53 1987   003   86-28373
ISBN 0-87835-224-4

10 9 8 7 6 5 4 3 2 1

Dedication

To
Charlotte and Jerry
and Helen and Bernard

CONTENTS

Chapter 4: THE ANALYSIS OF SYSTEMS ANALYSIS

Chapter 5: INVESTIGATING: RESEARCHING THE SYSTEM

Chapter 6: CREATING SYSTEM MODELS

Chapter 7: INTERPRETING: DESIGNING SYSTEMS AND THEIR ELEMENTS

Chapter 8: PRODUCING PRELIMINARY INVESTIGATION REPORTS

Chapter 9: WRITING PROCEDURES AND SYSTEM CHARTING

Chapter 10: PRESENTING LOGICAL SYSTEM SPECIFICATIONS

Chapter 11: COLLECTING AND ORGANIZING SYSTEM EXPERIENCE DATA: PART I

Chapter 12: COLLECTING AND ORGANIZING SYSTEM EXPERIENCE DATA: PART II

Chapter 13: PROCESS, DATA, AND MATERIALS FLOW DIAGRAMMING

Chapter 14: COSTING AND DECISION MAKING IN SYSTEMS ANALYSIS

Chapter 15: PROJECT MANAGEMENT: PLANNING, CONTROL, AND REPORTING

Chapter 16: SPECIAL PROJECT TYPES

Chapter 17: SPECIAL PROJECT CONTENT

Appendix A: AN EXAMPLE OF A USERS' GUIDE

Appendix B: AN EXAMPLE OF A OPERATOR'S GUIDE

Appendix C: AN EXAMPLE OF A TECHNICAL GUIDE

Appendix D: CASE STUDIES

Glossary

Bibliography

Index

PREFACE

Systems analysis, more than any other discipline in information resources management, has been a battleground of ideologies over the past two decades. Unlike programming and very much unlike hardware development, systems analysis sits "between two chairs," often uncomfortably so, between users and computer technology. In the past, textbooks have reflected this ambivalent position and have stressed technical skills and computers.

It is apparent to many that a new age of information systems development is beginning. In this new age, users, not technologists, will drive systems development; applications, not computer uses, will motivate projects; end-user needs, not analysts' predispositions and computer marketing plans, will determine what gets built and how. It is for this new age that this textbook has been crafted.

DISTINGUISHING FEATURES

Fundamentals of Systems Analysis with Applications Design departs from traditional textbooks in a number of ways important to both instructors and students:

Emphasis on User

The emphasis is on the user and **user application needs**; technology is an important player, but computers, *per se*, are never analyzed in detail.

Structured Systems Analysis

The book was developed using the tools of structured systems analysis and reflects the goals of information resource maintenance.

dBASE III Exercises Throughout

A popular microcomputer database management package, dBASE III™, is used in end-of-chapter exercises. These exercises serve three purposes: (1) to familiarize the student with a useful system development tool that can help analysts involve users, (2) to emphasize and exercise skills and concepts developed in the chapter, and (3) to use a single live case throughout the book.

Alternative System Development Methodologies Stressed

There are important alternatives to large DP-driven projects such as prototyping, end-user software, application generation and expert systems. These system development methodologies are stressed in the book.

Readable to the Non-Technologist

The author and editors have attempted to produce a book which can be profitably read by general management and business students alike. We have tried to make the book readable, interesting, and relevant to the non-technologist.

User-Driven Methodologies Emphasized

Because many business students who take a course using this text will *not* become systems analysts, it is important to keep the work of the analyst in the context of user needs. The text stresses user-driven methodologies such as **prototyping** and **end-user software**. While Chapter 15 and parts of Chapters 5, 8, and 10 look at traditional project management, most of the book attempts to see analysts as individuals who, like contractors, have customers to satisfy. Customer needs, not the prejudices and dispositions of technology-hounded analysts, come first; that is the lesson North American business is learning painfully. "It's a nice technology...what can we do with it?" will have to be replaced with "It's a nice application, how can we build it?" Complementing this is a lesson that users and systems analysts' customers have never had to learn: analysts are also human beings who have needs, pride, prejudice, background, deadlines, and aspirations. Underlying all of this is a new cooperative business spirit that increases the productivity, relevance, and satisfaction of workers, managers, and customers.

"Why" and "How" Questions Asked

To underscore the fact that this same cooperative business spirit extends to the classroom, this book is written in a non-threatening style. On the other hand, the book is also written in a high-minded spirit. Systems analysis is not all nuts and bolts; there is theory. There are reasons why things are the way they are, and there are forces working to change things. This means that students will be asked to think about why as much as how. Where terms do not exist, even in the current literature, new ones have been manufactured. I have not been hesitant to borrow ideas and metaphors from other fields to illustrate points.

Strong Emphasis on General Systems Theory

One final word about philosophy. The book leans heavily on concepts from **General Systems Theory (GST)**. GST is really the theoretical underpinning of all systems sciences, including systems analysis. I've tried to develop a General Information Systems Theory (GIST) from these concepts in a business context. I hope that both instructor and student get the GIST of the book's message.

Exercises and Discussion Questions

Most chapters of the book are augmented with three kinds of exercises. First, every chapter has a series of **discussion questions**. Many of these questions reflect hard issues in systems analysis and more generally, in the field of business data processing. Some of the questions have no ready answer, but should be asked of students and instructors. Too often systems analysts accept their own work uncritically while discouraging or leaving unanswered their clients' questions. The discussion questions included in this book probe the assumptions and golden calves of the field. They also exercise the skills students acquire through reading the material.

Short Cases and dBASE III Exercises

Most chapters have a **short case** for discussion to put the concepts in the text into context. Chapters 1 through 14 are also complemented by a series of **practical exercises** in dBASE III.

Free Case Materials and dBASE III Diskettes

An educational version of Ashton-Tate's popular dBASE III database manager is available free from Boyd & Fraser to text adopters. An additional Case Materials diskette includes a set of dBASE III-written programs and data files for the exercises. The programs and data illustrate a case ("HAIRLOOM STYLISTS") in which a small but aggressive hair-styling boutique has purchased a poorly functioning microcomputer-based information system. Students must analyze, understand, and represent the system, and interpret their findings through fourteen extended exercises. By providing a live (well, kicking, anyway!) system the text gives students a palpable context for the material in the book. As a bonus, students learn the kernel of dBASE III, a skill easily transferred to other database management packages.

Glossary

A glossary is also included at the end of the book for quick reference to key terms. All terms are printed in **bold face** when they first appear in the text.

ANCILLARY MATERIALS

A comprehensive instructors' support package accompanies *Fundamentals of Systems Analysis with Application Design*. These ancillaries are available to instructors upon request from Boyd & Fraser Publishing Company.

Instructor's Manual

Material in the Instructor's Manual follows the organization of the text. Chapters of the Instructor's Manual include:

- Lecture Outline (keyed to Transparency Masters)
- Chapter Objectives
- Vocabulary List
- Summary
- Rationale
- Teaching Notes
- Objective Questions
- Discussion Questions
- Answers to Objective Questions
- Over 150 Transparency Masters

ORGANIZATION OF THIS BOOK

The organization of the book is **hierarchical**, rather than traditionally linear. More technical details on material in each section appear in the chapters of the following sections. In this way the text is structured much like a well-functioning information resource. The book is divided into five sections:

Context of Systems Analysis

The first four chapters introduce the **context** of systems analysis including its theoretical bases (Chapter 1), its roles and functions (Chapter 2), its workers (Chapter 3), and an analysis of systems analysis itself (Chapter 4).

Processes of Systems Analysis

The three major **processes** of systems analysis are discussed, including investigating for knowledge (Chapter 5), representing knowledge as models (Chapter 6), and interpreting models into designs (Chapter 7).

Products of Systems Analysis

Chapters 8 through 10 examine the **products** of systems analysis: preliminary investigation reports (Chapter 8), procedures (Chapter 9), and specifications (Chapter 10).

Tools of Systems Analysis

The **tools** of systems analysis are examined including investigation techniques (Chapters 11 and 12), charting and diagramming (Chapter 13), and economic analysis for decision-making (Chapter 14).

Management of Systems Analysis

The last three chapters explore the **management** of systems analysis: project management (Chapter 15), alternative management techniques (Chapter 16), and special project content (Chapter 17).

USING THIS BOOK

This text is intended to be used in any of three ways. It is ideal for a thirteen- or seventeen-week course in systems analysis, looking at one chapter each week with exercises. It may be used in a full-year course in a data processing curriculum by spending more time on analytic, investigative, and reporting skills. In this case the three exercises provided at the end of the instructor's manual can be used as a six-week system design exercise, using the educational version of dBASE III distributed with the instructor's manual. Finally, the book can be used as it is titled: systems analysis plus application design in a two-semester sequence. I suggest that the text be augmented with exercises in application areas and live projects, to take advantage of the full two-semester sequence.

This sequence would not necessarily be intended for only DP, MIS, or CIS students since it would stress business applications. The goal of the two-semester sequence would be to produce students who would be informed, critical customers for systems analysts.

ACKNOWLEDGEMENTS

This text profitted from the helpful suggestions of a number of reviewers. Thanks go to Henry Etlinger (Rochester Institute of Technology), Peter Gingo (University of Akron), James Wright (Chadron College), Murray Berkowitz (Institute for Defense Analysis), Darell Gobel (Catonsville Community College), Michael Michaelson (Palomar College), Paul Ross (Millersville University), James Cox (Lane Community College), Christine Kay (DeVry University), Hooshang Beheshti (Roanoke University), and Darleen Pigford (Western Illinois University).

Very special thanks are offered to the staff at Boyd & Fraser in Boston and Brea, CA. Tom Walker showed a lot of faith in the project from the start; he worked hard to make the book more valuable to instructors. Donna Villanucci worked tirelessly and seemingly continuously to bring the project together. Pat Donegan and Peter Gordon worked the text over several times to make it into English and bore the brunt of working through the end-of-chapter exercises like a trooper. A number of editors provided guidance at all stages. Becky Herrington managed the production of the document and Ken Russo and Susan Reese produced the beautiful artwork that graces the pages.

Thanks to several of my colleagues, particularly Ron Murch for providing critique in the early stages and Tom Janz, with whom I've worked on the cultural aspects of systems analysis. Klaus Krippendorff of the University of Pennsylvania, introduced me to cybernetics and GST and provided the initial motivation for this book by making me aware of what "system" means. Praise, too, to my students over the years who have tried out all the exercises, read all these pages many times, and fed back their opinions in true cybernetic fashion. They were the pioneers and deserve applause for discovering and "settling" the territory that is this book.

To my wife Marilyn and my children, thanks for "released time" to write. It takes a long time to create a text and Marilyn knows what a "book widow" is. Her faith that the book could be written motivated me through the several years it takes to bring ideas from neuron to cellulose. She understands systems like no one else.

Calgary, Alberta
February 1987

Paul S. Licker

CHAPTER 1

SYSTEMS AND SYSTEMS DYNAMICS: BIRTH, GROWTH, MAINTENANCE

OBJECTIVES

1. Describe the components of a system's definition: elements, relationships, and goals
2. Describe specific system modes of functioning and the specific special units that arise in those modes
3. Distinguish three types of system environments with respect to predictability.
4. Distinguish and describe three system control architectures and the role feedback plays in each.
5. Describe and apply structural-functional design as both functional analysis and goal analysis.
6. Describe the components of a generic information system.
7. Detail and use the elements of a data flow diagram to produce and discuss several common morphs in information systems.
8. List seven steps in the System Development Life Cycle and describe each.

SYSTEMS AND SYSTEMS DYNAMICS: BIRTH, GROWTH, MAINTENANCE

1.0 COMIN' AT YOU FROM ALL SIDES

Ben and Iris are managers of a retail rental firm that specializes in commercial cleaning machines and supplies. Ben handles the retail outlets and customer sales; Iris takes care of the internal function of the business, orders supplies, and keeps track of the changing technology in cleaning equipment. Every Monday afternoon, Ben, Iris, and their general manager, Phoebe deWitt, meet to go over the week's plans. Usually the business has few peaks and valleys, but there are occasional runs on equipment. Sometimes it's lost or damaged and customers argue over responsibility; the equipment is still on the books but it can't be rented out. Phoebe makes some decisions during the Monday meetings. On other occasions, however, customers may place large reservations some time in advance; Phoebe's approval on these orders can be given only after examining the entire season's planned needs, the customer's past usage and the condition of the equipment, and inventory levels.

This Monday afternoon Cameo Cleaners, a small but aggressive establishment, has just received a major contract for cleaning several large downtown office towers. They would like four months' rental of several major pieces of equipment that Ben and Iris have a lot of call for. Phoebe wonders if having Cameo's account is as important as maintaining those items for regular customers.

"One thing's certain." Iris says, "Without those items, Tom's Quicklean isn't going to buy supplies from us. And probably Wilma's not going to either since she's Tom's sister-in-law."

"Anyway, Cameo could go buy their own. They've got two-year contracts with LeasCo," Ben points out.

"Not so, Ben," Phoebe replies. "Cameo hasn't the cash to do it. They need the equipment and we could get a good price on it for a four-month period, maybe longer. We should go with Cameo. They're money in the bank at no risk to us."

After the meeting, Ben isn't so sure he can face Tom Quiggly of Quicklean. Tom has rented equipment on an as-needed basis for years. Is Phoebe right? What do their customers want, need, or plan? Has anyone ever known, really? Is it a mistake to change policy from short to long-term rentals? The questions come at Ben from all sides. The only sure thing is that answers will come only after the equipment has left the shop.

Ben, Phoebe, and Iris are part of a system, a set of interrelated parts which function toward a goal. Their system is a particular type, a business. And it is a particular type of business system: a retail rental service. This book is about systems like this one: how they grow, maintain themselves, and progress toward their goals, what kinds of events they cope with, and how they manage that coping.

Chapter 1 examines systems as living organisms. As systems grow, they change in response to their own needs and to pressures from their environments. Certain systems function better; others, less well. Information plays a role in systems, which develop strategies for survival in unpredictable environments. Some systems feature information as a major product or component. We will examine those systems in detail in later chapters.

This chapter introduces concepts from General Systems Theory (GST) to show how all systems function. Systems consist of elements interrelated for a purpose. A system contends with an unpredictable environment through the development of specialized units, some of which control others. Information plays a major part in coping with unpredictability in the environment. System architectures develop complex decision-making mechanisms to select optimal actions in response to environments and the availability of resources within an environment.

Information systems are particular kinds of systems that "run" on information. A data flow diagram represents data relationships. Several laws governing the flow of information will be introduced and these laws assist analysts in understanding and designing information systems. The system life cycle of information systems (the System Development Life Cycle) contains a phase involving logical design in the life of an information system.

The chapters in this first section introduce you to systems analysts and their craft. Chapter 2 discusses the organizational milieu of systems analysis, while chapter 3 details the working life of the systems analyst. Chapter 4 analyzes systems analysis itself and prepares for later sections that discuss the products, procedures, and tools of systems analysis.

1.1   SYSTEMS

Ben, Phoebe, and Iris are parts of a particular kind of system, called Commercial Cleaning Rentals. Why is the study of systems important? Business systems like Commercial Cleaning Rentals work because of **information**. Information is the grease that lubricates the wheels of systems. Business systems require special kinds of information. Not only is there **business-oriented information** (profit and loss statements, sales levels, tax rates, discount rates, inventory levels) but also information about cleaners, customers and competitors to be managed. We call the system that supplies and manages information an **information resource**. Systems analysis is about the relationship between how systems function and their information resources.

Consider the simple system in Figure 1-1. Removing one of the middle blocks will cause the whole stack to topple. But why? How does the third block down "know" that all of a sudden the entire stack is unstable and must fall? Can instability be predicted and

controlled? When does that potential for instability become important? These are questions about a physical system and its relationship to the information resource that is "about it."

Related to every business system is another system that manages information about the business. Before we examine that relationship, we will discuss systems in general and what is known about how they operate. This area of knowledge is called General Systems Theory (GST). Later we will turn our attention first to information systems, then to the relationships, and finally to the job of maintaining that relationship (called "Information Resource Maintenance" or IRMaint).

FIGURE 1–1. An Unstable System

A **system** is defined as a set of **elements** that are related and that, through this set of **relationships**, aim to accomplish **goals** (Figure 1-2).

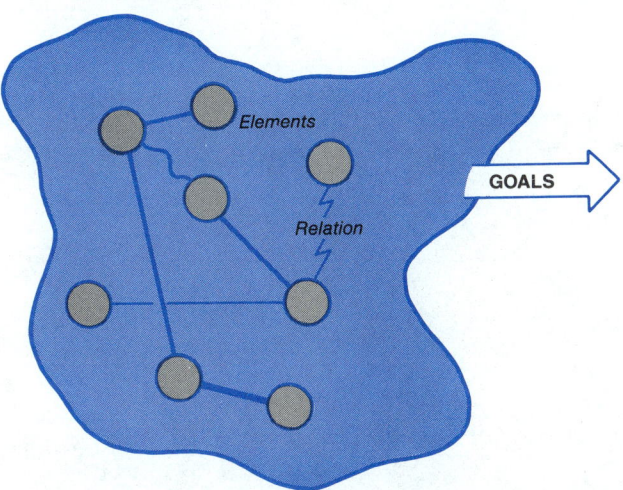

FIGURE 1–2. A System is Defined By Its Elements, Relations, and Goals

A system without a goal is merely a set of related elements. A set of unrelated elements will not do anything.

Relations among the elements limit the elements' behaviors. Consider children at a playground. Some run and jump; others cooperate in a game; still others merely stand or sit. Is there a system here? Yes, there are many.

We can isolate the systems by noting the restrictions that seem to exist on the behaviors of the elements. The game, for instance, places restrictions on those children playing it. Only certain moves are allowed at certain times. The game has a goal, to win. There is an essential binding together of elements, rules, and goals that we call "a game."

What if we abolished those relationships? Consider a team of students trying to build a bridge made of dried pasta (a yearly endeavor at the Engineering school at my university). It is highly unlikely that any bridge the students might come up with would stand under any load if they do not communicate with one another. The communication may be implicit, as when the students have all taken courses in structural design and are aware of structural design principles. In that case, their behavior is implicitly coordinated by their previous learning.

We require that a system have elements that relate to or communicate with one another to achieve a goal. We define systems by listing the elements, relationships, and goals. Discovering what those elements, relationships, and goals are, however, is not easy; in some circumstances it is impossible. This is so because *the nature of a system is, at least partially, a function of the observer of the system.* We call that the **Uncertainty Principle.** What it means is that in discussing a particular system, we can refer only to our observations, which are at least partly a product of ourselves rather than the system being observed.

This concern is well known in law and also goes by the term "objectivity." That is why we always ask for corroboration of a witness or replication of experimental results. In law and physics, we recognize that the observer is very much a part of the system report. Figure 1-3 is obviously a photograph of a bridge. Most readers could discuss its

COURTESY OF JEFF ALBERTSON/STOCK BOSTON

FIGURE 1–3. A Photo of a Bridge

purpose, function, and structure, but the bridge is somehow "different" for bridge engineers, automobile passengers, and drivers. Figure 1-4 on the opposite page illustrates this even further through a visual joke. The observation of a corner seems meaningful, but overall the drawing makes no sense. Remember this drawing when you report on something. Reports are only products of our limited visions, not the truth of what actually exists.

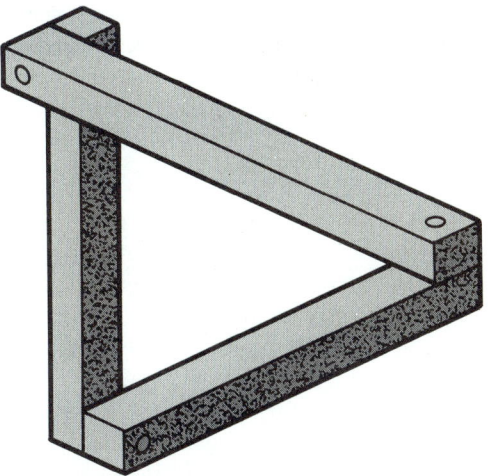

FIGURE 1–4. An Artist's Claim

Because of this we try to report as completely and objectively as possible, making clear our **observation framework**. For systems analysts, this means making clear what techniques are used for investigation. Senn's 1978 article, "A Management View of Systems Analysts: Failures and Shortcomings," points out that systems analysis has an inherent bias toward opportunities for automation, that is, seeing manual, noncomputerized systems as needing improvement. By reporting our biases as well as our data, we complete the valid observation as well as we can.

Given the restrictions that the Uncertainty Principle compels, a system is described in terms of its elements, relationships, and goals.

Elements

System elements are describable objects or events. Often the "elements" themselves are also systems. A classroom may be thought of as a system of students, a professor, desks, books, and so forth. Students and professors are themselves complex systems. When an element is also a system, we refer to the element as a **subsystem** of the larger system. Subsystems in turn may be composed of more subsystems. Where does this end?

Because our observations are at least partly functions of ourselves, it is practical to terminate the decomposition of systems into sub(sub-sub-...) systems when the observer cannot decompose them any further. We end up with systems that cannot be peeked into, called **black boxes**. We are all familiar with black boxes. As children, we were all curious about how toys, mechanisms, or our parents' watches worked. When we chose not to treat them as black boxes and attempted to open them up by looking inside, we usually destroyed them. Black boxes are essential building blocks of systems, or system "atoms." It is a curious fact that at the basis of each system lies a pool of ignorance we recognize and choose not to clarify.

The technique of decomposing systems into elements is called **functional decomposition** and is the basic tool of systems analysis; in fact, it is the "analysis" aspect. There are two procedures for functional decomposition. Goal analysis takes systems apart based on decomposition of their goals; behavioral analysis tries to isolate minimal units of elements related by their behavior. Both are discussed in chapter 4.

Relationships

Elements (that is, black boxes) within systems are related in a variety of ways. Some relationships are physical, others temporal (related in time), some logical (one is necessary for another), others transactional. Observation of these relationships is subject to the Uncertainty Principle and may depend strongly on the observer's background. An economist will look for monetary relationships; a marketer, for purchasing relationships; and an information systems specialist, for an exchange of information.

The relationships contribute to a system's goals by limiting the behaviors of the elements to those that do make a contribution. Societies are built on the premise that certain behaviors are allowed and others forbidden. A firm's procedures manual serves the same purpose. Sometimes relationships are explicitly spelled out in rule books or procedure manuals. At other times, they are implicit or unspoken, but very much in evidence. Consider how people dress and act in a business office — how often are the rules spelled out?

Goals

The curious aspect of system goals is that the Uncertainty Principle works backward here. It is impossible to have direct knowledge of a system's goals from "inside." Those within a system have no direct knowledge of its goals, which can be known directly only from the outside. That is, the elements of a system, conscious and intelligent, can at best only make relatively short-term inferences about why their behavior is being limited. Those outside the system can observe the interaction of the system with its environment and understand why the elements behave as they do (Figure 1-5).

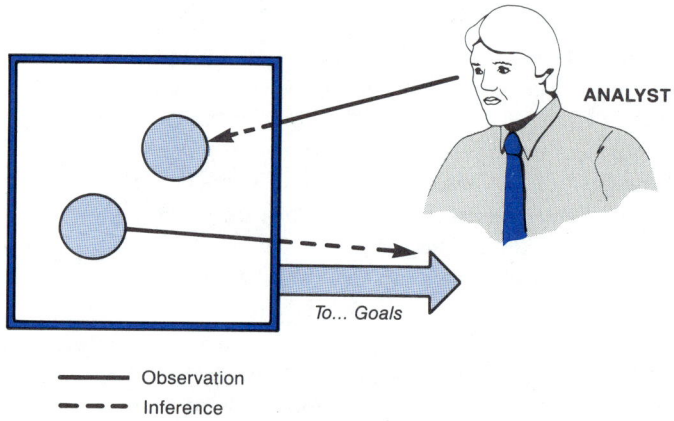

ANALYST

To... Goals

——— Observation

- - - Inference

FIGURE 1–5. The Uncertainty Principle at Work

Because goals are not known directly to the elements of a system, statements by system elements that "Our goal is X" or "We do this because of Y" are best treated as hypotheses or theories. This is not to deny that the company president is well intentioned when he says, "We will increase our market share by 20% over the next year." Those within the system the President administers have at best only tentative knowledge about actual goals by the time they are translated into directives at lower levels. In many firms, workers are thoroughly insulated from company interaction with customers through layers of bureaucracy. No direct relationship can be discovered by any worker. One of the strongest messages of *In Search of Excellence* (Peters and Waterman, 1982) is that knowing your customers (read here "observing your goals being met") is a primary determinant of excellence or even modest success. Low morale and poor performance result when workers have little conception of or stake in company goals and have to hypothesize why they are told to work as they do. Systems analysts, on the other hand, have a mandate to get to the system's goals from the outside. Their analysis is seen as more informed and objective simply because of their "neutral" and "outside" viewpoint.

Recap

Systems are identified by their elements, relationships, and goals. Elements are black boxes to analysts, known by name and behavior, but the Uncertainty Principle limits analysts' knowledge of the elements' contents and relationships. Analysts may have direct access to system goals, however, and are certainly in a better position to observe goals in action than are elements within the system.

Consider the systems shown in Figure 1-6. Each is identified by goals, relation-

TYPE OF SYSTEM	ELEMENTS	RELATIONSHIPS	GOALS
Physical (Clock)	Physical Objects (Mechanisms, Weights, Springs, Energy Transmitters, etc.)	Energy Exchange; Propinquity (Nearness)	Movement, Transferral of Objects, Rearrangement of Objects (Hands)
Intellectual (Naming of Variables in a Program)	Concepts, Ideas, Rules of Transformation, Procedures, Names of Programs	Name of... Letter in the Name of...	Having Similar Variables Named Similarly for Mnemonic Purposes
Information (Course Registration)	Students, Registration Files, Time-Tables Courses	Is enrolled in... Meets at Time... Is Eligible to Take... Is Registered in the Program... Is Closed...	Assigning Eligible Students to Courses and Sections of Their Choosing
Manufacturing (Pencil Making)	Raw Materials (Wood, Paint, Graphite), Semi-Finished Pencils, Machinery, Workers, Wrappers, etc.	Is Produced by Machines #... Works on Machine... Is in batch #...	Production of Several Kinds of Pencils

FIGURE 1–6. Examples of Types of Systems

ships, and elements. A mechanical clock is a *physical* system; the elements are physical objects and the relationships are mechanical in nature. The system for naming variables in a computer program is an *intellectual* one (another example is the Dewey Decimal System); elements are the names of things and the relationships are lexicographical. The third system in the figure processes *information*, specifically student course registrations. Elements here are names of students, course numbers, and files of information. Relationships are informational in nature; certain data go into certain fields and are processed in certain ways. The fourth system is like the third in that it *processes* something, in this case pencils, producing them from raw materials. The materials and machinery are the elements. The relationships are defined by the processes: what goes in, what comes out.

Physical, intellectual, informational, and manufacturing are four system types important to systems analysts. In subsequent chapters, we will discuss how to investigate and document each kind of system in order to understand and design information resources.

1.2 SYSTEM MODES

A system changes over time. The changes are referred to as **system dynamics**. They come from the relationships that limit elements' mutual interaction as well as interaction of elements with the environment in pursuit of the system's goals.

Boundaries

Systems interact with environments. We can conceive of an environment as elements **not** part of a system. The distinction between "in-the-system" and "environment" is drawn along a **system boundary**. A system boundary is defined as the set of system elements whose behavior is *not solely determined* by the behavior of other system elements.

In other words, elements within a system are influenced in their behavior by limitations forced on them by their neighbors in the system. Elements along the system boundary also have to react to the environment. What is the "environment" other than elements not in the system?

Two things characterize environmental elements. First, they seem to have "minds" of their own, from the system's viewpoint. Their behavior is influenced only slightly by system goals and interelement relationships. Second, some environmental elements are obviously more influenced than others: those "near" the system boundary are primary examples of elements prone to being influenced, while others remain more or less unaffected by anything the system does. The system must contend with easily influenced environmental elements, and it is at the system boundary that such contention usually arises.

In Figure 1-7, on the opposite page, A's behavior is totally determined by elements B, C, and D. We can predict A's behavior from data on what B, C, and D have done. The relationship may be complex, but we could ultimately analyze it.

But C's behavior is another story. What C does depends not only on what B has done, but also on what the environment has done and is doing. For example, C's job might be to write up sales slips for purchases; the unit cost comes from B and the actual item is purchased by a customer *outside the system*.

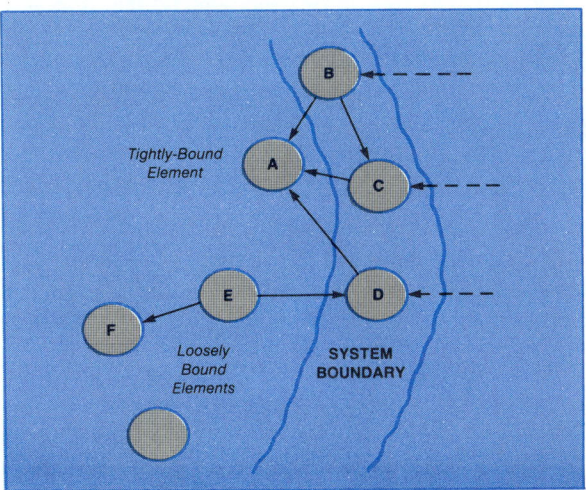

FIGURE 1–7. Binding: An element's dependence on the behavior of others

How much one element's behavior depends on others is called **binding**. We say that A is tightly bound, C is loosely bound, and B is *unbound*. Obviously, tightly bound elements cannot be at the system boundary; conversely, loosely bound elements are usually found *only* at the system boundary. The definition of binding depends ultimately on observers' being able to predict an element's behavior. What does the Uncertainty Principle say about determining where the system boundary might be?

Another term related to binding is **coupling**. We say a system is *tightly coupled* if most of its elements depend on many other elements in the same system. A system is loosely coupled if many of its elements are rather isolated, with influences restricted to only a few elements. A tightly coupled system is "knowable" from an understanding of only a few of its elements. A loosely coupled system has to be investigated in detail on an element-by-element basis because the element-to-element binding is itself loose.

Although elements away from the system boundary may be "predictable" in one sense, that predictability depends on knowledge of other elements. Ultimately, a system's behavior, as opposed to that of its elements, depends at least partly on the environment. The environment is always doing something with — and often to — the system. The system can act in one of three modes with regard to the environment: maintenance, defense, or growth.

Maintenance

System work that takes place entirely within the system boundary is called "maintenance." Maintenance is concerned with ensuring elements perform their proper functions in proper ways. For instance, one element of a computer system might be a storage device, charged with "remembering" information. If that information is subject to change without reason (that is, not according to the proper function of the storage device), it will have to be fixed or replaced.

Maintenance activities include the following:

recognizing when a problem has occurred;
having the knowledge to be able to fix the problem; and
having the time and resources to fix the problem.

Some systems have specialized units for maintenance. In our study of systems analysis, we will discover that systems analysis itself is a maintenance function, set up to keep the information resource running smoothly.

Defense

Every system has goals, goals that may conflict. Competition results from these conflicts as systems compete for the **resources** they need to achieve their goals. Systems are seen by other systems as resources and must therefore defend themselves. Defense activities take place at or near the system boundary.

Defense efforts are always critical. When the boundary is breached, the system is threatened with annihilation. The more tightly coupled the system, the more critical the boundary. Loosely coupled systems can take more stress, perhaps losing some subunits (groups of elements that are relatively more tightly bound to one another) but maintaining their integrity. Tightly bound systems, however, may crumble all at once if certain key elements are lost. Consider the stack of blocks in Figure 1-1 as an example of a tightly bound system that is very vulnerable. Many experts would say that the Western economy is also "tight" in this sense, because the major banks have bought heavily into the economies of third-world countries. A major goal of information system design is to produce "loose" systems that have layers of defense so individual failures do not prove catastrophic for the system as a whole. A system defends itself by "hardening" boundary elements to resist encroachment by neighboring systems.

Normally, defense is expensive, since defense activities do not contribute directly to achieving system goals. Systems without defense, however, soon find themselves treated as other systems' resources when those systems need to grow.

Growth

The third mode of system activity takes place at and *beyond* the system boundary. When systems grow, they add elements. By definition of "system," adding elements implies building relations to those elements. Another term for this is **organizing**. A system grows by establishing relations between elements already "in" the system and those in the environment.

Physical growth occurs when resources are taken from the environment and built into something useful to the system, either as a **nutrient** to supply energy or replacement material or as a new functional subsystem (or unit). Our food, our clothes, the houses we build, and our factories exist because we take "raw" materials (materials which have no "relationship" to the system under construction — vegetables, fibers, wood, bricks) and process them in some fashion (digestion, sewing, home construction, fabrication) and make them useful to us. Of course, it is important to remember that the resources we *assimilate* were useful in some sense to the systems from which we took them. Ecological principles dictate that the environment has nothing "raw," that all resources serve a purpose at every moment. We just are arrogant enough to imagine that our uses are more important. *Self-importance is every system's viewpoint* and thus the primary motivator for growth and the rationale for defense in conscious systems. In practice, growth is mutual and results in an exchange of resources in what general systems theorists term an "equilibrium state."

Equilibrium and Specialization

Growth implies competition and defense. When resources are scarce relative to needs, competition is vigorous; when resources are plentiful, competition is less noticeable. In mature environments, where all resources are organized into systems, there is constant competition. But there is also cooperation in that resources are exchanged among systems.

Figure 1-8 illustrates how resources are exchanged among systems at **equilibrium**,

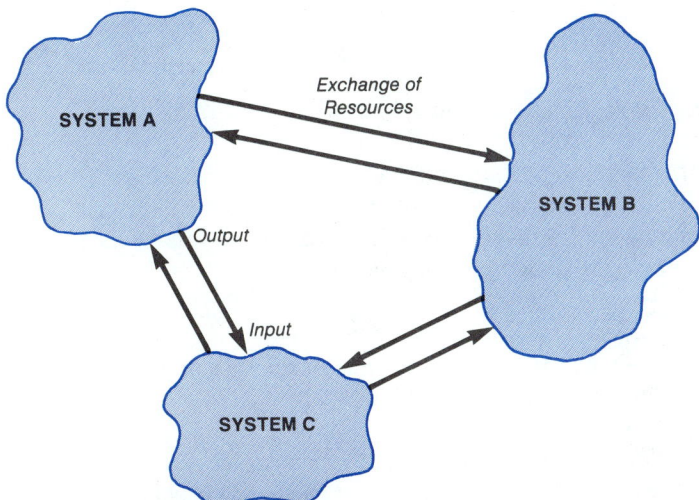

FIGURE 1–8. Systems Exchanging Resources in Equilibrium

a state achieved when competing systems have counterbalancing gains and losses. The net effect is a balanced exchange, the nature of which determines the behavior of those systems. Exchanges among systems at equilibrium are called **transactions**. The nature of the equilibrium (how much activity, how much fluctuation in the size and stability of individual units) depends on the nature of the transactions.

Maintaining equilibrium in the face of an active larger environment results in the development of specialized units called **cells**. Figure 1-9, on the following page, illustrates a system with cells that arise from maintenance, growth, and defense activities.

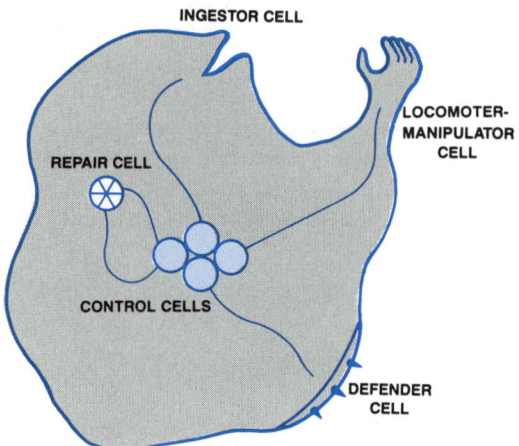

FIGURE 1–9. Specialized Cells

Repair cells are units that specialize in maintenance of system functions. They move to the site of a malfunction and quickly make repairs. Repairs are of two types:

> *R1 rebuilding a relation:* the repair cell acts to relate two or more cells whose transactions have ceased.
>
> *R2 restoring an element:* the repair cell rebuilds an element that has disintegrated. It may mimic the cell and eventually replace it, or it may rebuild the cell with raw materials brought along for that purpose. In some cases, prefabricated replacement cells may be brought along to be plugged in.

Defender cells are specially hardened cells that resist the disorganizing efforts of systems in the environment. Disorganizing would free elements of the system to act as free resources for other systems. Defender cells work in two ways:

> *D1 shielding the system:* defender cells utilize hardening to ward off disorganizing activities; the shields wear away after a while and defender cells may be lost.
>
> *D2 warning the system:* defender cells provide information about impending danger to the system. Defender cells that operate in this way are called **sense cells**, because they serve to make systems aware of, and thus able to counter, moves by other systems in the environment.

Defender cells work in conjunction with **locomotor-manipulator** cells, which move the system as a whole either toward or away from resources and manipulate the environment to make those resources more available. Stationary systems are far more likely to fall prey to other systems; inactive systems may starve for resources.

Ingestor cells specialize in acquiring and assimilating free resources in the environment or causing disorganization to make those resources available. Figure 1–10 shows the activities of ingestor cells (I) in acquiring resources (R) from a neighboring system

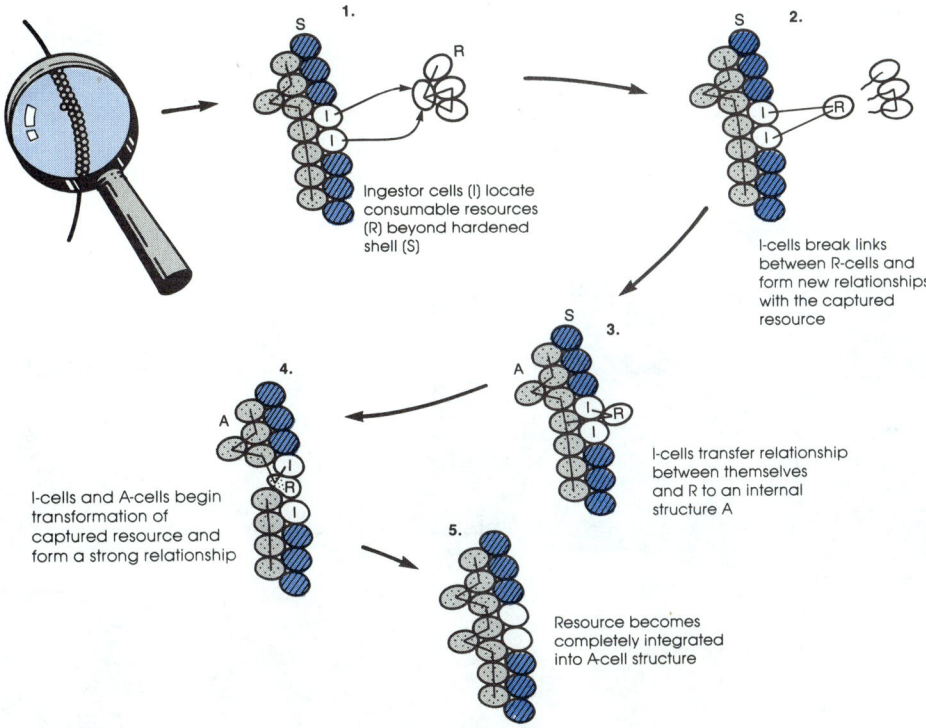

1. Ingestor cells (I) locate consumable resources (R) beyond hardened shell (S)

2. I-cells break links between R-cells and form new relationships with the captured resource

3. I-cells transfer relationship between themselves and R to an internal structure A

4. I-cells and A-cells begin transformation of captured resource and form a strong relationship

5. Resource becomes completely integrated into A-cell structure

FIGURE 1–10. How Systems Acquire New Elements

and transferring them to the interior of their own system A. Ingestion requires the following actions:

I1: breaking relations between located resources and the systems of which they are a part

I2: forming new relations with the freed resources; and

I3: transferring the relation to the internal structure.

Figure 1-11, on the following page, illustrates the actions of maintenance, defense, and growth as accomplished by repair cells, defender cells and ingestor cells.

By overseeing and directing other specialized cells, **control cells** coordinate other cells' activities towards the system goals. They do so by maintaining information on the status of the system relative to its goals. Status information is available from sense cells and from certain repair cells that store data rather than remember functions. Repair cells that merely remember without performing repairs are called **memory cells**.

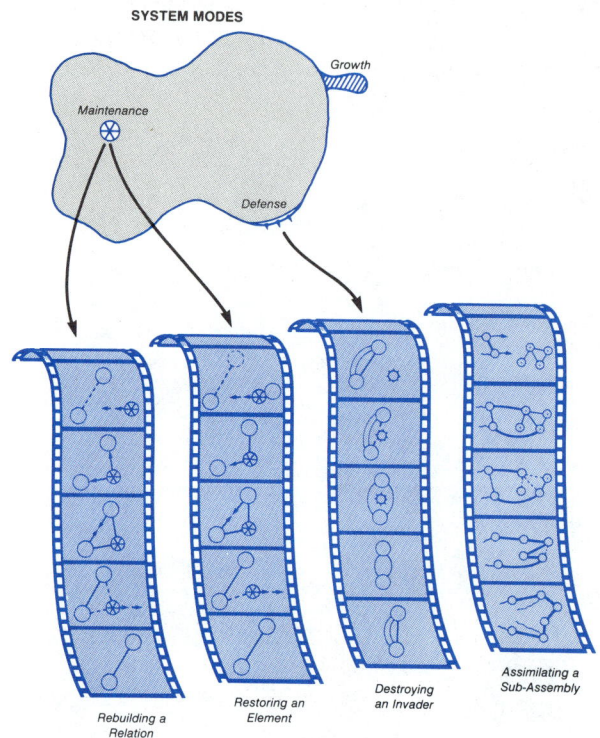

SYSTEM MODES

Growth

Maintenance

Defense

Rebuilding a
Relation

Restoring an
Element

Destroying
an Invader

Assimilating a
Sub-Assembly

FIGURE 1–11. Maintenance, Defense and Growth

There are four control functions:

C1: note and remember status information from sense cells.
C2: note and remember status information from repair cells.
C3: compare status information with goals (remembered in memory cells).
C4: select the appropriate responses (remembered in memory cells) and direct other cells to act.

These specialized cells (repair, defender, ingestor, locomotor-manipulator, control, memory, and sense) are the basic classes of specialization that develop in systems. The conditions under which such specialized cells develop depend not merely upon the system and its goals; they arise because systems have to contend with environments, the topic of the next section.

1.3   SYSTEM ENVIRONMENTS

The environment of a system contains all the elements not in the system. Emery and Trist discuss several types of environments in their thought-provoking book *Towards a*

Social Ecology (1973, pp. 38–56). The following discussion, with slightly altered terminology, is based on that book.

Let us distinguish three types of environments (Figure 1-12) by their **event textures**, that is, by how events in an environment relate to one another. This is best illustrated by the following example.

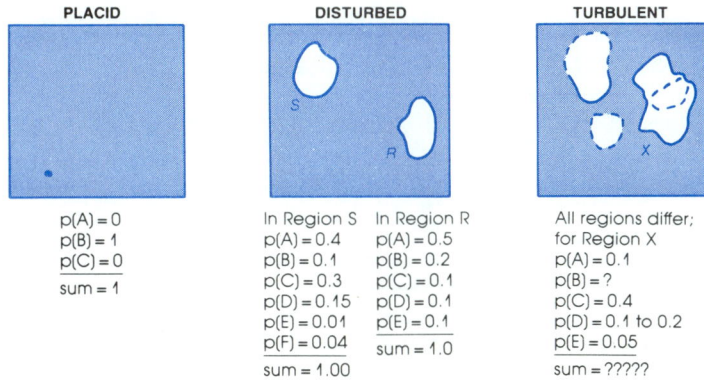

PLACID	DISTURBED		TURBULENT
	In Region S	In Region R	All regions differ;
p(A) = 0	p(A) = 0.4	p(A) = 0.5	for Region X
p(B) = 1	p(B) = 0.1	p(B) = 0.2	p(A) = 0.1
p(C) = 0	p(C) = 0.3	p(C) = 0.1	p(B) = ?
	p(D) = 0.15	p(D) = 0.1	p(C) = 0.4
sum = 1	p(E) = 0.01	p(E) = 0.1	p(D) = 0.1 to 0.2
	p(F) = 0.04	sum = 1.0	p(E) = 0.05
	sum = 1.00		sum = ?????

FIGURE 1-12. Three Types of Environments

Suppose you are playing billiards on a friend's billiard table. You note, with some envy, the smoothness of the felt top and its expensive slate underlay. Certainly any ball rolling across the top will follow a straight line on such a smooth flat surface. You feel that the texture of the surface is such that everything is predictable. In fact, whatever might happen on one part of the surface surely can be reproduced on another part; you think you'll play very well on this **placid environment**.

Now suppose your friend tells you that there are some small pits in the slate underlay. In fact, far from being absolutely flat and predictable, some sections of the table may cause the balls to veer left or right. Your friend says, "Don't worry, though. The chances are 90% that your ball will go straight anywhere on the table, except over here." She points to a corner and says, "Here the chances are 80% that it will go straight and 20% that it will veer right." Near that corner, in a **disturbed environment**, you will play less well but still effectively.

Near the end of the game, your friend's younger brother runs into the room, grabs two of the balls, and runs out. Your friend screams and runs off after her brother. "Give those balls back!" she commands. She comes back into the room with the balls and apologizes, "He's never done that before — it's weird what kids can do!" Now you begin to doubt whether you can play at all in this **turbulent environment**. Most of the time you know the probabilities, but you never counted on someone stealing the balls from the table...what else can happen?

A placid environment has a uniformly smooth and predictable event texture. You know everything that can happen and when and where it will happen. All events have a probability of 1 or 0. Of course, in practical fact, there are no placid physical environments.

An environment with a disturbed event texture is the most common one we encounter. A large number of different things can happen; we know what they are and their probabilities. In a disturbed environment, we can construct a list and assign probabilities to each event on the list. That is how we study for a test: we don't know the exact items on the test, but we do have an idea of the probability for each topic area. Thus, the first test will be "mostly" on chapters 1, 2, and 3, with a "little bit" from chapters 4 and 5 and "nothing" from chapter 6. We construct a table of probabilities based on our interpretations of the instructor's speech habits:

Event (Questions from each chapter)	Probability
1	30%
2	30%
3	30%
4	5%
5	5%
6	0%

Note that (1) we feel the list is complete (the probabilities sum to 100%) and (2) we know the names and probabilities for each chapter.

Imagine our surprise when the questions are mostly from chapters 7 through 9. Here is turbulence at work: we didn't even have a relevant list of possible events!

In fact, most physical environments are turbulent. We call them "disturbed" or "turbulent," depending on our confidence in the validity of the list of events and probabilities. When we are confident, we say the environment is (merely) disturbed. When we are shocked by events not on the list or by a "run" of one type of event or the "disappearance" of another, we think of an environment as turbulent. We can conceive of the three types as lying along a scale of turbulence:

Predictable, all probabilities are 1 or 0	Some probabilities are between 0 and 1		Many probabilities are unknown
Placid	Disturbed	Somewhat Turbulent	Quite Turbulent

Our conceptions are tempered — or caused — by observations of our experiences. Long experience with and careful observation of an environment allows us to fill in the list of probabilities with greater accuracy, lowering its apparent turbulence. Initial experience with a system we can not observe well may convince us that it is turbulent. Visits to foreign countries or cultures, starting new jobs, and dealing with upset individuals put us into apparently turbulent environments that seem to become less turbulent as our experience grows. Most business systems are highly disturbed; the job of the business analyst is to develop theories of how things go together to make things more predictable (that is, less disturbed, more placid). This knowledge helps us respond better to turbulent environments.

Responses to Environments

To survive, systems develop strategies to cope with environments, depending upon the system's ability to organize and control its own behavior and the degree of turbulence in the environment. In a placid environment, systems hardly have to do anything to cope; in deeply disturbed environments, however, most systems fail. In a disturbed environment, systems have to develop coping strategies based on an understanding of the environment — they have to learn or be told the probabilities. The procedure is roughly this:

S1: Establish a menu of possibilities through observation and categorization.
S2: Establish probabilities for each possibility.
S3: Develop strategies for coping with each possibility.

Consider the problem of responding to the marketplace with a product. Each season's production line depends on research and guesswork with respect to the competition and to consumers' buying habits and lifestyles. In each category are a number of variables to be discovered (the possibilities) and the odds established (the probabilities) for each description, long before the season begins. Wrong guesses are costly, so a great deal of effort goes into market research to discover all the factors that influence consumers' buying habits. Will consumers go for "deluxe" products this year? (What is the probability that an upper-end product will sell?)

Systems that survive in disturbed environments become adept at monitoring environments and playing against them. They develop specialized sense cells tuned to the particular events they want to note and measure probabilities for. That is why it is easier to cope in a (merely) disturbed environment. In turbulent environments, the system may not know what to sense.

A turbulent environment has a list of possibilities that may change capriciously; probabilities appear to change wildly from moment to moment. Of course, most environments are themselves large systems whose characteristics do not actually change rapidly — they only *appear* to do so based on our experience in a corner of it. The roller-coaster ride of oil prices over the past fifteen years is a good example. Larger forces may actually control the prices; they appear to change daily and capriciously because we do not know what those forces are.

Nonetheless, systems do cope over the short term by learning about the environment and by accumulating sufficient resources to cope with short-term violence brought about by that environment. To the extent that the system is strong, flexible, resource-rich, and open to information from the environment, it can cope with turbulence. Every system has a limit, however, beyond which it cannot cope with turbulence. How systems do cope with limited turbulence is the topic of the next section.

1.4 SYSTEM ARCHITECTURE

The term "organization" refers to how elements are related to one another. The way that units or subsystems are interrelated to cope with environments is called **system**

architecture. There is a spectrum of architectures that relates to how systems process information from their environments in order to cope. Three architectures of immediate interest are (1) simple, (2) cybernetic, and (3) learning.

The Simple System

Figure 1-13 depicts a simple system transacting with its environment. The particular kind of equilibrium it attains depends solely on the relative strengths and sizes of the system and the environment. The action-reaction loop implied by the transactions is called the **primary loop**, or **action loop**. Normally weak systems survive only in placid environments. They are effectively "hard-wired" directly to the environment in fixed ways; in fact, they do not vary their behavior at all. Major changes in the environment almost certainly overwhelm weak systems. Business systems analysis is not concerned with either simple systems or placid environments.

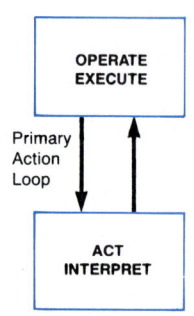

**FIGURE 1–13.
The Primary Action
Loop in a Placid
Environment**

Cybernetic Architecture

A more complex architecture is depicted in Figure 1-14. Note that the simple system reappears here within the primary loop. We say that this interaction lies in the **realm of action**, or deed, because the interaction can be completely characterized by the volume and type of transactions. But a cybernetic system (Wiener, 1948; Ashby, 1965, pp. 1–6)

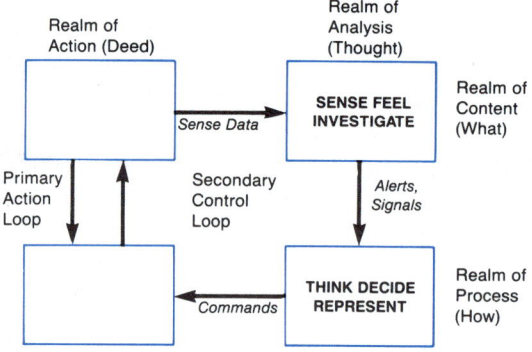

FIGURE 1–14. A Cybernetic System Operating in a Disturbed Environment

goes beyond mere physical interaction; there is an informational interaction in the **secondary**, or **control, loop**. Control loops dictate how control cells coordinate behavior of other cells to counter the environment. Sense cells are alert to particular events in the environment, prewired to "go off," to trigger alarms when certain critical events occur. The following are some typical events:

- The price of oil drops below $5 per barrel.
- The sales of widgets fall below 50,000 units in a month.
- Our delivery truck fails to operate safely.
- The monthly financial statements arrive after the fifth day of the following month.

In the above examples, certain **critical variables** have been pushed into a dangerous range. We are not concerned as long as oil is at least $5.01 per barrel or if we sell 50,001 widgets, or more.

Each alarm has to be treated based on a program or *policy* of action by a control cell that recognizes the critical situation and selects an appropriate and cost-effective action. This strategy is carried out in the **realm of analysis** or thought, going beyond the events in the realm of action and attempting to discover what the events mean to the system in terms of changes to the system's (usual) mode of action.

Events in the environment and the sensing of these events take place in the **realm of content** or "what is happening." Likewise, the interpretation of the events and subsequent counteraction take place in the **realm of process** or "how we will react." The four realms (deed versus thought and content versus process) characterize the complex cybernetic architecture necessary to cope with disturbances in the environment.

The control loop is information-based. Sense cells note what is happening and trigger alarms if critical variables are breached. The control cell notes what has happened, looks up an appropriate and cost-effective response, and issues commands to the system to take physical action. The "activities" in the control loop all involve information.

The control loop is fragile, however. There are at least ten ways in which the control loop can fail to meet a disturbance in the environment. Here are five. Can you list at least five others?

- Sense cells may not be sensing the right events.
- Sense cells may fail to operate correctly.
- Alarms from the sense cells may not be passed on to the control cell, or they may be distorted or garbled.
- The control cell may fail to heed the alarm.
- The control cell may be paying attention to other alarms and not have time for the incoming alarm.

If the connections and functions are correct, a cybernetic system may still fail because the selected actions may be inappropriate, too expensive, or impossible to carry out. Control cells act like computer programs: certain inputs (alarms) call for certain actions. If this relationship does not work, if the control cell fails to function properly, then the system must ultimately fail or else be very lucky.

That even cybernetic systems must ultimately fail in turbulent environments is axiomatic where the turbulence is itself critical to system functioning or if environments are extremely strong. This is so because ultimately an event will occur for which sense cells are not assigned or for which no effective counteraction is available. A cybernetic system cannot change its strategy, but the environment could change its tactics. A cybernetic system can only "read" the environment, it cannot change its "program." For that, a more complex architecture is needed.

Learning System Architecture

This increase in complexity is illustrated by the **learning system** in Figure 1-15 on the following page. The primary and secondary loops are augmented by yet another

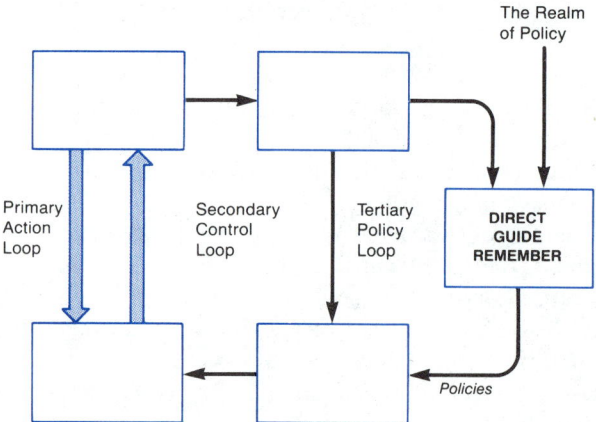

FIGURE 1–15. A Learning System Operating in a Turbulent Environment

loop, called the **tertiary** (third), or **policy, loop**. This loop allows a system to change its program of responses — its *policy* — in response to analysis of the success of previous policies. In the *realm of policy*, the learning system accumulates information on the value of specific policies and changes them (and the associated sense cells) in accordance with what it learns. In business, this is called "policy making"; in information resource management it is called "programming."

The policy loop enables the system to "reconstruct" or "reorganize" itself from time to time based on accumulated knowledge of past transactions. Policy making is therefore limited in effectiveness by the following (only a partial list):

- the actual turbulence in the environment.
- the amount of memory available to store past experience.
- the ability to change policy rapidly and accurately.
- the amount of memory available to store a new policy.
- the ability of the control cell to scan the new policy rapidly.

Sometimes the three loops are distinguished by terms derived from bureaucratic structures. The action loop contains the workers; the control loop, the managers; and the policy loop, the executives. This analogy is helpful in that the building of loops and the adding of "realms" can continue indefinitely. To cope with increasingly higher orders of turbulence, systems can add layers of control, evolving more complex architectures over time (Bateson, 1972, pp. 166–176). Figure 1-16 on the opposite page, illustrates the addition of a next layer. The costs of additional sense, control, and memory cells and the obvious decrease in relative speed of reaction to the environment are apparent.

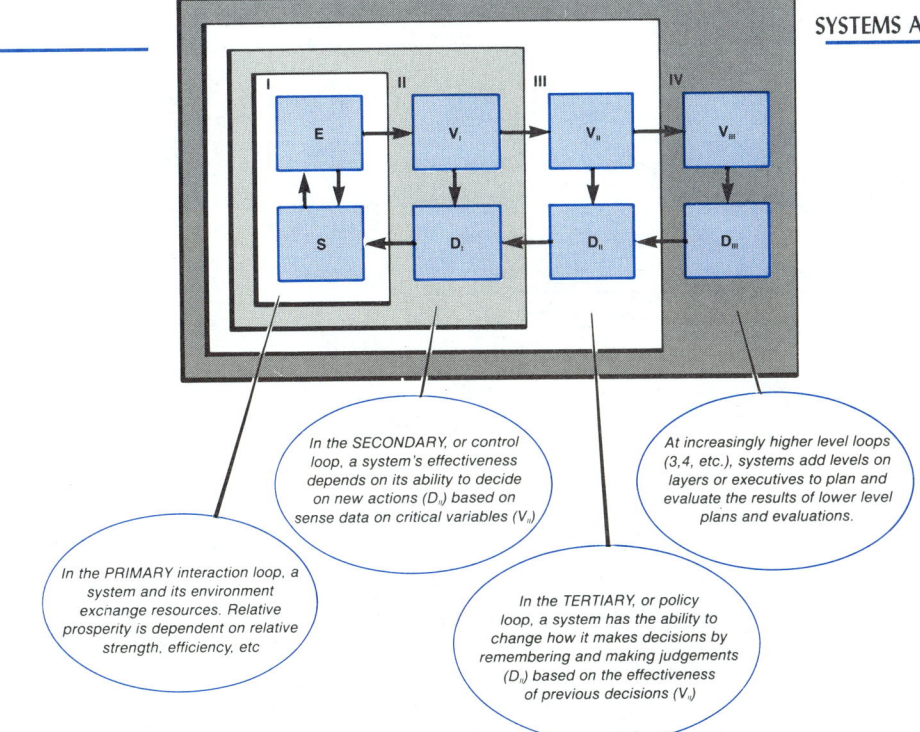

In the PRIMARY interaction loop, a system and its environment exchange resources. Relative prosperity is dependent on relative strength, efficiency, etc

In the SECONDARY, or control loop, a system's effectiveness depends on its ability to decide on new actions (D_{II}) based on sense data on critical variables (V_{II})

In the TERTIARY, or policy loop, a system has the ability to change how it makes decisions by remembering and making judgements (D_{III}) based on the effectiveness of previous decisions (V_{II})

At increasingly higher level loops (3,4, etc.), systems add levels on layers or executives to plan and evaluate the results of lower level plans and evaluations.

Figure 1-16. The Addition of a Layer of Control

Feedback in Complex Systems

The higher-level loops exhibit the property of **feedback**, information about past activities used to select future activities. For instance, the control loop transmits information about activities in loop I to the control cell to make possible changes in how loop I functions. The policy loop feeds back information about the value of previous policies to a control, or decision, unit (D_{II}) to change policies.

The purpose of loops is to correct differences between reported states and desired states. Sense cells send alarms when the difference becomes large enough; control cells select an appropriate action according to a fixed program to reduce this difference by "feeding back" information to a lower-level loop. This how **negative feedback** functions.

Negative feedback does not imply "nasty" or "nonproductive". The term simply refers to the use of information to reduce differences between measured and desired states of the environment. Negative feedback loops are important in many situations in daily life. A thermostat in a house or in an automobile engine is an example (see Figure 1-17 on the following page). When the difference between the measured temperature and a preset, desired temperature becomes large enough, the active part of the thermostat warms sufficiently to move to a position that breaks the contact and shuts off the furnace. When the temperature falls (as it will with the furnace turned off), the active part cools, moves back to a position in which contact is re-established, and the furnace is turned on again, heating the air.

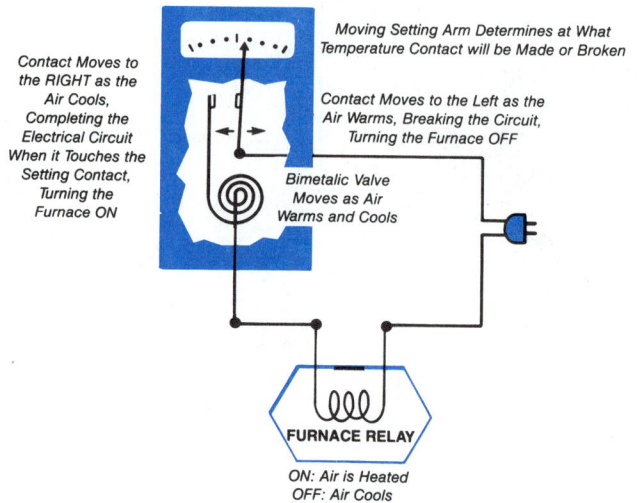

Figure 1–17. Use of Negative Feedback in a Thermostat

Negative feedback is commonly used in finance. Economists may recommend increasing interest rates to bring consumer and corporate borrowing down by making it more expensive. This decreases the money supply and, theoretically, reduces inflation by giving people less money to bid prices up. When the inflation rate has been lowered, interest rates may also be lowered, making credit more available, increasing the money supply, and, theoretically, perhaps increasing inflation.

As you can see, there is a relationship between the amount, kind, and frequency of negative feedback, the procedure for making decisions (the control cell), and the goals of a system. This relationship may require frequent tuning — something that may be difficult, especially where the environment is quite turbulent. Too much negative feedback can result in an **overcontrolled system** that is unwilling or unable to take risks. Even though we have not yet defined risk in systems terms, the message is clear. Negative feedback tends to motivate a system to operate within a narrow range of actions, decreasing flexibility and putting a damper on unique, novel, or creative behavior. Perhaps the word "negative" is appropriate after all.

Positive feedback operates, as you would expect, to increase deviations. When would that be profitable? Many systems have a "desire" to play, experiment, or be entertained by running risks. Since negative feedback strives to make the environment more predictable, positive feedback is characteristic in systems that strive to make the environment less predictable.

Positive feedback can also be used to explore the environment. Where turbulence is already under "control," systems may need to see what might happen if certain things occur. "What if we do this...or this...hey, that's interesting!" This sort of non-task-oriented (and non-goal-directed) "play" is important in determining a system's ultimate (not short-term) success. It is only in controlled situations that the system can safely "try out" new policies. Stated another way, a system cannot afford to wait until impending disaster to try out a new plan; it needs to "tickle" the environment from time to time.

Playing "what if" games is an important part of management-planning exercises, intended precisely to anticipate, in the form of problems, future turbulence, and to measure the probable effectiveness of various plans. We may therefore view each level of positive feedback as a **decision-support system** for the lower levels. In this way, a system with a complex architecture, lots of resources, a great deal of time, and a fairly well-controlled environment can prepare for later contention with the environment with lowered threat.

Resource Distribution and Architecture

Resources in the environment, like actions of the environment, are not predictably distributed. In Figure 1-18, resources are **uniformly distributed** and **nonreplenished**.

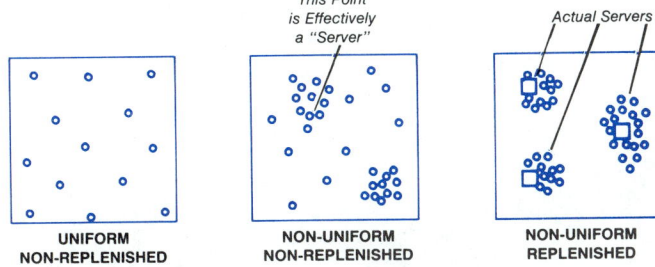

FIGURE 1–18. Resource Distribution Situations

Each area of the environment has the same probability of containing a resource as any other, but the resources are not replenished. As they are used up, new ones do not appear (are not manufactured by the environment) unless the system transacts to replenish the resources from its own stock.

The next example is a nonreplenished, **non-uniform** distribution. Different areas have different probabilities of containing resources. The distribution may be concentrated at one or more centers, called **servers**. Systems needing resources may be forced to queue up at these servers, which may be as mundane as vending machines, as sublime as altars, as rare as the ends of rainbows, or as ridiculous as back-scratching machines.

The distribution of servers determines many aspects of system behavior. One such behavior is **resource seeking**, or **homing**. Systems needing resources seek out the needed resources, which are located at servers. There they may have to contend with other systems or provide specific elements in a transaction. Cybernetic systems are more well equipped than simple ones to seek and home onto servers; learning systems can learn which servers are best and which to avoid.

Resources can also be replenishable. Replenished resources may be manufactured in the environment or merely transported from elsewhere. In either case, the growth of a system depends on locating and assimilating these resources through locomotor-manipulator and ingestor cells.

System growth is therefore dictated by resource distribution and replenishment. Note that in nonreplenished situations (Figure 1–19), systems reach a maximum size and then "die" for lack of resources. Where resources are replenished, growth can be unlimited.

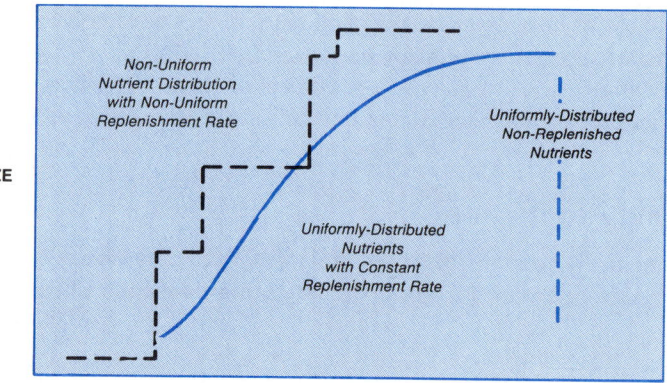

FIGURE 1–19. A Nonreplenished Situation

Where there is a balance between resource needs and replenishment, systems reach characteristic sizes determined by two factors:

1. As systems grow, the distance between ingestor cells supplying resources and those needing nutrients within the system interior increases faster than the size of the boundary (where ingestor cells must be to function). This naturally limits growth, whatever the resources are, and interior cells eventually starve.
2. As systems grow, resource requirements for maintaining the supply units drain the supply of resources away from other units. In the trade-off, ingestor cells must decrease in number, or the number of other cells, some of which are essential to maintenance, growth, and defense, must decrease.

Growth has some characteristics that apply to every system:

- Young systems are small and relatively unspecialized.
- As systems grow, specialized units arise.
- Growth rates rise, then decrease as resources become less available.
- Without replenishment of resources, system growth eventually ceases and may become negative.
- With replenishment, system size reaches a maximum characteristic of the type of system and replenishment rate.
- Distance and supply factors can affect system integrity for loosely coupled systems if resources are limited.
- Feedback levels may develop to enable a system to retain its structure and function as a hierarchy of policy levels.
- In systems with feedback, however, overhead costs eventually override any increase in safety or adaptability.

As systems specialize, functionality becomes more important, because systems with specialties rely on those specialized units (other units simply can no longer perform the specialized functions).

Now, armed with knowledge of how systems function in general, we turn our attention to a specific kind of system, one whose elements and relationships involve information: the information system.

1.5 THE INFORMATION SYSTEM

In an information system, elements involve data. The relationships among the elements are based on data transactions. The goal of an information system is to process information. When an information system is a subsystem, this information-processing goal probably contributes to control activities of the larger, at-least-cybernetic system.

The terms "data" and "information" are not interchangeable, but they are related. Data are values that refer to descriptions of things: the temperature is "seven degrees Celsius"; the wind is "from the northwest." Data is an abstraction about a system and reflects some aspect of that system's state.

Speaking informally, information is "processed data," data that somehow has been made purer or more useful. The practical test for "information" is one of usefulness to someone, reminding us of the Uncertainty Principle. We say that "information is in the eye of the beholder" because data informs people seeking the data to solve problems or make decisions. Consider the temperature. Normally we are not much concerned with the temperature. But when we have to decide how to dress in the morning, we seek data on the predicted temperatures for the day to make our decisions: we want to be informed. The data becomes information at that moment.

The academic study of information and data in the abstract lies in a field called **semiotics**. We are concerned here, however, with the description of systems that use information rather than the distinction between data and information. Therefore, this text will not normally distinguish between the two. Note, however, that the purpose of an information system is to create higher-quality information out of lower-quality data for decision makers who control other systems (Figure 1-16).

Elements

Following Kroenke's (1981, pp. 25–33) definition, the elements of an information system fall into five categories:

1. *Hardware*: terminals, cables, processors, memory units, storage devices, multiplexors, and other equipment
2. *Data*: files, records, fields, characters, digits, bits
3. *People*: users, programmers, analysts, vendors, instructors, system managers
4. *Programs*: applications, operating systems, communication packages, spreadsheet generators, tutorials, demonstration software
5. *Procedures*: manuals, documentation, how-to-do-it guides, startup, takedown, backup, restore, and emergency shutdown

These categories can be abbreviated as PUSH-D (procedures, users, software, hardware, data).

Relationships

An information system is a collection of elements that exchange data with one another and with the environment. The particular equilibrium reached in an information system is determined by the exchanges of data among the elements and with the environment. Figure 1-20 illustrates this exchange:

*An information system receives **input** data from its environment and directs **output** information back to the environment.*

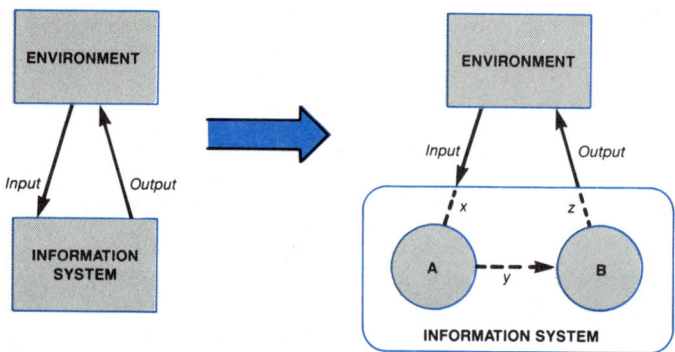

FIGURE 1–20. An Information System Exchanges Resources with Its Environment

This exchange is not to be confused with the physical exchanges that physical systems require. For example, information has to be *encoded* and *transported* on a medium (for example, on a diskette), but the medium is not the information. The exchange of media may be necessary for the information exchange to take place, but it is the information exchange (input-output) that defines the relationship. This exchange is a **data relationship**.

Let's look inside the information system in Figure 1-20, which depicts two elements, A and B, and three relationships. A provides input to B; B receives input from A. A receives input from the environment; B sends output to the environment. Note the names of the data elements that are exchanged on the lines representing the relationships. Data element x comes to A from the environment; y goes from A to B; z is sent from B to the environment.

Elements in the environment that originate information (Figure 1-21 opposite) are **sources**; those that receive information are **sinks**. They are often the same element and the transaction is two-way (as when a manager requests and receives information on sales). In other instances, the transaction may be one-way (a manager requests that information on sales be sent to another manager).

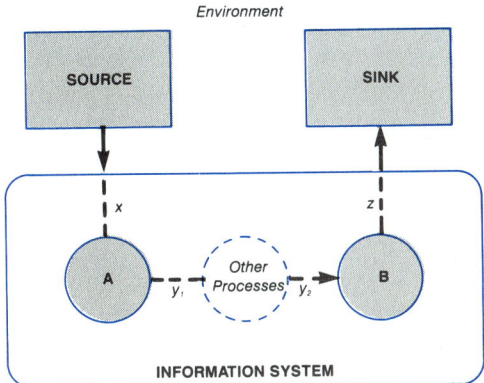

Figure 1-21. Specialization in Information Processing Interaction

The value of an information system lies in its ability to transform source data into information. Information systems do not create data. In this sense they are not **intelligent**; all intelligence lies outside the information system in the sources, and all output information can be traced to sources outside the system. Information systems are valuable because they transform, reorganize, and even beautify data, but they cannot create it.

Given that, what can information systems do? They add, subtract, multiply, divide, sort, align, select, channel, highlight, and even appear to make decisions. Is this not creating data?

No. The processes of an information system **enrich** the data. That is, they refine raw data and remove extraneous, unwanted data to make the wanted data more informative. By digesting data, an information system makes it more palatable for the sinks in the environment that consider this system as a server. These sinks include managers, supervisors, workers on assembly lines, clerks handling airline reservation systems, and consumers receiving their monthly statements by mail. But data is not created in information systems. When that happens, managers are misled about monthly sales, supervisors get incorrect orders, workers are assigned to work on the wrong batches, travelers end up without seats, and consumers complain about their bills.

Information System Goals

Because information systems exist to upgrade data, the analyst can easily begin logical design by employing the tools of **goal analysis**. We start with the overall (control) goals of the system expressed in "needs" statements. The sinks, the receivers of information, may say this:

- "I need a monthly report of sales by region by product code."
- "I'd like to see reports on production slipups by machine and shift."
- "I've got to decide how much product to make. I need to see a trend analysis of production versus sales for each product on a three-month rolling-average basis."

These needs statements really describe the things for which the sinks line up at the information server. An analysis of each goal results in a design that, when implemented, will meet the goal. The goals of an information system are implicit in its output.

Hence, information systems analysis generally runs *backward* from the output rather than forward from the input. In this way, we implicitly recognize several "laws" of information (Figure 1-22):

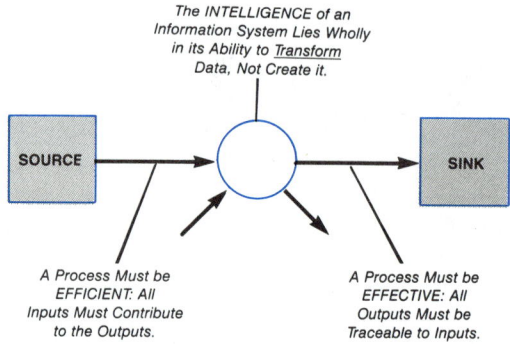

FIGURE 1–22. The Information Resource Must be Efficient and Effective

The Law of Conservation of Information (LCI):

Information cannot come from nowhere. (Corollaries: Information must be data based; all information produced at a sink can be traced to sources outside the system; all information produced at the sink can be traced to noninformation sources producing the raw data.)

The Law of Utilization of Information (LUI):

Information should not be ignored. (Corollaries: No information should be brought in from a source unless it is essential in producing information at a sink; information requirements can be determined solely from output needs.)

The first law guarantees system effectiveness. The system will produce the information required, and that information will be based on data, not created imaginatively by the system. The second ensures system efficiency. Processing will not be wasted on information that does not contribute to required output.

Where these laws are violated in system design, a number of harmful things occur. Information that cannot be traced to specific sources is called **noise**. Noisy information is inaccurate or irrelevant. Consider a system that supplies inventory reports. If the reports are not based on the actual inventory, then they are inaccurate. A clerk who merely guesses at inventory levels while producing report is introducing noise, rather than processing data.

Processing information unnecessarily wastes system resources. This situation is common in older organizations that used to need certain information but no longer do. In some cases, alternate, less expensive, and more reliable sources for information have been located.

Information System Functions

Information system data relationships dictate a **logical structure** among the elements, just as the rules of a game dictate structure among the players. The data relationships are themselves determined by how data flow among elements of the system, how inputs and outputs are "connected." What happens between the connections are data processing activities, or **processes** for short.

A **data flow diagram**, or DFD, illustrates these connections. Figures 1–20 and 1–21 are DFDs. Figure 1-23 depicts three elements (X, Y and Z) and three data relationships (a, b, and c). The data relationships are called **data flows**; the elements are called **data processes** or merely **processes**. We read Figure 1-23 as follows:

Process X accepts b and c as input and produces a as output.
Process Y produces b as output; process Z produces c as output.

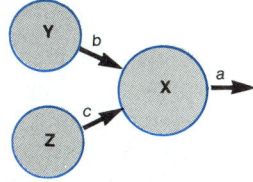

The Law of Logical Data Flow (LLDF):
X completes only after *both* Y and Z complete. (Corollaries: X completes only after B and C are completely examined; A is complete only after B and C are completely examined; A completes only after both Y and Z complete.)

FIGURE 1–23. Law of Logical Data Flow

As drawn, the diagram violates both laws of information, but let us ignore that for the moment. Data relationships obey an information law called the **Law of Logical Data Flow** (LLDF): no process can complete until all its inputs are complete. By this law, X can not complete until both b and c are complete. By the LUI, the only reason for waiting for b and c to complete would be because a depends in some sense upon b and c. Thus, a corollary of the LLDF is that the outputs of a process are complete only after all the inputs are complete. A further corollary could be that no output is complete until all processes contributing input are themselves complete, and processes cannot complete or reach their goals (that is, produce their outputs) until all processes supplying input are themselves complete.

Now we have relationships among elements as well as data flows. We read Figure 1-23 as an analyst might:

> *The output a is not completely available until X completes; until both b and c are completely available; and until both Y and Z are completed.*

In practice, the LLDF is the analytic tool used to perform the "backward" analysis referred to in section 1.5. Even more curious, this backward analysis corresponds exactly to the technique of goal analysis:

> *The goal of having a requires that we achieve the goals that function X is designed to achieve. That requires having produced b and c, which in turn requires having met the goals that functions Y and Z are designed to achieve.*

The LLDF has a number of implications. First, it does not require inputs and outputs to occur in any special order, only that outputs cannot all be complete before all the inputs are complete. Thus, a data flow diagram does not refer to strict time order. DFDs are **asynchronous**, or unclocked. On the contrary, all processes in a particular DFD may be working at the same time.

The second implication is that an information system terminates when the output processes terminate and not before. If output processes appear to be "quiet" while others are going, the system still is not complete. In other words, if some part of the system is still working, however loosely coupled the system, the whole system is working.

To illustrate timing in a system, a DFD is not a good representation. Chapter 6 discusses modeling and the kinds of representations we may choose to build. DFDs are valuable for depicting data relationships; other models do better with time.

Data Flow Diagrams

A data flow diagram shows basic data relationships as and processes that are related by the data. DFDs also depict other things (Figure 1-24 on the opposite page):

- Sources and sinks (collectively called **external entities**) are depicted through square boxes. By definition, they are outside the system being represented.
- A **data store** or file of data is depicted by an oblong box.

The distinction between data flow and data store is simply that a data flow is data moving rapidly between processes (or data stores), while the data store is data at rest. Generally, data stores depict files, complex structures, or sets of data, while data flows represent single data elements or structures such as records.

Data can be changed only at low speeds. Imagine trying to dress a small child who is running. "Organizing" the child's clothing is a near impossibility. Most parents learn how to dress a child moving at speeds below a fast walk. Similarly, information systems

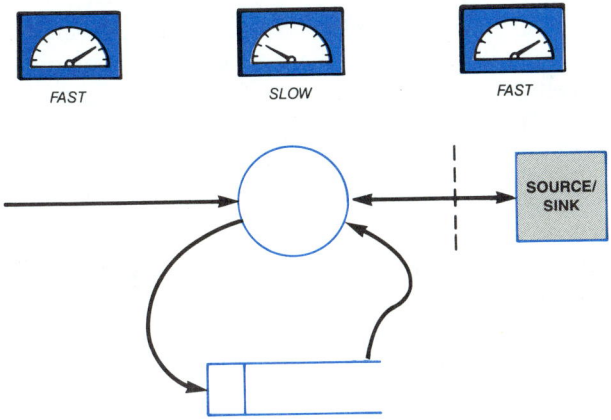

Data can be manipulated only at <u>low speeds</u>.
Processes and data stores slow data down for
examination. Computation and data handling
take time. We assume that all data flows are
infinitely quick. When they are not, we designate
a <u>data store</u>. Processes <u>outside</u> the system take
indefinite time. We accept them as infinite-speed.

FIGURE 1–24. Timing Data Flows

cannot process rapidly moving data. Processes and data stores slow data for examination, transformation, or accumulation.

We assume that data flows, on the other hand, are infinitely quick, when in practice this is not so. When we have to slow data down — to accumulate it or reorganize it into files — we draw the symbol for a data store. A data store is to a data flow as a file is to a thought. Conceptually, a file is just slow moving data. In DFDs, data stores usually represent files that are being accumulated, or **batched**. Data flows represent the movement of such data, and processes represent the transformation of the data.

Because sources and sinks are outside the system, we don't care how rapidly or slowly they process data. We imagine that they are infinitely fast. This gives rise to a two-way classification of **data entities** (see Figure 1–25). Based on this chart, we can

		Does Data Change?	
		YES	**NO**
How Rapidly Does Data Move?	**SLOWLY OR NOT AT ALL**	Process	Data Store
	RAPIDLY OR INFINITELY FAST	External Entity	Data Flow

FIGURE 1–25. Classification of Data Entities

also think of a data store as a "dumb" process that does not do anything or a "slow data flow"; an external entity as a "fast" process or a "smart" data flow. If we find data flows that seem to change, there must be an external entity or process.

The final symbol appearing on a DFD Figure 1–24 is the dotted line, which denotes the *system boundary*. Elements connected to sources or sinks across this line lie on the system boundary and are by definition either ingestor (input) cells, locomotor-manipulator (output) cells, or defense cells (either input or output). Specific configurations of these cells are commonly used in the design of systems; it is to these that we now turn our attention.

Morphs

A configuration is called a **morph**, the Greek word for "form." We use it here to mean structures of elements that are common, recognizable, useful, or worthy of discussion. Of the infinite number of morphs, we will discuss about a dozen.

Think of a morph as a "molecule," a meaningful set of "atoms" or symbols. While a process or a data flow alone does not have much use, in certain combinations they are quite useful. Figure 1–26 illustrates two morphs that are basic to other morphs. The first

MORPHS

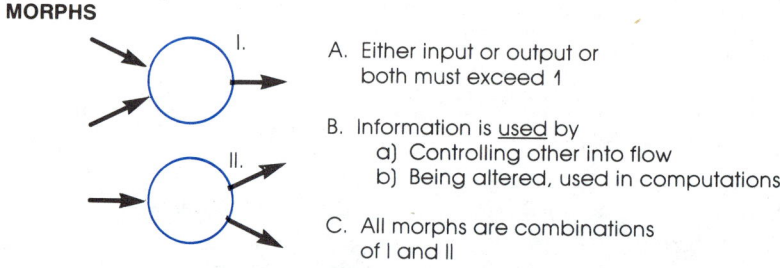

A. Either input or output or
   both must exceed 1

B. Information is <u>used</u> by
   a) Controlling other into flow
   b) Being altered, used in computations

C. All morphs are combinations
   of I and II

FIGURE 1–26. Two Elementary Morphs

shows a process with two inputs and one output; the second, a process with one input and two outputs. The first morph uses one input to direct the transformation of the other input into the output. The second morph uses one output to report on the results of the transformation of the input to the other output.

Eight more morphs are depicted in Figure 1-27 on the opposite page. They are:

Input batch: collect input *a*'s until informed to stop by *b*.
Output batch: empty the batch (*a*) upon signal (*b*).
Input validation: incoming data (*a*) is compared against valid *b*'s and sent along (*c*) if valid; otherwise, an error message (*d*) is sent to the source.
Output validation: a request for data (*c*) is validated against valid *b*'s and data (*a*) is sent (*d*) if the requestor is validated.
Split: Incoming data (*a*) is split into two classes (*b*) and (*c*) of data; unclassifiable data results in an error message (*d*).
Merge: Two sets of data (*b* and *c*) are merged together and sent (*a*).
Selection: One of two sets of data (*b*, *c*) is sent (*a*) upon signal (*d*).
Transformation: Results (*a*) are determined by operation (*c*) upon original data (*b*); error conditions are noted in (*d*).

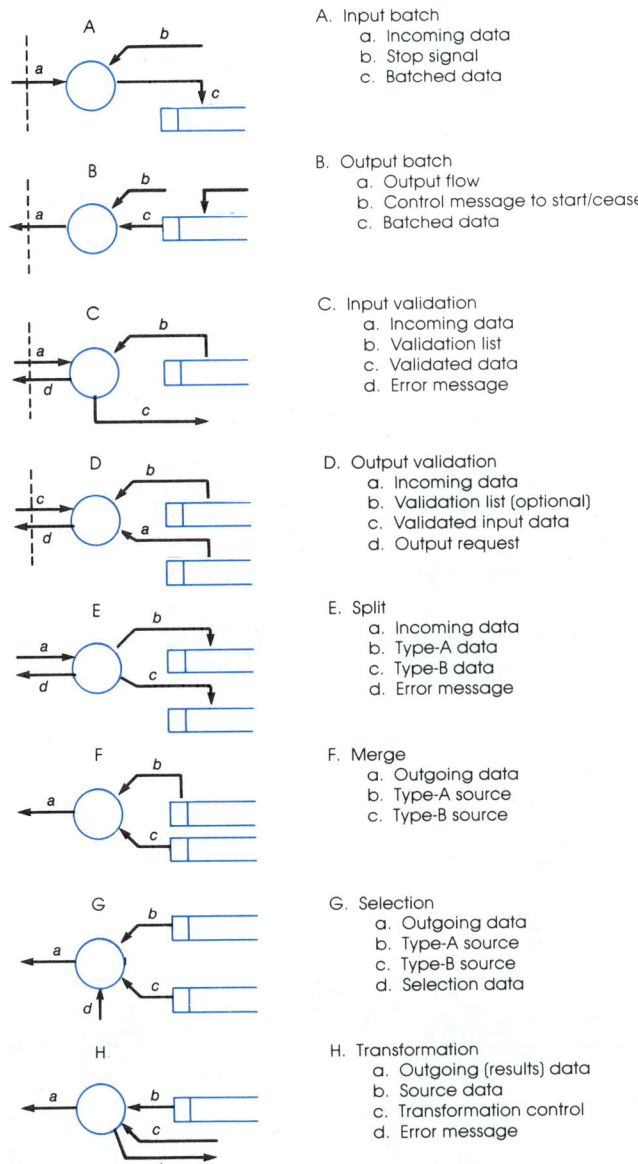

A. Input batch
   a. Incoming data
   b. Stop signal
   c. Batched data

B. Output batch
   a. Output flow
   b. Control message to start/cease
   c. Batched data

C. Input validation
   a. Incoming data
   b. Validation list
   c. Validated data
   d. Error message

D. Output validation
   a. Incoming data
   b. Validation list (optional)
   c. Validated input data
   d. Output request

E. Split
   a. Incoming data
   b. Type-A data
   c. Type-B data
   d. Error message

F. Merge
   a. Outgoing data
   b. Type-A source
   c. Type-B source

G. Selection
   a. Outgoing data
   b. Type-A source
   c. Type-B source
   d. Selection data

H. Transformation
   a. Outgoing (results) data
   b. Source data
   c. Transformation control
   d. Error message

FIGURE 1–27. Eight Important Morphs

A complex general query combining several of these morphs is diagrammed in Figure 1-28. Here, a query for information is satisfied as a response from input data based on control information (a list of valid requestors, for example, or restrictions on which data can be accessed). A report, or **audit trail,** on the request is generated, as well as output. A typical query of this nature is the following (expressed in the dBase III language):

SUM EMPWAGE TO ALLMANAGERS FOR POSITION = "M"

The processor sums employee wages to a variable called ALLMANAGERS only for records for which the POSITION field has the value of "M" (for manager). Control information indicates whether or not any file is in use, and whether EMPWAGE and POSITION are fields in this file. A report is displayed on the operation and, under certain conditions, the sum is printed on the screen as a response. EMPWAGE is changed after this operation as the "output" of the operation.

Morphs are the starting place for information systems analysis. We look for certain familiar **boundary morphs** if there are problems with security or relevance. We examine certain familiar splitting and merging morphs if data stores seem to end up with the wrong data. If reports are incomplete or inaccurate, we trace missing data back to their possible source(s) through **processing morphs**. And if systems are complex, slow, or hard to learn, we try to find redundant or unused morphs.

Our discussion of information system functions (now expressed through morphs) will be complete only after looking at how system functions change. The discussion in the next section addresses the development of information systems.

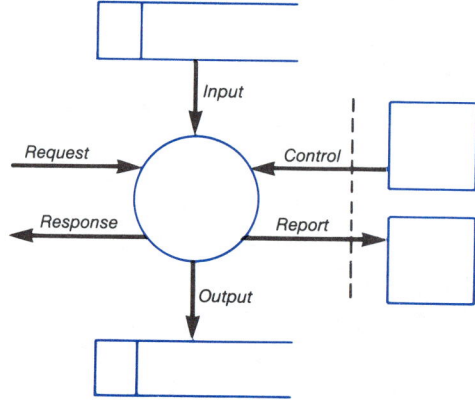

FIGURE 1–28. A Complex General Query

1.6   THE SYSTEM DEVELOPMENT LIFE CYCLE (SDLC)

When we speak about the life cycle of an information system, we use the special term **System Development Life Cycle**. Information systems are developed through a process of investigation, design, construction, and recycling. Most texts use the abbreviation SDLC and stress the "development" aspects. Let's examine this traditional view and add commentary from a GST perspective.

The Stages of the SDLC

Although there is no standard description of the SDLC, most writers specify between five and ten stages or steps. These steps resemble those found in any construction project:

1. system study, problem definition
2. preliminary investigation; feasibility study
3. logical (functional) design
4. physical design ("blueprinting")
5. implementation (construction)
6. installation
7. operation, maintenance, system audit, system study (back to #1)

The **system study** is a formal or informal examination of system functions. Questions are asked about how well the information system is functioning at the present time. Usually a study is initiated by a memorandum from a user of the information system. Sometimes there is a planned management audit or review of the system on an annual or quarterly basis. The result of the system study is a report recommending that something be done to improve the functioning of the system. Chapter 5 discusses the role of system studies.

The **preliminary investigation** takes the system study a step further by detailing the nature of the problems and opportunities for improvements. Chapter 4 outlines the ways such improvements can be viewed: growth, maintenance, and reconstruction. A preliminary investigation (Chapter 8) takes the logical view of the system and defines system development goals in terms of existing system goals using structural-functional design and analysis techniques (detailed in Chapter 4).

A **feasibility study** begins with a set of proposed "solutions" and attempts to build a case for the feasibility of at least one of them, based on costs and benefits. For a variety of reasons (detailed in Chapter 8), feasibility studies are usually inappropriate in systems analysis. Chapter 14 discusses cost/benefit calculations, but frequently it is impossible to make accurate estimates of either.

While the preliminary investigation moves from system goals to system functions, the feasibility study presumes that the solution that meets the goal is a combination of known units whose costs can be computed. Our approach is to avoid feasibility studies. They are intended for implementation of systems that are already fairly well understood and for which it is easy to calculate costs and benefits in the early stages. Given the complex nature of information systems today, such assumptions are rarely justified.

Because the preliminary investigation begins with goals and works toward functions, analysts are less likely to miss logical solutions than they would be with predefined, "canned" solutions of the feasibility study. The risk, of course, is that many proposed logical solutions may be infeasible.

Logical design is the complete analysis of the proposed or improved system. It is obtained by functionally decomposing the DFD until physical elements (black boxes) are obtained. These are units whose descriptions depend on the operating environment. For example, the function "write reports" can be decomposed only into actions involving ink and paper, with knowledge of where each item in the report will go on each sheet. This is not a functional consideration; instead, these factors can be determined only with knowledge of which machines will do the printing, how wide the paper is (or even whether paper, as opposed to displays, will be used), and so forth.

Logical design begins with the preliminary investigation report (Chapter 8) and results in logical specifications (Chapter 10).

There are three components to logical design: investigation, representation, and interpretation. During **investigation**, the analyst gathers data using a variety of investigation techniques (Chapters 5, 11, and 12). These data include operational details, wish lists, complaints and expectations from staff and users, goals and goal analyses, and collections of documentation.

Representation is the activity of depicting systems graphically as **models**. Chapters 6, 9, and 13 review techniques for creating and using models. **Interpretation** compares the data collected by the analyst with the analyst's prior experiences in implementation, analysis, training, or reading. This procedure merges both "inside" and "outside" views of the system, "interpreting" the system's goals and functions in more real-world terms. Chapters 7, 10, and 14 discuss the analyst's interpretation activities in more detail.

Following logical design, **physical design** puts clothing on the store dummy, so to speak. The physical design specifies how the black-box functions are to work in the real world. Physical design is concerned with costs, schedules, and feasibility. Whereas logical design ends with the full functional decomposition of a system, physical design is concerned mainly with specifying the wires, boxes, and programs that will work to carry out the functions to meet the goals. Logical design may have specified that a particular process produce a report of sales by region. Physical design will specify how the sorting will take place, what the report will look like on paper, and precisely how software and hardware will work to do that.

The logical design might look like this:

```
PRODUCE SALES-REGION REPORT:
   DO WHILE MORE REGIONS
      DO PRODUCE REGIONAL-SALES-LINE
      SUM REGIONAL-SALES TO SALES TOTAL
   ENDDO
END
```

Physical design, on the other hand, would look like this:

```
Page Header: SALES FIGURES BY REGION.....Page xxx
Page Trailer: REGIONAL TOTAL   $ xxx,xxx,xx
Line Format:
rrr''''/'ddddd.dd/*10'ssssss.ss
rrr = region number ddddd.dd = sales figure ssssss.ss = sum
```

Physical design looks inside the black boxes; logical design is not concerned with physical descriptions of how the processes will work. If physical design determines that each report will be two hundred pages if double spaced and one hundred twenty-five if single-spaced, a decision might be made to single-space the report for reasons of mailing, reading, and paper costs. These considerations are ignored during logical design.

Implementation is the actual construction of the system or subsystem. This includes coding and testing any software, purchase or lease of the hardware, writing procedures, and designing the physical facilities (layout; landscaping; furniture; and interior decorating, including lighting, heating, and traffic planning). Approximately 30 percent of all life cycle costs are incurred before implementation, which spends another 20 percent. The rest are spent in installation, operation, and audit. By pushing costs as far back into the system life cycle as possible, expensive choices made later have a greater chance of paying off.

Installation means bringing the new or improved system on line. Installation activities include wiring of components, training of staff, organizational changes, public relations efforts, placement of furniture and interior decorations, final integration of software with operating procedures, and switchover from existing operations, including file conversions (from one format to another or from one medium to another). Installation can take place in one of five ways (the terminology is again based on Kroenke's definitions):

> *Pilot:* one user group is selected for installation. After that group has been tried out, the entire user population is given the new system. One accounts receivable group may try the system first, then the others.
>
> *Piecemeal:* one **function** is installed, then another, then another until all functions are "converted" to the new way. For example, accounts receivable may be changed first, then billing, then sales management.
>
> *Phase-in:* the old system is removed while the new one is slowly phased in; for instance new accounts may go on the new system as old accounts are converted from the old one.
>
> *Parallel:* both the new and the old systems may function simultaneously for a period of time, after which the old one is dropped.
>
> *Plunge:* Over the weekend, the old system is carted out and on Monday morning, everyone finds terminals on their desks.

Parallel installation is the safest and is used in on-line systems for which constant readiness and accuracy is important. After a few weeks or months, bugs in the new system have been shaken out and users know how to use it. This is the most expensive technique, however, requiring two complete systems.

Phase-in, piecemeal, and pilot installation are less expensive but difficult to manage, and the intermediate results may not always be easy to interpret. Just because the system works well on one function or for one group does not mean that other functions or groups will be successful. In terms of technology innovation (Chapter 2), however, pilot and piecemeal introduction are quite favorable.

The plunge is almost never advised, but because apparent costs are low, many small businesses, professionals, and underfunded organizations such as clubs purchase off-the-shelf systems and just install them, hoping they will work. Sometimes that happens; sometimes the firm's reputation is damaged and morale is severely affected. The trend toward microcomputer-based "user-friendly" business systems only exacerbates this tendency, often with no actual improvement. Results often depend more on training and conditioning of staff than the quality of the installed system.

Operation and maintenance take up 50 percent or more of the life cycle budget. Maintenance functions include repair and replacement of system elements, data backup and restore, recovery from disasters, and routine repair and regular upgrading of hardware as well as growth through adding functions. **Post-implementation Audits** will be taken at various times following installation of the new system. Generally a complex system will be audited three to six months after installation and may be annual thereafter. A new system is like a new employee; regular performance reviews are both necessary and fair. Smaller systems working in smaller organizations might never be audited. Operational auditing of information systems is still more art than science, although the trend toward controlling costs will only speed development of rigid and reproducible auditing procedures. Chapter 14 discusses some techniques for putting cost and benefit values on systems. When costs start to exceed benefits, it is time for a system study and the cycle then repeats.

The SDLC as a Multi-Loop System

The term "cycle" is appropriate, because the SDLC is a negative-feedback-guided multiloop learning system (Figure 1-29). At the physical level, **operations** take place. The information system acts in production mode to print paychecks, track production of widgets, and produce monthly financial statements. Our focus here, however, is just on the information system. But within each functional area, the information produced controls the function (payroll, production, finance) in a cybernetic way.

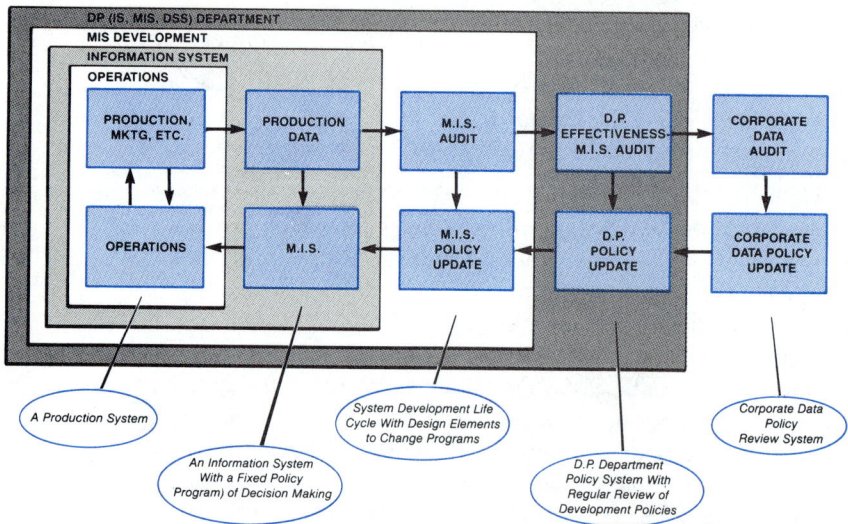

FIGURE 1–29. Information Resource is a Learning System

At the next level, the cybernetic level, the control loop uses data about how well operations are being performed to select **tactics** based on Information Resource Unit policies. Decision makers use these policies to control projects or production schedules.

These management information systems (MIS) policies are subject to change from higher organizational levels. The MIS audit (the term may vary from firm to firm) serves to fine-tune operational policies to keep them consonant with those of the firm. For instance, a previous development effort may have overspent its budget by 50 percent because it used a database manager that had poor documentation. Tool procurement policy may be changed to weight documentation higher in selection procedures.

MIS is one aspect of data processing (DP), or, to use a more modern term, **information resource management** (IRM). At this higher level, still other policies may be implemented to change policy update procedures. For instance, policy updates may only concern new system development. But in an era of more sophisticated system development methodologies (Chapter 16), such a fixation may be obsolete. IRM or MIS policy updates may need to include new techniques such as prototyping, reusable software, and end-user software development.

Finally, the firm ultimately controls the information resource function, which can also influence how IRM policy updates are to be performed. Corporate data processing, information resource management, and management information systems policies are becoming increasingly sophisticated as management uses computers more extensively and comes to depend more on up-to-the-minute, accurate data and decision-making assistance. This makes IRM policy making increasingly a concern of upper management.

This multiloop analysis shows that the activities of the systems analyst (falling in lower-level loops) are a component of an evolving, active, complex system reacting to the corporate environment. This type of system is even more complex than a learning system. Our analysis is made even more complex because at each level, information resources are constructed and used to assist in complex decision making.

1.7 SUMMARY

A system is composed of **elements** that are related to each other to achieve a **goal**. Systems' behaviors are determined by those relationships and the environment the systems are in. Descriptions of systems are limited by the **Uncertainty Principle**, which dictates that it is impossible to know precisely what a system's elements and relationships are from outside the system and what a system's goals are from within it.

The **system boundary** is composed of elements whose behavior is influenced from outside the system. Systems may be loosely or tightly **coupled**, depending on the strength of influence among elements. Systems function in three modes, perhaps simultaneously: **maintenance**, **growth**, and **defense**. Maintenance means fixing an element or restoring a relationship among elements. Growth means assimilating **resources** or free elements from the environment and building relationships with the new elements. Defense means resisting such assimilation attempts from other systems in the environment. Systems reach **equilibrium** with their environments through a relatively static **exchange of resources** in **transactions**. The nature of the equilibrium is determined by a number of factors, including the kind of transaction, the strengths of the environment and system, and the degree of specialization of the system.

A number of specialized subsystems, or **cells**, arise as systems grow. These include **repair**, **defender**, **ingestor**, **locomotor-manipulator**, **control**, **sense**, and **memory** cells.

System environments vary along a dimension of predictability from **placid** (completely predictable) to **disturbed** (unpredictable, with known probabilities) to **turbulent** (unknown probabilities).

Systems respond to turbulence by attempting to build strategies or policies matching action in the environment with their own reactions.

Systems also respond in terms of their **architecture**. A **simple** system has a fixed pattern of behavior. A **cybernetic** system can sense environmental changes and choose appropriate responses. A **learning** system is more complex in that it can change the menu of choices based on computed effectiveness of the existing menu. As systems grow in complexity, they become more flexible at a cost of slowness of response and overhead. These architectures, which are based on **negative feedback**, try to decrease the difference between actual and desired states of the environment.

System growth and interaction with the environment are determined by resource availability. Resources are **distributed** throughout the environment, clustering in **servers**. They may be either **replenished** or **nonreplenished** and the rate of replenishment and distribution pattern limit system growth, as do certain other factors characteristic of each system.

An **information system** is composed of data-manipulating elements that are related to each other by exchanges of data. Information systems are not **intelligent** since all data is ultimately traceable to sources outside the system. Instead, the apparent "intelligence" of an information system stems from its ability to **enrich** data by making it more understandable or usable.

Information obeys several laws. The **Law of Conservation of Information** states that information must have a source. The **Law of Utilization of Information** states that information must have a destination. The **Law of Logical Data Flow** dictates relationships among data processes in terms of logical dependencies. We diagram these dependencies using **data flow diagrams** which show how information is manipulated between **sources** outside the system and **sinks**, also outside. Commonly occurring configurations of **processes**, **data flows** and **data stores** are called **morphs**.

Information systems evolve by creating and modifying lists of activities or programs. This process is mirrored in the **System Development Life Cycle**, which consists of at least seven steps: **system study**, **preliminary investigation**, **logical design**, **physical design**, **implementation**, **installation**, and **continuing operation and maintenance with periodic audits**. Using the **General Systems Theory** framework introduced in this chapter, one can see that the SDLC participates in a multi-level learning system within an organization.

Discussion Questions

1. Consider a drinking fountain as a *system*. Describe this system in terms of (1) elements, (2) relations, and (3) goals. How do you know you are right?

2. Section 1.1 points out the impossibility of completely "objective" observation and states: "By reporting our biases as well as our data, we complete the valid observation as well as we can." Criticize this approach to system reporting. What is the obvious practical difficulty in doing this? How much help will those biases actually be? Is there another alternative?

3. Consider the following "system":

    If black came up on the last draw, bet on black. If red came up, bet on red only if it came up twice in a row.

    Otherwise, don't bet on this draw.

    In what way(s) is this system like any system discussed in Chapter 1? What important differences are there? Is it still a "system"?

4. Elements are bound and systems are coupled. Provide one example of each of the following:
    a. a system that is tightly coupled
    b. a loosely bound element within a tightly coupled system
    c. a loosely coupled system
    d. a set of tightly bound elements within a loosely coupled system

    What can you say about the set in 4d? What characterizes the element in 4d? Consider social systems that are tightly or loosely coupled; what do you call sets like 4d? Elements like 4b?

5. Equilibrium and specialization go together. Why? Describe how specialization arises in non-equilibrium situations. How does an exchange of resources bring about an equilibrium condition?

6. Consider the following kinds of specialized units (termed "cells") and provide a common example of each for a computerized order-entry system:
    a. ingestor cell
    b. defender cell
    c. repair cell
    d. sensory cell
    e. control cell
    f. memory cell

7. All "real" environments are turbulent, but what does that mean in practice? Describe a typical door-to-door sales situation in terms of turbulence from the point of view of the salesperson. How does the salesperson go about trying to reduce turbulence in this situation?

8. Cybernetic systems are more capable of surviving in a disturbed environment than simple ones; likewise, learning systems are better equipped for survival in turbulent environments than in cybernetic ones. Why? What is the cost of increasing the odds of survival?

9. Information systems "enrich" data, but they are not intelligent. How is data enriched? Does the appearance of intelligence in an information system enrich the data in that sense?

10. Provide examples of the laws of Conservation of Information, Utilization of Information, and Logical Data Flow within a forms-completion process. Do the laws also apply to a complex decision-making process? Why or why not?

DESIGN EXERCISES
Using dBASE III

1. Insert your dBASE III diskette in the A: drive of your microcomputer; put the Case Materials diskette in the B: drive. Turn on the CAPS LOCK function (usually a key labeled "Caps Lock"). Enter the following (the symbol ~ should not be typed; it means press the ENTER key):

        A: ~
        DBASE ~

When the screen fills up with information, a small prompt ("(DEMO).") will appear in the lower left-hand corner with a blinking cursor (_) waiting for your command.

The following commands allow you to explore dBASE III as a program. After you type each line, press the ENTER key:

CLEAR ~	Clears the screen
DIR ~	Lists all dBASE III data files
DISPLAY FILES LIKE	Has a similar effect. (At this point we will no
*.DBF ~	longer indicate ~ at the end of lines except as explicitly needed to make a point.)

How many dBASE III files are there? Which is the largest? The letters "*.DBF" indicate "any file with the 'last name' 'DBF.'" "*.PRG" refers to dBASE III program files. Obtain a list of these. How many are there? Do there appear to be any consistencies in how the programs are named?

One of the dBASE III data files is called CLIENTS. Here are some commands to let you examine this file:

DESIGN EXERCISES
Using dBASE III *continued*

USE CLIENTS:	Lets you use this file (you can USE only .DBF files, so you should not type the .DBF)
DISPLAY STRUCTURE:	Tells you the names and characteristics of all the **fields** in CLIENTS
DISPLAY ALL	Displays all the **data** in the CLIENTS file

NOTE: If your computer is connected to a printer, you can add the phrase TO PRINT after DISPLAY statements to get a **hardcopy printout**. For example, to print out the structure of the CLIENTS file you can say

DISPLAY STRUCTURE TO PRINT

How many fields are in CLIENTS? How many **types** of fields are there in this file? How many **records** are there in the file? Notice that dBASE III shows only one screen at a time. Now, using the above commands, discover the structure and content of the other data files.

2.   You have examined the HAIR system a bit using dBASE III's tools. Now we are going to attempt to explore the behavior of the system. Use this command:

DO HAIR   Runs the program called HAIR

The system prompts you for a "Control Key." This is a code which authorizes you to access the system. Certain codes will let you change data. For now, just press ENTER and use the menus shown on the screen to explore HAIR. What seem to be the elements, relationships, and goals? What limitations of your own do you observe? Note what happens each time you press keys. Any time you wish to leave the program, just press the ESC (escape) key. This will bring you back to the prompt.

Consider HAIR as an environment that you have to contend with. To what degree is it turbulent? What kinds of resources are you exchanging with it? What "transactions" occur? Draw a very high-level DFD of the system based on your observations. Consider the specialized cells within HAIR. What cells (you may identify them by module or menu name) are sense cells? Defense cells? Ingestor cells? Locomotor-manipulator cells? Can you really observe the workings of other cells?

When you have accumulated enough information to complete the report outlined above, press ESC and then enter

QUIT   End the use of dBASE III

CHAPTER 2

SYSTEMS ANALYSIS AND THE ORGANIZATION

OBJECTIVES

1. To understand the role, function, and structure of the Information Resource Unit within a modern business organization and be able to list several goals
2. To describe how the Information Resource Unit utilizes human resources in pursuit of its goals
3. To describe the implications of several alternate structures of the Information Resource Unit in terms of functions, effectiveness, stability, and long-range planning
4. To describe the interface between the Information Resource Unit and other major units and departments in an organization
5. To plan the introduction of an Information Resource Unit to several kinds of organizations, varying in size and mission

SYSTEMS ANALYSIS AND THE ORGANIZATION

2.0 INTO EVERY REIGN....

Reign King, Inc., is a medium-sized manufacturer of rainwear and related garments, headquartered in a major Eastern city. For forty years it has produced quality raincoats, rainproof outerwear, and accessories under the leadership of Benny Millerhoff, who started the firm as a young man. Reign King does about $4 million in business and sells its products throughout North America to a steady, reliable market.

Until recently, Reign King completely ignored the automation revolution, concentrating its efforts on styling and marketing of its apparel. Five years ago, however, Raul Costa, one of the designers, was introduced to computerized design and cutting of fabric during a trip to a fashion show in Japan. Raul immediately saw the benefits of computerization. Not only would the computers keep labor costs down, but they would ensure a consistent line of garments without variation. Raul also could not ignore the implications of further computerization — computerized grading (sizing of garments), sewing, and cloth inventory control.

Benny resisted for a while. The initial cost was high for a firm the size of Reign King, and the benefits, while substantial, seemed hardly worth the risks in terms of labor relations, training, and public relations. Benny finally gave in and bought a number of computer-aided design and manufacturing (CAD/CAM) components, which were installed over a two-year period.

For a while, fascinating things happened. Although samples still took just as long to make by hand, the manufacturing cycle was reduced from six months to three or less. Inventory control was so tight that ordering of fabric could be done by the stylists directly, and on-hand inventory (and costs) were reduced significantly.

But beyond the obvious benefits lie problems. There are now new people around, people who understand computers but not the apparel business. Programmers and vendors flit in and out, taking direction from Raul. Benny feels like he is on the outside looking in. There is pressure from Seymour Stanger, his son-in-law and bookkeeper, to automate payroll, finance, and personnel records. Benny's partners know that Reign King's competitors have automated their marketing, ordering, and shipping. Will all of Benny's seventy-two employees now start demanding computers? Will these computer people, cut from different cloth than Benny, take over the needle trade from Benny and his generation?

Reign King may weather the storm if Benny, Raul, Seymour and the others can agree on a corporate information resource strategy. But first they have to understand a number of things about how and why an information resource functions in the larger organization. Raul seems to feel that because CAD/CAM will revolutionize the manufacture of garments, it is benign, easy to adjust to, and fits naturally into an organizational frame. In taking a hard-nosed dollars-and-cents view of the application of computers to accounting, Seymour ignores the management aspects of information resources. No one has looked at the long-range implications of computerization. Only Benny cares about what speeding up the metabolism of Reign King will do to its viability as an organization.

In this chapter, we will look first at a firm's information resource function, its role, and its goals. Next, we will examine how the resource functions and what it does to meet its goals. Third, we will reflect these functions onto the larger organization and examine several ways to structure an information resource unit (IRU), depending on the firm's needs and environment. Since the information resource often serves the rest of the organization, relations with other subsystems is the topic of the fourth section. Finally, we will examine ways of planning for a formal information resource unit within an organization and study several phenomena involving organizational maturation and technology innovation. Then we will return to Reign King to see what precipitates what.

The information resource is people as well as machines.

2.1   THE INFORMATION RESOURCE UNIT

Figure 2-1 illustrates the interplay of a number of subsystems within an organization. These interactions range from simple (distribution acts as a conduit between manufacturing and the environment in a simple exchange of resources) to complex (marketing both affects and is affected by manufacturing, sales, and long-range planning), but only the information resource management subsystem interacts with every other subsystem.

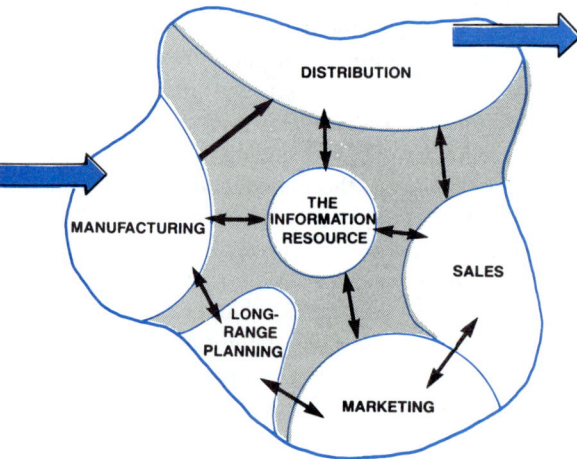

FIGURE 2-1. Function Units in a Business Organization

Organizations run on information. Figure 2-2 illustrates a systems view of a typical business organization. Organizations exchange financial and human resources with their

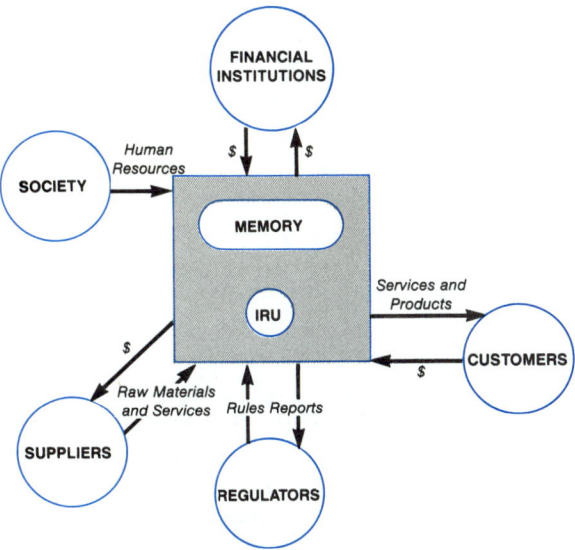

FIGURE 2–2. The Firm as an Open System

environments. Several specialized environments include customers, suppliers, the financial community, and regulators. Central to control of these exchanges is an internal information subsystem, which acts as a learning system for developing managing interaction strategies with the environment to achieve goals.

Chapter 1 illustrated several of these informational functions carried out by the information resource. As **automation** of these functions proceeds, the volume and speed of information supplied increases, sometimes beyond the capacity of the other subsystems to cope. In effect, the well-timed sequence of events established over the years to control interaction with the environment is thrown out of kilter by the rapid growth of information supplies. Such a situation, called **information overload**, is only a symptom of a poorly synchronized and poorly structured information resource.

Consider the following example:

Ralph Condon manages a team of custodians in a large building complex in Bigtown. Every day of every week, Ralph receives memos from the general manager of Housing Managers, Inc., which manages twenty-five such complexes. The memos detail new regulations from the authorities concerning housing conditions. While not all the information concerns Ralph and his team, he has to read it all to find out what does concern him. In addition, each week he receives several magazines about building management and a variety of brochures from manufacturers of cleaning, repair, and building supplies. Every day he hears from five to ten tenants about their plumbing, paint, or parquet, in the form of angry notes and phone calls. Ralph ignores most of it, because he's got six custodians to supervise and little time to do it. "Overloaded, underpaid, overwrought, and underground" is how he summarizes it.

Ralph's situation is this: too much information and not enough time to assimilate it all. The underlying problem is that Ralph's "information resource" is poorly organized and managed. Instead of all building information, Ralph should get only what he needs. He feels he has to make decisions about everything: what to buy, when to buy it, what to repair when and by whom, what do do first, and how to respond to city directives, county imperatives, and national requirements. In truth, Ralph need respond to little of this; the decisions should come from a plan, and the information supplied should be consistent with that plan and in a form Ralph can use. In other words, if Ralph's information resource worked properly, Ralph could just supervise his custodians and not worry about information.

Information overload is just one of the symptoms of a poorly functioning information resource. There are others:

- People do not make decisions or pass this responsibility along to others who can not make decisions either.
- Either nobody knows what information is available, or information is available but no one knows where it is or whom to get it from.
- No one will take responsibility for supplying information to those who need it.
- Everything runs late, inefficiently, or with great waste — and no one can document it.

These are all symptoms of an information resource that needs "tuning up." Its functions are obviously not being performed well and its goals are not being met.

The goals of a properly functioning information resource are as follows:

1. Information is made available in a **timely** fashion. That is, the supply of information is synchronized to the needs pattern of those who use the information.
2. Information is **accurate**, **precise**, **relevant**, and **reliable**. This means that there are no errors ("14" means four more than ten, not one more than forty); there is sufficient detail ("about $10,000" may be $14,999 or $5,001; "next week" may be Monday or Saturday); the information delivered can be used to make the required decision; and the three previous considerations can be counted on.
3. Information is **understandable** and **useful**. Merely accessing information is valueless if the information cannot be understood and assimilated to the task at hand, which is usually decision making. Furthermore, information makes a difference. Knowing something allows the recipient to do something about a situation. It is not only relevant, it is critical.
4. The resource is **cost-effective** (the cost of obtaining, understanding, and using information is less than the cost of not having it available). Costs here may be both tangible and intangible. If there are political, social, ethical, or moral risks involved in getting information, these have to be factored in as significant costs.

In other words, the goals of an information resource are to make information available at the right time to the right people in the right way at the right cost. To do this, the information resource must carry out several functions (see Figure 2-3):

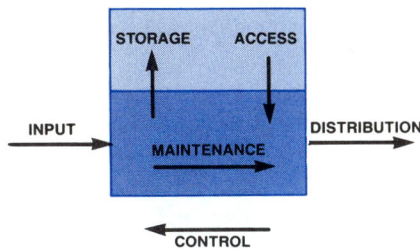

FIGURE 2–3. The Systems Functions of an Information Resource

1. **Input**: data are captured from the environment and from within the organization and translated if necessary.
2. **Storage**: data must be reformatted for efficient storage and retrieval.
3. **Access**: stored information must be made easily available, yet secure from intrusion; there are obvious tradeoffs here.
4. **Maintenance**: sometimes information is lost or damaged and has to be replaced or repaired. Schemes for ensuring the integrity of data are complex.
5. **Distribution**: information is distributed to appropriate subsystems within the organization or the environment in exchange for other resources.
6. **Control**: because the information resource subsystem is a learning system, several layers of control structure are required to regulate the other functions.

In short, an information resource is just like any other well-developed, mature learning system, growing, maintaining, and defending through a variety of specialized subsystems, as detailed in Chapter 1. Its resource is essentially informational in nature (although financial and human resources are, of course, necessary), and its transactions involve information rather than physical materials. Its "clientele" is the whole organization, since everyone in the organization needs information, and its structure (depicted in Figure 2-4) follows that of the generalized systems described in Chapter 1. Ingestor cells bring information into the resource, locomotor-manipulator cells handle distribution and access to data by other cells. Sensor cells assist control cells in managing the subsystem to achieve its goals. Repair cells maintain the resource internally. Defender cells

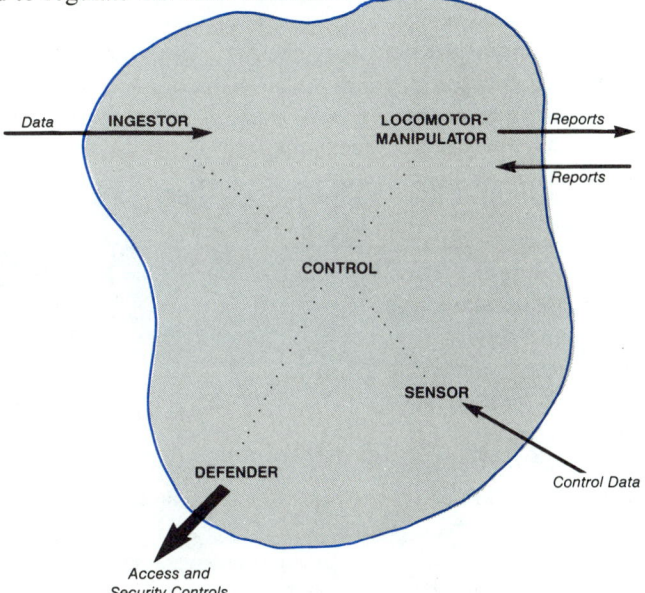

FIGURE 2–4. An Information Resource's Specialized Functional Subunits

ensure integrity from outside access to the system, and memory cells store both data and data about the information resource.

In a **computer-based information resource (CBIR)**, most of these functions are accomplished through the use of computers. The next section discusses the function of the **information resource unit**, a specialized group of individuals who work to achieve information resource goals.

2.2   THE FUNCTIONS OF THE IRU

The subsystem that performs the functions of maintaining and administering the information resource within an organization is called the **information resource unit (IRU)** (although it may go by a variety of names). It serves to meet the goals outlined in section 2.1 by performing the following functions:

- data gathering, entry, checking, and filing
- data processing in a batch (off-line) or real-time (on-line) mode
- information resource subsystem (application) design
- information resource application development
- information resource application maintenance
- information dissemination
- site planning, management, and maintenance
- long-range technology and financial planning
- human resource administration
- financial administration

These ten functions are illustrated in Figure 2-5. Note that the "business" of the IRU is to transform corporate data into corporate information for its users. The administrative functions relate to the other resources needed to perform this function: human, financial, physical, and technological.

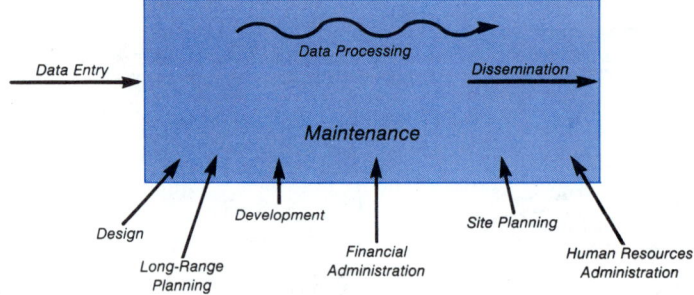

FIGURE 2–5. Functions Which Meet IRU Goals

In most IRUs, especially in small organizations, these functions are not so neatly parceled. Often IRUs consist of one or two individuals who administer small minicomputer installations on relatively low budgets. In turbulent corporate environments, however, IRUs grow to become complex organizations in their own rights. How do these functions operate? Each is illustrated in the vignettes below, which illustrate general principles through specific cases.

Data Entry

Daniella Brown supervises nine women and three men who enter mail transactions into a mail-order firm's database through terminals connected to their mainframe computer. The data entry program, in operation for a year and a half, checks the entered data and computes totals. Unknown to Daniella's staff, the program also keeps track of data entry errors, keystroke rates, and periods of more than five seconds between keystrokes. Printed reports on entry summaries are sent to management on a weekly basis; Daniella gets a weekly productivity report, too.

Data Processing

Across town, three mainframes work hard for the Intermountain Insurance Company. Over one hundred separate applications run nightly on the Intermountain Bit Busters, doing everything from running an electronic mail network to riding herd on sales agents in eighty-six districts throughout the mountain West. Jack Caldwell supervises eleven operations staff members to keep Intermountain profitable and functional, handles calls from irate users, and watches while programmers and engineers sweat over the installation of new equipment.

Application Design

Meanwhile, across the border in Alberta, Canada, Eleanor Borden is awake at her terminal, writing a report on the design of EXCELL, an "expert" system she and two others on the staff of Canergy, Ltd. have designed. EXCELL helps petroleum landmen decide the best possiblities sites for future drilling in Alberta's rich oil lands. Eleanor is a systems analyst who has worked with a number of database management packages on her "super" microcomputer (2 megabytes of memory, one 60-Mbyte hard disk, two 1.2-Mbyte floppies). She designed the specifications for EXCELL, which will be written in PROLOG later by her colleages.

Application Development

He is called "Bits-for-Brains," but Leon Price just ignores the jibes. He's the hottest programmer ever to hit town. Leon is building a structured COBOL application for the marketing department of Sensorium, a manufacturer of industrial control equipment. Leon is three days into the coding of the application. He has taken the systems analyst's specifications for a number of modules of the larger market-analysis program and rewritten them in pseudocode, a kind of pared-down English that looks a lot like programming but still captures the flavor of what the analyst wants. He has already translated one specification into COBOL and is doing some preliminary testing using test data the analyst has provided.

Application Maintenance

"What a pain!" programmer/analyst Agnes Ramsey says to herself as she ushers Steve Diamond out of her office for the third time this week — and it is only Tuesday! SOSH doesn't work, and it is up to Agnes to make it work. Her job is to see that SOSH, a social services package, continues to provide analyses of case management practice for city social workers. As supervisor of nine workers, Steve has a right to bug her until she either fixes the package or writes a memo describing how to get around it. Agnes had better find out how to meet Steve's needs soon. You can blame problems on the computer for only so long.

Information Dissemination

Tom Montalbano walks around his Information Center, feeling justifiably proud of his work and that of the three staffers who assist him. The Information Center occupies a suite of offices on the third floor of corporate headquarters for Urban Infrastructures, Inc., a multinational construction company. The highly computerized engineering department uses the latest in CAD/CAM (computer-aided design/computer-aided manufacturing) software and a large mainframe computer. Project managers worldwide communicate with HQ through satellite links to filter in data on expenses and projections, accomplishments, and crises. Tom's group provides some design and financial management packages to be run through terminals and on micros at the Information Center, where project managers use canned programs or generate their own using powerful "application generation" software. Tom is also in charge of disseminating system announcements, distributing manuals, maintaining on-line system help facilities, and producing the annual report of the HQ MIS group.

Site Management

"Where are we going to fit three hundred square feet of terminals into a room already crammed with equipment?" asks Bernie.

"Well, we'll move the wall, run some cables up through the existing cable shaft from the computer below, install some extra power under the floor, and, with some luck, people will just be able to stand up and type!" quips Vaughn, site manager for the Information Resource Department at Sandyford Technical College. Vaughn's got to figure out how to run power, additional lighting, and ventilation, and how to manage the traffic flow into and out of the room to prevent theft, asphyxiation, and fire. The sprinkler system is surely inadequate, and where is the student monitor going to sit to provide help for confused COBOLists and frenzied Fortraners?

Long-Range Planning and Forecasting

It has been nineteen years since Henry Bigsky quit ranching, but he can still sense a stampede and a technology stampede is one of the most dangerous. As a long-range planner working for the director of MIS services for Lonestar Forestry Products, Henry is working on a long-range planning document, forecasting technological opportunities and needs for the department and trying to get a handle on where Lonestar needs to be positioned to get the most for its computing dollar in 1990. Do they go with another mainframe? **What's** going to happen to the cost of micros as IBM forces standardization and stabilizes software? Can Lonestar use cellular radio handled from their computer to communicate with logging teams? Are staffing costs going to continue to rise as rapidly? If so, should Lonestar rely more on applications generators, or should they contract out system development?

Human Resource Management

As assistant director of computer operations, Andrea Wallace is in charge of staff training and development at a medium-sized trucking firm headquartered in a southern city. As such, she administers performance reviews annually to each of twenty-three employees in the computer operations department. Lately, Andrea has noticed that the turnover in programming staff has decreased significantly, perhaps as a result of the specialized training in management she gave supervisors through a consulting firm last year. She is also pleased that the reviews are being done on schedule. Meanwhile, Andrea has to get busy on the Human Resource Plan that the personnel department has her complete every year.

Financial Management

"Eighty-six dollars an hour! You charged me eighty-six dollars for one hour of a trainee analyst's time? Where do you get this figure, Sandy?" demands Blair. Sandy responds, "We use a chargeback algorithm. We estimate all our expenses for a given period and then allocate them to categories of costs for our users. Each analyst, buck private to general, charges eighty-six dollars an hour. A programmer works out to fifty-nine, additional operators to thirty-four. Each CPU cycle costs you 0.0000002 cents. I can show you the figures. But Blair, you know as well as I do that we've got to pay for our little cousin here." Sandy waves toward the computer. "Our analysts don't work for free, even if you ultimately don't decide to go ahead with your project.

"If we're going to be able to justify our budget, we've got to account for everything in this shop. Besides, think of it this way. If you hadn't spent that eighty-six dollars on Trent, you might have spent fifty-nine hundred on programming you really didn't need. The way I figure it, you should give us back, say $581.40 as ten percent of your savings. What do you say?"

2.3   STRUCTURES FOR THE IRU

As the above vignettes illustrate, work in an information resource unit occurs within a variety of structures and employs a variety of individuals. In the past, the IRU hired only operators (who ran the machines), programmers (who produced the software), systems analysts (who designed whole systems), and system managers. In the 1980s, however, the information resource "business" is far more complex.

Today's IRU is likely to employ a public relations coordinator and a full-time trainer. Trends in technology percolate technical skills upward (technology diffusion) and inward from the marketplace (technology invasion), thus requiring people with new specialties and new skills unrelated to the technology itself. Typical of these are the following:

1. **Information center coordinator**. The IC coordinator runs a store-front data processing shop, a technical college, or a newspaper. Like Tom Montalbano, the IC attempts to bring computing power to the non-technical staff of a firm without impairing "normal" operations. This is done by providing hand-holding, training, very simple access to corporate data, and a variety of easy-to-learn tools.
2. **User-analyst** (computer services representative). This member of a user group acts as a mediator between the group and the IRU. The user-analyst is the user counterpart of the systems analyst. In fact, the user-analyst may well be the client for the systems analyst's efforts. Combining in-depth knowledge of the user group's function, role, and needs with the ability to converse in information resource terms, the user-analyst can be an effective player in information resource management in certain structures.
3. **Officeware analyst**.
4. **Telecommunications analyst**.
5. **Disaster and recovery specialist**.
   The above three individuals specialize in particular applications but these applications are not merely single programs (like a statistics program) that are run once in a while. They are permanent features of the information resource function. Although disasters do not occur frequently, the disaster and recovery analyst maintains vigilance over system function and plans for recovery.
6. **Data administrator**. The data administrator is directly responsible for the integrity of the corporate database. The DA carries out this responsibility in a variety of ways, often using a computerized database administration package.
7. **Business manager**, lawyer, public relations advisor, accountant, marketing manager. Most businesses employ specialists to handle the affairs of the "business" (as distinct from production of the product or provision of the service). Information resources have been slow to respond to the market nature of software and information, but many firms are participating in the marketplace, firms that until the 1980s were content to concentrate on banking, retailing, and transportation. They realize that, under the proper circumstances, their software is saleable; their data, desirable; and their on-line access, attractive.

Regardless of the external market, however, IRUs now come under the same budget scrutiny as other departments. The term "cost center," which stressed the drag that the IRU was seen as having on cash flow, is being replaced now with "profit center," emphasizing the ability of the IRU to bring cash into the firm, or at least assist other departments and units in doing so.

With this professionalization, specialization, and commercialization has come a reorganization of the human resources within the IRU. Several alternate structures appear in today's IRU in medium- and large-sized firms:

- the task force
- the mini-bureaucracy
- the project group
- the consulting group
- the diversified conglomerate

The **task force** is commonly found in organizations just beginning to systematize their information resource. This group might have a programmer or an analyst borrowed from another firm or a consulting group and representatives from user organizations. The task force is charged with creating a permanent structure for the IRU (see Figure 2-6).

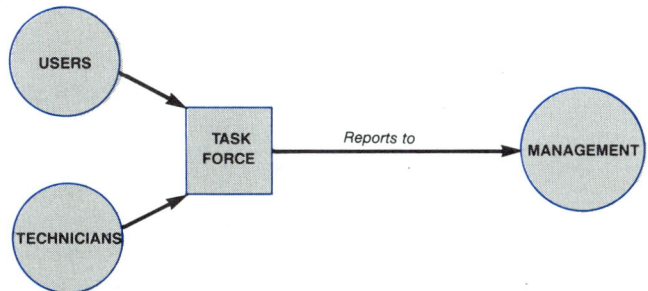

FIGURE 2–6. The Task Force

Most firms develop a **mini-bureaucracy** within the IRU, as illustrated in Figure 2-7. The breakdown reflects the three disciplines traditional to the IRU: operations, pro-

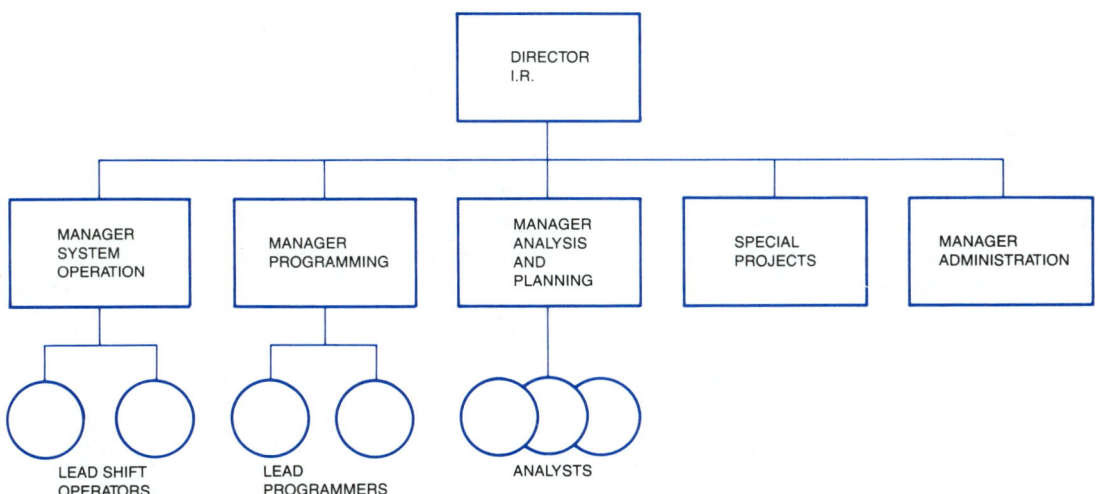

FIGURE 2–7. The Mini-Bureaucracy

gramming, and analysis. Individual career flow is generally from right to left, ending with administrative responsibilities. The bureaucratic structure makes the division of labor and responsibilities easy and natural, but it does not lead to rapid implementation of work, excellent relations with users, or easy cooperation among highly specialized workers. For that reason, the mini-bureaucracy has undergone a number of changes to make it more responsive, user-friendly, and cooperative.

One alternative is the **project group** organization. Here projects are formed in a "grid management" style by tapping individuals from each discipline to work on a single project. In Figure 2-8, Allen (from systems analysis), Pat (from programming), and

PROJECTS	MARKETING	INVENTORY	NEW VENTURES	FACILITIES PLANNING
Systems Analysis				
Allen		X		
Twyla	X			
Geoffrey			X	X
Programming				
Pat		X		X
Bob	X			
Ansel	X			X
Support				
Jim		X		
Reggie				y
Laura	X	X		
Sally	X		X	

FIGURE 2–8. Grid (Matrix) Management

several support staff are assigned to an inventory project. They report administratively to their respective managers (in analysis and programming) but they receive their assignments from the project leader (who is almost always a senior systems analyst). The project structure and grid management work well in theory, but they do little to improve relations with users (since project responsibility still rests within the IRU group). They may also serve to overspecialize staff members if the project lasts a long time.

The **consulting group** (Figure 2-9) is another alternative. Here, individuals or teams are assigned to work within user organizations on a temporary basis to plan, design, implement, install, and sometimes operate and maintain specific systems. The consulting model is a direct response to the need to improve relations with users. In one variation, the assigned workers are seconded to the user group and report to it administratively during the life of the project.

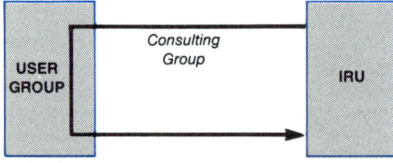

FIGURE 2–9. The Consulting Group

The final organizational model is that of a **diversified conglomerate** (Figure 2-10). This mature IRU consists of a number of semi-independent units that provide information-resource-related services throughout the firm. These units include consulting services, projects, an Information Center, long-range corporate data planning, and a

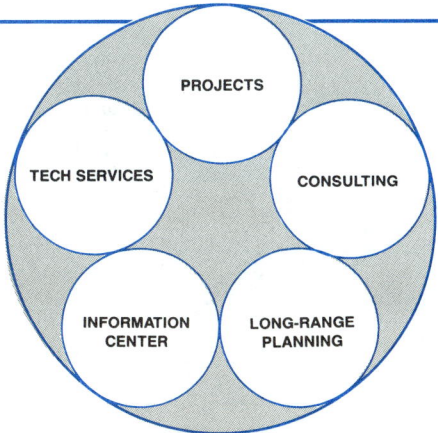

FIGURE 2–10. Specialized Functions in a Mature IRU

variety of technical services, including forms, documents, records, and data management.

The organization of the IRU strongly influences both user relations with IRU workers and the appearance of the information resource to the user. Consider a library. Where the library's policy is "open stacks," users can go find what they want, browse, explore, and generally get themselves into all sorts of trouble. The "closed stacks" model, while preventing problems, limits access; it is very good at keeping books on the shelves, however. Similarly, where the IRU organization is bureaucratic and rule-ridden, access to information tends to be rigid and user perceptions cold. When the IRU has the flexibility to respond to a variety of user needs in a variety of ways, access is easier, but users need more training, advice, and counseling. Traditionally (Figure 2-11), IRUs rely on procedures (policy) and the inherent, detailed, technical knowledge

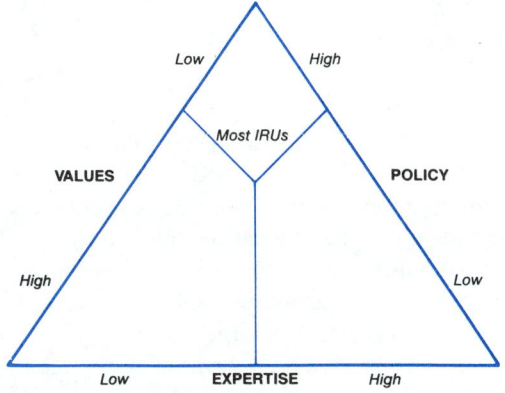

FIGURE 2–11. The VALUES-EXPERTISE-POLICY Triangle

of the staff (expertise) to the detriment of empathy (shared values). Recent research shows that such a situation builds a wall between user and analyst. As Chapter 3 will emphasize, acquisition of empathy-related skills is a must for the analyst who wants to leap over the wall (Janz and Licker, 1987). Figure 2-12 illustrates the typical current and desirable situations with respect to these three dimensions. Now we will turn our attention to that relationship.

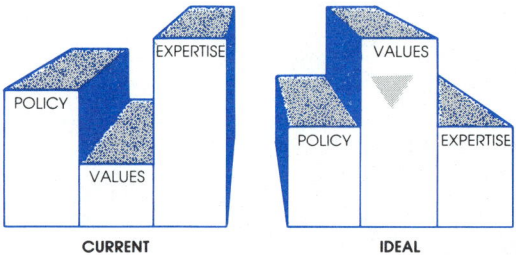

FIGURE 2–12. Current vs. Ideal IRU Profiles

2.4   THE IRU AND THE REST OF THE ORGANIZATION

The IRU can have one of three organizational relationships with other departments (Figure 2-13).

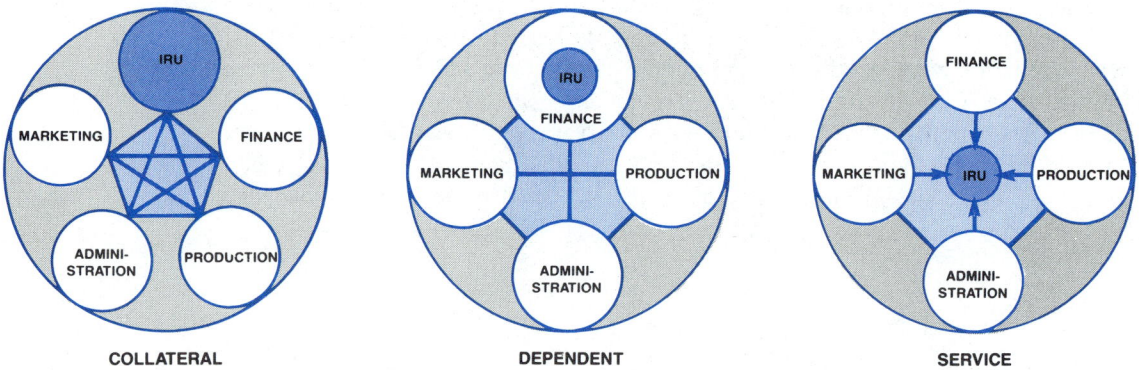

FIGURE 2–13. Three Relationships between the IRU and the Rest of the Organization

1. **Collateral**. The IRU is a subsystem that interacts with other departments, also subsystems, on an equal basis. This view can also be called "functional" — it sees the IRU as a function, like marketing, administration, production, and finance. As software, information, and access become profitable products, the collateral relationship should become more prevalent in larger firms.

2. **Dependent**. In this model, the IRU is either located totally within a user department (often accounting or finance, sometimes engineering) or clones of the IRU are found in several departments. Many firms begin their IRUs this way, only later amalgamating the pieces and building a formal organization. With the appearance of small, cheap, powerful microcomputers, however, even larger firms with mature, collateral IRUs are discovering the reappearance of the dependent "microIRU" within user departments. Universities, in particular,

seem prone to this evolution as professors and students acquire microcomputers that the academic computer group does not want to administer and maintain.

3. By far the most common relationship is that of a **service organization**, which provides information resource services to all departments on a priority basis (that is, first-come first-served after political considerations). This model naturally evolves from the dependent relationship, since user groups do not want to be dictated to by those who provide necessary services. Like a service organization, the IRU is functionally similar to the mail room, building services, and shipping. As the IRU grows in budget and technical expertise and as the firm comes to depend more and more on IRU services, the service relationship becomes dissatisfying for all players and evolution into the collateral model (which is more professional but riskier) or the multi-IRU model is likely. The information center is an attempt to preempt the trend toward small user-based IRUs by bringing the IRU directly to users, bypassing the mini-bureaucracy commonly found in IRUs.

There are three other possible relationships between the IRU and the rest of the organization. The IRU can be **centralized** or **distributed**, **functional** or **service-oriented**, and **in-house** or **contracted out**.

When the IRU is centralized, all facilities are available at a specific location, although terminals may be placed elsewhere in a wide network. The centralized structure provides a single door for user access to personnel, projects, and influence. By decentralizing into coordinated units that have discretion with regard to projects (subject to technical coordination and general organizational considerations), the IRU can open many doors to users in a variety of ways, depending on user skills and resources. Thus, the novice user can go to an information center, the experienced user can contact a consulting group, and the political, powerful user can put together a task force.

The functional IRU works on its own projects, while providing the information resource function to the firm. Projects may be initiated by users, but only if they are consistent with the corporate Information Resources Plan (IRP). In the service mode, the IRU works *only* on user-initiated projects. There is obviously a tension between these two models. The functional approach requires a politically skilled IRU and an organization whose executive appreciates the value that the IRU adds to the firm.

Many firms either fail to develop in-house IRUs or actively choose not to. Sometimes the purchase of outside information resource skills is more cost effective, especially for small businesses that cannot afford to staff their own microcomputers. A $7,500 computer purchase is cheap; a $30,000-a-year programmer is not. Others simply install terminals to a service bureau and contract system development to the bureau's technical staff. Paying an inflated price for those services rendered on a per-minute basis means they do not have to bear the price of human and technical resource management. Others have only a skeleton staff and access their computer through a power fourth-generation language (4GL), which gives them the ability to create, access, query, and report on their organization's data without ever "developing" an application. Chapter 4 sheds some light on this approach.

Thus, there are three distinct kinds of relationships the IRU can have with the rest of the firm. In each case, there can be changes in these relationships over time. A dependent IRU may become an independent service group and then evolve into a collateral functional organization. There is a natural association of the collateral organization with functional responsibilities, the service-provision organization with service responsibilities, and the dependent organization with specific departmental responsibilities. Most IRUs are still the departmentally dependent. As conglomerate IRU organizations appear in collateral relationships with other functional departments, the IRU will find itself a major player in the firm.

It is to these trends that we turn our attention for the final section of this chapter.

2.5   PLANNING AN INFORMATION RESOURCE

The previous sections dealt with the IRU as though it were a static element in the system of the organization. However, the IRU has origins and characteristic growth patterns. These patterns result from influences from the technology, from management, and from workers. There are several phenomena at work; two of the most important are "technology innovation" and "technology diffusion."

Technology Innovation

Technology innovation is the introduction of a new way of working, thinking, or transacting. Technology is a way of accomplishing work. One could say, as Hannah Arendt does in her book *The Human Condition* (1958), that work is the task of building something permanent, of making a change that matters. Technology dictates the tools through which we work. Modern management theory strongly stresses the role of tools in work as mediators between the skill of the worker and the product: to have good products requires good skills and appropriate technology.

A technology innovation always influences the product of an organization. How a new technology is introduced depends on a number of factors, including the following:

1. Where an organization is complex (that is, a large number of specialties or professions), innovations are more likely to be suggested. High complexity, however, makes it more difficult for innovations to be adopted because of conflicts and the inability to reach agreement among the complex web of interests.
2. Where an organization is formal and emphasizes rules and procedures, there is less likelihood that innovation will be sought. During the actual construction or installation of the innovative technology, however, work proceeds more quickly and with less confusion.
3. Innovation occurs more rapidly from the top down. Not only is management support valuable, management innovation is more adoptable. However, highly centralized organizations (with low levels of participation in decision making by lower levels) make it more difficult for information to be available to innovators at the top.

Innovations that are revolutionary in nature are less adoptable than those that are merely evolutionary. Bringing in a computer on Monday morning may be a blueprint for disaster. A more reasonable approach is a phased-in implementation.

The characteristics of the innovations have strong effects on the probability of their adoption. Figure 2-14 illustrates a model of technology adoption put forward by Zalt-

FIGURE 2–14. Technology Adoption (Based on Zaltman, Duncan and Holbek, 1973)

man, Duncan, and Holbek (1973). They hypothesize that innovation adoption requires the completion of a five-step "program":

1. **knowledge-awareness**: understanding the innovation
2. **attitude formation**: adopting a stance toward the innovation
3. **decision**: deciding whether or not to adopt
4. **initial implementation**
5. **continued implementation** and sustained employment

Different characteristics of the innovation are important in different phases of innovation adoption. For instance, initial cost (which is quite high for computerized technology) may form a barrier in the attitude-formation stage, whereas continuing costs may come out only during decision-making.

Other factors during early stages that inhibit innovation are perceived low return on investment, risk in charting new territory, difficulty in understanding or finding out about the innovation, lack of compatibility with existing methods, the apparent complexity of the innovation and its implementation, and the perception that the innovation is "forced" from outside rather than being a natural outgrowth of the business (called a "technology stampede").

Later in the adoption process, other factors come into play: whether the innovation appears to be state of the art, continuing costs, the number and the strength of the perceived advantages, and the nature of organizational commitment ("Can we move the computer out if it doesn't work?").

During implementation, the question "Will the end ever be in sight" comes into play. This is a strong factor because information resources seem to grow constantly and require continuing maintenance. Other important factors include the perception of widespread, rather than localized, value ("Is this only for the Veeps? Who needs it?"), the availability of continuing innovations related to the existing one ("Can we keep up-to-date and state-of-the-art with this system?"), and how the innovation affects interpersonal relationships.

The IRU and its paraphernalia represent an innovation in most firms. The factors mentioned above influence both the initial acceptance of an IRU as well as its subsequent implementation. Mere implementation, however, rarely produces a wide clientele for the services of the IRU. That depends on the phenomenon of **technology diffusion**.

Technology Diffusion

Gibson and Nolan (1974), among others, have commented on the diffusion of the innovation represented by computerized information processing throughout a firm over time. According to Gibson and Nolan, diffusion follows a pattern that is typified by a "growth" or "contagion" curve. In this pattern, illustrated in Figure 2-15, information

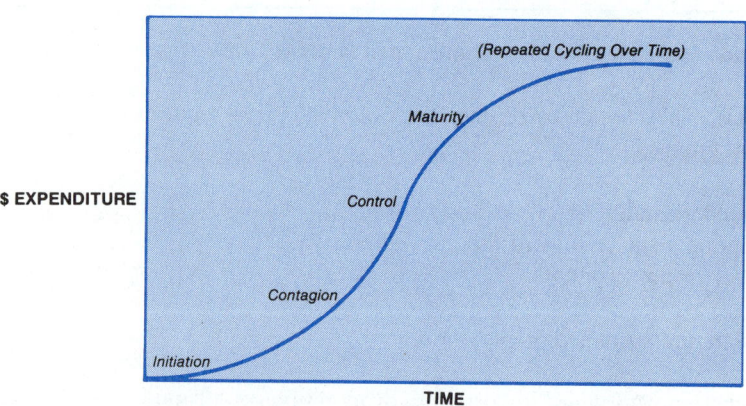

FIGURE 2–15. Gibson and Nolan's Stages Hypothesis

processing expenditures are slight in the early stages because the level of commitment is low as isolated groups "try out" the innovation. As these pioneers interest those with whom they interact in the value of the technique, those people, in turn, influence others. This second stage is the "growth" stage, and it is in this stage that growth is most rapid. Individuals seek the new techniques that they know work for others. Their resistance is relatively low and their interest, high. Also, because expenditure is still relatively low, risks (that is, costs that might not be paid back) are also low.

Expenditure does rise, however, and as it does, perceived costs rise. Also, individuals whose interest is high become more scarce. The remaining non-users (termed "non-adopters" in the literature of technology diffusion) have an inherently higher resistance, have more difficult applications, or are more difficult to approach. Perhaps the non-adopters are working on projects that cannot be changed, or they have long-term commitments to other techniques or workers. In some cases, their work simply may not be compatible with the information resource as it currently exists and at current resource levels.

The administration of the firm begins to notice as expenditures become considerable and corporate investments in equipment, people, and physical plant become significant. Typically, in this third stage, controls are placed on use of the resource. Sophisticated "charge-back" schemes (through which users of the information resource pay for the actual employment of each element of the resource — say, computer time, paper, connect time for terminals, use of disks or tapes) are developed and installed, and the information resource begins to spend a significant proportion of its resources on overhead such as a newsletter, training, bookkeeping, and long-range planning. In other words, in this "control" stage, the IRU becomes like any other cost center in a firm — accountable for its expenses.

Finally, Gibson and Nolan hypothesized a fourth stage, called "maturity," in which growth is minimal. The IRU provides maximum service for the existing demand. Controls make it difficult to innovate; hence, new applications that might stir new demands are not forthcoming.

Research has demonstrated two things. First, most firms have not reached even the third stage in Gibson and Nolan's model. Second, in many firms, stages may be skipped and growth may not be smooth like the growth curve, but rather choppy instead. By purchasing a major computer, many firms are parachuted into full-fledged information resource operations for a variety of reasons other than diffusion. One might be a company-wide directive to switch to computerization to appear state-of-the-art, to compete better, or to comply with regulations. Gibson and Nolan's concept is helpful, however, when we discuss how particular systems offered through the IRU (such as an accounting package, a market- research package, or a billing/order-entry system) are adopted throughout a firm or within a department. The growth curve is actually a model of resource utilization, as outlined in Chapter 1. Growth peaks when the availability of new users is still high and there is a "critical mass" of existing users. Later, non-adopters are more likely to be refusers than users for whom the IRU has not yet demonstrated advantages.

Technology diffusion, particularly through the IRU, can be facilitated in the following ways:

1. Education: Where a system is to be introduced among nontechnical people, prior training in "computer literacy" may assist in combatting resistance, prejudice, and ignorance.
2. Top management support: This is a perpetual motivator for employees.
3. Decentralized services: Offering the services through a variety of locations is better than forcing potential users to come to a distant place to access the service.
4. User friendliness: Although systems are probably neutral with respect to "friendliness," it is not uncommon for people to feel put out with systems they feel uncomfortable using. Careful attention to display formats, order of screens, wording, reading speed, color and graphics, and the unambiguous meaning of terms can assist.
5. Transparency: To the extent that potential adopters do not have to learn new procedures to get their work done, initial resistance will be lower. Many systems can be phased in using existing paper forms; those forms can be duplicated on the displays and on-screen reports can appear identical to printed ones.

Planning in a Turbulent Environment

As pointed out in Chapter 1, turbulent environments are characterized by extreme unpredictability; not only are the possibilities difficult to predict, but the *nature* of the possibilities is often unknown. Probabilities change radically at the whim of unknown and unpredictable relationships. To the degree that a turbulent environment is strong and smart, organizations will find coping with it difficult.

Planning an IRU is difficult precisely because of the turbulence in the immediate environment. This turbulence arises from three sources:

- technology changes
- social and economic changes
- user changes

The first is the well-known "march of progress," which is neither predictable nor forecastable. Specific advances are hard to anticipate, and those that have been anticipated are difficult to forecast in terms of impact, price, capacity, and so forth. As recently as 1982, the 64Kbyte microcomputer was the coming standard; now 640Kbyte machines are standard and 2-megabyte powerhouses seem poised to muscle their way into the minds, hearts, and offices of corporate America. Other "trends" did not develop as expected, including office automation, the portable microcomputer, and cottage industry software. Programmers, who were supposed to be obsolete by 1985, are now needed in increasing numbers.

Technology forecasting is a tricky business in the best of times. Several techniques have been developed, although it is apparent that their predictions have not been of high or lasting quality. Planners who need technology forecasts for IRUs may require as much in the way of courage as they do in knowledge of statistics.

Economic and social currents in society also produce waves in the turbulent IRU-planning environment. Each change in taxation law that affects depreciation of equipment, each employment-related regulation, every educational white paper — all influence the way people see information resources both as sources of employment and as contributors to the national balance sheet. To cite only one example, consider the effect of government regulations on the hiring of minority groups into highly specialized units. Because tests for programming ability have been suspect for many years, there exist few objective instruments by which to select programmers. Hence, justification of individual hiring decisions may prove difficult under a court challenge.

Finally, users are more "literate" and, with that knowledge, more demanding of IRUs to deliver convivial and useful services. Simultaneously, lasting management, having awakened to the power that the information resource provides, seeks and receives increasingly sophisticated decision-support aids, both at the low end in microcomputer spreadsheeting and at the upper end through integrated, distributed corporate database management. In addition, clientele realize that they, too, have a stake in the information resource. In recent years, legislation has been passed concerning access, privacy, and accuracy of databases, as well as the rights individuals and groups have to data about themselves. These developments only serve to increase the turbulence in the IRU's environment.

An IRU introduction plan must thread its way carefully about these sources of turbulence. The five principles outlined in the previous section concerning technology diffusion apply equally to the IRU as an organizational "technology" (that is, a method of meeting goals):

1. Education, public relations, and consultation are far more important now in setting up and "diffusing" the IRU throughout the organization than they were even ten years ago. A plan for the introduction of the IRU should include an educational plan, a public relations policy, and a procedure for consultation with user groups. While a **steering committee** has usually been the formal guiding

body for IRU introduction, such bodies are naturally limited in responsiveness and educational capabilities, since they are usually composed of higher-level individuals, technically minded users, and others who are already naturally inclined to know about and support the IRU. Others who are less likely to understand the IRU and its purpose are likely not to have any participation in the planning stages. Because of this, steering committees should be augmented with resource persons who are familiar with public relations and education. For instance, along with the IRU, there should be a formal training program in IRU fundamentals for affected individuals long before the actual applications are implemented.

2. Top-management support is obviously necessary for expensive ventures like the IRU. Beyond this, the support should be visible and effective. Top management must understand information resource technology, environmental influences, and the stakes not only within the firm, but outside, too. Again, a training plan before actual IRU planning begins may pay off.

3. IRU layers should be decentralized so that ordinary mortals can understand and use some IRU facilities without comprehending the whole unit. That translates into having IRU staff who interpret the IRU for others or training individuals within the organization to do so. Otherwise, technology diffusion relies on the informal organizational channels of keeners and stakeholders who will certainly discourage timid and hesitant potential adopters. In the extreme, presenting the IRU as a monolithic take-it-or-leave-it whole polarizes staff into a small cadre of confident collaborators and a far larger group of non-adopters, producing more turbulence.

4. An IRU should be user friendly, which means approaching the user at his own level, with tools and techniques already familiar to the user or that can be learned easily. An IRU that is isolated and develops its own organizational motive power, vocabulary, constituency, and vested interests is one doomed to be misunderstood and disliked. The IRU that is seen as an easy extension of work patterns (because it fits in easily with how individuals work), understandable, on a par with other organizational systems (many workers find that the special constituencies that grow up around technical services like the IRU are impenetrable, leading to the perception that you have to be an insider to get service), and sharing the stakes of the users rather than pursuing its own arcane or selfish goals is going to be well liked, supported, and well used.

5. Finally, the ease of use of the IRU should extend to its relative transparency within the organization. Growth of the IRU should parallel and complement growth of the information resource, not some arbitrary plan dictated by technology stampedes. Coupling IRU growth to specific measured needs eliminates the necessity to anticipate much of the turbulence in the technical environment. By making the IRU transparent (that is, well integrated into the organization and its goals), support for the IRU becomes implicit in management's support and the organization's goals as a whole.

Growth Plans for the IRU

IRU growth may be influenced by the initial introduction plan. In small firms, the IRU may actually be "bought" in turnkey fashion from a vendor or consulting firm. In these cases, the computer arrives on Monday and the head of the task force that planned the computer becomes the director of the department. On Tuesday, two programmers are hired, and on Wednesday, full-scale production begins.

There are difficulties with such a plan, stemming from lack of user and staff training. In very small firms, owner/operators simply do not have the time to become familiar with the technology before it is introduced; this usually leads to a considerable time period in which everyone has to "get used to" the computer. In larger firms, a specialized department may quickly become a vested interest, and a costly one at that. Often the turnkey introduction of the IRU is the result of a technology stampede (see the first case, below).

Growth may be steady, but more likely it occurs in spurts, with staff added to match either new application demand or new technological opportunities. This push-pull relationship (a push from the users, a pull from the installed technology) often characterizes the entire growth pattern of IRUs. Managing this relationship becomes a major responsibility of IRU management. As the work technology provided by the IRU diffuses throughout the organization, the stress of push-pull becomes noticeable and important (see the second case, below).

Following are four cases in IRU introduction and growth. The first, called "**technology stampede**" documents how a small organization becomes stampeded into setting up an IRU. The next describes a typical **quantum diffusion** process. The third looks at a **technology invasion**, or what happens when one firm acquires another and forces technology into the organization. Finally, the fourth case looks at **uncontrolled growth**, its fruits, and its difficulties.

Four Cases in IRU Planning

Merv's Menswear is a small, three-outlet retail men's clothing store that has been operating in a town of half a million for seven years. Merv McInerny has built Merv's into a class operation serving the better-dressed men in town; his clientele are loyal and steady customers. About eight months ago, Merv read in a menswear trade journal that microcomputers are making inroads into the small retailing industry, handling purchasing, inventory, billing, accounting and even some point-of-sale activities. Within a month after reading the article, Merv saw three turnkey vendors and two computer salespersons and spoke to six of his competitors about computerization. Only one had actually tried computerizing, but three used a service bureau to handle billing and accounting. There was a lot of interest, though, and Merv was intrigued with the computer as a way to cut costs, keep better track of sales, and perhaps reduce "shrinkage" (theft).

Six months ago, Merv brought in Balsam Business Consultants, who, after a quick look around Merv's, convinced him to purchase their business system, built around a small microcomputer. From the day it was installed, Merv has experienced difficulties; within two weeks, his bookkeeper had quit. His inventory is now so thoroughly mismanaged that visual checks have to be taken weekly. In effect, the computer has been junked.

While it prints checks nicely, it is not easy to work with. Several different diskettes have to be located and mounted to run each application, and there is not enough capacity to expand any of them. Also, no one knows how to reprogram it except the folks from BBS. Their monthly maintenance charge of $200 seems excessive for the little help they offer, but neither Merv, his bookkeeper, nor his sales people have the time to learn about computers!

Hiway Robers, Inc. is a supplier of robes and dressing gowns to Merv's Menswear. At Hiway, Hiram Conway, the owner and general manager, used a different tactic to build an information resource within his firm. First, he educated himself and his staff about computers and information. He came to understand his own business in information terms before he contacted any consultants or vendors, even taking a course in computers at the local junior college. He then spent the money to employ a consultant to teach him about computers and to examine his business, to give him an independent view.

When he contacted vendors, he sent them a request for proposal (RFP), which asked vendors to spell out hardware and software they felt would be required in his small manufacturing environment. He also stressed education, continuing suppport, and expansion into other application areas. He had some reservations about the process, especially since he recommended only a range of prices rather than a fixed maximum. He asked each vendor to spell out the potential tradeoffs (price versus capacity, maintainability versus functionality, speed of operation versus conviviality) and requested a specific implementation plan from each.

Hi chose two vendors to give presentations to him, his accountant, and his hired consultant. Present also were his wife, who handled the books, and the shop foreman, who filled out most of the forms. While he had decided beforehand that the quality of the presentation was not as important as the material presented, he was surprised to discover that he responded to hype. He was glad that his sessions with the consultant and his training prevented him from acting immediately.

Hi chose Probusiness Systems because their clear presentation spelled out costs as well as benefits and their training and continuing maintenance plans were superior. He also appreciated that their implementation plan included training one of his own staffers in the "internal workings" of the system so that he would not remain forever dependent on Probusiness. They demonstrated their willingness to build the system beyond the basics they planned to introduce and, further, to advise Hi on hiring. They also showed their trust in him by allowing him to reserve certain aspects of the customizing of their package further down the road, but they also made sure that they understood his current needs.

After eighteen months, Hi has expanded his computerized information resource twice, adding inventory and supplier data bases, expanding his computer capacity 50 percent, and adding two staffers, one a part-time programmer and the other, a data-entry clerk. Hi is satisfied with the steady growth so far and is unafraid of adding subsystems when they seem necessary.

When Merv left his office late one afternoon, he walked down Main Street to the Bootique to look at some shoes. Jim Fawcette, the operator and, until recently, owner was there as usual. Jim sold his store a month ago to Footware International and although he has stayed on to operate the store, he is not particularly happy tonight:

"Oh, Merv, say, how are things going with your new computer?" asked Jim.

"Not so new, and not so good, thanks. Why do you ask?"

"Well, if you look over at the sales booth, you might notice a new POS terminal. The systems people at Footware brought that little honey in last Wednesday, and now we're linked into their mainframe in Bigtown for orders, billing, inventory, and so forth."

"Great, Jim. I guess your experience is better than mine, no?"

"Merv, for ten cents I'd throw it out. Now, I'm not saying we shouldn't automate. Footware bought the place and as far as I'm concerned they can put robots in the stockroom. But what this terminal is doing to my staff is almost criminal. You know, we used to be able to get inventory information by going back there and counting. Even new clerks knew the stock in a few days. And I could phone my accountant and get up-to-date information quickly and pleasantly. But this new terminal is the only way we can find out about shoes, sales, and stock and, well, there's simply no way to learn it. Those systems guys are four hundred miles away and they don't return phone calls. And when they do, who can understand them? The operator's manual is written in Martian, the people at Footware know less than we do, and I'm pulling my hair out trying to use it. One salesman, a guy who's worked here for seven years, just quit this morning. I'm dreading having to train a new guy on this system without any good manuals, and just when I'll get a stock movement report is a mystery. I saw one example when they installed this thing and it's almost impossible to understand. It wouldn't be so bad, but I've phoned other stores that Footware's acquired and they all say the same thing: just wait a month and things will get worse."

"Say, Jim, you got any size 9C black loafers?"

"Who knows, Merv?"

Four hundred miles away in Bigtown, Footware International is having its problems, icebergs that Jim Fawcette has only seen the tip of. At the stylish headquarters of the largest footwear retailers in the North, Debby Bannister, director of systems development, is upset. For over six years, CSD (the cash systems division) has been steadily, and without much foresight, growing. From a minicomputer that handled accounting, Debby has built an institution around a pair of mainframes, a satellite network, and over one hundred and fifty remote terminals in every retail outlet, three warehouses, and two office buildings. She heads a staff of fourteen, developing and maintaining not only accounting software but also network control, management decision support, facilities planning, market research, and point-of-sale terminals for on-line entry of sales — the last application includes inventory, billing, and efficient ordering and bill payment. Yet for the most part, Debby has only responded to needs rather than anticipated them. As Footware has grown, so, too, has the need to provide computer-based information resources to sales personnel, managers, store operators, and executives. As the need has risen, Debby has bought, built, and borrowed the necessary technology in an economical way. But still, she has acquired machinery and staff without much planning. The results have been unpleasant. Newly acquired stores have no technical support, manuals are poorly written, training is non-existent, and delivery time for applications is long and getting longer. Debby has not planned for the introduction of a computerized information resource, so she cannot anticipate the next crisis. And the most critical factor, user involvement, is missing as technology invades the stores. The corporate policy of acquiring small, quality outlets means that almost every new user of the POS system is also a new employee of Footware, unfamiliar with CSD and its procedures. The phones ring constantly and, with that last call from Smalltown's Jim Fawcette, Debby has about had it.

These four cases show how technology stampede, quantum growth, technology invasion, and uncontrolled growth can affect organizations and their abilities to cope with turbulence. As Hi's experience demonstrates, quantum, controlled growth, lead by good corporate planning and a good knowledge base, point to success. Being stampeded or forced to automate and to set up or cope with an IRU without planning can lead to disaster.

2.6 CONCLUSION

The **Information Resource Unit** provides access to the information resources of the organization. It is charged with managing and maintaining that resource. In order to do that, the IRU must carry out several functions. Foremost among these are new system and system element design, development, and installation; critical in this chain is systems analysis.

The IRU employs a variety of individuals in pursuit of this goal. Programmers, operators, systems analysts, managers, and librarians comprise the technical, professional staff. In addition, clerks, secretaries, receptionists, and accounting staff supplement these efforts, especially in large firms. Budgets for IRM may run into many millions; even small budgets require careful control. In the larger firms, the interface with users and other departments may be formalized and require trainers, public relations staff, and marketing representatives.

The structures through which the IRU can achieve its goals include the **task force**, the **mini-bureaucracy**, the **project group**, and the **consulting group**. Often the functions of the IRU are dispersed across a number of departments or divisions. The IRU can be **centralized** or **decentralized**. It can operate along **functional lines**, or it can be a **service organization**. Finally, it can employ **in-house personnel**, or it can be partly or almost completely **contracted out**. Many small firms work through a consulting firm.

Many suborganizations within a firm utilize the services of the IRU. Relationships can be **dependent, integrated,** or **collateral**, depending upon how closely IRU personnel work with others in the firm. The general model pursued in most larger firms is that of a dependent IRU as a service organization within the larger organization. In smaller firms, the IRU is completely integrated into another department, usually accounting or finance. The third model is usually the result of a system's rapid growth and the need to set up an integrated information resource in a new subsystem from the ground up.

IRUs grow in a number of ways. Growth can be either rapid or slow, phased-in or turnkey, steady or in quantum leaps. Controlling growth of the IRU is a major concern for many organizations. Just as important is keeping up with technology changes. Major strategies for IRU growth include **technology stampede, technology diffusion** from the top down, **technology invasion** from the outside, and **uncontrolled growth** as a default.

Many large firms and most small ones work without the benefit of an information resources plan and thus pursue unplanned, uncontrolled growth as their only goal.

2.7   CASE: THE CASE OF THE LOST QUAYS

Over Seize Shippers is based in an Eastern port city and handles a large volume of overseas shipping. Basically, OSS deals with damaged cargo, which it buys it off boats at dockside in as-is condition and resells to jobbers. Most of the goods eventually make their way to "budget" or "military surplus" stores in town; often, merchandise is exceptionally fine and needs only cleaning or minor mending. Over Seize makes a reasonable profit from its operations, headed by Sandor DiMaggio and his sister, Marina.

Sandor does most of the dockside haggling, while Marina handles all the bookkeeping. Over Seize also employs five warehousemen, two clerks to keep track of the paperwork, two sales representatives who collect from the jobbers (who buy on credit), and a number of casual laborers. In addition, Over Seize contracts out its hauling and cleaning.

Until recently, Sandor and Marina kept information in their heads, and their dockside dealings were based on wits and bluff. When merchandise is mellowing with mildew and frozen strawberries are self-pureeing in summer heat, the DiMaggios cannot afford to consult complicated books. But there has been some recent competition from others who specialize in food, clothing, and nonperishables. Prices offered for salvage lots at dockside are rising and Over Seize cannot afford to keep bidding them up.

Sandor knows a little about computers and wants to set up a computer group in his company of eleven people. He knows that computers can do his payroll (which is currently handled manually by Marina), and keep track of shipping arrivals, prices paid, and the movement of goods in lot quantities. On the other hand, he doesn't just want to buy a computer and let *it* run *him*! He has read up a bit on inventory systems and microcomputers, and he knows some programming from a community college course. He'd like to go about planning in an intelligent way.

1. From what you know so far, what (a) positive and (b) negative comments would you make about how the DiMaggios have gone about planning?
2. Given the size of the firm, its annual gross revenue of $1.5 million, and its profit of $125,000, what recommendation would you make about the introduction of an information resources unit?
3. Consider these alternatives and evaluate each:
   a. a turnkey installation supplied by a local consulting company
   b. hiring a programmer who would buy a computer and set up a computer department
   c. having Marina learn about computers and start to use one
4. Sandor decides to hire a programmer with some knowledge of business (his nephew Tony) to set up a computer department with a budget of $40,000. What criticism would you offer.

5. Imagine that Over Seize were twenty times as large as it is, with over 200 staff members and a yearly gross of $30,000,000. What planning considerations would be different, assuming that everything is currently done manually.
6. The computer seems central to this case, yet there is already an information resource unit functioning. What is it, how does it work, and how does it fit into the firm now?

DISCUSSION QUESTIONS

1. Consider the IRU that supplies computing services to an academic community. How are these services "partitioned" into various roles such as administrative computing, research, academic records, registration, and so forth?
2. Suppose you were hired as a systems analyst to work on the design of a system to track personnel career data in a large bank. With what sorts of individuals would you interact to find out what is needed? How would you gain access to these people? What organizational barriers would you likely encounter in your quest?
3. Why do you suppose the IRU has to be run on a business-like basis? What conflicts can you see arising from this goal? What is the "business" of an IRU, and how would you propose it to be carried out in the following three examples?
   a. A farm cooperative, owned by the members, for which the IRU would supply marketing, agricultural supply, and distribution information
   b. A government department, staffed by four hundred individuals, that is charged with maintaining commerce, labor, and resource need information and for which the IRU handles regulatory information on small businesses
   c. A large medical clinic of eleven doctors, eight nurses, and three receptionist/clerks that uses a manual information system for billing, patient-record maintenance, recalling and scheduling, and supply ordering
4. What "public relations" functions can you see as necessary for a large IRU within a large organization? What is likely to happen when public relations go sour? Can you see the technology itself assisting in this function?
5. Consider the task-force, mini-bureaucracy, project group, consulting group, and their ability to respond to the following typical "crises" that an information resource unit might face:
   a. The computer fails and is down for three days during peak season
   b. The VP Accounting decides that she is not getting enough service of the appropriate type from IRU staffers and puts a lot of pressure on the president to do something
   c. The union president says that there is no way his people are going to use the new POS terminals unless they are thoroughly checked out for environmental hazards and unless the POS operators get an immediate 5 percent pay increase for doing "technological" work

d. The manual system is going to be junked within one year and some new computers brought in

How is each structure suited to respond to each situation? Which is best? Considering that these situations are not common but are representative, how would you "evolve" a structure to cope best with a series of such situations?

6. How should Miriam respond to Laurie in the following conversation?

Laurie [L]: Listen, I'm the vice president of finance. Your department started here so we should be getting much better service from you down there.

Miriam [M]: Laurie, I can see your needs, but we have many other groups to respond to, and we've got our own initiatives to pursue as well. For instance, we're planning to upgrade to a mainframe in the fall and half our staff is planning that transition now.

[L]: That's great but irrelevant. We need service. You provide the service. Your initiatives are unimportant to us in finance. We need better response and better-quality work in a shorter period of time. You can't just go your own way without providing service to us.

[M]: ?

7. Indicate the forces that might create a collateral IRU from a situation in which the IRU is completely dependent within another organization.

8. Can an IRU ever *shrink*? What might cause this to occur? Can an IRU ever become dependent from a collateral position? What might cause the IRU to shift from a service organization to a functional one? What does the IRU have to sell?

9. Consider Merv, Jim, and Debby. Prescribe an organizational "liniment" that could "cure" their problem at this point. Begin by understanding what could have caused the problems in each situation and then point out what each organization can do to get growth on a proper trajectory.

10. What causes a technology stampede? How can managers avoid them?

11. Information technology has a particular way of innovating and diffusing within an organization. Conduct an investigation into information technology innovation and diffusion in your school or firm. Where did the new ideas come from? Who nurtured them? Who advertised them? Why were they attractive? How did growth occur: steadily, spasmodically, in quantum leaps, or through technology invasion? What was the "price" of growth? Who paid? What are the existing growth plans for the short and long terms?

DESIGN EXERCISES
Using dBASE III

1. The files that are on your data diskette have few records in them. The command in dBASE III that adds data is called APPEND. Bring dBASE III up, type USE CLIENTS and then...

APPEND   Tells dBASE III to add records to the end of the file

DESIGN EXERCISES
Using dBASE III *continued*

Fill in the **form** with information about yourself. Note that dBASE III will not tell you if you enter the wrong codes; it will let you know if you try to enter a letter into a numerical field (but not a number into a character field!). When you have finished, press ENTER; otherwise you can continue adding records (the version of dBASE III distibuted with this text limits you to 32 records in a file, however).

2. It is apparent from the name of the CLIENTS file and its fields that HAIR serves a function for a beauty salon. In fact, the system has been developed for HAIRLOOM Stylists, Inc. Bring up dBASE III and use these commands to explore CLIENTS and other files in more detail:

USE CLIENTS	Tells dBASE to examine CLIENTS.DBF
DISPLAY ALL NAME	Displays the names of all customers
DISPLAY ALL NAME, PHONE	Displays a table of names and phone numbers

How many different names are there? How many different combinations of name and phone number?

DISPLAY ALL NAME, COLOR	Displays a table of names and hair color (natural or otherwise!)

How many different hair colors are there?

DISPLAY NAME FOR COLOR = "BK"	Displays the names of all customers with BK (black) hair

How many customers have black hair? Grey hair (GR)? Red(RD)? Blonde (BL)?

DISPLAY NAME FOR COLOR = "GR" .AND. AGE < 30	Displays the names of all customers with prematurely grey hair

How many prematurely grey customers are there? Note the use of ".AND." (this means exactly what it seems to) and the mathematical comparison " < " (less than). You can use ".OR." and the symbols " > " (greater than), " > = " (greater than or equal to) or " < = " (less than or equal). What do these mean and how many are there:

DESIGN EXERCISES
Using dBASE III *continued*

DISPLAY HAIR FOR STYLNAME = "FRANK" .OR. STYLNAME = "NANCY"
DISPLAY STYLNAME FOR HAIR = "CU" .AND. AGE > 60

As you can see, DISPLAY finds a record for you and displays the information you request. The LOCATE command works like display, but does not display the record. LOCATE finds the next record meeting certain conditions.

LOCATE FOR HAIR = "CU" .AND. AGE > 60
DISPLAY STYLNAME

You can move from record to record within a file like this:

SKIP   Go to the next record
SKIP + 3   Skip over 2 records
SKIP –3   Back to the same record

Notice that SKIP does not DISPLAY. If you prefer not seeing the names of the fields displayed along with the data, enter this:

SET HEADING OFF   Use ON to turn it on again

3. Referring to the table on the opposite page, USE PRODUCTS and APPEND records for the data listed. This data is needed for the other exercises. After entering the data, you can DISPLAY all the records and check this against the printed list.

DESIGN EXERCISES
Using dBASE III *continued*

ITEM	ITEM NAME	DESCRIPTION	PRICE	SALE PRICE	REORDER	ON HAND	SUPPLIER	ON ORDER
1201	MOUSSE	BEAUTY-BAR	7.55	21.00	30	66	MALLORY	N
1101	GEL	BEAUTY-BAR	5.16	12.00	10	3	JONES	Y
3005	NAIL POLISH	BLACK	4.50	8.00	12	5	MASTER SUPP.	Y
1601	BRUSH	CONFRAC BUSH	12.95	24.00	5	3	DON ELLIS	Y
1003	SHAMPOO	DRY	9.43	16.00	100	148	JONES	N
1308	RINSE	DRY	5.55	10.00	60	60	JONES	N
1450	CONDITIONER	DRY-BEAUTY-BAR	6.14	12.00	100	144	MALLORY	N
3004	NAIL POLISH	ELECTRIC ORANGE	3.50	6.00	12	24	PARTICULARS	N
1306	RINSE	FOR OILY HAIR	5.55	10.00	60	40	JONES	Y
1001	SHAMPOO	FOR OILY HAIR	9.43	16.00	100	210	JONES	N
1501	SPRAY	HAIR NET SPRAY	3.18	6.00	50	180	MALLORY	Y
2005	BRUSH	HARD BRUSH/LONG	7.44	18.00	10	12	DON ELLIS	N
2150	DRYER	HIGH PWR DRYER	16.60	26.00	10	8	DON ELLIS	N
1002	SHAMPOO	NORMAL HAIR	9.43	16.00	100	21	JONES	Y
1307	RINSE	NORMAL HAIR	5.55	10.00	60	63	JONES	N
1450	CONDITIONER	NORM-BEAUTY-BAR	6.14	12.00	100	80	MALLORY	Y
1450	CONDITIONER	OILY-BEAUTY-BAR	6.14	12.00	100	101	MALLORY	N
3003	NAIL POLISH	ORANGE	3.50	6.00	12	16	PARTICULARS	N
3007	NAIL POLISH	ORANGE 3	3.50	6.00	12	12	PARTICULARS	N
2151	DRYER	PORTABLE DRYER	12.60	20.00	10	9	DON ELLIS	N
2201	CURLING IRON	POTTER	9.45	16.00	5	8	DON ELLIS	N
3001	NAIL POLISH	RED 1	3.50	6.00	12	19	PARTICULARS	N
3002	NAIL POLISH	RED 2	3.50	6.00	12	12	PARTICULARS	Y
3006	NAIL POLISH	RED 3	3.50	6.00	12	35	PARTICULARS	N
2003	BRUSH	KIDS SOFTBRUSH	6.50	10.00	10	3	DON ELLIS	N

4. Hairloom Stylists, Inc. was started four years ago by Douglas van Vliet after six years' employment in a competitor's salon. Doug wants his salon to become known as THE place to get your hair styled in Bigtown. To get there, he has invested a lot of money in decoration and staff skills. To insure he stays there, he asked Mercurial Software, Ltd. to create a microcomputer package to keep track of his business. Originally, HAIR served only to record receipts and handle inventory. Mercurial recommended the hardware (which Doug subsequently purchased through them) and, over a three-month period, built a software system around dBASE III. Doug paid almost $6,000 for the system two years ago and has since spent another $2,500 expanding the memory and adding modules to keep track of personnel and customers. Trish

DESIGN EXERCISES
Using dBASE III *continued*

Drinnan, his receptionist, has spent a great deal of time learning how to use the system and Byron Jarvis, his bookkeeper and brother-in-law, has also become interested enough to take a course in dBASE III. Trish and Byron are the team that overlooks the system and calls in Mercurial when there are problems. Originally, Doug and his wife, Kelly Jarvis, spent a few hours with Mercurial's programmer/analyst, Larry Parsons. But the system itself was built at Mercurial's office and delivered when it was completed.

Having looked at the system a bit, consider the following questions:

a. What **functions** does HAIR carry out?

b. What **structure**(s) does (did) the "unofficial" IRU take (during recent history)? How does it relate to other areas of this type of business?

c. How would you describe the **growth** of the system? How has the technology been introduced?

CHAPTER 3

THE CRAFT OF THE
SYSTEMS ANALYST

OBJECTIVES

1. To describe the role of the systems analyst in the organization
2. To identify the activities of the systems analyst within the systems department
3. To list the major skills necessary to succeed in carrying out these activities and meeting the requirements of these roles
4. To describe the attitudes and experiences that contribute to the portfolio of systems analysts
5. To forecast the roles, activities, and products of the systems analyst in the near and mid-term future

THE CRAFT OF THE SYSTEMS ANALYST

3.0 KENT CLARKSON, SYSTEMS ANALYST

7:30 a.m. Kent Clarkson, systems analyst for a great metropolitan newspaper, reports for work. Actually, he doesn't really "report" for work, he logs in, then examines his electronic mail from the previous night's work on NEWSYS, a major news abstracting project he's leading. He directs the work of four programmers, one project secretary, and two novice coders on this project, which is intended to abstract news articles from each of the four daily editions of the *Monitor*. As lead analyst during the systems analysis phase, Kent had an important role in the specification of this project. As a "reward," he was placed in charge of his team of seven, given six months less than optimal, and told to go to it. Work is going well, but Kent has some misgivings about one of the programmers' skills, and his relationships with the newsroom staffers are less than wonderful. Thinking back over events of the past year has Kent pensive and doubtful that this is entirely what he was looking for when he took his first course in BASIC.

That course was twelve years, two degrees, two jobs, and four projects ago. While it would not be truthful to say that Kent has experienced only successes in twelve years, the failures have been few, unimportant, and generally easy to repair. Graduating from Metro Polytech with a B.S. in computer science and a minor in business, Kent moved quickly into a programmer-analyst position with a local warehousing firm, that was eager for young talent and desperate to computerize. Kent parlayed his technical skills, his understanding of business jargon, and a great deal of empathy into a redefinition of his own job: lead analyst.

At the same time, he acquired an M.B.A. with a specialization in information systems. Because he designed the inventory system for WhereHouse, Inc. almost single-handedly, he gained a reputation as a doer. Unlike many of his cohorts, he didn't lose his sense of humor, his patience, or his ability to explain things without jargon. In a few words, Kent was well liked, well read, and well heeled.

The *Monitor* shared these feelings when they went to a headhunter four years ago to look for a lead analyst for their NEWSYS project. Kent was by then almost technically obsolete because he had concentrated almost exclusively on the technology of the inventory system at WhereHouse, Inc. His *people* skills appealed to Louise Lyon, director of DP at the *Monitor*. Louise was fed up with the narrow-mindedness of her staff, most of whom regarded editors and reporters as necessary evils to be tolerated. She liked Kent's personality and his obvious leadership skills. Kent spent an afternoon with Louise and the associate editor of the *Monitor*, who was also impressed — impressed enough to offer Kent the position, with a considerable increase in salary and a great deal of freedom to influence information resource policy.

First on the agenda was revitalization of the computerized news-gathering system, primarily by updating the programs that communicated with reporters. By examining other systems and by extensive interviewing, Kent was able to redesign the "protocols" that governed conversations through the terminal, making them more sensitive to reporters' and editors' styles. The result was a widely acclaimed user-friendly package that others in the newspaper chain were quick to adopt. Louise then decided that Kent could begin work on NEWSYS, a system designed to abstract articles, producing summaries for the editors (a kind of one-page newspaper for the morning set-up) and the *Monitor* librarians, who handle queries from the public. Since both editorial and public relations goals were important, Kent's approach, which involves management and librarians as well as other analysts in a prototyping mode, is genuinely appreciated.

Kent's promotion to the NEWSYS project has not been completely without a down side. Kent hears a lot of complaints from programmers about work schedules, and he realizes that he doesn't have the resources to meet his delivery schedule. The NEWSYS project and the *Monitor's* production share the same computer; the three micros Kent wanted just haven't materialized due to the realities of cash flow. New players have shown up on the stage, too. Barnett Barett, owner of the Barett chain of newspapers with outlets from Portland to Shreveport, spent four hours with Kent last week finding out about the design details for NEWSYS. Barett repeatedly told Kent how important the project is to increased sales and the new ideas Barett has for minipapers, twelve editions daily, continental distribution of stories over microwave, and on and on. Barett dreams big on small dollars — circulation has dropped overall in the chain and he's looking for a morale boost. Meanwhile, Louise has accepted a position as VP Information Resources for a New York bank. With her position opening up, Louise is pressuring Kent to apply for the job. The other programmers, too, are subtly (and sometimes not so subtly) pressuring him to move up, over, out — anywhere away from them. Meanwhile, Kent has to learn more about local area networks and become more current in technology. No sir, this is not at all what Kent had envisaged twelve years ago.

3.1   SYSTEMS ANALYST: SUPERHERO OR GRAY-FLANNEL DRONE?

Kent Clarkson reminds us immediately that there is a *person* behind the jargon. People have people-concerns, regardless of the nature of the tools they employ in their work.

Kent does not know if he is fish or fowl, superhero or drone. He is at a point where being at the cutting edge of technology is a bit like being at the cutting edge of a sword. In other words, Kent is unsure of his role.

This section concerns the roles of the systems analyst in the organization. There are three roles that stem from the three functions an analyst serves within a modern firm:

- **entrepreneur**, change agent, salesperson
- **internal consultant**, teacher, mentor
- **employee**, desk jockey, contributor to profit

Each role is derived from a unique function the analyst serves (see Figure 3-1):

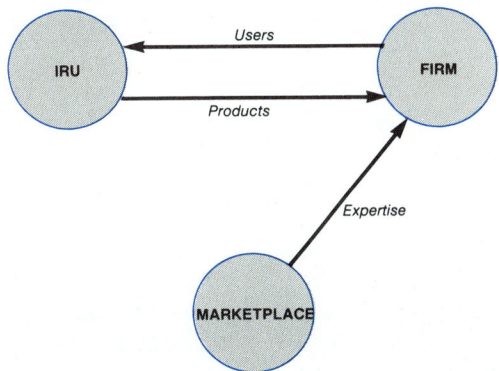

FIGURE 3–1. Information Exchanges

1. Analysts sell the services of the information resource to other employees of the firm to change the way people work toward their own goals and provide others in the firm with the latest and the best in technology. Analysts are simultaneously vendors, promoters, and impressarios because of their unique access to information resource technology.

2. On the other hand, analysts serve a more neutral role as providers of information, assistance, and advice about the technology to those who may already recognize their needs. It is common for analysts to recommend retaining a good manual system rather than attempting an expensive conversion to a poor computerized system. Analysts provide this source because of their training and because they often have spent the bulk of their careers as programmers outside the firm. It is also likely that analysts are *politically neutral* within a firm simply because they may spend far more time working with technology than dining with vice presidents.

3. Finally, and ultimately, analysts are employees of some organization. With the exception of consulting systems analysts brought in from software houses and DP consulting firms, analysts are like other workers. They rise in the morning, go to work, put in their eight or more hours, go home, and one day out of thirty, get a check. They are subject to similar pressures to complete their work, have similar fears for their jobs, and react the same way to poor ventilation. When their bosses recommend actions they do not agree with, they, too, wonder whether they have the right or the responsibility to speak up. Once a year, they attend a company Christmas party and behave like mortals faced with temptation. In short, they are people who work for a living.

The analyst must meet the expectations of a number of people, as well as career expectations within the context of late twentieth-century economics. That there are conflicts is less surprising than the fact that, for the most part, analysts (and programmers and others in the broad area of electronic data processing) have been able to avoid or ignore these conflicts for the past forty years. As a number of important trends accelerate, however, the conflicts become more apparent, more pressing, and more costly. This chapter discusses the variety of roles analysts fill both within the organization in general and within the systems or information resource management department specifically; what skills, attitudes, and credentials analysts should present to meet the challenges of these roles; and the nature of these trends and how they will affect analysts.

3.2   THE ANALYST AS SEEN BY OTHERS

Systems analysts are often seen against a backdrop of organizational considerations that cause others to view them in a number of ways. In addition to performing systems analysis tasks, analysts are often charged with carrying out administrative duties. Because they generally have a lot more experience than programmers or operators and because their work with the organization has exposed them to a variety of supervisory situations, analysts are often seen by their supervisors as capable of handling many administrative assignments. Such assignments may include supervising other analysts (the "**lead analyst**" or "senior analyst" position formalizes this), supervising teams of analysts and programmers as a team leader, handling the technical direction of a project as a project leader, or moving into administration directly as a data administrator or director of an information center.

As an informal or formal team or **group leader**, the analyst will have responsibility for one to ten other workers and, in addition, may continue to have technical responsibilities. It is not uncommon for systems analysts to be shifted into project management positions in their second or third year of analysis work as a reward for consistent and reliable efforts on projects. Project management requires special skills beyond systems analysis. It is frequently the policy in larger organizations that systems analysis is necessary, but not sufficient, for project management work. An analyst should have seminars, university coursework, and experience as a team leader to qualify to head up a project (Chapter 15 discusses project management).

As a senior or lead analyst, the systems analyst has administrative responsibilities without necessarily controlling the work of others as a project leader might. Lead analysts may head a team of analysts on a large project, while senior analysts may become what are termed "discipline directors" in a grid management scheme. In any event, these lead positions are usually rewards for service and experience. An analyst will typically spend one to three years as a junior analyst, two to five years as an analyst, and then become a lead or senior analyst after a total of five to eight years' experience. Project leadership is conferred on those who have the appropriate experience, formal training, and knowledge.

An additional internal role is that of consultant. In addition to informal consultancy mentioned at the start of this chapter, analysts may work literally as consultants, either through consulting firms or as consultants to user departments. Formal consulting may be recognized in a larger firm as a reward for service and as an alternative to entering management.

Analysts, therefore, can work as analysts, supervisors, managers, directors, and consultants without leaving their field. Whereas others in the firm may see analysts in complex and conflicting ways, within the IRM group systems analysis has a relatively clear role and presents consistent occupational options to workers.

From the outside, however, the analyst's role is less clear and more obscure, less consistent and more contradictory. Because analysts are typically the first wave of the flood tide of change and because systems analysis has its roots in industrial engineering (with the associated patina of efficiency expertise), analysts are usually seen negatively by outsiders and users. For instance, where a system fails to meets users' needs, the analyst may be sent in as a detective. If the interaction with users is not managed well, the analyst may appear overbearing, pushy, or insensitive. Early interviews may be complaint sessions which have to be handled with tact and sensitivity. Otherwise, the analyst comes across as an efficiency expert who will find fault with users who want to blame the system. Unfortunately, most analysts have learned far less about human nature than about file management. It is not surprising that the user-analyst gap is large and growing, and that users almost automatically see analysts as stony-faced threats who want to bring computers in to solve every problem that users don't recognize while ignoring problems users really have to have handled.

The situation is hardly improving. Most systems analysis textbooks have at most three to ten pages on the art of interviewing. Few analysts have taken courses in human relations; programmers are not required to take such courses, and programming is seen as the first step on the analyst's career ladder. Analysts are often promoted without having had a day's practice in interviewing. Therefore, it is one of the major challenges of the second forty years of computer systems development to provide analysts with human relations skills, especially since several of the newer techniques require intense interaction with users. Let us now turn to the skills, attitudes, and credentials necessary for success in systems analysis.

3.3 SKILLS, ATTITUDES, AND CREDENTIALS FOR SYSTEMS ANALYSIS

The work of a systems analyst is complex, challenging, and fulfilling. Obviously, a variety of skills, attitudes, and credentials are required. Analysts must develop patience, judgment, and tact to a far greater extent than any other group in IRM. Analysis work is highly attractive, especially to programmers, who naturally see it as a step up the employment ladder. This section discusses the skills, attitudes, and credentials that analysts must have or develop to succeed in analysis work.

The Skills of the Analyst

Figure 3-2 lists skill areas put forth in a recent paper by Scharer (1984). The author

- Effectiveness in performing functional specification and general design
- Ability and willingness to communicate with users
- Effectiveness in performing initial data collection and problem definition
- Effectiveness in training users, writing procedures
- Ability to analyze and resolve problems
- Ability and willingness to establish and meet project goals
- Ability to prioritize, organize and control project assignments
- Acceptance of responsibility
- Ability and willingness to communicate with other information resource specialists
- Ability and willingness to follow policies, procedures and standards
- Innovativeness and ability to recommend new techniques or methods

FIGURE 3–2. Eleven Criteria for Systems Analyst Performance (adapted from L. Scharer)

recommended that these skills be weighed in the performance appraisals of working SAs; later we will examine a similar list for entry-level analysts.

Scharer's list is supported by a great deal of recent research in the area of DP training, traits, and personnel requirements. A typical report is that of Cheyney and Lyons (1983), which indicates that SAs need skills in system design, technology, and some areas of information resource management.

The list of skills can be divided into three classes: **leadership**, **technical**, and **political**. Under leadership skills, we find **communication**, **project management**, **assumption of responsibility**, and **innovation**. Technical skills include **system design**, **system investigation**, **technical communication**, and **documentation**. Finally, political skills are represented by **teaching** or sharing of skills and **problem resolution**.

The list serves to remind us of two important principles in information resource management. First, the "problems" worked on are problems that others present; the measure of success of a project is how satisfied the *user* is, whatever the technicalities of the solution. The other principle is that solutions must be feasible; they are constrained to a great extent by the existing technology. Thus, "success" means both pleasing the user and achieving an economic, practical solution that all parties can agree to.

How does one acquire these skills? First is formal training. Most analysts receive at least some training at the post-secondary level. A computer science curriculum is valuable from a technical standpoint, but the MIS (IS, DS, DP) curriculum found in most business schools stresses the business and management aspects of information resource management and provides the management background necessary to work effectively in a modern organization. A course in BASIC or Fortran is more useful if that knowledge can be reflected against the backdrop of a business environment. Courses in management of human resources, management science, and market research offer that backdrop.

In addition to these formal curricula, analysts receive regular training through university extension programs, technical training institutes, and private offerings by consultants. In many organizations, extensive in-house and contracted educational programs run continuously, stressing programing languages and packages, system development methodologies, and some human relations and management skills. Under normal circumstances, a junior analyst in a large firm can expect to spend two weeks or more per year in the classroom.

Other skills are not so easily acquired by formal learning but come rather from experience and on-the-job training. Most analysts learn interviewing by doing it. Interviewing is difficult; it only seems easy because almost anyone can talk. Wherever possible, junior analyst training should include formal exposure to interviewing by experts.

Many analysts are given basic courses in "the firm's business" to familiarize them with procedures and policies before they investigate any information resource activity. It is far more likely, however, that they will learn this from informal conversation with co-workers and from their own on-the-job experiences.

Likewise, project management may be taught in structured courses, but most firms have their own semi-formal way of conducting projects that is usually learned on the job. Some methodologies are formalized in manuals taking up five to ten feet of shelf space of manuals — but does anyone *really* have time to read all that?

Other skills, such as communication and taking responsibility are closely tied to the personality of the analyst and are acquired through experience. Here, too, the *quality* of the experience dictates how effective an analyst becomes: bad experiences often engender poor attitudes and poor skills.

Skills must be practiced to keep them sharp. Unfortunately, maturity often brings promotion into administration, where technical skills cannot be regularly exercised. Analysts must resign themselves to a continual narrowing of expertise as they rise in the organization.

Attitudes for Systems Analysis

The fundamental attitudes that systems analysis requires are inquisitiveness, patience, flexibility, and a general appreciation of others, known as empathy.

High scores indicate attitudes consistent with those shown to be effective in system analysis success. Recent research by Janz and Licker (1987) has shown that attitudes toward information resource management workers are based on three factors: **perceived expertise, reliance on policy**, and **empathy**. Most users perceive analysts as relying too heavily on their technical expertise and company policies to the detriment of empathy.

Other important attitudes are tolerance of others' views, especially the views that users have of themselves (which is part of empathy), and courage. Courage has two aspects. First, there is the courage to put forth one's own views and back them up. Analysts frequently feel constrained by the technology and their own narrow views. They are often afraid of innovative approaches, especially those that may obviate the need to automate, because they have the impression that, as technical experts, they *must* recommend computers. The other kind of courage is restraining oneself from making suggestions where they obviously are not wanted — an aspect of tact. Analysts are commonly known as brash, pushy people who come into an area talking jibberish (or

COBOL) and make suggestions left and right about what can be improved where. Recent research has shown that being perceived as agents of change is enough of a burden for analysts; being perceived as inflexible, loud boors makes their job impossible. Keeping ideas bottled up, *even when they seem great*, requires courage of a high order.

How does one acquire these attitudes? First, the training of analysts should be broad, not narrow. It should include exposure to a wide range of thought. For instance, a course in sociology would not hurt, while a third course in programming might be redundant. Second, analysts need to experience their firms as well as study them. This author has proposed that analysts rotate through the firm in their early years as junior analysts ("DP/Company Interaction: Is Job Rotation the Answer?"), getting exposure to users, managers, customers, and regulators and a broader understanding of how people work. Third, as Cheeseborough and Davis (1984) have pointed out, **mentoring** is an important aspect in information resource management. As the field matures and as the older workers actually become older (until recently, old meant over thirty), mentoring relationships between experienced workers and new recruits can provide both an ear for listening and a voice for advice. The common view that users have of analysts and others in information resource maintenance as computerniks run amok can be reduced easily when older, more experienced and more sensitive workers can curb some of that enthusiasm for the technology acquired through years of intensive training.

Credentials

Currently the only credentials one needs to become a systems analyst are a degree of some sort and one or two years of programming or related experience. Some graduates of undergraduate MIS programs move directly into junior analysis positions. This system has developed because most larger firms have their own in-house training programs which "certify" the junior analyst, who does little real analysis work for the first year or so of employment. Others, however, are thrown to the wolves on their first day of employment. Obviously, formal "credentials" are not yet the critical requirements of analysis employment.

On the other hand, informal credentials *are* necessary. For instance, having worked on successful projects provides one with a good track record, a sort of "practical" credential. Naturally, if the project succeeded by luck or because there were no other alternatives or if one's contribution to the project was to draw data flow diagrams, such credentials are worthless. Most firms recognize the value of experience as a credential; that's the basis for the requirement of one to three years' experience.

Recently, the Association for Systems Management developed the **Certified Systems Professional** (CSP) certificate to fill the gap. The CSP is granted upon passing of an examination; thereafter, recertification is required every three years after demonstrating continuing education contact of 40 hours. These requirements are easily met in most firms; the CSP is not meant to be exclusive, but to provide standards for the entire community.

One doubtful credential is a degree in computer science. The battle rages unabated with proponents on either side insisting either that CS coursework is irrelevant to business or that business students are ill prepared for the technicalities of their rapidly changing universe. Like other disageements over exclusiveness, there is no resolution. Analysts must understand programming and the technology, but there is ample evidence that overemphasis on technical aspects to the detriment of leadership and political skills produces an individual who has little chance of success in the business world. While practicing analysts all wish they had more technical courses, the analysts who fail are those who have no understanding of the firm, business, or management. Thus, extensive training in programming, for instance, is not as valuable in information resource management areas as one might think.

Critical Success Factors

Analysts like Kent Clarkson succeed because they have technical, leadership, and political skills; attitudes that contribute to flexibility, empathy, and tact; and credentials in the technology as well as in professional life. Evidence is mounting that the critical success factors for analysts include:

- appropriate formal training
- exposure to the organization
- empathy for the user
- brainpower
- a sense of humor

3.4 CAREERS IN SYSTEMS ANALYSIS

Because systems analysts work for all types of organizations, there are many different approaches to systems analysis work. However, given the skills, attitudes, and credentials needed to succeed in SA work, the paths into and out of systems analysis are limited by experience and training.

Usually, analysts move into systems analysis after a stint as a programmer (Figure 3-3). Since programming skill is a prerequisite for much analysis work, it is logical for employers to desire a strong background in and fluency with programming.

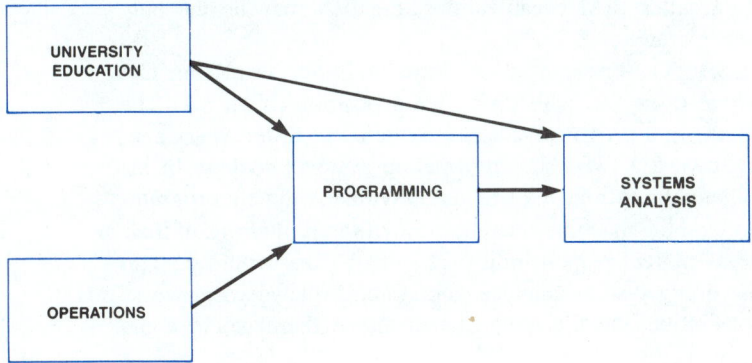

FIGURE 3–3. Sources of Systems Analysis Human Resources

In many firms, an analyst begins as a **programmer/analyst** (P/A). Clearly the P/A is neither wholly a programmer nor quite an analyst. In some firms, P/A is a job classification intended to "professionalize" programmer positions without adding major analysis components to the job. Other firms expect the P/A to do both programming and analysis work. Maintenance of existing packages and systems, for instance, requires both sets of skills. The lists in Figure 3-4 shows how difficult it is, however, to be expert in both. These lists, derived from one prepared by a special task force of the Data Processing Management Association, attempting to classify programming and analysis work. Only in the area of specifications (analysts write them, programmers read

CONTRAST OF PROGRAMMER/ANALYST ACTIVITIES		CONTRAST OF PROGRAMMER/ANALYST SKILLS	
PROGRAMMER ACTIVITIES	**ANALYST ACTIVITIES**	**PROGRAMMER SKILLS**	**ANALYST SKILLS**

Understand Module Specs	Investigate Facts and Data	Program Design Concepts
		   Logic Structures, Top-Down
		   Construction, Modularity
			Information Gathering and Evaluation
			   Researching, Interviewing,
			   Observing, Judging Relevance

Derive Coding Plan	Identify Problem		
Coding	Interview People		Understanding Human Behavior

Clean Compilation	Arrange Facts and Data	Program Design Tools
		   Flowcharts, HIPO,
		   Pseudocode, etc.
			Understanding Organizations
			   Principles of Administration,
			   Organizational Behavior,

Design Test Plan and Data	Create "System" and "Procedure"		
Develop JCL	Develop Specifications	Program Development Tools	Accounting, Finance, etc.
Run Test Procedure	Present Specifications	Program Coding	Environmental Influences and Trends

Debug	Modify Specifications	Language Syntax, Coding
		   Conventions, Efficiency
		   Considerations
			Problem-Solving
			   Developing Alternatives, Evaluating
			   Alternatives, Decision-Making

Continue Testing Until Correct	Obtain Approval of Specifications		

	Develop Time/Cost Schedule	Solving Logic Problems
		   Arithmetic,
		   Report-Generation, Control
		   Breaks, Data Validation, File
		   Management,
		   Table-Handling, On-Line
		   Applications
			Communicating
			   Reading, Writing, Speaking,
			   Listening

	Develop Development and Implementation Plan		
	Design Detailed Specifications for Modules		Persuading Others
	Supervise Programming and Testing		Project Management
	Monitor Implementation	Program Modification and Enhancement	
		Program Testing	
		Job Control	
		Documentation	
		Working in Groups	

FIGURE 3–4. Programmer and Systems Analyst Skills and Activities Compared

them) do the lists overlap, and then in a hierarchical way. P/A work is often very demanding because of this. In smaller IRM organizations, the P/A may be the only employee other than a computer operator.

Because programmer and SA positions overlap so little, it is not surprising that experience in programming alone does not qualify one for a position as an SA. The opposite may well be true, in fact. In a provocative essay written almost ten years ago, Philip Kraft warned of a drift toward a two-class information systems society. In his vision, one class would be populated by technically trained individuals, mostly programmers and operators, who would tend the machines but, because of the limitations of their training, never rise into positions of real responsibility. The other class would contain university-educated elites, those doing systems analysis and systems management work. In this class promotion, responsibility, and the rewards that attend them would come easily.

That Kraft's prediction has not come true may be best explained by the remarkable democratization that the march of technology forces on individuals once their formal training ceases. Managers (and analysts) cannot afford to rest on their training, which rapidly becomes obsolete; the pursuit of knowledge keeps everyone huffing and puffing after the technology. The rapid growth of the industry and the continuous realignment of responsibilities into new jobs (such as telecommunications specialists, database specialists, database administrators, information center staffers, office automation specialists, CAD/CAM, CAI, graphics) have served to "keep the pot boiling" and to provide new promotional and vocational challenges for all IRM staffers.

On the other hand, it is clear that to become an analyst requires basic training and experience. A rule of thumb is that analysts require knowledge of two programming languages, one of which is used in the firm, and two years of programming experience, not necessarily in business programming (although business programming is, in the terms used in job advertisement, an "asset"). A university or college degree, preferably in business, is almost universally a prerequisite. A programmer who shows aptitude and skill in dealing with others and some detailed knowledge of an important major application may be promoted into systems analysis, but most firms will require coursework in systems analysis either in-house or at a university, often through an extension program.

After a year as an **analyst trainee**, the SA may acquire the title of "**junior analyst**." Salaries are typically commensurate with the upper ranks of programming within three years, by which time "junior" may be replaced by "senior". The senior analyst then may become either a permanent project manager, a senior consultant, or a manager or supervisor of a programming or analysis group. Because most firms prefer their supervisors to have experience in systems analysis, the career paths for those who want to remain in programming are still severely limited.

Beyond supervision, analysts can move into management, although analysis skills and experience alone are usually not sufficient to run modern information resources. A strong business background, experience in working with paying customers, and a feeling for the marketplace usually contribute to business success, and information resource management is no exception. As IRM becomes a **profit center** rather than merely a line item on the cost side, organizations are demanding that their information resources be run on at least a cost-recovery basis internally. Often IRMs are expected to show a paper profit if not a real one.

As a field, data processing has shown a remarkable tendency to reward its employees with increased access to technical resources rather than stressing promotion to increased managerial responsibility. In line with this, a two-tiered or "parallel" career path model has often been used. In this model (see Figure 3-5), technically adept employees who do not want increased responsibility for tasks or projects are given paper promotions and positions of internal consultancy. Edwin Shore (1983) recommended these titles: information control specialist, technical specialist, senior technical specialist, senior consultant, and advanced consultant. Highly-qualified employees can then "advance" to positions of increased technical responsibility without having to

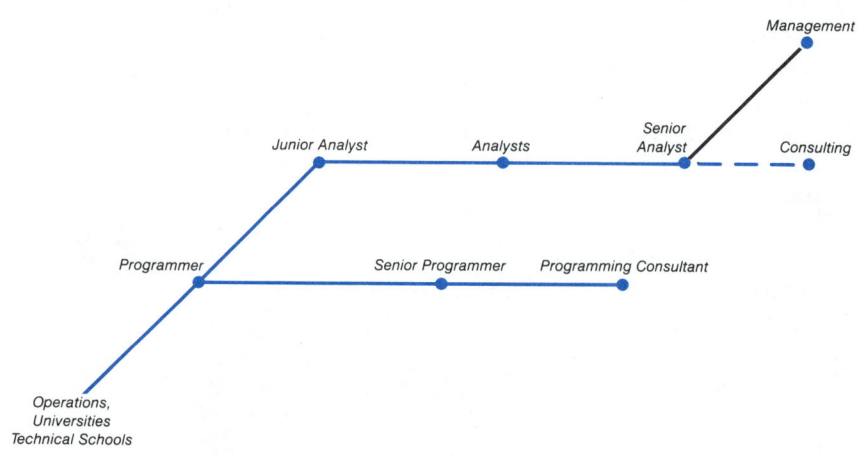

FIGURE 3–5. Career Paths for IRU Specialists

employ, or hone managerial skills. In this way, organizations can develop what Shore termed a "pentagonal" structure (Figure 3-6), which is more appropriate for the

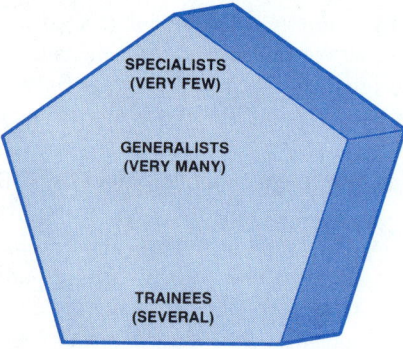

FIGURE 3–6. A Pentagonal Model of Career Structures in Systems Analysis

information resource field than the traditional triangular or pyramidal structure of the bureaucracy (still found in many firms and illustrated in Figure 3-7).

Contrasting with the **technical tier** just described is the **managerial tier**, which stresses managerial skills such as planning, delegation, negotiation, and communication. Managerial skills are not directly selected for information systems training in universities and colleges and rarely appear in systems analysis and design courses (except under the heading of "project management"). Because of this, a recent trend has been to move user-managers (often from accounting or engineering) into IRM because

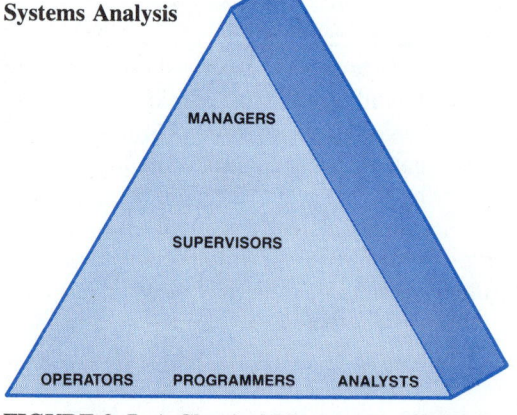

FIGURE 3–7. A Classical Management Triangle of IRU Positions

they posssess critical and valuable knowledge of the firm and its products and have acquired technical skills along the way.

So we see that systems analysis is part of two different kinds of career paths, one heading for increased specialization in the technology ("technical tier") and the other for increased managerial responsibility ("managerial tier"). Several trends are working to change these career opportunities and, as well, the work systems analysts perform. That is the topic of the next section.

3.5 THE FUTURE OF SYSTEMS ANALYSIS

CORVUS SYSTEMS

The nature of the IRU's work is changing as technology trends accelerate.

The rule of thumb in computer work is that knowledge has a half-life of four years (coincidentally the length of undergraduate schooling). Four years out of the university and freshly promoted into a senior analyst position in an accounting firm, half of what an analyst knows is obsolete (some of it even hazardously so, being six to eight years old). Another four years later, without additional training or involved experience, only 25 percent of the newly appointed manager's technical knowledge is immediately applicable.

This simple obsolescence profile is but one of the powerful trends in IRM that affect the way analysts are trained, employed, and promoted. Four other trends in the technology are apparent:

1. **Technology infusion** is bringing computer knowledge to staffers through information centers and to management directly through seminars and various end-user software-development methodologies.

2. **Technology dilution** is making technology less distasteful and jargon ridden. The appearance of user-friendly interfaces to complex software is completing the process of "utensilization" by which detailed, powerful computer facilities are toned down, and made accessible and less threatening to untutored users. This process began with compilers that, while limiting the flexibility and optimization available to the machine-language programmer, made machine language skills unnecessary. Each layer of software has homogenized the facilities of lower layers and made each successive layer more accessible.

3. **Technology adaptation** is creating new forms and formats for information resource management within which to work. New employment opportunities, such as database administration, are appearing with each change in software. The appearance of microcomputers as office machines has introduced a variety of jobs for analysts, as have office automation and telecommunications. Along with the adaptation of the technology has come a change in the way projects are developed and managed and a new need for special analysis skills.

4. Finally, **technology evolution** is removing some arenas of information resource maintenance and ending career paths that used to attract analysts while more sharply defining others that are becoming more important.

Figure 3-8 illustrates how these four trends (infusion, dilution, adaptation, and

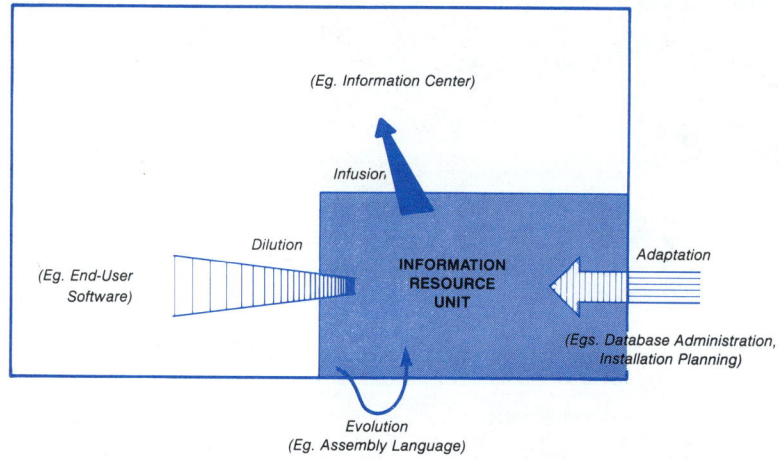

FIGURE 3–8. Technology Transfer Processes in the IRU.

evolution) have shaped the job marketplace for programmers and analysts over the past twenty years. Note that opportunities for assembly-language programmers have moved into the systems area almost exclusively while analysis jobs have moved to higher "layers" such as data administration, installation planning, security and disaster planning, and specialized application analysis.

Infusion means that more non-data-processing jobs are utilizing information resources in integral ways. Dilution means that the "version" of that technology employed by workers at all levels has begun to resemble "programming" less and less. Adaptation and evolution imply that jobs in information resource management are themselves changing to meet the opportunities of the latest technology, causing specialization within the IRM fields and the disappearance of some vocations. Although Kraft predicted the demise of the programmer and others have predicted similar fates for analysts (see James Martin's 1982 book, *Applications Development without Programmers*), the opposite is more likely: the job class will continue, but the work will change and acquire a set of specialized names.

Three important results of these trends have been the appearance of information centers, the use of prototyping in system development, and the development of several database-related vocations.

In summary, the four technology trends are changing not only the content (with its half-life of four years) but also the form of information resource management, forms that provide new career paths for analysts and increasing managerial challenges to IRM.

3.6 CONCLUSION

The systems analyst may be either an **entrepreneur**, an **internal consultant** or an **employee** within an organization. These three roles dictate three distinct relationships among an individual and others in the organization. Recognizing that these roles are distinct is important in understanding how analysts work and how they function within larger domains such as departments and firms. Within an information resource management group (DP department, MIS group, DSS team, etc.), the analyst usually plays a **lead position** role, directing the activities of others in the group. Often, analysts are given administrative duties; in some departmental structures, analysts may be **team leaders**, **project leaders**, or **consultants** to management. On the other hand, employees outside the IRM group see analysts as quasi-official **representatives of the IRM group**, "**efficiency experts**," internal **vendors of computer services**, and **detectives**. These varying roles put a great deal of pressure on analysts to perform well and meet conflicting images.

Analysts need a variety of skills: **leadership**, **technical**, and **political** skills rank highly. In many ways, analysts also need **managerial** skills to work in the administrative areas they often find themselves. Skill alone, however, is not sufficient to meet the needs of systems analysis in a modern organization. Certain *attitudes* stand out as helpful: *empathy* for users, *understanding* the client-consultant relationship, *flexibility, patience* with others' lack of familiarity with information technology, *tolerance* of others' views, and *courage* both to put forward one's own ideas and, conversely, to restrain oneself from making suggestions where they are obviously not wanted. Finally, analysts need to have accumulated certain experiences to, first, attain analysis positions and, second, move into positions of leadership. These experiences include education, project work, political exposure, and practical exposure to the firm's business.

Careers in systems analysis generally follow two distinct paths. The **technical tier** begins with technology-intensive work in computer operations and programming, leading to systems analysis through years of experience and aimed at internal consultancy. The **managerial tier** usually begins with education in business and computers, moves immediately into programmer-analyst or junior analyst positions, and passes through to supervisory positions within IRM. Many analysts today find themselves in positions outside IRM proper, working in such diverse fields as organizational development, training, information centers, and technology planning. Other analysts have begun their careers outside IRM totally, moving into systems analysis because they possess critical knowledge of the firm and its products, as well as having acquired some technical knowledge along the way.

The future of systems analysis is difficult to predict. However, certain trends are clear. **Technology infusion** is bringing computer knowledge to staffers through information centers and to management directly through seminars and various end-user software-development methodologies. **Technology dilution** is making technology less distasteful to the populace in general and is spawning knowledge bases of computer technology with which the analyst must contend. **Technology adaptation** is creating new forms and formats for information resource management within which to work and new employment opportunities for analysts that did not exist in large numbers before this decade. Finally **technology evolution** is removing some arenas of information resource maintenance and ending some career paths that used to attract analysts.

3.7   THE CASE OF THE CAUSTIC CRITIC

Ty Runnals has worked as a marketing analyst for Market Fax, Inc. for two years. In this job, he gathers information for paying clients on markets for everything from fingernail brushes to financial paper and to be blunt, he is tired of the work. What really fascinates him is the computer. He has fallen in love with Lotus™, gone bananas over benchmarking, and lost his heart to hardware.

Ty would like to move from marketing analysis to systems analysis for Market Fax, a move that he feels would suit his personality and new career goals. The latter includes running his own marketing consulting firm in five years. Ty has a good, solid knowledge of existing software packages; he has taken a course in programming, and has spent a lot of time learning the packages. He feels he is ready for more.

A confrontation with Calvin Coldheart, his manager, was less than comforting. Calvin pointed out that Ty has had no experience other than using microcomputers. He further indicated that if Ty kept his nose clean and his achievements high, he would hold several accounts for the firm within a year or two. Ty pointed out that this wasn't what he wanted, and anyway, he had shown how fast he could learn. Calvin sent him to Barney Bitz, the head of DP, for a talking to.

Questions

1. Are Ty's goals consistent with his skills, attitudes, and credentials? What, if anything, is missing?
2. Are Calvin's criticisms on target? What else might Calvin have pointed out to Ty?
3. What might Barney say to Ty? Will Ty like what he hears? Role-play this interview. What are the possible outcomes?
4. Assume that Barney takes Ty as a trainee analyst. What kinds of initial assign-

ments should Barney give Ty? Will Ty succeed? What will Barney and Ty learn from these trials?

5. Outline a better career plan for Ty over the next several years. Include training, experience, and conversations he might have in trying to change his career.

Discussion Questions

1. An analyst is often seen as "the representative from the great state of computers" by non-technical people in the organization. What are the negative implications of this image? What can analysts do to counter this image? How can management prepare its staff properly for working with analysts given this tendency? Are there any positive implications of this image?

2. Analysts often work in conflict situations where their goals of investigation and design run up against inertia, fear, and a genuine feeling that things are fine just the way they are. How might an analyst prepare to meet and counter these feelings? What aspects of public relations work (such as advance publicity, training, and conferencing) can be applied here?

3. The following is a typical conversation:

   Joe: I'm really afraid that when your analyst comes down to the sales group there's going to be problems.

   Mary: Really? Why do you think that?

   Joe: These guys from the data processing department really have to show off what they know. They're always telling us what we're doing wrong and assume that just because there is room for improvement that we don't know what we're doing here. They are really conceited and rude.

   Mary: Is that it, just public relations?

   Joe: Public relations???? Really, Mary, your people walk in and right off the bat they say, "This is wrong," and "Here's a problem," and "Why on earth do you do it that way?" That's not public relations, it's just bad manners!

   Discuss why feelings like Joe's arise. Are they justified? Analysts are often accused of having a "problem orientation." John Gall, in an amusing book called *Systemantics* advised that "should you come across a system that seems to work, tread lightly." Others might say, "If it ain't broke, don't fix it." How can these lessons be applied to the training of analysts in school to prevent the arrogance Joe perceives?

4. Three employees have applied for an opening at the junior analyst level. Examine their resumes. How would you go about selecting the best candidate?

   Name: Roberta Smith

   Position: Senior Accounts Clerk

   Education: Business Certificate, Lamont Junior College (2 yrs.); courses in COBOL and systems development methodology in our firm.

   Experience: Three years accounts clerk; 2 years senior accounts clerk handling routine bookkeeping assignments with increasing responsibility for direction of junior clerks. Supervises three clerks now. Extensive knowledge of ACCOUT, our computer package for handling ordering and billing. Worked as user liaison with IRM department for 6 months on that package.

Name: Howard McMullen

Position: Programmer

Education: Computer Science Degree, Philpott College; advanced courses in CICS, ADABAS, and microcomputer packages.

Experience: Junior Programmer 2 years. Programmer for 2 years. Major programming responsibility for the ACCOUTS package modules. Some maintenance work on this and other packages. Has had increasing technical resonsibility over the 4 years and is eager to learn new packages.

Name: Carla Chan

Position: Marketing Analyst, Distribution Division.

Education: B.B.A., Philpott College. Majored in Marketing with a minor in Information Systems. Some knowledge of COBOL and BASIC. Has 6 years' experience using our marketing packages, SAS, and microcomputers for market analysis.

Experience: Marketing analyst for six years. Has extensive knowledge of our products and markets. Is responsible for supervising four market researchers and one junior marketing analyst. Was the client for MARK8, our marketing package, and worked closely with analysis staff for the year it took to develop the package.

5. Assume that each of these applicants is to be interviewed for the position. Present at the interview are a personnel administrator and you, manager of systems planning. Assume that you know the following:
   a. You are a COBOL shop; all development is in COBOL.
   b. A major responsibiity of the successful candidate will be implementing and operating an information center.
   c. You would like to move to more prototyping and less full-scale development from scratch.
   What questions would you ask of the candidates?

6. Here is a partial transcript of the interviews. How would the answers change your judgment?

   You: What special skills would you bring to the job?

   Roberta: I learn fast and get along well with those who work for me. I've always been fascinated by computers and the way they do simple but repetitive tasks well. I've always thought that if we could computerize everything around here we'd all be a lot happier.

   You: Do you think computers will solve everything?

   Roberta: No, but they sure do make things run better. In my group, I've made sure we transfer everything possible to computers. At first there were some complaints, but I smoothed that out — everybody was scared of computers but they only needed to be told about their benefits. Now everything runs fine.

   You: What special skills or background would you bring to the job?

   Howard: Well, I'm a crackerjack programmer who can learn just about any language there is. I can find bugs others swear don't exist. There's not a program around I can't debug. And I write very efficient code, too.

   You: Have you had any experience talking to users?

Howard: Yes, I spend a lot of time with them showing them how to fix errors in ACCOUT. Actually, I tell them that what they've done is use it wrong, because there really aren't any errors there. You know, most users come in irate, but if you just point out their errors, they go away pretty satisfied.

You: Ever thought about doing something about this situation?

Howard: No. There is no way you can get users to stop making errors or get them to read the manuals they're supposed to read.

You: Carla, do you bring any special skills or experience to this job?

Carla: Yes, in two areas. First, I've got a very good idea about this company, what it does, who its customers are and where it is going. Second, I have a strong user orientation, having been a user myself for six years. I understand why system development takes so long, and I've got some ideas about what to do about that.

You: But you have little formal work experience with computers?

Carla: That's not nearly so much of a handicap as you might imagine. I do have courses in programming, but more important, I've worked with both programmers and analysts in the MARK8 system development project. I had them go over not only the design, but also the code. When my people have a problem with the system, they come to me first; I think I can guess when they've misunderstood the manual and when the system doesn't work.

You: Is that often a problem?

Carla: Sometimes. But I've got an idea that the problem is in constantly retraining people every time there's a change of personnel or a change to the system. I'd move to see some changes in the way users are trained and in how liaison is maintained. We should move more into high-user-involvement development; you know, users aren't dummies, but sometimes they feel inferior around systems people.

7. Make a survey of newspaper and journal ads for IRM personnel. What seems to be stressed in the hiring of systems analysts? What seems lacking in terms of other skills? How many of the jobs seem to be among the "new" professions in systems analysis?

8. Look around your university or college. In what ways have technology infusion and dilution become evident in either the administration of the schools or the daily lives of students? Consider the alternative: "hard" technology without dilution and a "wall," in terms of access, between those who know the technology and those who don't. What aspects of university community life would be different? Would these differences be positive or negative?

9. Consider the evolution of technology. The process of dilution implies that technology becomes more palatable but less flexible and more standardized. Can you see evidence of this in the software marketplace, particularly in microcomputer offerings? Examine recent issues of *Byte* or *PC Week* Magazine and look at the advertisements. How are products "pitched" to consumers like you and like managers of small and large businesses? Do they seem more like utensils than specialized scientific or business instruments? What are the implications of this process?

10. The history of information technology began with simple programming of sequences on punched tape and moved into the programming of computations as a stored program on the ENIAC. From there, we have developed a series of layers, mostly in software. These layers make dilution and infusion possible. Examine software systems you use in your courses and complete the diagram on the next page:

Layer	Advantage of Layer
Basic hardware, machine layer	n.a.
Data management programs	Access to data without having to know much about the hardware
Operating system	
Application program (eg. _____)	
Application generator (eg. spreadsheet)	

DESIGN EXERCISES
Using dBASE III

1. Examining the contents of a file is not the only way of looking at it. dBASE III allows you to count the number of records meeting a condition, and average or sum numerical fields for specific records. These values can be saved in *memory variables* and displayed later or used in computations or reports. Bring dBASE III up, USE the file named PRODUCTS, and employ these commands:

COUNT TO SIZE	Count the number of records
COUNT FOR SUPPLIER = "JONES" TO JSIZE	How many for JONES?
AVERAGE SALESPRICE FOR SUPPLIER = "JONES" TO JAVG	Average sales price for Jones's products
SUM SALESPRICE FOR SUPPLIER = "JONES" .AND. ITEMCODE > "1000" .AND. ITEMCODE < "1500" TO JSUM	Total value of Jones's hair products
? JSUM	Display the sum
? JAVG	Display the average
? JSIZE	Display the count

[You can use almost any name for a variable other than *key words* (like SUM, COUNT, DISPLAY, TO, etc.) and names of fields in active files. dBASE III will signal *"Syntax Error"* if you try to use a key word as a variable name.]

DESIGN EXERCISES
Using dBASE III *continued*

Using these commands, find out the average recall period for hair coloring (ACTIVITY between 1030 and 1039) activities (USE SERVICES); the number of male (GENDER = "M") stylists (USE STYLISTS); the number of stylists who work Tuesday afternoons (the value of the TUESDAY field is "A" or "N").

2.   Doug has recently become disenchanted with his system. For one thing, Mercurial offered a poor consulting service and refused to work closely with Doug to improve the system, insisting that the system as delivered was "complete." Whenever Larry comes out, he seems to have little time for Trish and none for Byron. Trish received two days' training from Larry, but Larry does not really understand the hair business and only wants to fix the programs as quickly as possible. Trish seems to want to learn more about programming in dBASE III, but Larry certainly does not want to teach her programming, dBASE III or how HAIR works.

Doug, on the other hand, has become infected by the computer virus. He has just asked Larry this question:

Say, Larry, how do we go about finding out which of my stylists perform the expensive (over $50) activities the most frequently? And what items are moving fastest among which types (age, gender) of customers? And what days of the week are the busiest? What days do teenagers come in? What services do the older customers (over 60) prefer? What's the true average recall period for hair treatments (ACTIVITY between 1022 and 1025)?

Can you answer these questions either using dBASE III or by examining the file? What skills would Larry have to have to be able to work with a client like Doug? What is happening to the technology that might have an effect on Larry's usefulness, given Trish's expressed interests?

CHAPTER 4

THE ANALYSIS OF SYSTEMS ANALYSIS

OBJECTIVES

1. To describe the functions and structure of Information Resource Maintenance
2. To describe how an information resource functions and what its components are
3. To analyze information resource maintenance into its component elements and to describe the contributions of systems analysis to IRM
4. To list and provide examples for eight alternative activity structures within which to achieve information resource maintenance goals
5. To describe the process of logical system design within the environment of systems analysis in general systems terms

THE ANALYSIS OF SYSTEMS ANALYSIS

4

4.0 WHERE TO BEGIN? WHERE TO END? WHAT TO DO IN THE MIDDLE?

Polly Perrault is the VP, Customer Relations for Shoes Unlimited, a major discount footwear retailer in the Midwest. Shoes Unlimited sells shoes to the public through thirty-four outlets, found exclusively in suburban shopping centers. Founded nine years ago by Simon Brewster, the company has grown rapidly. Now, after having defined a niche for itself in a highly competitive market, SU is finding that it is not enough merely to change the ideas customers have about the shoes they wear. Originally the idea of a cafeteria-style shoe store without traditional shoe sales staff was pooh-poohed by the banks Brewster went to for capital. After demonstrating that people had a low opinion of shoe store staff and were uncomfortable in these stores, however, Shewster convinced the Bank of the Midwest to put up the money. Things went well for over eight years. Accustomed to dreary, uninspiring, unexciting shoe emporia, customers fell in love with the upscale, upbeat idea of a shoe cafeteria, where they could pick what they wanted after a computer measured their feet. No longer did they have to wait for half an hour while an obsequious clerk rummaged around in the back for a 5-1/2 AAA — the computer always knew what was in stock.

The cafeteria idea worked well until manufacturers realized that shoes are a popular item and started producing a bewildering variety of shoes, many poorly made or imperfectly sized. Although Shoes Unlimited is tied into several manufacturers by computer, quality control is poor. The wrong shoes, poorly made, arrive in record time, and customers bring them back in record time. And Polly is the person who has to field the complaints. What she'd like is some way to use the computer to keep track of returns, then she could analyze them for manufacturer, style, customers' comments, and customer characteristics so that she could anticipate — perhaps forestall — some of the worst batches of bad shoes. But Shoes Unlimited, which uses computers in a very modern way to conduct its retail operation, contracts out its management computer work. Polly has no idea how to get a project going. Her colleagues in the upper-management team never anticipated using the computer in any way other than to assist the actual sales function. Where does Polly begin — and where does she go after that?

An Introduction to Information Resource Maintenance

Polly's pressing problem is one of "maintaining an information resource." A wealth of information is available to her, but she lacks the organizational resources to use it in a constructive way. While Shoes Unlimited is a pioneer in the use of computers for retailing, it has no organizational mechanism in place to change, upgrade, or refine the information resource *per se*.

As discussed in Chapter 2, an information resource is the accumulation of information, the machinery to access and use that information, and the people to assist in maintaining the information. The information itself usually has some organization: sales information, personnel information, customer information. The specific organization, however, varies. The most important thing about an information resource is that it is a resource for the organization, just as are human and financial resources.

Figure 4-1 illustrates the equal status the information resource has with its financial and human counterparts. A business organization can be seen as an engine that turns environmental resources into equity. In these terms, a business organization is an open system (see Figure 2-2) in equilibrium with its environment, growing by transforming lower-quality **input resources** to higher-quality **outputs**.

The information resource naturally grows out of the two major aspects of system function: physical transaction with the environment and control of that interaction (as detailed in Chapter 1). Information can therefore be about the *environment* (arising from transactions) or it can be about the *organization* (stemming from the control of the interaction). Where the environment is placid, the information

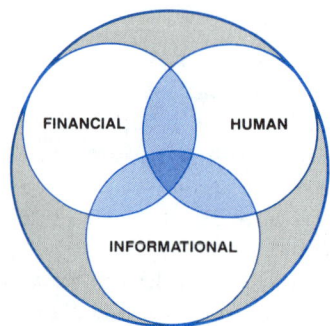

FIGURE 4–1. Three Classes of Resources

resource can be relatively simple, and the information quantity can be low; its complexity, limited. Where an organization deals with a turbulent environment, the information resource must be large, well articulated, and complex. The information resource is itself a subsystem that mirrors in structure and function the environment within which the organization functions.

As organizations grow in a turbulent environment, information resource management becomes an expensive, resource-consuming function. As Polly is discovering, Shoes Unlimited's rapid growth in a changing marketplace has left the information resource aspect of the firm far behind.

This chapter is about a firm's information resource and the roles that systems analysis — and logical system design in particular — plays in maintaining it. We begin by defining in more detail what an information resource is, how it functions, and how it changes. We next introduce an analytic technique — structural-functional design — that is used to describe and analyze systems. Then we apply this technique to information resource management and maintenance. Activities within each subfunction can be organized in a variety of ways — this chapter looks at eight important organizations of work.

4.1 A GENERAL MODEL OF INFORMATION RESOURCE MAINTENANCE

Systems analysis seeks goals as one part of Information Resource Maintenance (IRMaint), which, in turn, contributes to the goals of Information Resource Management (IRM). They are all concerned with an organization's information resource. This resource may be a well-organized subsystem or a poorly developed and barely communicating set of functions. In many firms, IRM is sophisticated, handled by its own department, and staffed with well-trained individuals. Few firms experience smooth growth as Gibson and Nolan (1974) (see Figure 2-15) predicted, and many would describe the evolution in even the best-managed situations as "chaotic" and "unpredictable." Models such as this try to capture in a few lines the complex forces that mold any organization. Given the rapidly changing, turbulent environment for computer-based work, it is surprising that growth ever occurs in a well-behaved fashion. Several articles have recently attacked the notion of a set of stages through which information resource development must flow (Benbasat, Dexter, Drury and Goldstein, 1984; King and Kraemer, 1984). The essential "systems" nature of information resource introduction, growth, and control, however, is amply apparent from studies of the role of IRM in organizations.

Even where management of the information resource is poor, certain common aspects are apparent:

1. The components of the information resource include people, machinery, and information in a variety of forms, formats, and media (Chapter 1).

2. The resource contains information necessary for the functioning of the organization and may be a byproduct of the function. It can be used, therefore, either to conduct the affairs of the organization at a primary or secondary level or to plan for change, through historical analysis, of the data.

3. The resource processes information. Programs (in a computer-based resource), procedures, and equipment handle the information, refining it from low-grade "data" to higher-grade "information" and passing it from place to place. Knowledge of these processes is not always immediately available. Since the resource is like any other system, specialization and resource conflicts arise *within* the resource itself, potentially hiding the system's functioning and obscuring this knowledge.

4. The resource *is* a system. It has locomotor-manipulator cells, control elements, sensor elements, and a system life cycle (as outlined in Chapter 1). Furthermore, it interacts with every other system within an organization; as it grows and specializes, its interaction with other systems becomes increasingly sophisticated and entrenched. The implication for IRM is that "technology is not enough." As pointed out in Chapter 3, it is precisely because of this complex and sophisticated interaction that the skills of the analyst must include a portfolio of management skills and organizational savvy.

The information resource is unique in one important way. The major visible component — the information — is *about* the organization, but the resource as a whole is *within* the organization. Conflicts, therefore, often result, and important organizational issues such as access, confidentiality, and authority are commonplace.

Chapter 2 mentioned some of the criteria for a properly functioning information resource. Below are more details on the needs dictating these criteria.

Information must be made available in timely fashion. It must be collected quickly and move quickly from its source to the sinks. A **responsive** information resource delivers the required information rapidly upon request or explains why there are delays. Speed, however, can be excessive. Information must be timely, meaning that it arrives within the scheduled time frame. A common situation is the receipt of information out of order: sales summaries arrive before the details, raising negative feelings that may be prematurely acted on. Who has not had the experience of trying to pass someone who is trying to pass you? Sometimes it is necessary for one system to slow down its reactions. In systems terms, an organization must try to avoid **oscillation** (illustrated in Figure 4-2) which may result from overreaction.

FIGURE 4–2. An Illustration of Oscillation

Information received must be **complete**. Otherwise, decision making is difficult and risky. The worst situations occur when decision makers wrongly assume completeness. Then, decisions are made with a false confidence and usually result in disasters. Insecurity mounts when decision makers suspect that information is missing. System modeling (discussed in Chapter 6) is a tool that analysts use to examine completeness. There is often a trade-off between timeliness and completeness: it takes time to assemble complete data and, on the receiver's side, it takes time to assimilate complete information. Overly complete data can cause information overload.

Inaccurate information is valueless. One of the greatest benefits of computerization has been the reduction of errors in routine data processing. In business data processing, the information resource should be able to diagnose errors, prescribe corrections, and sometimes itself correct the errors, as close to the source as possible. The term "maintenance" is often applied to this kind of work, although in this text we will use a wider definition.

In any information resource, **consistency** is a criterion of correct functioning. A report should not change in format without notice. Figures should not show dollars and cents one day and only dollars the next. Input procedures must be "frozen" for a period of time so that data entry clerks can learn them and feel comfortable with them.

Finally, the information must be **available**, that is, "accessible" and "usable." Accessiblity requires that those who need information can contact the resource easily. As systems grow in complexity, specialized access mechanisms develop. Maintaining those mechanisms for accessibility is a difficult task, one which is commonly taken on by systems analysts.

The information resource provides information in a timely, complete, accurate, consistent, and available fashion. It is the job of information resource maintenance to make sure that the components of the information resource meet their functional requirements. The specific procedures through which this job is accomplished are described in the section 4.3.

4.2 STRUCTURAL-FUNCTIONAL DESIGN

Let us turn our attention for a moment from how systems function to how to describe systems we would like to build. Systems analysis is aimed not only at understanding and describing systems, but also at designing them. This section introduces a technique for designing systems based on an understanding of their specialized functions and their goals. The technique is called **structural-functional design**. The term, which exposes three aspects of systems analysis, is particularly apt because:

1. The technique is **structural**: it breaks items down into structures of structures. The tree that results captures the interaction of parts as they contribute to higher-level wholes.
2. It is **functional**: it concentrates on functions performed by units within systems and their contribution to the functioning of the higher-level units to achieve goals;
3. It is a **design** technique: it can be used, subsequent to analysis, in redesigning an existing system.

Structural-functional design is performed in two different ways: functional analysis and goal analysis. They are, in fact, ultimately equivalent in systems that meet their goals. The former concentrates on the activities and behavior of a system; the latter, on the system's goals. Naturally, activities should contribute to goals.

Consider a system with the goal of planting a field. This goal can be analyzed into subgoals (Figure 4-3):

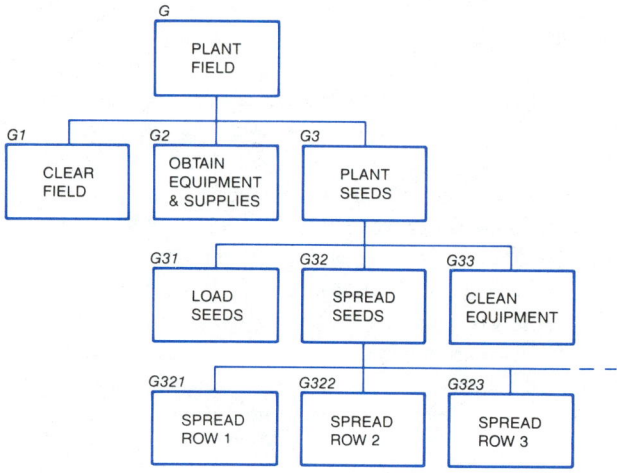

FIGURE 4–3. A Goal Analysis of Field Planting

    G1: clearing the field
    G2: obtaining the needed seeds and equipment
    G3: planting the seeds

Each of these goals can be tested for completion and further analyzed into sub-subgoals. Consider G3:

    G31: loading seeds into the seeding equipment
    G32: spreading seeds on the field
    G33: cleaning the equipment when finished

Spreading seeds is done on a row-by-row basis

    G321: spread row 1
    G322: spread row 2
    G323: spread row 3, and so forth

Functional analysis looks at each activity. Since a goal can be easily equated with the activity needed to meet that goal, translation between functional and goal analysis is straightforward. Functional analysis is more productive in analyzing observations of working systems; goal analysis is useful in deriving new designs for systems that do not yet exist.

The value of either type of analysis in design is that the design of a system is equivalent to the implementation of each function in the structural-functional design so that the function actually accomplishes the required sub(sub-sub-...)goal.

Hence, the design process follows the steps listed on the next page.

1. Begin with system goals as you observe or infer them.
2. Subdivide into subgoals and continue this subdivision until goals are clearly identified with specific actions.
3. Equate each such goal with an activity that accomplishes that goal or performs the specific actions at each level.
4. The design is now equivalent to the indicated **structure of functions** (activities).

The resulting design is not optimized, however. That is, the design is functional in the sense that it works and is guaranteed to accomplish system goals, but it may not be the most efficient design.

What is certain is that, from the standpoint of logic, implementation of the activities that appear at the lowest levels of the design will accomplish the system's goals.

To distinguish this design from optimized designs, the product of a structural-functional design exercise is a **logical design**; the others, based on optimization of activities, are **physical designs**. As will be discussed in chapter 7 and in section 4.4, physical designs require detailed, accurate knowledge of the interaction of the system with its environment to perform the optimization; a logical design makes no assumptions about the operating environment of the system. It really is "logical" in the sense that it is derived from logic; any given implementation might be horrible, however. For each logical design, there can be many valuable, as well as many useless, physical designs.

Logical designs are not unique, by the way. They are derived from observers' data and intuition. According to the Uncertainty Principle, observations must reflect the observer as well as the system observed and, therefore, are not unique. Furthermore, the use of goal analysis does not make reference to sequence. All that is required is that goal G1 be accomplished when goals G11, G12, and G13 are met. The order that the goals have to be met in is not specified.

Structural-functional design and its products go under several names. The resulting tree diagram is called a **hierarchy chart**, a **structure chart**, and a **Visual Table of Contents** (VTOC). This text refers to it as a structure chart. Chapters 6, 9, and 13 make reference to structure charts as basic elements of systems analysis.

4.3 MAINTAINING THE INFORMATION RESOURCE

Figure 4-4, on the following page, illustrates the structure of the functions of information resource management. Four functions contribute to the goals of IRM: operations, planning, promotion, and maintenance.

Operations (IRO) include routine operation of the resource and provision of service to its users. You probably are familiar with computer centers and how they function. IRO is a business function that provides information and information processing services to customers who pay for the services and supplies.

Planning (IRP) is a management function that forecasts needs in service provision, maintenance, and promotion. IRP employs a number of tools, primarily economic forecasting and technology forecasting. The goals of IRP include producing immediate, short-term (1-2 years), long-term (2-5 years) and very-long-term (5 years and more) plans for human, financial, and information resources. An information resource plan lays out the

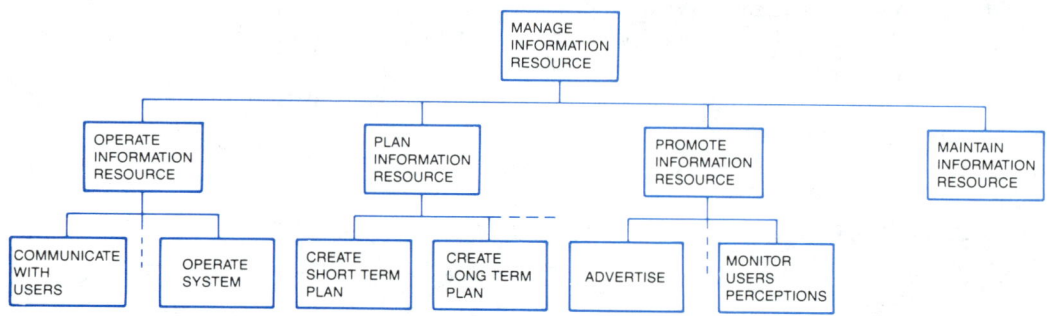

FIGURE 4–4. Goal Analysis of Information Resource Management

general requirements for the IRM function, what funding is needed, what technology likely will be required, and what skills will be needed in the stated time frame.

Promotion (IRProm) is a marketing function. If this function is carried out appropriately, information resource products and services are offered to the correct clientele in the correct time frame and in the correct way. Of course, "correct" is an arguable term. As with any marketing function, not everyone can be pleased. Consider a university computer center. Its "market" may include undergraduates in fine arts who need to produce graphic artwork and thus require the latest laser graphics printers and color terminals. Graduate students in physics and chemistry require hours-long runs of exquisitely complex mathematical models. Business students need access to microcomputers and the latest in spreadsheet and financial planning software. Administrators are concerned with very large data bases, access security, and on-line query of data. Untenured professors may be cranking up the "paper mill" and in one week need seven word-processing revisions to a scientific paper. A sophisticated professor holding a chair in medicine "absolutely requires" access to national networks for on-line or computer-based conferencing with esteemed colleagues. Finally, information resource managers themselves need to be able to configure the computer system without taking it down for a week.

Is it hard to satisfy such a diverse group of individuals? Yes. Each facility offered generates yet another vested interest group that resists using something new. Many computer center operators need to maintain two or three versions of the same operating system simply because an important "customer" cannot be bothered with converting programs or files. Even small projects can be hamstrung because a key individual has learned one word processing package and refuses to learn another.

Maintenance (IRMaint) is concerned with four subgoals (see Figure 4-5). These are:

- **Substitution**: automating existing processes
- **Refinement**: improving or correcting existing processes
- **Enhancement**: adding new functions to existing processes
- **Development**: building new processes

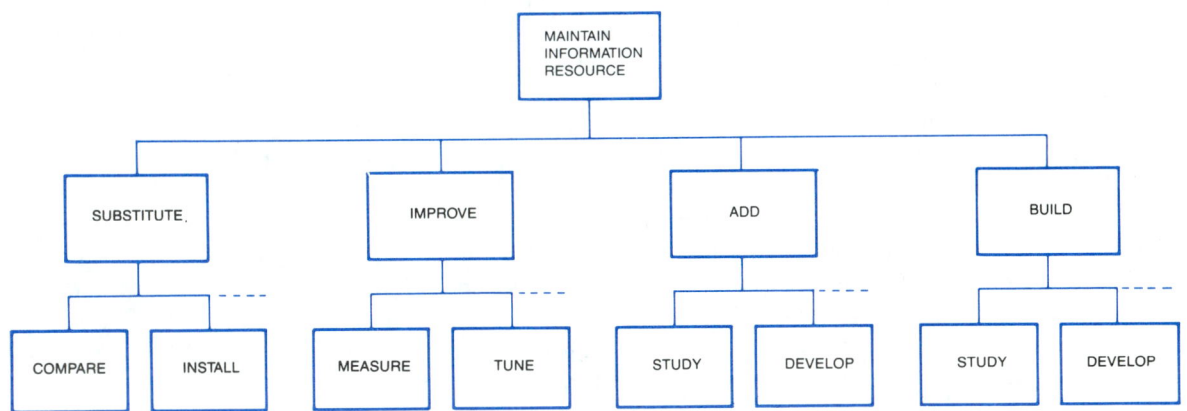

FIGURE 4–5. One Goal Analysis of Information Resource Maintenance

Although most work in systems analysis is refinement or enhancement, analysts are commonly trained to seek new opportunities. Because of this, the IRMaint function has the unwelcome reputation of always seeking "computer solutions" to problems that do not exist. As John Gall stated so well in *Systemantics* (1975, pg. 92), "When investigating a system, tread lightly; you may be disturbing one that works well." Or, as Thoreau reputedly said, "If you see someone coming toward you with the obvious intent of doing you good, run for your life!"

The overwhelming bulk of IRMaint work is precisely that — maintenance of an existing information resource. There are estimates that between 50 percent and 100 percent of all IRMaint work is in the areas of improvements and enhancements, most of which is either repair or small improvements. As the trends discussed in Chapter 3 continue, maintenance of existing systems will begin to approach 100 percent in most organizations, especially when end users begin to create (and damage!) their own software.

In *substitution*, one kind of system is substituted for another. Almost always this implies the substitution of automation in one of its guises for manual or mechanical techniques of providing information resources. The entire system may be replaced or one new element simply revised. The fields of micrographics, reprographics, and records management are mainly concerned with the development of improved mechanical systems to substitute for existing manual, pencil-and-paper systems.

Commonly, however, computer systems are pressed into service. An example that is becoming the rule is the replacement of manual typing systems with computerized or semi-computerized word processing. The functions remain the same: collection of raw documents, transcription, production of final copy, revisions, and repetition of the process, ending in reproduction and distribution.

With the introduction of rapid, typically verbal, command-oriented automation, however, procedures have to be rewritten. Typically, many peripheral procedures and activities remain the same (particularly raw document production and distribution, as well as most aspects of keyboarding), while central activities are drastically altered and many new procedures introduced (for instance, procedures to combine documents, provide automatic abstracting, and spelling checking). These new procedures require training programs. Sometimes these programs can be delivered through *computer* programs. Retraining is required when procedures are changed. For example, logging procedures for documents are changed, but the concept of logging is not — a change of procedure can cause confusion; documents may be lost if retraining is not done well.

Refinement or improvement of an existing system leaves the system intact and makes repairs or improvements that "tune" the system to work more appropriately. There are three reasons for tuning.

First, the needs of the organization with respect to the information resource may not have been correctly assessed in the first place. A medical billing system designed to work in a one-doctor office will not work efficiently if a second physician is hired. Smooth procedures for shifting from one physician to two need to be constructed.

Second, the needs of the organization may have changed. Organizational growth, changes in regulations, changes in policy or markets, or rises or falls in supplies and costs may require changes in how information is maintained. A common example is expansion of a product line. Whereas a three-digit product code may have sufficed for 600 products, four digits (or more) will be required when the one-thousandth product is introduced. An interesting problem occurred in the reporting of National Hockey League statistics in 1983 when Wayne Gretzky scored his one-thousandth goal: the computer that tracked points worked only to three significant digits (000 to 999) and thus began registering Gretzky in the rookie class with 001 goal!

Third, the original logical design specifications may not have been implemented correctly. Such errors, or bugs, comprise much of the traditional software maintenance of a data processing shop. The trend toward end-user programming makes much of this redundant, of course, as users create their own bugs. Recent research (Alavi, 1985) points out that many users fail to document their work, rapidly change their design conceptions, and lose track of what they were trying to do in the first place. They then rely on IRM professionals for assistance. Policies on responsibility for refining users' work are not yet consistent in the IRM community.

Enhancement of an information resource means adding system functions to an already complete and working design. Usually these additions are hardware and software, although new procedures may be designed to incorporate existing facilities in new ways. Typical enhancements include the addition of new files, the extension of records to include new fields, the addition of security-checking layers to prevent intrusion, the creation of new reports, and the speeding up of all processes.

Installation of faster hardware requires changes to the operating system to take advantage of the new speed. Expansion of a corporate database may require building or buying a database manager to provide on-line access to the data (here, of course, "enhancement" begins to look like building an entirely new system). In an enhancement, the bulk of the information resource is unchanged, but interfaces between subsystems may be reworked extensively.

One such interface is the procedures through which the resource is accessed. Consider, for instance, an on-line insurance records system that provides a listing of accounts by renewal date *or* insurance type, but not by date *and* type. An enhancement would allow easy specification of this two-way breakdown of accounts. If many such reports were required, an enhancement would provide a menu of such reports and the facility to specify a variety of breakdowns, moving toward a full database management/query facility.

New system development is the most extreme of IRMaint goals. Because of this, it is also the most expensive, the most risky, and the least common. Not only is there usually an existing system in place (which is to be scrapped rather than replaced on a one-for-one basis), but the new information resource is expected to accomplish functions not in the specifications of the old system. Some old system functions are to be discarded, perhaps because they were there to get around the limitations. In addition, new system development has strong organizational implications, since organizations often build around existing resources, implying the need to reorganize, assign new responsibilities, and revamp the flow of authority. Often, a new information system requires a new management system just to handle the information resource. In the extreme case, an information resource department is built from scratch.

The four goals (substitution, refinement, enhancement, and development) dictate four functions within IRMaint, functions that sometimes are stongly segregated. Usually, substitution and development work is performed by a **project team**, either composed of information resource professionals within a firm or brought in from a consulting group. There may be a separate **maintenance group**, or, more commonly now, maintenance may be assigned on a system-by-system basis to take care of refinements or enhancements. Technology infusion and adaptation are disturbing these neat alignments, resulting in multidisciplinary task forces, analyst prototyping of new systems, and information centers. The next section discusses the role of systems analysis in support of the goals of IRMaint.

4.4 SYSTEMS ANALYSIS ROLES IN IRMaint

The four basic components of IRMaint (Figure 4-6) are:

1. **design:** planning and specification
2. **implementation:** building or changing a system according to the design
3. **installation:** putting the system into operation
4. **audit:** measuring the function of the system

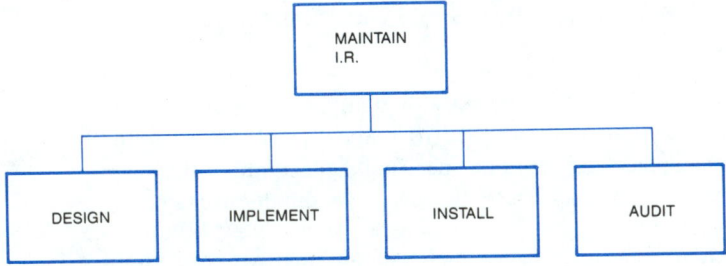

FIGURE 4–6. An Alternate Goal Analysis for IRMaint

(Chapter 1 introduced these components as the Information System Development Life Cycle). Within system design there is a strong distinction between *logical* and *physical* design mentioned in section 4.2. The best way to illustrate that is through an example.

Jim wants a new car. His accumulated savings from summer jobs amounts to $1,500, and his part-time work in the university library gives him an additional $200 monthly. Jim does not have an exact make, model, or year in mind; he only knows that:

- It has to get him from home to the university.
- It has to get good mileage.
- It has to be as attractive as he can afford.

He goes to Harry's Lots-a-Cars Lot and tells Harry his four criteria. Harry goes right to work and brings out a 1970 Mustang, very sporty, but also very squeaky and smokey. Jim says, "No way" so Harry brings out a 1980 Toyota, without squeak or smoke, and, frankly, without any class, given that it is missing part of the front bumper and has obviously been in at least one, if not several, serious accidents. Jim says, "Not this one either." Harry now wheels out a bicycle. "What? That's a bicycle! I want a car!" Jim howls. "No way you want a car. I give you class, you want quiet. I give you quiet, you want style. These cars meet all your requirements, but you obviously don't want a car. Here's sure transportation in all kinds of weather (never fails to start, yes?), very good gas mileage, and classic, absolutely classic lines. And it's only $200. It's not a car, but it does what you want."

The bicycle meets the logical design requirements that Jim specified, but not the unspoken physical requirements. Each of the requirements is met (and thus the overall goal will be achieved), but Jim obviously does have some sort of physical image in mind. And Harry does not have any of those physical objects on his lot.

In **logical design** of information systems, the structure of the functions to be met is laid out and logical tests are performed to see if, for instance, the Law of Conservation of Information is adhered to. We make no statements of how a function is to be carried out. Only in physical design do we say things like:

- We will do sorting on a computer.
- The customer name will be in the upper left-hand corner.
- We will program in COBOL.
- Employee ID numbers will be six digits followed by a classification code of A, B, or C.

These decisions are made after logical design because they imply that functions will work in a specific way (1) on a computer as opposed to manually; (2) in a certain position on the form rather than anyplace else; (3) in a specific language rather than any other one; and (4) in a certain physical format. These are specific assumptions about the environment the system will operate in.

Experience shows that delaying physical design decisions has two payoffs. First, changes to design can be done independently of other physical design decisions. For example, we may decide to add fields to records. This may conflict with the existing physical record layout, thereby invalidating the whole physical design. The second pay-off is that top-down goal analysis brings out the structure of functions, thereby ensuring that the system will meet its overall goals. Bottom-up physical design, on the other hand, may send us down a design path with dead ends. Jim, for instance, is out there looking for a cheap, reliable, classy car, something that obviously does not exist (at least, not on Harry's lot). Note that top-down, logical design guarantees only *some* solution; it does not specify what that solution might be. If we fail to design, we have "proven" logically that the system goal is unattainable; we have not described how the system will act in the physical environment — that is the job of physical design.

Physical design specifies the following (among others):

- specific hardware boxes and specific physical, electrical linkages among them
- layout of machinery
- physical procedures for operation
- training of staff
- implementation procedures
- installation procedures
- audit procedures

The first three items specify what the system does and how it will work. The fourth item is part of physical design because the skill portfolio necessary to do the work is not apparent until physical design is well advanced. The final three items are part of planning the other elements of IRMaint and can be done concurrently with physical design. Commonly, large teams are involved in physical design, including management, analysts, chief programmers, interior decorators, vendors, and physical plant managers, not to mention union representatives, head office personnel, and secretaries and clerks.

Systems analysis proper is concerned only with logical design. It contributes to IRMaint in six ways.

First, it provides information on **user needs and capabilities**. These needs may relate to an existing problem, plans for expansion, lack of understanding of the function of information resources, or just plain fear of computers. The analyst investigates and discovers what these needs are.

Second, systems analysis provides information on **user experiences** with the information resource. Often this is of a **critical incident** nature. Comments such as "I remember three days at the end of May when the system was down and we couldn't get our payroll out" or "I never get the information I need to prepare my month-end reports before the fifth or sixth of the next month" relate to specific events or chronic situations users encounter. This information may relate to system failure, the habits of the system, the way a user accesses the information resource, or what use is made of reports. Data on these experiences are used to make a model of current system functioning.

The third role is to provide guidance on **priorities** in maintenance. What should be done first? What should be done immediately? What can be postponed? What is too expensive to do? What is so vague that it does not make sense to proceed? What can never be done? These and other evaluations are a critical function of systems analysis and depend not only on analyst skills but also on the past experience analysts may have had with a certain class of users.

Systems analysis also **helps create plans for maintenance activities**. This is implied by the other functions. While such plans may be quite vague in the early stages, certain magnitude judgments can be made about expense, scale of project, and political vulnerability based on data the analyst collects. If this is an entirely new approach to a new problem, maintenance work in the form of improvements and enhancements will be almost certain.

Along with plans for IRMaint activities, systems analysis should provide plans for **information resource audit**, based on precise specifications. Because systems analysts often perform system testing and the post-implementation audit, it is reasonable for them to consider how the system should be audited even during logical design.

Finally, systems analysis supports IRMaint by **providing a link into the information resource's management**. Analysts "have their fingers on the pulse" of the information resource. By having a broad view and deep understanding of the organization, they can easily interpret IRM goals to users. For instance, the system operator may realize that system throughput is increasing and may be quite eloquent about how nice that is. But the analyst, who understands performance data, may realize that increased throughput at the expense of inquiry response time for an important class of users is not a desirable outcome at all and can present this to IR management as well as users. (This political role of systems analysts was detailed in Chapter 3.)

In providing this support, systems analysis operates at six levels, as shown in Figure 4-7. At the highest level, analysis contributes to the goals of IRM directly, usually in an organizational or political way. At the design level, systems analysis "proper" creates logical system designs. At the next level, systems analysts investigate, represent, and interpret data information resource functionality.

The fourth level is called the level of **products**, because goals are specified in terms of tangible products: investigation reports, logical models, logical specifications. At lower levels, systems analysis follows known **procedures** to create those products. For instance, conducting an investigation is the procedure necessary to create investigation reports. At the next level, the **tool** level, we specify the tools analysts use to follow the procedure to produce the products. One such tool is interviewing. Another is simultaneous verbal protocol analysis. A third is direct observation.

There is a pattern to analysis work. Systems analysis is part of the control loop in managing an information resource. The pattern of **investigate-represent-interpret** occurs frequently as analysts understand the environment, analyze it, and recommend changes. In fact, investigate-represent-interpret is the basic building block or "paradigm" in analysis work.

"Investigation" stems from the Latin meaning "to track." Like a hunter pursuing quarry, the analyst examines the area for clues, hints, and hidden meanings. Information is collected systematically and organized for later representation and analysis. Simultaneously, the "third eye" of the analyst is open for special signs. Managers often tell analysts one thing when they mean precisely another. Comments like, "Everything is going well here," may mean that everything is

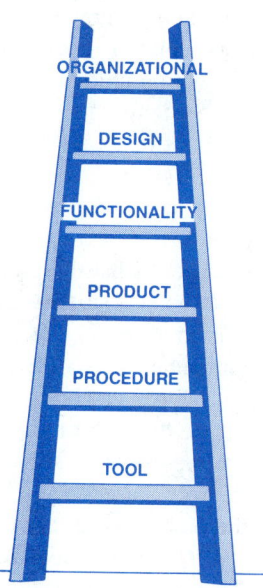

FIGURE 4–7. Six Levels of Systems Analysis Support to IRMaint

kiboshed. People beat around the bush, talk in circles, obfuscate, and are often confused; the analyst must be aware of this while investigating. People may work better and more efficiently when being observed; they may suddenly understand why a procedure is necessary just because they have never thought about it before.

At any level of this analysis of systems analysis, sensitivity, awareness, and attention to detail are needed. However, the analyst is neither a perfect mirror nor an informational vacuum cleaner. It is the analysts' own awareness of the fallibility of key procedures and tools that keep them honest.

"Representation" means to "re-present" data uncovered through investigation. Appropriate information is selected, analyzed, abstracted, and then placed into system *models* which are valuable to users, clients, operators, implementors, vendors, and managers. Modeling is the most visible product of systems analysis. Traditionally analysts have produced flowcharts. More modern techniques have greatly enlarged the repertoire of models to include over two dozen common kinds of charts, diagrams, systematic text, and so forth. (Modeling is detailed in Chapters 6, 9, and 13). The ability of the analyst to *present* these models to appropriate audiences is crucial to the achievment of any information resource maintenance goal.

It is the *interpretational* abilities of the analyst rather than the more mechanical investigation and model-building skills that distinguish information systems analysts from other professionals. The analyst must bring past experience to bear on present problems and must tap into a personal resource of knowledge, values, and intuition about systems, machinery, software, people, and management, synthesizing recommendations for action in the form of logical specifications.

This synthesis is evaluative. That is, it is not a value-free array of incontrovertible fact. On the contrary, it represents the analyst's unique viewpoint.

The synthesis is judgmental. It makes specific recommendations to do x and not to do y. It is also motivational in that it attempts to persuade others of its value.

Finally, the synthesis is personal and idiosyncratic. It represents not only clean things like data, well-learned techniques and professional attitudes, but also not-so-clean things like individual personalities, the need for groups of analysts to reach consensus, the career aspirations of the analyst, and momentary interpersonal concerns and conflicts.

The synthesis is the product of human beings working to put forth their own ideas, in the best interests of the ultimate customers — management and users. Students often ask if analysts should make recommendations that managers seem to want when the data point elsewhere. Like other ethical decisions, the answer lies beyond the realm of systems analysis and in the area of organizational behavior. The trend toward professionalization of information resource personnel has made it easier to address these concerns because analysts can fall back on professional ethics as well as techniques, knowing that there are others who share those same values. But the products of the analyst may well reflect organizational pressures the analyst feels. (Creating products by synthesizing fact in response to organizational realities is discussed in some detail in Chapters 7, 10, and 14).

Thus, IRMaint is supported by systems analysis at many levels, each of which requires investigation, representation, and interpretation. Products (reports, models) are created following procedures (conducting investigations, analyzing data) and employing tools (interviewing, flowcharting, writing specifications).

The student of systems analysis should not come away with the impression that IRMaint is a monolithic, unified structure of activities that everyone agrees with and behind which sits a body of proven theory. On the contrary, there are almost as many ways of structuring these activities as there are people performing them. Chapters 15, 16, and 17 discuss some of the issues and alternatives in structuring IRMaint activities into projects, for example. Chapter 15 concerns one particular structure, the development project, the "centerpiece of the system development life cycle." Chapters 16 and 17 discuss some alternatives that arise because of unique requirements of time, content, and resources. The formalization of system development methodologies is itself a growth industry, with several annual conferences and many researchers actively and productively at work.

4.5   ALTERNATIVES IN IRMaint ACTIVITY STRUCTURE

Eight alternatives common in the organization of IRMaint activities have evolved in response to increasing sophistication among users, increasing hardware capability, and the increased costs and decreased availability of expertise relative to the need. These alternatives are:

1. project
2. application generation
3. prototyping
4. end-user programming
5. the developmental approach
6. maintenance
7. package modification
8. personal computing

Figure 4-8 illustrates the organization, personnel requirements, and strengths of each of these approaches. Several alternatives (primarily end-user programming and the developmental approach) have engendered reorganization of the information resource function in the organization and are the subject of continuing research. Adopting any structure implies a number of organizational issues (many of which are difficult) and a series of related activities (many of which require good planning).

STRUCTURE	ORGANIZATION	PERSONNEL	USEFUL FOR
PROJECT	Linear or a Network of Activities; Deterministic	Analysts, Programmers, Designers, Testers, Project Managers	Large Systems, Long Time Horizons, Well-Understood Areas of Work
APPLICATION GENERATION	Single-Shot	User, Sometimes With Analyst Consultation	Extremely Well-Known Areas; Users Need not Know Programming
PROTOTYPING	Iterative, Somewhat Probabilistic	Analyst, Coder, User	Tailoring Specifically to the Needs of one Non-technical User
END-USER PROGRAMMING	Cyclical, Pre-Determined	End User, With Assistance of Information Center	Well Defined Application Areas, Savvy Users With Information Center Backup
DEVELOPMENTAL	Iterative	User, Analyst, Coder, Toolsmith	Quick, Responsive Development of a Tool for one Person or a Small Group
MAINTENANCE	Single-Shot	Programmer or Programmer/Analyst	Correcting Errors in Code, Investigating Recognized Anomalies, Enhancements
PACKAGE MODIFICATION	Like a Short Project	Analyst, Programmer, Users,	Changing an Existing Purchased Package to Tailor to Installation Needs
PERSONAL COMPUTING	None (Random)	User = Operator = Analyst = Programmer	Responsive Exploring of a Problem by an Involved User who is not a Trained Professional

FIGURE 4–8. Alternative Structures for Information Resource Maintenance Activities.

The Project

The **project** structure is the traditional way of organizing new-system development. The project is organized into a network of tasks (see Figure 4-9), which are performed

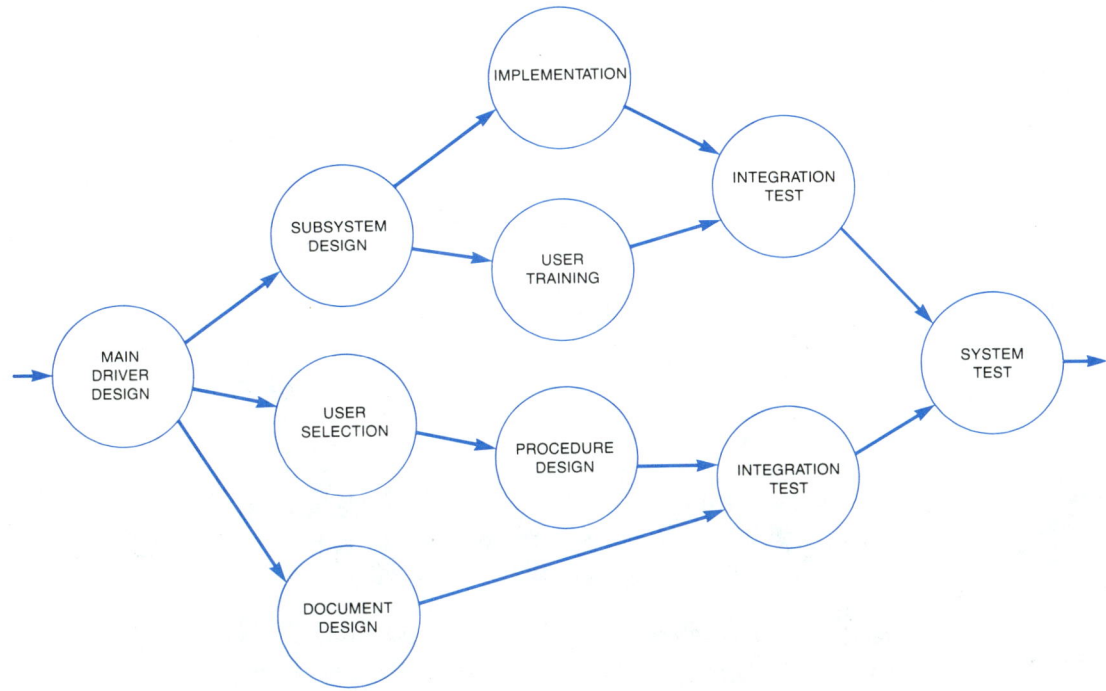

FIGURE 4–9. A Project Task Network

under the direction of a **project leader** who has overall responsibility for project activities. The network organization implies that work proceeds *linearly* in time, that is, from task to task as each is completed. This, in turn, implies that scheduling of these tasks is crucial in meeting overall project goals on time; delay in a critical task will delay many subsequent tasks. A variety of mathematical scheduling techniques have been developed to enable project managers to plan, execute, and track the progress of projects as well as forecast completion dates and resource requirements. Foremost among these are PERT and CPM (discussed in detail in Chapter 15).

The early years (1945-1970) of information resource management were preoccupied with building new applications on expensive machinery using relatively few technical staff who were overworked and often inexperienced. Project work dominated in that period. As a result, much of the terminology we use to describe IRM work is derived from the now unnecessary emphasis on project work characteristic of that era. Emphasis then was on creating an **informational infrastructure**, and little thought went into improving project work or looking for alternatives.

Project work is typically governed by the **System Development Life Cycle** (SDLC) (discussed in detail in Chapter 1). This concept specifies that after performing the logical system design, the analyst hands these specifications to the system designer. The designer creates the physical design and hands these specifications to implementors. At each step, value of some sort is added in the realization of thoughts and ideas.

A typical SDLC-governed project involves analysts, programmers, IRM supervisors and management, and user "clients" who are responsible for keeping up-to-date on the progress of the project toward realization of their request. Project staff must be ready and able to work on their respective tasks when the time comes; a typical project can last from one month to several years.

Long projects have a cost, however. Future human, financial, organizational, and computer resources have to be planned for well in advance. Imagine that you are about to plan a two-year project that has as its main goal the introduction of a microcomputer-based local area network into all offices at the headquarters of a multinational energy firm.

The difficulty is that very few resources can be allocated accurately before a project begins. The simple question "Who is going to do the work?" taxes imaginations, given the rapidly changing technology. Workers who are pleased to work for $25,000 annually this year may be demanding $40,000 in the second year of the project. And there remains the difficult question of how to predict technology, a question left to other texts to ponder. Nonetheless, project managers must estimate accurately before they begin.

The reasons that projects work less well now than they did twenty-five years ago are twofold. First, environments within which projects function are more turbulent now than before. Economic, technological, and social forces create deeper, more frequent disturbances, which make forecasting of resource needs and availability almost impossible in some cases.

Second, most of the "easy" work has already been done. The first business IRM work was in accounting, order processing, and billing in cash — areas well known, easy to control, and based entirely on accurate historical information of past organizational activities. To construct a management-decision-support application, however, IRM personnel have to understand ergonomics, the psychology of decision making, and the social psychology of organizations, as well as the firm's business. As we move from "easy," number-based, historically grounded *reporting* applications to "hard," knowledge-based, poorly grounded *decision-making* applications, the appropriateness of large projects becomes questionable.

As Chapter 16 will discuss, projects will continue to prosper. However, for new "projects" intended for single individuals or small groups, for "throw-away" software, and for exploratory enhancements of existing systems, other formats are more appropriate.

Application Generation

Not all information resource maintenance has to begin from scratch, rebuilding expensive wheels. Often analysts or users employ "kits" and construct pieces of the information resource from prefabricated parts. Sometimes this construction can itself be automated. The term for this is "**application generation**," the automated construction of software on command by another piece of software, called an "application generator."

Application generation can best be seen through an illustration such as in Figure 4-10.

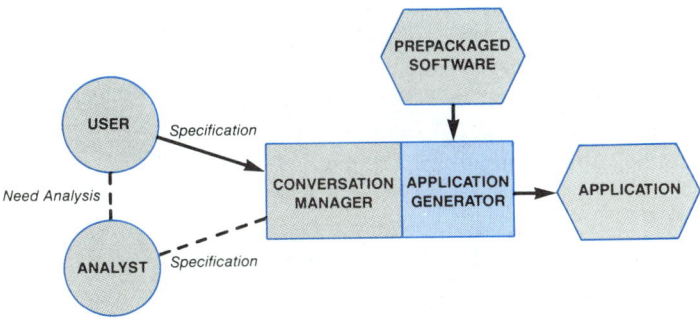

FIGURE 4–10. Application Generation

Consider the needs outlined in the following interview:

> *Tom:* Sure, we have a lot of data analysis to do, but it's usually just the same stuff over and over again.
>
> *Analyst:* Can you give me some examples?
>
> *Tom:* Well, at the end of each month we have to produce a report of sales by region by product. Every quarter we have to produce a quarterly report the same way. And there's the annual report. On December 31, we have to make three reports that differ only in the period being accumulated. It seems silly to write three programs to do that!
>
> *Analyst:* Perhaps. Are there other examples?
>
> *Tom:* Try this one. We also produce exception reports, you know, where a product is selling less than 90 percent of forecast, where a salesperson is selling less than 85 percent of quota, where a region is underselling, and so forth. And these reports have to be amalgamated on a monthly, quarterly, and annual basis, too.
>
> *Analyst:* So you have a lot of reports to produce that use essentially the same data, but that are, let's say, "keyed" to certain needs, like the need to control a month or a quarter or a product or a salesperson. Is that right?
>
> *Tom:* Yeah. Y'know, when you look at it that way, most of what we do is really just turn on a switch and say, "Go do the report."

Here is a perfect example of the value of an application generator. A single set of data is used to produce a variety of reports based on **parameters** that control either the way the report looks or routine differences in how data are handled, accumulated, or compared. Tom's work seems amenable to the "generator" approach. In effect, Tom can command one of a number of applications by uttering "magic words" (that is, parameters).

An alternative is to produce sixty-four programs and have Tom request them by name or number through a menu. This is a viable approach until Tom wants a sixty-fifth program or loses interest in sixty-three of them. Rather than have an individual programmer create or generate sixty-four (or -five or -six) programs, the application generator actually recreates the program on command through a conversation.

Doesn't this approach "waste" resources in recreating the application every time?

Yes, it does. If an application is going to be used often, it may be more efficient to "hard-code" the application after Tom realizes how useful the program is. There is another side, however. If the number of applications is unpredictable, if the usage varies, and if there is a need to change the application often, it becomes less practical to apply expensive human resources, to applications that differ only parametrically from others. It is better to let the computer do it.

Added benefits of the application generator approach are these:

1. A "front-end" conversation manager poses questions to the users about their desired application. This manager accumulates information on application needs from users. It also keeps user consciousness high about the actual needs and provides at least a modest level of user involvement, a factor in system development that research has shown is critical to the success of application development.
2. Expensive specialized talent in information resource maintenance does not have to be applied to development of "small" applications. Such talent can be better directed toward understanding the underlying principles behind all the applications.
3. Users can make minor changes to formats simply by "regenerating" the application, thereby further freeing professional systems staff from routine work.
4. The generator may itself be a saleable product.
5. Application generators have a way of "growing up" and generalizing into full-fledged information resources.

Several forms of application generators are available. **Architecture-based** software development and **report generators** can create permanent applications based on either input or output characteristics. The former is typified by forms generators, common in database managers. These generators create display-based forms onto which operators enter data. Popular among the users of microcomputers, such generators are highly integrated into databases and can serve to simplify data capture and thereby increase the value of the database manager.

Prototyping means beginning with the reports the user wants to see

Report generators have been around since the middle 1950s. Grandchildren of RPG (report program generator), such programs are essentially application generators that accept commands from users about how data are to be displayed or printed, generate stored specifications for the reports, and then, when supplied with data, create the reports as desired. Using a report generator, users can create a number of variations of the same report, experimenting with layout and simple calculations. Again, report generators are integral to most microcomputer-based database managers and have widespread usage in mainframe business systems.

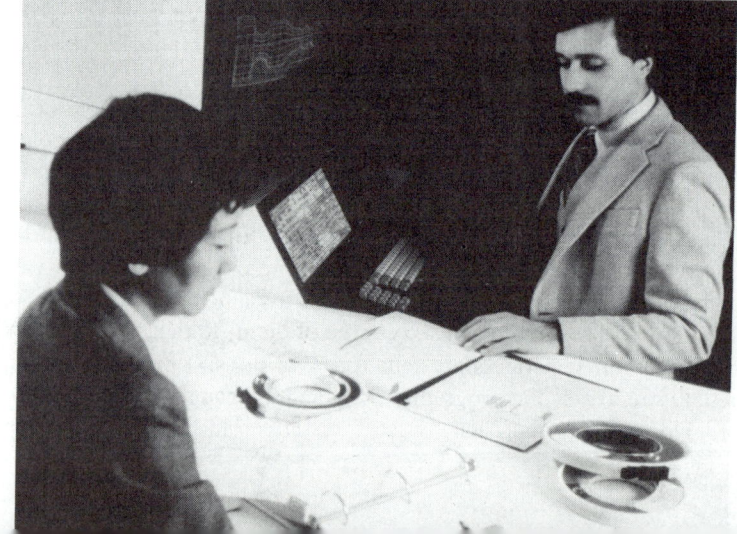

Prototyping

Prototyping (Figure 4-11) turns system development on its head by beginning with the final product and working down to its constituent parts. Prototyping is based on the premise that most users have little need for or patience with the parts of development. That is, a user can neither use nor understand the value of a small module within a larger system. On the other hand, the user is the best source of criticism about a whole application

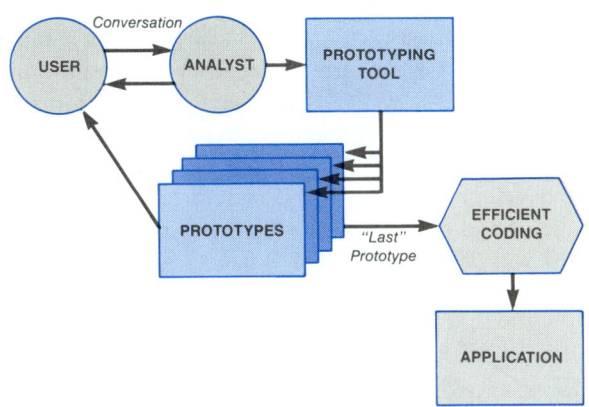

FIGURE 4–11. Prototyping

and how it appears to work. If these premises are true, then users should never see anything other than the entire system in operation.

The major shortcoming of prototyping is the inability to prototype very large systems and the relatively inefficient results that have to be rebuilt in an efficient fashion. The advantage is that systems can be produced that seem to *simulate* an operating final product. Although the prototype may work slowly and may not actually process any data, there is always something for the user to relate to, thereby keeping this important resource involved in the system's development. Prototyping is discussed in more detail in Chapter 16.

Several products are now available to produce prototypes; many of the more interesting ones are aimed at the micrcomputer market. Spreadsheet generators are ideal for use in a prototyping mode, since simple applications can be made "apparent" at early stages on screen.

The difference between prototyping and application generation is that the former is guided by an analyst in consultation with users as critics. The latter is created by the IRM group and then handed to the users to generate and maintain all applications. In both cases, "final products" may be hard-coded in more efficient ways by professionals once the actual value and cost are determined and it is decided that the costs of "professionalizing" the product are worthwhile.

End-User Software Development

Application generation and prototyping lead to the development and maintenance of information resource elements by the users themselves. Users can craft their own systems in three ways. First, they can learn programming languages and build their own programs using fourth-generation languages like FOCUS, CONDOR, and NOMAD that are far less difficult to learn than Fortran, COBOL, and PL/1. James Martin (1982) thinks that this is the start of a trend toward what is termed "programmerless programming."

Second, some database managers are easy to apply to simple goals. In particular, dBaseIII™ has been remarkably successful because it allows users to create their own applications quickly on their own databases.

Third, spreadsheet generators are a form of **end-user programming**. The astonishing success of VisiCalc™, Lotus 1-2-3™, Symphony™, and Supercalc™ has demonstrated that "untutored" users can, in fact, become accomplished programmers of their own business applications.

End-user software development is cyclical (Figure 4-12). Recently researchers have noted that the technique of end-user-developed software is beginning to show some of the problems associated with adolescence. For instance, many users get themselves in

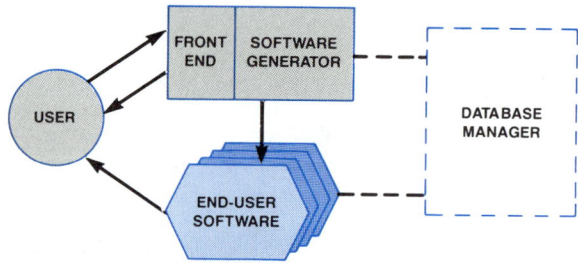

FIGURE 4–12. End User Software Development

too deep and find that their "next" application is just too hard. When they ask for professional assistance, the professionals may rebel when faced with a lack of documentation, poor, unsystematic coding styles, or an ignorance of specialized techniques to perform sophisticated sorts, computations, and queries. On the other hand, users understand the application itself. Whether end-user software will realize its potential as applications become harder is arguable. Trapped between the attractive economics of application generation and the power of prototyping, end-user software may end up as a museum piece by the next century as users find that the simple applications can all be automated and the more complicated ones require the intervention of an analyst.

The Developmental Approach

An interesting combination of prototyping and application generation is found in the developmental approach fathered by Ralph Sprague (1980) in the area of **decision-support systems** (Figure 4-13).

A decision-support system (DSS) assists management in making decisions. To do this, a DSS needs the requirements on the following page.

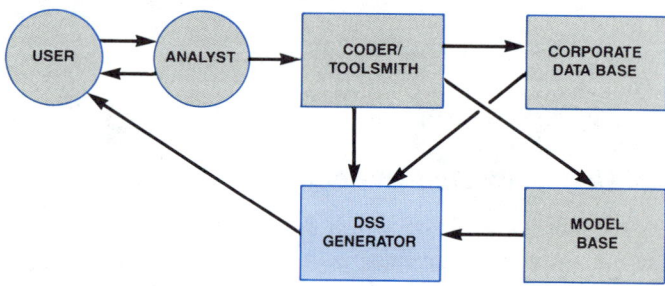

FIGURE 4–13. The "Developmental" Approach to DSS Development

a corporate database
tools to query the database
tools to generate reports on the results of computation
a set of models

The database is usually set up independent of the DSS. A DSS generator creates specific DSS applications (called the specific DSS or SDSS). The SDSS assists the decision maker in locating data and specifying how the data are to be manipulated. To answer questions such as "What if we try this...," models are "run" against the data to determine the outcomes of specified courses of action. Managers receive graphical or textual reports showing the results. They can then choose the best course of action based on their trials.

A DSS is only as good as the database and the models used to pose what-if questions. A significant amount of literature exists on how to build, use, and evaluate DSSs. Of importance here is the team approach to building DSSs.

The team works with the manager. On the team are a decision analyst, a coder, and a DSS "**toolsmith**," who constructs the models needed by the decision maker. As the DSS "develops" (hence the term "developmental" approach), more and more of the system is hard-coded by the programmer. A database manager, graphics support packages, and report generators are important sub-units that have to be integrated. Users employ models and then want them refined. This approach closely resembles prototyping.

The developmental approach is ideally suited to building a system for an individual. It is expensive, but the result is a highly flexible product suitable for supporting a range of decisions in an area covered by the models and the database. Once the basic DSS is in place, it can easily be cyclically improved.

Maintenance

The **maintenance** project is usually the result of a development project. The work involves making repairs or minor enhancements, tuning the system to run better, or adding new files, fields, or devices. Rarely will the maintenance project construct a whole system. The term "maintenance" is unfortunate, because although it is derived from the true meaning of "maintenance" (related to the phrase "hand [main] holding [tenance]"), the connotation is of uninteresting, unchallenging, and unrewarding "dirty" work, which it is not.

Maintenance is usually a single-shot piece of work involving a single activity, a single worker, and often a limited time resource, classified in the following ways:

1. Clean up: correct the results of bugs in the data.
2. Fix up: remove errors in software or hardware.
3. Spruce up: make things run faster or look better.
4. Wash up: remove system functions, files, or records that are no longer useful.
5. Do up: make up a special run, a small application, convert a file to a new format, adapt a package or piece of equipment, train someone how to use a changed package or piece of equipment.

In many ways maintenance is like project work. Before the work begins, estimates are made of the resources needed to grapple with the question of whether or not the work is worth the cost. Next, the analyst (or more likely, a programmer-analyst where software is concerned) tries to understand the current system and its difficulties. Then repair or enhancement work begins, and the product is tested and released.

On the other hand, there are important differences. First, it is not always possible to maintain a running system. Often an "Out of Order" sign has to be placed on software. This naturally forces users to rethink their strategies and may lead to conflict — it is the maintenance staff that will bear the brunt of complaints while the software is out of commission. Second, project work is often performed in teams, while maintenance work is usually the domain of the ace super-sleuth, working alone or with an able assistant. Because of this, the maintainer has to have not only the skills of the detective, but usually the stamina of the original construction team. Maintenance work is hard and demanding, especially since documentation is not the "jewel in the crown" of programmer skills. Maintenance personnel may have to work from incomplete specifications of how the system was intended to function coupled with users' often-garbled descriptions of what the system *actually* did.

Maintenance work has traditionally been passed down to novice programmers and trainee analysts. A more recent trend has been to employ crack squads of maintenance commandos, specially trained in maintenance work. The development of software engineering as a field of study and practice and the growing body of literature on **maintainability** and reliability in software systems have fostered the education and employment of maintenance specialists.

Another trend is **life-cycle engineering**, the creation of teams that have complete responsibility for a project before, during, and after installation of a system. In this model, the individuals who designed and programmed the system also maintain it. "If they are not going to document, then let them remember what they did" is the philosophy here. Life-cycle engineering recognizes that systems have life spans. Estimates at this date are that for traditionally developed software-based projects, over 50 percent of all expenditures on system development actually occur during maintenance of the software after it has been installed. Other evidence from recent surveys shows that 50 to 90 percent of all software expenditure is on maintenance. If the trend toward end-user programming accelerates, it is likely that few programmers will ever work on new-system development.

Package Modification

After the Second World War, land developers and manufacturers in North America realized that building houses and other complex goods from scratch wasted scarce resources and was far slower than using prefabrication techniques. Communities such as Levittown, and a wide variety of consumer goods were built from factory-fabricated units and assembled as needed at the work site. It took almost twenty years for the software development field to catch up to this trend, but **package "tailoring"** is now a way of life for many organizations that cannot afford to purchase or build hand-crafted software.

There are many examples of packages that users buy and then tailor to their own needs. Complex accounting packages may be bought from consulting firms for a fraction of initial development cost. The consulting firm tailors the package to the specific needs of the client for an additional fee, perhaps also supplying training and maintenance under contract.

As in the consumer goods areas, an original idea is developed and documented by a vendor and made as flexible as possible. After each sale, the vendor works in a maintenance mode. From the customer's viewpoint, the software is "full of bugs", that is, it does not work exactly as the customer would like it to.

Some packages are built in prefabricated fashion, so that the blocks the user wants can be constructed on the spot. Lotus 1-2-3™, has a number of auxiliary files (called "drivers") that are built during a system installation phase. Most complex microcomputer software is distributed in like fashion these days. The training and continuing maintenance aspects, however, are less widely available. Vendors usually rely on third parties to do the training and maintenance.

Personal Computing

Finally, the use of **microcomputers** is a way of packaging IRMaint work. The introduction of microcomputers into the business environment is only just beginning to have an impact. Personal computing is based on a **stochastic** work structure that we might call "play." The goals of personal computing are enjoyment and displacement of work onto the computer. The process is linear, moving more or less in a straight line toward some goal, but highly probabilistic. We are never really sure how far we are from the goal and the nature of the goal may change from time to time.

Personal computing is governed by the criterion of enjoyment, meaning anything from addictive hacking to productive analysis of data. Often, the introduction of a personal computer into an environment has a greater effect on personnel than on work itself.

The model, bluntly speaking, is finger-painting. When an attractive design is achieved, the work is complete. In many ways, personal computing is a throwback to the earliest days of computing when scientists and engineers operated their own vacuum-tube and relay computers. They were the analyst, the user, the client, and the coder all in one. It was, in fact, from those days that computer work gained its unfair aura of being a mysterious kind of wizardry — one needed a degree in engineering to use a computer in 1945. Even in 1960, a coder had to know about bits, latches, and switches. This feeling of technical competence is recaptured in personal computing, but with the support of today's software. As with end-user computing, it remains to be seen whether personal computing, having captured the imagination of a subset of managers and professional workers, will substantially alter how IRMaint work is done.

4.6 SOURCES OF IRMaint

Our discussion of IRMaint so far gives the impression that all IRMaint work is handled by an organization within the consuming firm. That is not the case. There are alternatives specified on the next page.

- contracted piece-work
- turnkey system purchase
- package tailoring

Contracted piece-work is available through software consulting firms, which provide something like a mobile IRMaint group willing to work for clients. Working with a software house in this mode requires the purchasing organization to present some level of expertise in both IRMaint and the application area. The most critical success factor is the involvement of informed users in the work. The overwhelming variety of software and the continuing shortage of expertise in the face of the rapid growth in the field means that client organizations should not rely exclusively on hired expertise. Stated another way, the personnel department that contracts with ABC Software to develop a personnel system must first build up some level of expertise in its own ranks. Many firms employ user representatives who negotiate with the IRMaint group for development work; this same model should be applied to outside contracting.

Turnkey systems purchase involves more than software. Turnkey suppliers are one-stop shopping centers for all IRM needs: project planning, analysis, design, installation, training, and maintenance. The vendor in this case sells a package of hardware, software, and services far more complex than that offered by the software house. At certain times, a system appears, people are trained (sometimes the vendor assists in hiring operating personnel or supplies these workers under contract for a period of time), and new procedures are put into place.

The third source is to purchase packages and then **tailor** them in-house to meet specific needs. Package tailoring is becoming one of the most important ways businesses are entering the IRM area without the high initial costs of setting up development "shops." The trend can only continue as labor costs continue to rise relative to hardware costs.

4.7   CONCLUSION

Systems analysis as an activity contributes to the goals of maintaining an information resource within an organization. This chapter concerns the **products**, **procedures**, and **tools** of systems analysis and how they fit into this broader organizational goal.

Systems analysis is a component of **information resource maintenance**, which in turn is a component of **information resource management**. Other subsystems of information resource management include **operations**, **planning**, and **promotion**. At this level, goals are essentially managerial and business-oriented; technical considerations exist only to support business goals.

Information resource maintenance subsystems include **system design**, **implementation**, **installation**, and **audit**. These four goals are technical in nature and are concerned with *understanding* and *measuring* a system's functions. Within system design are **logical system design** and **physical system design**. These two sets of activities are commonly performed by the same individual or team — the systems analyst. At this level, technical matters are paramount. The major activities (logical versus physical design) are distinguished by their emphasis on understanding the structure and functions of a proposed system or system change ("logical" design), or the specification of the boxes, wires, and physical activities that, when put together properly, carry out the tasks needed to achieve those functions in that structure. **Structural-functional design**, as outlined in Chapter 1, is the major tool of logical system design; physical design principles are as varied as the fields in which design must be carried out: architecture, electronics, office landscaping, and ergonomics.

Information resource maintenance activities can be carried out in a variety of organizational formats. The most common way is to put together a **project**. Other alternatives include **prototyping, end-user-developed software, decision-support systems** (also called the developmental approach), **personal computing**, and **maintenance**. In addition, IRMaint can be accomplished in a number of "packages," including **in-house development, contracted piece-work, turnkey systems purchase**, and **"package" tailoring**.

Regardless of the structure of activities or the packaging of the work itself, systems analysts conduct their work in a pattern of recurring processes. There are three parts: **investigation, representation**, and **interpretation**. Investigation is the uncovering of facts about the existing system. Representation means the production of system models, which are tables, charts, drawings, or text that describe both the existing system and the alternatives that represent improvements. Interpretation is the "art" of systems analysis, in which the arrayed, represented facts are evaluated to determine a new logical system design. The result of this three-part process is the *logical design* or **logical specifications** of the new system or improved system element.

Information resource maintenance is undergoing transformations (as discussed in Chapter 3). Most important, much of the maintenance work is being transferred to users, which implies that the more difficult and challenging work is becoming even more difficult and more challenging. Several of the trends discussed in Chapter 3 are changing the way organizations approach information resource activities. In particular, the phenomena of the **information center** and **distributed processing** are causing analysis work to change in nature and structure. The development of "fashion" in IRM has brought about a variety of organizational specialists and vested interest groups, as well as an enduring set of intellectual challenges to systems analysts in performing their craft. Ideas such as "decision-support systems," "expert systems," and the introduction of office automation into the ranks of executives have changed not only the structures of work, but the market for systems analysis and the environments within which systems analysis is carried out as well. Turbulence, rather than placidity, is the order of the day.

The first four chapters have spotlighted the context of systems analysis and discussed the fundamentals of General Systems Theory on which the rest of the book depends. The remaining chapters of this book are organized in a structural-functional way. Chapters 5, 6, and 7 discuss procedures for logical system design. They concentrate on investigating, representing and interpreting, respectively. In these chapters, we will look at the theory behind logical system design and explore how investigation efforts are planned and organized, how system modeling works and assists in the design effort, and how the analyst goes about interpreting these models into logical system specifications.

Chapters 8, 9, and 10 respectively look closely at the products of the three procedures: reports, charts, and specifications. Practical guidelines are presented for organizing and writing reports, selecting and drawing system charts and diagrams, writing specifications, and making presentations. We will see how these products fit into IRMaint generally and how managers in particular should and can pay attention to them.

The following chapters go one level deeper and take a close, nitty-gritty look at the tools the analyst uses. Chapters 11 and 12 detail techniques of gathering data, concentrating especially on the choice and employment of techniques and their effects on the workers investigated. Chapter 13 details the use of specific diagramming techniques employed by analysts and system designers. Chapter 14 explores the use of economic analysis in the service of decision making in systems analysis.

The final three chapters take a close look at project work. Chapter 15 scrutinizes project management, especially planning, control, and reporting, since these are aspects that concern user management, too. More detail on prototyping and information centers is presented in Chapter 16. Chapter 17 looks at the types of technology the analyst has to consider across various fields such as records management, microcomputing, office automation, and consumer products and games. Management perspectives, rather than technical skills, are stressed in these last three chapters.

4.8   A CASE OF MISTAKEN IDENTITY

Jean Kilpatrick is a "front-line" social worker for a municipal social-work agency. Her primary concern is working with families who abuse their children. Most of her work is interviewing families and recommending temporary placement of abused children into foster homes. She then refers the families to private counselling agencies in the city. The designation of her job as "front-line" is appropriate; many times she feels as though she were on the front battle lines in a war.

Yesterday Jean received a computer-printed report from the data processing department in response to her request for a search through the records. Jean wanted a list of past contacts her agency has had with a particularly troublesome family, the Bollingers. Today Jean is confused. There is a Bollinger family with prior contact; in fact, there are *three*. Two of them have two children and one has four. No ages are recorded for the children. The family with four children has no addresss; the other two families' addresses do not match that of Jean's troublesome family. The information is five years old, however, and her Bollingers have moved several times. Jean knows that her family has two children. One of the listed Bollingers has had a contact with the welfare people; the other two are middle-class families like Jean's clients.

The information is so confusing that Jean will ignore it, just as she has ignored other search results in the past. Besides, the request took three weeks to satisfy and Jean had to fill out two complicated forms. One was for access to the information itself, the other to approve her request to have access. Maybe she will just assume that her clients are not any of the listed families. Anyway, her contact information form (BTC-3-82) does not have any place to indicate knowledge of prior contacts and the Bollingers aren't saying anything. If only data processing could just once give her something useful in return for filling out all these forms.

1. Consider the information resource Jean is using. Does it meet all the criteria set out in this chapter? If not, which ones does it not meet and how do you know?

2. What would be a proper role for a systems analyst who is called in to determine whether or not the information resource is in need of change?

3. Using the information above and assuming other facts about people who work in organizations, decide on a structure for systems analysis activities in support of IRMaint to change the system Jean has come to ignore. What would some appropriate goals be? What is the scope of the effort? What are some of the risks involved?

4. Suppose the analyst's initial recommendations are to go ahead with a redesign. Examining the nine alternatives in packaging IRMaint activities, evaluate each with respect to this project. Which would be the quickest? The costliest? The most disturbing to workers? The easiest to manage?

5. If the project is to go ahead, Jean's supervisor, Sandy Mahoney, will be the "client." What will Sandy have to know? What skills will Sandy have to have? What should Sandy demand to be kept informed of during the work? How can Sandy measure the progress toward goals in this work?

6. Suppose Jean's employer decided to scrap their information system, which is only one of several applications run on the city computer, and build or contract their own. Make a recommendation for the best source of new systems. What are your criteria? What are the risks and benefits of each source? What skills, knowledge, and attitudes must the client have? Should Sandy still be the client on this work?

Discussion Questions

1. Consider an information resource such as the student course registration system in your college or university. Discuss the differences between information resource *management* and information resource *maintenance* in this system. Do they involve the same people? In what ways are you as users of this system affected by decisions others make in IRM and IRMaint for the student registration system? What specific resources are involved in the operation of these systems?

2. Five criteria for proper information resource functioning were mentioned in this chapter. Is it possible to rank them in terms of general importance? Can you think of exceptions to any ranking scheme? What are some of the trade-offs? Consider as an example, the student course registration system from Question 1. Can there be trade-offs there, such as inaccuracies or inconsistencies to allow for, say, increased availability?

3. Why do you suppose that IRMaint traditionally has been organized along vocational lines (analysis, programming, operations) rather than along functional lines (investigation, representation, interpretation)? What would be the advantages and disadvantages of each? Can you think of an example of IRMaint for which vocational organization would be a requirement?

4. Provide an example of logical and physical specifications for the information resource provided to you by this book. What corresponds to the logical specification? What corresponds to the physical design?

5. Consider education in systems analysis. Describe building knowledge in this area in terms of substitution, refinement, enhancement, and development. Can these same terms be applied in other areas, such as management in general?

6. Three functions are described in this chapter: investigation, representation, and interpretation. Consider the task of providing information about parking regulations to visitors to a college campus. Describe the goals of investigation, representation, and interpretation as they relate to this goal.

7. The JKL company manufactures fittings for pipe used in drilling operations. Drilling technology is changing rapidly and drilling operations are increasingly computer-directed and -controlled. An inventory system already exists on a mini-computer through which receiving and shipping personnel maintain a stock of fittings. Describe how you might organize structures of work (project, maintenance, prototype, etc.) to accomplish the task of improving this inventory system.

8. The trends discussed in Chapter 3 and alluded to in this chapter are going to affect how user management interacts with IRMaint personnel. Consider the JKL company in the previous example and forecast how each of the technology trends (infusion, dilution, adaptation, and evolution) will change how managers in the JKL company will approach or be approached by the IRM group to meet the same improvement goals for each trend. For example, how would technology infusion alter the way managers in the JKL company interact with the IRM department? How would technology dilution make a change?

9. The JKL company is also going to build a new order-entry system, one that has to interface with the inventory system. Discuss four plans for such development: in-house, piece-work contracting, turnkey installation, and package tailoring. Which carries the most risk? Which has the least cost? Which has the shortest pay-back period? Which has side benefits, and which has hidden costs? Develop the plans based on the analysis of IRM presented in this chapter.

10. Consider the goal of going out to a concert with a friend. Perform a goal analysis on this activity, functionally decomposing it at each level. How do you know when to stop decomposing? What can you do with this goal analysis of a practical nature?

11. Why does structural-functional design work? On what assumptions is it based and, if carried out correctly, why should the designs produced actually work?

DESIGN EXERCISES
Using dBASE III

1. Let's turn our attention to the programs which make up the system. The following enables you to look into the programs:

TYPE HAIR.PRG   Print out the commands stored in program HAIR

Note that the commands look a lot like BASIC and that they go by and **scroll** off the top of the screen. You can get the screen to **pause** by pressing the **control** (CTRL) **key** and the NUM LOCK key at the same time (do this by holding CTRL down and then touching NUM LOCK briefly). Some keyboards have a PAUSE key that performs the same function. To get the listing going again, press any key.

Two commands you will find in the programs are these:

DO p   run a dBase III program (p stands for the name of the program)
RETURN   go back to the command just after the DO that brought you to this one

Using the TYPE command, record the ways that various programs DO others. When you've finished recording this, transcribe it into a structure chart.

2. Consider the conversation that Doug had with Larry previously. What are the pros and cons of each structure discussed in the text?

Work in pairs and role-play the conversation between Doug and Larry. As you converse, use the dBase III commands and your common sense to **prototype** sample report lines for Larry's needs. Can you actually locate a small set of reports that Larry might need? If so, can you imagine how you'd build an **application generator** for him?

CHAPTER 5

INVESTIGATING: RESEARCHING THE SYSTEM

OBJECTIVES

1. To understand the necessary role of knowledge in logical system design
2. To detail three different techniques for obtaining knowledge in logical system design
3. To describe the role and function of investigation reports in the logical design process
4. To detail how to set up an investigation effort given the goals of the investigation

INVESTIGATING: RESEARCHING THE SYSTEM

5.0 "WHO SAYS?"

"Who says?" demanded Odette Bergman. "Who says the program's not usable?"

"Derek," responded Irene Takaoka, "and Saul and Monica. And everyone else who uses it. Odette, it just doesn't work right."

"Well, how do you know that?"

"Because I asked. Look, I'm their manager. I can't let something as important as this company's accounting go down the tubes while I listen to my workers complain. They should know. After all, they use it every day."

"Still, Irene, we worked six months on that package and you signed off on it. It worked then and we've heard nothing since then. Why all of a sudden isn't it usable?"

"Odette, I can't tell you. It's just built up. The package is hard to use, y'know, and when Monica joined two weeks ago, we discovered that we couldn't explain to her how it worked. Then we looked carefully at some sample runs and discovered that the balances didn't balance. Maybe you tested it wrong; maybe we didn't check it closely enough last March. But it's wrong. There are lots of other problems, but I haven't written them down. I know your group invested a lot in this, but it's got to be re-examined, and quickly. Where do we begin?"

5.1 UNDERSTANDING AN EXISTING SYSTEM

Irene and Odette have already begun. They know something is wrong and neither is happy. What they do next is investigate the accounting package and its users. In a way, they are detectives. The system life cycle of the information resource "recycles" for a number of reasons:

- Changes to the information resource itself: failure of an element, loss of data, changes in equipment
- Changes in the users of the resource: increased needs that result in increased usage, changes in perception, changes in expectation, new users or a new kind of user

- Changes in the environment of the resource: new requirements to be met, increased competition, lack of funds to support the information resource, changes in corporate strategy
- Changes in the nature of information resources: new equipment available, obsolescence of existing equipment, changes in industry pricing, trends in technology

In Irene's case, there was a change of personnel along with a number of changes in perception of how the package ought to work. We do not know yet why these changes caused Irene to confront Odette, but certainly that has to found out. But before anything can be done, the existing system must be understood.

Why "understood"? Won't the existing documentation do? The documentation is very important, but existing formal information is only part of the story. Users' experiences (Figure 5-1) are a significant factor, and they will not appear on charts or specifications. And the actual performance of the package may differ remarkably from recommended or assumed performance standards. Workers may fail to use the system correctly or for correct purposes. Finally, the system just may fail altogether.

Why must we understand the system before doing anything about it? In addition to the obvious problem of not even knowing which system we are talking about, there is the difficulty of determining what, if anything, is wrong. "Complex systems fail in complex ways," implies John Gall in *Systemantics* (1975, pg. 62). Just because a system appears to be ineffective or unusable does not mean that it actually is. On the other hand, subsystems that appear to be functioning correctly may actually be failing, as Irene discovered long after the fact.

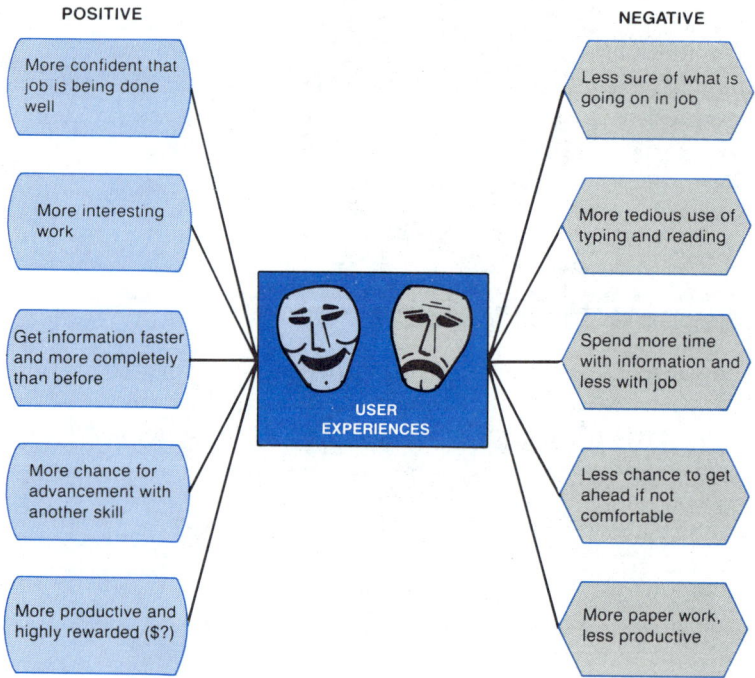

FIGURE 5–1. Varieties of User Experience (p. 5.7)

This chapter is concerned with discovery and knowledge. The analyst is not only a knowledgeable person, but a person who can learn new facts. The analyst is a critical part of the learning system to which IRMaint contributes. Thus, an analyst's major responsibility is to investigate existing systems in an attempt to understand them, because understanding is key.

The plan of this chapter is to look first at how existing systems are "understood" and what this means. Next, we will examine three formats for investigation: the preliminary investigation (PI, examined in more detail in Chapter 8), the feasibility study, and the detailed investigation.

Then we will turn our attention to the reporting function, how to make investigation results understandable and usable. Finally, there will be a detailed look at how to put an investigation into gear. Chapters 11 and 12 present details on the most important investigation techniques; here, we will be concerned with the administrative aspects of setting up an investigation.

The Problem of Knowing about a System

Chapter 1 presented the basic problem with systems — unless one is inside a system, it is theoretically impossible to have first-hand knowledge about the system's elements and relationships. On the other hand, from within the system, exact specification of the system's goals is difficult. In each case, knowledge is created indirectly through an understanding of experience with the system and its environment. In fact, knowledge depends on **systematic examination of relevant experiences**.

"Systematic examination" implies that the analyst's method for observing a system is objective, repeatable, and noncontroversial. Others can, and would, find the same facts if they systematically examined the same way.

Experiences should be relevant. In the case of the accounting package, the experiences that count are those that involve the package and/or the preparing of accounting reports. The politics of the accounting department are less relevant. How the firm's products are shipped is irrelevant. If a shipping clerk discovers that the accounting package is useless for printing packing slips, that is interesting, but probably irrelevant.

Finally, the analyst's investigation has to be experience based. It is the workers' experience with the information resource as they try to meet organizational goals that counts. Working from outside a system, the analyst builds a model of the system from user experience (Figure 5-2).

To summarize, an analyst applies observation methods to relevant experience with the information resource. The result is a set of facts. Knowledge arises through examination of these facts for significant patterns.

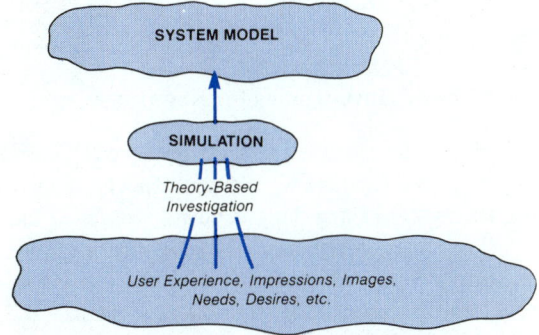

FIGURE 5–2. Building System Models from User Experience

These patterns owe their appearance in part to the already accumulated knowledge of the analyst. For example, if the analyst already knows how the accounting package is supposed to be used (perhaps through interviews with Irene and Odette), then the fact that Derek and Saul cannot get the accounts to balance tells the analyst where a problem lies. If the analyst finds out later that Derek received on-the-job training from Saul, the analyst may see a typical pattern: the published "correct" way to use an information resource is not the actual way it is being used. The analyst is on the way to "knowing" that the problem lies in the way the use of the package is taught - or not taught.

Of course, the analyst may examine experiences unrelated to people. Downtime, throughput, transit time (the time it takes to process a record, form, or file), and a variety of formal measures of efficiency and cost are also observed, recorded, and analyzed.

Investigating to Obtain Knowledge

The investigation is cybernetic in the sense that it is guided by a policy and responds to its environment (that is, the system being investigated). Observations are made, recorded, and then analyzed to obtain knowledge (Figure 5-3) that is a theory of the user/system. This theory influences later observation, recording and analysis until a stable model is developed.

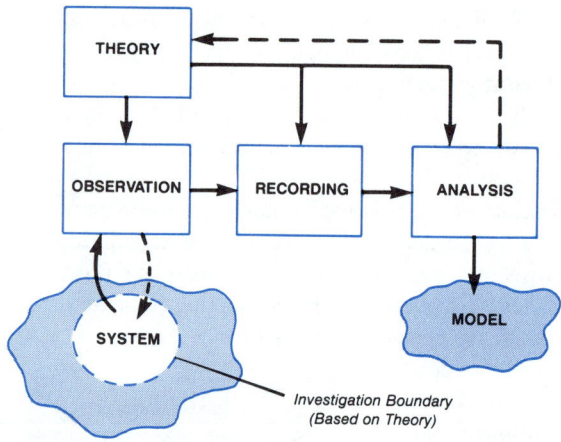

FIGURE 5–3. Investigating for Knowledge

First, the system must be observed. One of the major tasks facing an analyst is limiting investigation to relevant aspects of the system and its environment. **Determining the system boundary** is best approached in a step-wise fashion.

The initial motivation for investigating may come from complaints or audits of an existing system. Beginning there, the analyst notes every subsystem implicated in the problem or opportunity mentioned. Certainly the accounting package is included in our example. Is the inventory subsystem included? Maybe. Whose accounts don't balance? Who handles them? Maybe these individuals should be included. What about the individuals who use the system; should they be included? Yes. Is the personnel department involved? Maybe — who recruited these individuals?

Next, this large set is reduced by testing. If the system worked correctly and no one complained, which individuals and subsystems would not be involved? In other words, which elements are loosely connected?

The result is a diagram showing what is included and what is not included (Figure 5-4). The preliminary investigation contributes primarily by setting this boundary and limiting the scope of the detailed investigation.

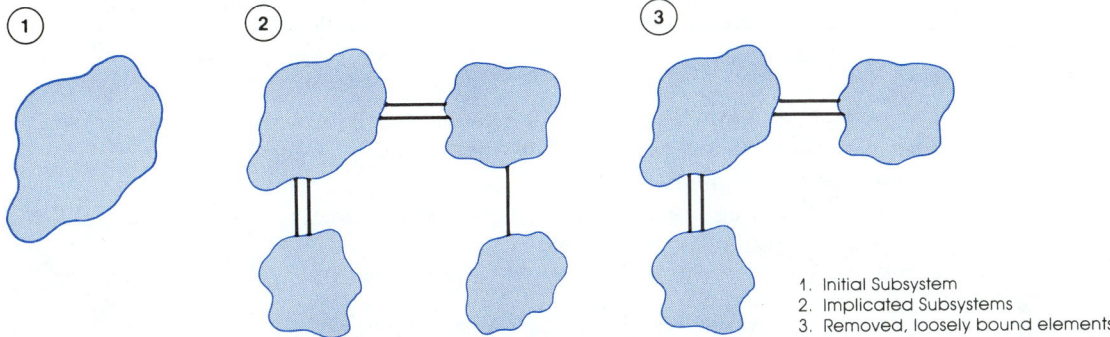

1. Initial Subsystem
2. Implicated Subsystems
3. Removed, loosely bound elements

FIGURE 5–4. Determining the Investigation Boundary

Observations are made only within the boundary. The major criteria in observation are reliability, validity, and usefulness. The observations must be recorded and then analyzed. The result is a set of statements pointing out how the system works, what needs to be improved, where these improvements should go, and how to make the improvements.

Knowledge-Based Design

The philosophy of this text is that systems analysis contributes to **knowledge-based** design. This means that the goals of system analysis include the translation of knowledge about the system into coherent, relevant, useful designs. The designs will be so only if the data collected reflect the analyst's knowledge about the system. Because knowledge is systematic, facts can be profitably debated. Because knowledge is relevant, facts can be effectively included or deleted from consideration in the design. Because knowledge is based on experience, the system that is designed fits into the same environment as those having the experiences.

Knowledge-based design succeeds if information collected by the analyst comes exclusively from observation rather than from other sources. "System knowledge" is obtained through observation of the system (Figure 5-5). Logical design depends on two sources: system knowledge and design knowledge. Thus, system knowledge depends exclusively on the state of

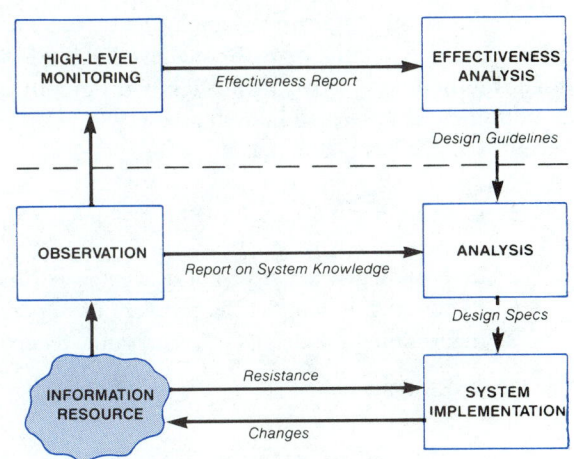

FIGURE 5–5. Observation in the IRMaint Process

the system being observed plus noise introduced in the observation process. Keeping this noise to a minimum is the goal of analyst training and the IRMaint management.

The logical design can come from only three sources: system knowledge, knowledge about design creation, and the logical design process itself. If the logical design process is not noisy and if design knowledge is a constant (for example, if we can expect all analysts to know basically the same things about designs and how they are arrived at), the logical design is a function of only the system being investigated. Since the data collected during investigation are experience based, the logical design merely reflects the experiences (desires, expectations, hopes, fears, and so on) in those subsystems investigated.

We want knowledge-based design to produce **analyst-independent** designs, designs that do not depend on particular analysts' methods, knowledge, skills, power, or appearance. The practical fact, however, is that knowledge-based design is an ideal. Architects are fond of putting their "imprint" on their designs, and as long as systems analysis is a proud art, we can expect signatures even in logical design.

5.2 TECHNIQUES FOR OBTAINING SYSTEM KNOWLEDGE

Knowledge-based design depends crucially on getting high-quality information about the system under study within the boundaries set for observation. Analysts rely on three stuctures to obtain this kind of knowledge:

1. the preliminary investigation (sometimes called an "initial investigation" (Leeson, 1985), "preliminary survey" (Awad, 1985) or even just "problem/opportunity definition" (Wetherbe, 1984))
2. the feasibility study
3. the detailed investigation (often referred to as "information needs analysis" or "requirements definition")

There is little agreement among textbooks and handbooks as to the exact names, natures, and functions of these studies. Each commercially available **system development methodology** labels them slightly differently and assigns different goals to them (a good review is presented in Wetherbe (1984, 128-130). Regardless of the differences in terminology, investigating has three distinct goals:

1. determining the nature of the problem or opportunity within some boundary of study
2. determining whether or not it is even worthwhile (usually read "cost effective") to continue looking at the problem or opportunity
3. determining the details of the system, its environment, and its functions

The Preliminary Investigation

The **preliminary investigation** (PI) has four purposes:

1. to outline the high levels of the structural-functional design of the system
2. to define the problem or opportunity
3. to determine the study boundary
4. to evaluate the need for further study and recommend ways to continue

The PI is "preliminary" in the sense that a full-scale effort to collect data is delayed pending the definitions and evaluations. The techniques employed to meet these goals resemble those used in the detailed investigation.

Most IRMaint efforts begin with a request from a potential client. Consider Zircon Industries. In expectation of new regulations concerning plant closings, the VP of manufacturing and the director of employee relations want to be able to merge their files to create reports on justification and execution of orderly layoffs and rehiring for specific plants, shifts, and jobs. They have sent a memo (Figure 5-6) to the manager of information resources outlining their desires.

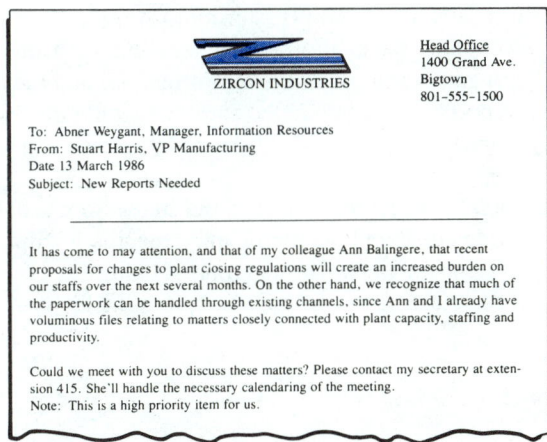

FIGURE 5–6. A Request for Assistance

Note that their memo speaks not only about their wishes, but it seems specific about how these wishes are to be met. Stu and Ann want to be able to use their files together. It is not clear to the manager of information resources what they want to do with the files. Nor does he immediately understand which files they are referring to, what kinds of reports they would like, and who would have access to the information and the reports. On the other hand, the request cannot be rejected just because it is written in technologically vague terms.

The manager of information resources asks Dominic Valle, a senior analyst, to speak with Stu and Ann about their memo. Dominic knows that the first interview should settle two points: "where" the problem or opportunity is (which subsystem in the organization), and "what" does that problem or opportunity seems to be.

The "where" question is important in beginning to set limits on the investigation. The "what" question begins the structural-functional design. Let's listen to the interview:

> *DV* [seated in the office of Stu Harris, VP of manufacturing. Ann Balingere, director of employee relations, is also present]: I understand from your memo that both of you see a problem on the horizon. Who is going to have that problem?
>
> *AB*: Well, I always take the heat first when we close a plant, but Stu here eventually has to close it. Then I have to relocate or lay off the workers.
>
> *SH*: Yes. You see, the reason we're concerned is the paperwork, you know. It's already awful, but we see new regulations on the horizon requiring us to give ninety days' notice of a closure. That's going to mean challenges and that means documentation.
>
> *DV*: So that's the "paperwork" you refer to in your memo?
>
> *SH*: Right. Now Ann's got all the personnel information in her files and I've got plant efficiency and performance data in mine. We'd like to be able to put this information together somehow when the Feds want the report.
>
> *DV*: There are two different areas, then, as I see it. You [gesturing to Ann] already have the personnel information. You, Stu, have information pertinent to plant closures. The first "problem" is getting access to two sets of information, and the second is figuring out what kind of reports would satisfy the expected regulations. Is that right? [AB and SH nod approval.]

In this excerpt, Dominic is trying to put a "geographic" limit on the study. So far, it seems so far to involve only employee relations data and plant efficiency records within Zircon, but obviously some information is needed about what the report should say.

Next, Dominic tries to define the goal structure of the proposed effort at the highest levels:

> *DV*: OK, now let's look at what you'd like to see from us. Ann, what do you want to give to the Feds?
>
> *AB*: Really, we need a report that justifies the plant closing on economic grounds and, at the same time, states how we'll handle the employee dislocation. What the Government and the public don't understand is that we don't want to hurt people who have worked for us for decades and that we always try to put them into another job if we can.
>
> *SH*: And the economic argument is always based at least partly on labor costs and the opportunity to move assembly to more automated lines and plants. So we need the labor figures and, at the same time, a report on retention, retraining possibilities, and, of course, the cost.
>
> *AB*: It's not easy to make these decisions, as you might imagine. So we also want something to help us make the decisions about plant closings and relocation.
>
> *DV*: What I'm hearing is that you'd like something that uses data: first to convince yourselves; next, to help you decide how to go about closing a plant; and third, a rationale in the form of a report [scribbles a few notes; see Figure 5-7].

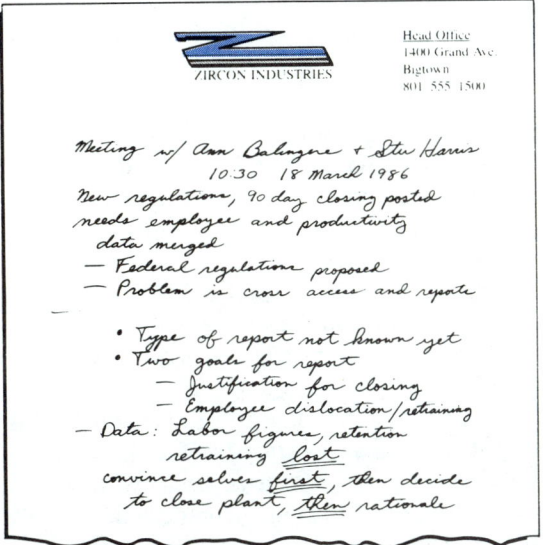

FIGURE 5-7. Dominic's Notes

SH: Yes, but I'm not sure about that order of importance. We don't know yet about the precise details of the regulations.

DV: When will you know?

SH: Oh, we know in general what's required, I guess. We've got a labor lawyer working full time in the chairman's office on this legislation, so we'll have details. I guess you're right, though. Rather than just documenting, why not have something that will really help us make the best decisions?

The scope has grown before Dominic's eyes. He may have to cut the problem down to size before continuing. In fact, after an hour of discussion with Stu and Ann, Dominic is almost certain that the report will have to wait until more information is available and that the really important subgoal is to assist plant-closure decisions.

Another goal of the initial meeting is to obtain a list of individuals to speak with and permission to speak with them. Eventually, Dominic will have a good idea of which subgoals he will be able to investigate, which are pie-in-the-sky dreaming, which will be politically unadvisable, and which are already already achievable but perhaps not well documented.

The PI terminates when the analyst understands enough about the nature of the problem or opportunity to state its definition, to indicate the scope of the investigation (that is, the study boundary), to detail at a high level the goal structure of the proposed new or improved system, and to document the users' needs for information. These form the bulk of the PI report and justify the analyst's recommendations either to continue with a feasibility study or detailed investigation or to quit now.

If Zircon goes ahead with the project, a feasibility study may follow if the problem or opportunity is relatively familiar. More often in a structured approach to systems analysis, a detailed investigation is performed, delaying questions of feasibility.

Feasibility Studies

Based on an engineering approach to problem solving, the **feasibility** study assumes the definition of a problem as a given and attempts to put a price (in time and dollars) on various ways of achieving the solution.

The solution is usually a construction based on known parts. For example, if Dominic is familiar with the kind of problem facing Ann and Stu, he may propose using existing facilities, such as a report generator and a corporate database, to produce the required report. He could easily determine the cost of doing this by looking at past attempts to do the same thing.

Figure 5-8 tables the major differences between feasibility studies (FS) and PIs. The FS is a bottom-up, solution-oriented, construction-based approach, while the PI is a top-down, goal-oriented, analysis-based one. What are these distinctions and why are they important to analysts and clients?

First, as Figure 5-8 shows, an FS assumes that the structure of the functions needed to meet a goal is already known. The major effort is in locating the functions among "parts lying around a shop" and putting them together. Because each existing part (called a "module") is well known, it will be the linkage or construction that determines how well the ultimate result works. Estimating the costs depends on how rapidly and effectively the parts can be put together, tested, and installed.

	FEASIBILITY STUDY	PRELIMINARY INVESTIGATION
Basic Design Approach	Bottom-Up, Incremental, Synthetic	Top-Down, Holistic, Analytic
Orientation -Solution	Constructed, or to be Determined From Candidates	Consideration Delayed—Some Solution Assumed Later
-Problem	Taken as Given, Goal Here is Select Best Solution	To be Determined in P.I.; Goal is to Define the Problem Structurally
Techniques	Cost-Benefit Analysis Schedule Analysis Technology Assessment	Structural-Functional Design Goal Analysis
Indicators	Well-Understood Technology Common Problem With Intense Past Experience High Emphasis on Cost Control	Technology not of Immediate Concern Unknown Area Sufficient Research Funds Available
Drawbacks	High Cost if Wrong; Need to Scrap Whole System	Could Design a System With Less Than Maximum Feasibility or Performance

FIGURE 5–8. Feasibility Study and Preliminary Investigation Compared

The PI, on the other hand, aims to define this structure of functions from the top down, beginning with the system's goals.

While feasibility studies are oriented toward demonstrating the (degree of) feasibility of a number of alternative solutions, the PI seeks to define the problem. Because the FS presumes that a number of solutions exist, little effort is made to *prove* that the solutions are, in fact, solutions.

These are then compared in terms of a number of *feasibility factors*. Three are considered important (Kronke, 1981, pp 100-101).

1. **Schedule feasibility**: do we have enough time to put it together?
2. **Technical feasibility**: do we have the skills and is the technology available to build it?
3. **Economic feasibility**: do we have enough dollars to put it together and operate it?

"Time, tools, and dollars" is a shorthand expression for this sort of feasibility quest. The various alternatives are weighed on these three scales.

Feasibility studies are most valuable when problems and opportunities are well understood and familiar to the IRU. Otherwise, time-tool-and-dollar-feasibility prognostication is a dangerous endeavor.

The Detailed Investigation

The **detailed investigation** is intended to gather data to fill out the structure chart begun by the PI. The data bear on both the existing system and its environment and a proposed "new" system and its environment.

The detailed investigation proceeds until the structure chart is complete to the stage of physical implementation. In other words, the analyst knows the design is complete when, to understand how a module functions, physical information on speed, volume, format, form, or layout is needed. Since this can be determined only by relating the logical design to the physical environment (that is, buying equipment, drafting forms, laying out physical records, specifying terminal types), the logical design is complete and the detailed investigation is over.

Investigating While Prototyping

Prototpying is a relatively recent innovation in system implementation. (Chapter 16 discusses prototyping in detail.) Of interest here is how the analyst gathers information during prototyping. Prototyping is an output-driven development technology, focusing on the output (usually a report) and making modifications based on needs to change the output. The prototype itself is a kind of "question." In effect, the analyst asks, "How closely does this come to satisfying your needs or conforming to your image of what a report should look like?" Because prototyping bypasses the traditional information-gathering phases of the SDLC, we may be tempted to think of it as lacking a knowledge base. In fact, however, it is exclusively determined by the needs of the user, as filtered through comments about the prototype.

Prototyping can be considered a kind of rapidly iterated SDLC with the IRMaint goals of enhancing and refining an existing system (the first prototype). That being so, data collection is limited to comments about each successive prototype. The analyst will keep notes about design decisions, which will form the "knowledge base" for the design effort. It is a happy coincidence that when the knowledge base is complete, so is the application.

5.3 THE ROLE OF INVESTIGATION REPORTS

Investigation itself informs the analyst, but the products of investigation have to be put into a form from which others can profit. Thus, there is a report for each kind of investigation: preliminary investigation report, feasibility study report, and detailed investigation report (usually part of the logical system specifications). This section is concerned with the role these reports play in presenting an image of a system to the outside world.

The System as Seen from the Outside

The Uncertainty Principle dictates that the views outsiders have of a system depend at least in part on their frames of reference. "Frames of reference" mean the terms with which an observer describes a system and the categories in which the observer perceives and distinguishes events, elements, and relationships. Because the observer's frame of reference is probably different from that of anyone inside the system, the description by an observer probably differs from "truth," whatever that may be.

In practice, we have come to rely on a philosophical trick. We define a system merely to be the description derived from overlapping several observers' descriptions and call it "objective." If one observer's description differs from all others, we tend to label that description as "subjective" and "false" in some sense. We require corroboration of an observer's view. Frequently, in carrying out our analyses, we do have access to elements of the system we are observing — the people who work in or with it — and we use their testimony to provide this corroboration. We call that "evidence"; the more massive the evidence, the more credible the investigation report.

The "system as seen from the outside" becomes a mixture of the detached observer's notes and statements taken from others closer to the scene. But the problems of observation cannot be solved merely by accumulating evidence.

In most cases, our observation techniques require the analyst to *code* observations into terms, categories, or sets based on some scheme best known to the analyst. In other words, even corroboration may be "slanted" because the analyst has a fixed scheme for writing down observations. This can lead to some problematic situations:

1. The analyst asks, "How often does the screen just go blank?" The data entry clerk replies, "Frequently, at least once a day." The analyst writes down "daily." In fact, the screen goes blank several times during a single session, which the clerk mentally lumps into a single "going blank" event.

2. The analyst notes that the *average* transit time for a particular form — the time it takes a form to be processed through a department — is four days. Any analyst would observe this. But there is no *typical* time. The major problem is the *unpredictability* of the transit time; customers do not have a clue how long it will take to get their orders processed — sometimes it takes a month!

3. In a group interview, the analyst notes, as would any observer, that the participants frequently seem upset at the response time their terminals provide them. He concludes that they are "frustrated" by the slow terminals. If he were more flexible, he would find out that it is not knowing how *long* the terminal keyboard

will be frozen that frustrates them — they don't care about the *speed* at all. If they knew they had five minutes to get a cup of coffee, they would be quite happy.

These are some examples of what might be called "jumping to the wrong conclusion"; using a term like that, however, is precisely the problem we are referring to here. In fact, it is using a particular reporting frame of reference that is the problem.

Terms of Reference

One of the major contributions of the observation report is a list of terms of reference, a kind of dictionary that defines observational terms. Such a list allows the readers of the report to judge for themselves whether certain errors in observation and recording are being made.

For example, the terms may include the following:

1. Data entry: entering data through a terminal
2. Terminal: computer workstation employed by a data-entry clerk
3. Keyboarding: entering data through a terminal's keyboard
4. Data-entry clerk: the person who keyboards the data at the terminal

Disregarding, for now, any circularity among these terms' definitions, consider that they tell us a number of things about the analyst's frame of reference:

1. Data come from outside the computer system; none of it is supplied by the computer.
2. The terminal is a work tool of the clerk.
3. Data enter only through the keyboard; any preparation of the data prior to keyboarding is not considered "data entry," including checking performed by the supervisor and logging completed forms.
4. Any person who keyboards must be a data-entry clerk — or only data-entry clerks' keyboarding can be considered to be data entry. For example, if a supervisor corrects a trainee's keystrokes, this is not considered data entry.

Presenting Data

Generally data are presented either as **aggregated results** or as **case examples**. Consider the following two excerpts:

A: Aggregated Format
Over half of the executives spoken to consider accuracy of data the primary consideration. Only a few are concerned with very high speed, and none thought that being able to access data immediately at all times was a criterion. When asked whether or not the quality of their work would increase if more data were available on sales and production, 80 percent thought it would. Most of these (75 percent of the 80 percent) warned that too much data would be worse than the current situation, however.

B: Case Example Format

Smith (not his true name) is typical of the respondents. He likes to be on top of things, but he's not willing to run the risk of inaccurate data in order to get it a little faster. He claims that he could use more data, but he's not sure precisely which data are important. He worries that because of this he'd get an avalanche.

Neither of the above reports is laden with conclusions of the form "...and therefore we ought...." In systems terms, the process of creating a data presentation is "unintelligent."

Presenting Investigation Results

Investigation reports such as those of the primary and the detailed investigations are not merely informative; they also try to persuade. Generally, the knowledge gleaned by the analyst is the basis for a series of recommendations. The format of the report should contribute to the credibility of these recommendations.

Here is a recommended general outline for any investigation report (details are in Chapters 8 and 10):

1. Executive summary: one to three pages summarizing the following sections, with recommendations highlighted
2. Problem statement: short paragraph describing the problem or opportunity as it was understood before the data were collected
3. Rationale: justification for collecting the data; why the data bear on the problem statement
4. Data collection procedure: A description of how the data were collected, including the specific techniques used, the sample selected, the time period required
5. Data-handling procedures: how the data were transcribed, stored, and protected; where the data are now available and in what form for which individuals
6. Analysis of the data: reduction of the data to comprehensible proportions; aggregations or case examples
7. Conclusions based on the data: recommendations for action or termination; rationale for these recommendations

Even though the **problem statement** will change during the IRMaint effort, it helps the reader understand the analyst's frame of reference. Consider these two "different" problem statements:

A. The volume of information that counter clerks have to handle to respond to a customer complaint seems to be interfering with their ability to satisfy customers.
B. Customer complaint forms are difficult to fill out while the customer is present. This detracts from the interpersonal relationship and lengthens the amount of time it takes to handle a complaint while not contributing to the quality of the encounter.

The first problem statement arose from the initial request for assistance (see Figure 5-9); the second problem statement evolved later to be the basis for the detailed investigation.

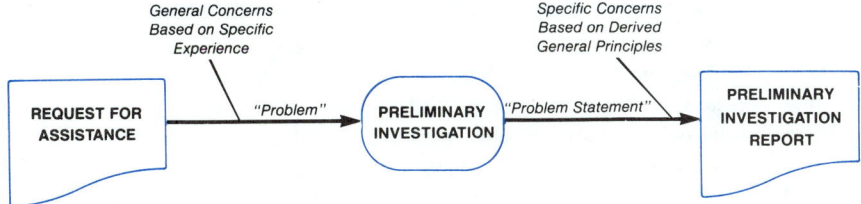

FIGURE 5–9. The Source of the Problem Statement

Recommendations can appear at two levels. On one level (Figure 5-10), the analyst recommends particular aspects of design based on knowledge. At a higher level, the analyst also recommends continuing with the project, cancelling it now, or redoing part of the data collection effort to bolster conclusions.

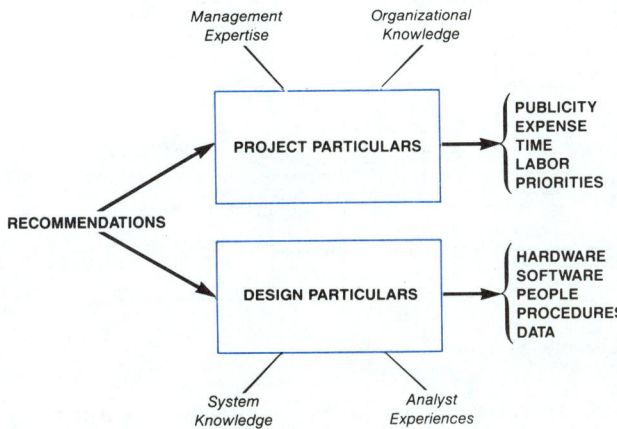

FIGURE 5–10. Two Levels of Analyst Recommendations

There are three reasons for cancelling projects based on investigation data:

1. The data are impossible to gather; knowledge will never be sufficient to complete a design without great risk.
2. The data show that no design will work for the cost limitations given.
3. The data show that benefits will, in fact, be slight for any effort put into the work.

Using Investigation Reports

Investigation reports provide others with an outside view of a system based on observations by an expert. The knowledge that results contributes to both design and project decisions. In the case of the feasibility study, a Go/No-go decision is based directly on the conclusions, while the other reports, especially the detailed investigation, contain design recommendations.

Because the reports are used by a variety of people, their language should be as nontechnical as possible, except where specific design points are made. Readers include system designers and programmers, but the more immediate audience is clients, managers, and users. Systems analysis is a mirror that speaks. If it speaks enigmatically, those who look into it may see nothing or they may see only what they want to see. The iterative nature of structured systems analysis is such that each look into the mirror is more illuminating than — and built on — the last.

5.4 SETTING UP AN INVESTIGATION

An investigation effort is itself a system. Because it contends with at least a disturbed environment, the investigation is conducted using the feedback loops found in a cybernetic system. Figure 5-11 illustrates this structure.

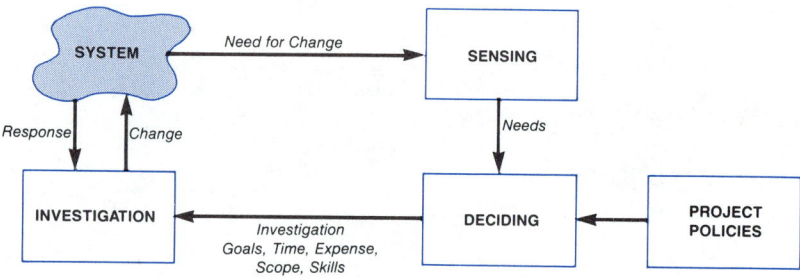

FIGURE 5–11. Investigation as a Cybernetic System

There are three major cybernetic functions served in the investigation: sensing, deciding, and acting. The analyst begins the investigation based on a sense that something needs to be done. Regular or *ad-hoc* system audits bring about this signal. A system may be functioning poorly or it may fail dramatically all at once. This brings the analyst to the scene.

The decision to investigate has a number of implications in terms of time, expense, scope, urgency, and skills. These, in turn, imply a set of limits on subsequent actions and make up the set of decisions that the analyst has to make early in the investigation.

Other decisions include determining the investigation techniques to be used. Along with that decision, the analyst has to obtain permission to use the techniques to collect the data. Building credibility is an important aspect of setting up the investigation. Finally, an investigation team has to be selected and the investigation begun.

System Audits: Sensing

As Figure 5-12 on the opposite page illustrates, a system investigation "begins" when a member of the IRU detects that there is a need for a change. Sensing the need for a change requires a sense cell dedicated to sensing that need. In many organizations, this is implemented formally by a regular system audit.

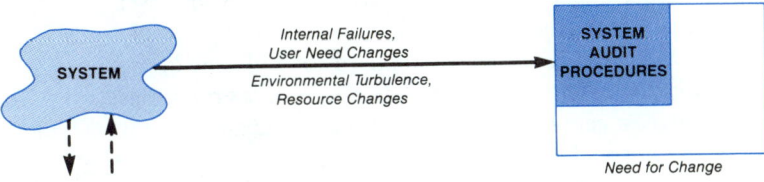

FIGURE 5–12. Beginning a System Investigation

The **system audit** simply determines whether the information resource is working adequately. There are four ways in which the information resource can prove inadequate:

1. internal failure, implying a need for maintenance activities (repair or regeneration)
2. changes in the nature of the required interaction between the information resource and its environment (usually because the users or user needs change), requiring growth in the system by addition or enhancement of functions
3. turbulence in the environment, implying a need for defense activities to maintain system integrity
4. changes in the nature, availability, or popularity of particular resources used to construct the information resource, implying a need for maintenance of a replacement nature

A system audit may be simply a measure of performance against specifications, concerned only with internal operations and the need for repair or regeneration. Or it may be a complex evaluation of the interface, concerned with user perceptions of adequacy. The IRU itself may conduct ongoing evaluations of hardware and software trends, while a special security group may be attuned to integrity of access and data.

Jerry Fitzgerald's book, *Designing Controls into Computerized Systems* (1981), outlines six aspects of control-based performance auditing. As illustrated in the diagram in Figure 5-13, these kinds of control are **deterrent**, **preventive**, **detective**, **reporting**, **corrective**, and **recovery**. Errors, omissions, and fraud should be deterred or prevented within the system itself. When such problems do occur, they should be detected and reported, and correction or recovery attempted. The audit system should track these events and create its own report, usually summarizing the control situations. While primarily concerned with internal failures and intrusion from the environment, a control-based audit is helpful in focusing attention on the need for on-going audit, one

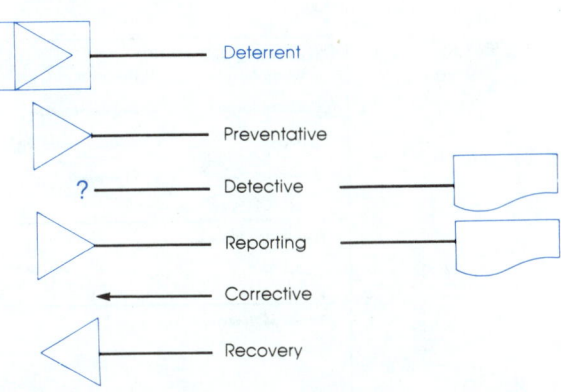

FIGURE 5–13. Fitzgerald's Six Aspects of Performance Auditing

that the information resource itself can provide automatically if it is computerized or tracked on paper.

User polls, suggestion boxes, and computerized surveys are ways in which user perceptions of adequacy can be sensed. Unfortunately, this class of change is normally not sensed in a rigorous, systematic way. The result is that the need for a change has to build up (as in Irene's case in section 5.0) without supporting data, or a catastrophe has to occur to motivate individuals to request action. Often, and again unfortunately, this is done in an atmosphere of anxiety and in an adversarial spirit.

That is why a cooperative audit should form part of every IRMaint effort. At the very least, system performance and user perceptions should be regularly measured and evaluated. Otherwise, the sensing aspects of IRMaint may not function, and a problem that should be looked at may be hidden until it is too late.

Setting Limits on the Investigation

Whether the investigation is prompted by a regular audit or a catastrophic system failure, a **request for services** will be created, usually on a form. Figure 5-14 illustrates a typical format for this form. Sometimes the request for services may go no further than a phone call because either the client's requirements cannot be met or the reason for making the request is mistaken or invalid. Such requests are often gathered at specific,

REQUEST FOR SYSTEM SERVICES

Date_____   Source_____

Requesting Organization_____ Local_____

Brief Description of Problem

Type of Request
- [ ] Education   [ ] Recovery   [ ] Other (Please Describe Below)
- [ ] Enhancement   [ ] Documentation
- [ ] Repair   [ ] System Study
- [ ] Special Run   [ ] Planning

Known Limitations

Terms of Reference

FIGURE 5–14. Request for Services

sometimes annual, intervals and prioritized before further action.

After an initial phone call or interview, the analyst and client may agree to proceed. The next step is to determine the limits of the investigation. The result is another form, called a **project brief**.

The project brief identifies the client and IRU contacts; the nature of the problem as noted by the system audit or user perception; and dollar, time, personnel, priority, and scope limitations as they are seen at this time. The project brief is a letter of agreement between the IRU and the client, although at this early stage the only committment is for a preliminary investigation. Figure 5-15 illustrates a project brief.

Limits are often set by policy and expediency. That is, it may be the policy of the IRU to limit preliminary investigations to worker days and to give such investigations only to senior staff, with priority determined by the number of affected users and the severity of the sensed difficulty. On the other hand, priority in this imperfect world may also be determined by the organizational rank of the requester. Determining the investigation boundary is one of the major goals of the preliminary investigation. Arbitrarily limiting scopes while setting up the investigation can result in difficulties later.

Deciding on Techniques

The project brief makes no mention of the techniques to be used by the IRU to gather data, because the brief is an agreement between two parties to commence. The decision about techniques comes later. Several factors enter into this decision:

1. The time available to research the system
2. The skills available on staff
3. The breadth of the investigation may bring it into politically sensitive areas in which interviewing is either required or forbidden or in which even the most unobtrusive observation methods may not be allowed. Obtaining permission may prove difficult; therefore, either the scope has to be modified or an indirect observation method employed, such as interviewing only top executives.

```
PROJECT BRIEF

Date _____          Seq # _____
Department _____
Client Contact _____  Local _____
                            Local _____

Nature of Problem as Seen by the
Requesting Organization




Limitations:
   Expense:


   Time:


   Personnel:


   Priority:


   Scope:


IRU Contact _____  Local _____
IRU Project Mgr. _____  Local _____
```

FIGURE 5–15. Project Brief

The basic model that this text uses for data collection is illustrated below:

$$A \rightarrow O1 \rightarrow R1 [\rightarrow O2 \rightarrow R2]$$

An **actor's** (A) activities in a system are noted by an **observer** (O1), who may also be the actor. The observer's observations are recorded by a **recorder** (R1), who may also be the observer. In the diary technique, for example, A, O1, and R1 are the same person. The recordings are then collected by another observer (O2) and compiled, tabulated and summarized by another recorder (R2). For example, in the diary technique, O2 and R2 are the analyst who collects and analyzes the diaries. In an interview, A and O1 are the interviewee, while R1 is the interviewer (there is no second observation and recording step).

The decision about which technique to use is critical, not only because of the expense involved in collecting data but because of the problem that arises from over-investigating if an effort has to be redone. People remember from interview to interview, so there is increasingly larger danger of contamination of one set of data by the previous set. Most investigation efforts do not attempt to re-investigate. It is likely, however, that several people interviewed in a preliminary investigation will be re-interviewed during the detailed investigation.

Obtaining Permission and Building Credibility

In most organizations, systems analysts must obtain permission to collect data. This usually means requesting such permission from department or division heads. Since these are often the persons requesting services from the IRU, there is usually little difficulty.

The project brief may detail precisely where data can be collected and when. Additional documentation may go along with this in the form of a **project summary** (Figure 5-16), spelling out how the analyst sees the work going, which techniques are to be used, and which specific persons or class of persons are to be observed or interviewed. Once approved by the client organization, the investigation can proceed.

The credibility of the IRU is on the line during any investigation. While the analyst is becoming knowledgeable about the users, those interviewed or observed are gaining knowledge about IRU and its values and skills. Building credibility is an important aspect of the investigation. It is during the investigation that analysts come into close contact with those who have little knowledge of information systems. In this interaction between two at-least-cybernetic systems, each is attempting to learn the other's "program" of policies; the impression the analyst promotes will be difficult to *un*-learn. Research has shown that the analyst is strongly identified with the proposed system or product and therefore the impression will rub off.

Techniques for building credibility include the following:

1. documenting every interaction, decision, and agreement with clients and those from whom data are gathered
2. making no promises about the ultimate system or product
3. breeching no vows of anonymity, privacy, or security; also making no such promises if they cannot be honored
4. briefly summarizing your observations in individual memos back to interviewed executives and managers and soliciting amendments from them
5. never disclaiming responsibility, never hiding your true purpose
6. obtaining permission to use client organization time and resources

PROJECT SUMMARY

Date__________ **Seq #**____
Project Title:________________
Client Dept.________________

Description of Problem

Nature of Effort to be Put Forth

Recommended Budget:__________

Recommended Time Allotment:________

Recommended Man-days:__________

Project Manager:____________

FIGURE 5-16. Project Summary

Selecting the Investigation Team: Acting

Having sensed the need for an investigation, defined the limits, decided on techniques, and obtained permission to gather the data, the last link is to select an investigation team and set the team into action. The team should have the necessary skills or be available for training.

In many IRUs, there is a policy that specific individuals operate in specific phases of IRMaint work. A novice or junior analyst may be apprenticed to work on a project with a more senior person. Just as likely, investigation work may be the job of a skilled, seasoned veteran. A good policy is to match rank-for-rank the interviewee and the interviewer. One concern of students in an MIS program is going head-to-head with the company president on their first day of work. A rank-matching policy usually alleviates this fear.

In many IRUs, a *project team* is established for each project that is approved by IRU and user management. This formal approval is provided before a project brief goes back to the client organization. The project team (Chapters 15, 16, and 17 discuss project management in more detail) is together during the life of the project. In small firms, a programmer/analyst may do all the work, from the initial meeting with a potential client through programming and training. In larger firms, teams of ten or twenty people have responsibilities determined by a project "book" or management system. Here, selection of the project team is a long-term decision that requires complex assignment procedures. Project leaders may be technical people from the IRU or users who are "seconded" to the project.

The team begins by meeting with the client subsequent to the approval of the project brief. Next, specific investigation assignments are made by the project manager based on recommendations from the client, statistical sampling requirements, geographic concerns, staff availability, and IRU policies. At this point, data are collected and the IRMaint activities begin in earnest.

5.5 CONCLUSION

This chapter began a section on the products of logical system design. Each product is the outcome of a major set of activities. **Knowledge**, the result of **investigation** activities by the analyst, in necessary in logical design.

Understanding a system is necessary before redesigning it, which can be difficult because the analyst is usually outside the system. To understand a system, then, the analyst must formally investigate it. Acting like a scientist, the analyst collects data and attempts to see patterns in those data. The logical design has to be based on this knowledge. **Knowledge-based** design is an attempt to fashion an information resource based exclusively on factual information collected during investigation.

There are three different techniques for obtaining knowledge. The **preliminary investigation** constructs a structure chart of information resource goals. The **feasibility study** evaluates a number of proposed techniques for meeting those goals. The **detailed investigation** collects in-depth information on current or proposed information resource functions based on the goals detailed in the preliminary investigation. In addition, there are techniques for collecting data during prototyping of systems.

Investigation reports serve as the basis for design decisions during logical design. Therefore, the terms of reference are important. Data have to be presented to interested parties, including clients and physical designers.

The process of setting up an investigation consists of these steps: **sensing** the need to collect information through an audit; **setting limits** on the goals and scope of the investigation; deciding on **techniques** for investigation; obtaining **permission** from those who will be asked to cooperate; building **credibility** for investigation goals; and **selecting** the investigation team. The investigation can be seen as a cybernetic system embedded in the larger IRMaint system.

Chapter 6 scrutinizes the next major activity of the analyst: representation. Representing the data gathered in a fashion such that designs can be produced is called "modeling." The models built comprise the most visible aspect of the analyst's work, an aspect most closely identified with systems analysis itself. Chapter 7 then looks at interpreting these models and creating specifications for new or improved information resources.

5.6   THE CASE OF THE TWO HATS

Tracy Appleton works for Camrose Holdings, a real-estate management firm responsible for thirty-seven buildings in Bigtown. These include apartment, office, and commercial properties. Camrose is a wholly owned subsidiary of Rosebud Properties, a major developer in the area. Tracy's responsibilities are to handle data processing for Camrose Holdings (billing, accounts payable, accounts receivable, materials and inventory management, payroll) on a minicomputer owned by Rosebud. Tracy is trained in data processing, but until now, has used only canned programs provided by the minicomputer vendor.

Yesterday, Tracy's manager, Ed Leary, and the general manager Flora Malashevsky spoke with her about the possibilties of using a microcomputer to handle her responsibilities on-site at Camrose, so she would not have to go over to Rosebud's offices three times a week. Also, if the micro were available at all times during working hours, rental queries probably could be handled on the phone. A predicted rental shortage could force demand and prices up and change the way space should be marketed. That would mean some kind of spreadsheet program to balance rentals in anticipation of block moves.

Ed and Flora now have high expectations of micros, Tracy, and the marketplace, but who is Tracy to question their optimism? Now, if only she knew how to proceed from here....

1. What should Tracy do first?
2. Tracy works alone, although she has access to Rosebud's data center. Assuming she can obtain some assistance from entry-level junior analysts there, what is Tracy's next step?
3. Ed and Flora are only two of the potential sources of knowledge. Whom else should Tracy speak with? What other kinds of data can they gather?

4. Ed and Flora want to know how much it will cost. How should Tracy answer that question? Should she answer that question?

5. Role-play an interview between Tracy and Stan Lannigan, director of marketing for Camrose Holdings. Flora recommended that Tracy speak to Stan about possibly including some sort of decision-support system for Stan on the microcomputer. Tracy really wonders if that should be part of the project at this time.

6. Step back from this situation a bit. What obstacles will Tracy experience here in observing and reporting? What barriers will exist between her and those who have to make decisions about continuing this project? Can she anticipate and make provision for some of these problems?

DISCUSSION QUESTIONS

1. This chapter begins with a philosophical challenge: how is it possible to "understand" a system when one is outside its boundaries? What are the dangers of *mis*understanding because of this problem?

2. The text states that investigations are based on theories the analyst holds about the system under observation. But doesn't theory come from observation? How do analysts get around this chicken-and-egg problem?

3. Marilyn has been contacted by the manager of quality control of her plant to look into improving the existing manual, paper-based quality control system. Quality control is essentially a manufacturing concern, but improving it could have repercussions for marketing, sales, and customer relations. How should she go about determining the investigation boundary?

4. How does the knowledge obtained from a preliminary investigation differ in type, quality, quantity, and value from that obtained from a feasibility study? From a detailed investigation? What causes those differences?

5. The manager of quality control, Phil Hodges, is discussing with Marilyn O'Keith the possibilities of automating the quality control system mentioned in Question 3. Phil wants to know whether computerization is feasible. When and how should Marilyn answer that question? What does she need to know to be able to answer it honestly?

6. A detailed investigation has as a major goal the completion of a structure chart that shows the structures and details of the functions. Given a problem such as the one Marilyn is working on, how will she know when to "quit" designing?

7. The investigation report is a message from the analyst to a number of people. Since they have to "observe" this report and make sense of it, what problems might such a report face that resemble any system observation? How can the analyst prepare for these potential problems?

8. Discuss the apparent relative advantages of aggregated data and case example presentations.

9. An investigation report is both descriptive and persuasive. How does such a report differ from the following reports:
   a. an end-of-term grade report at a college
   b. a football coach's report on why the team lost 65-0
   c. an on-the-spot television report of a disastrous flood
   d. a year-end report to shareholders

10. An analyst's recommendations are two-fold: specific logical designs and an evaluation of the advisability of continuing with the work. Are these two basically incomparable, or can one set of recommendations have something to do with the other? More specifically, can an analyst present a fully developed logical design for a project she recommends not be continued? Can an analyst recommend a project be continued without a fully developed logical design?

DESIGN EXERCISES
Using dBASE III

1.   Let's look again at the files and their structures. First, compare the fields of CLIENTS and SESSIONS and note that "Name" and "Phone" appear in both. The combination of name and phone number uniquely identifies each Hairloom customer. Look into each of the other files and list the **unique identifiers** in each file, whose use is described below:

CLIENTS	Describes each customer
STYLISTS	Describes each stylist
SKILLS	Skills that stylists may have
STSKILLS	Skills that stylists actually DO have
SESSIONS	A visit by a customer resulting in a receipt
RECEIPTS	Each charge (for activity or sold item)
SERVICES	Each activity that can be performed
ACTSKILL	A skill that goes into an activity
PRODUCTS	Products that can be sold
SUPPLIER	Suppliers of products

   We can discover the actual unique identifiers for each file by **sorting** the file and then listing the data in the field or combination of fields. Sorting is accomplished with the SORT command. Try USEing RECEIPTS and entering these commands:

DESIGN EXERCISES
Using dBASE III *continued*

SORT ON RCPTNO TO RCPTN	Sorts on the receipt number field to the file named RCPTN.DBF
SORT ON RCPTNO, TYPE TO RCPTNT	Sorts on receipt number and type to RCPTNT.DBF
SORT ON RCPTNO, TYPE, ITEMACT TO RCPTNTI	Sorts on receipt number, type, and item or service item code to RCPTNTI.DBF

Now you can examine each of the sorted files to find out which of the three sets of fields contains unique identifiers. How can you accomplish this? Which do?

2. Doug's dissatisfaction with Mercurial Software has lead him to contact you, an independent consultant. In Doug's initial phone call, he mentioned a need to plan his business, to keep control of costs, and to understand his clientele and staff. Continue with the role-play from Chapter 4, performing an initial interview with Doug, looking at the system as it now exists. How would a **feasibility study** at this point differ from a **preliminary investigation**? Given what you've discovered about Hairloom and Doug's needs, create a **request for services**, a **project brief**, and a **project summary**. Your time and budget estimates may be considered "informal" now.

CHAPTER 6

CREATING SYSTEM MODELS

OBJECTIVES

1. To define the term "system model" and provide an example
2. To list seven properties of models and provide examples for each
3. To list six criteria for system models and evaluate a given system model against those criteria
4. To describe the three major design problems that modeling assists in overcoming
5. To describe the modeling process and its role in producing logical system specifications
6. To characterize and evaluate available modeling tools
7. To analyze information to be used in models and to employ analysis techniques in the production of models
8. To use system models in particular contexts to produce logical systems specifications

CREATING SYSTEM MODELS

6

6.0 THE CASE OF THE DATA SWAMP MONSTER

"Wait a minute!" shouted Delia. "The way I see things, we're going to be eaten alive unless we find some way to make some sense out of this."

"Now you wait a minute, Delia," Neil puffed, his face red, his voice rising. "This is four weeks' work you're throwing away. Twelve interviews, hours of replaying tapes. Speaking of tape, miles of red tape getting into the files room." Neil was angry.

"Neil, I understand your disappointment, but we're swamped with data and if we just slop it on Elvira's desk, she's going to have our heads. We've got all kinds of ToDo paragraphs here, but I'm sure that neither Elvira nor our programmers are going to know what to do from our report." Delia put on a soothing voice.

"So what? They've always used our written reports in the past, so why not now?"

"For one thing, this is really a monstrous project. Just two minutes ago you mentioned all your interviewing. It's not just a few fixes, they want a new records management system. And for another, the programmers you refer to aren't hired yet.

"We're not just walking down to Alesandro, who's worked here for eleven years, and asking him to imagine what the new system should look like in COBOL. Elvira's interviewing new graduates and they aren't going to walk in the door and write records management COBOL, not at the salaries we offer. They expect neat flowcharts and data flow diagrams, detailed data definitions, and — oh yes — Elvira's got to be able to understand them, too. Alesandro doesn't care — he knows the company inside out and backwards. But Elvira's new, too, and I'm sure that your interviews uncovered her plans for change. What are you going to do when Elvira decides to change a file spec three weeks before release?"

"Do what we've always done — jump in and change the code. What do you think we ought to do?" said Neil.

"Well, we can't just change code. That's the programmer's job. We've got to record and document the proposed change — and if necessary, talk Elvira out of it. But if that's what she's going to argue for, we need to be able to test out her idea to see how much the change will cost and what effect it will have on system performance. We don't just want to shout at her things like 'It will be too much work' or 'It is really pretty now. Why change it?' We need good arguments. That's where models come in."

"So you think a picture's worth a thousand words?"

"A million, Neil. And maybe your job, too."

In this chapter, we will look at the systems analyst's most visible product: charts, graphs, and diagrams. When most students of information resource management think about analysis, that's the "picture" that comes to mind most often. Just as architects produce blueprinting, analysts produce pictures that guide builders, in this case, system designers and implementers.

Chapter 6 looks first at the role of **modeling**. While it may not be immediately apparent, the job of the analyst is mostly creating images of what a system "looks like" for a variety of people: customers, designers, programmers, users. If these images do their job, the audience will understand the system and agree to work toward building and using it. If the pictures do not readily come to mind or if they are confusing, precious time is lost and tempers may flare.

The following chapter also examines the definitions, the uses, and the details of the process of modeling. For an analyst, that process means first culling data gathered during the preliminary investigation and later. When the appropriate data have been located, the analyst abstracts the data into forms that are useful in making diagrams and tables. Finally, the analyst draws the models and documents them.

The chapter ends with a guide to the use of system models by analysts and others.

6.1   THE ROLE OF MODELING IN THE DESIGN PROCESS

As the conversation at the start of the chapter points out, modeling serves three purposes. First, there needs to be communication among individuals whose roles, functions and backgrounds differ dramatically. Second, the information resource needs to be represented in a systematic fashion. Finally, representations to be used directly in design need to be produced.

The needs imply three roles for modeling: communication, documentation, and translation (see Figure 6-1).

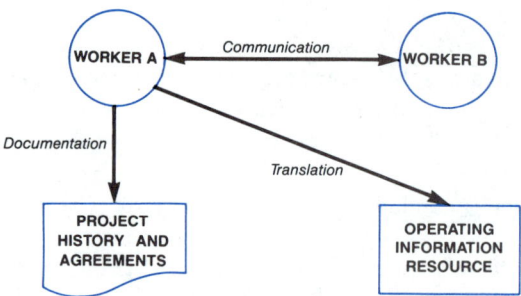

FIGURE 6–1. Three Roles for System Models

The Communication Role of Modeling

Those involved with computers are often surprised that those in the non-computer world don't understand what computer professionals are talking about. Specialized language (**jargon**) and work-related speech (**argot**) do not affect just the computer professional's interaction with those outside the IRU; often there is a sharp split within the group itself.

For instance, there is the gap between COBOL programmers and computer users in terms of expressing what the computer can do. A programmer is comfortable speaking of "procedures" and "data divisions." The user, on the other other hand, understands the computer in terms of the applications: "registration" and "account files." This difference is usually between **internal requirements** and **external requirements**. System managers have little desire to become involved with programming, but will speak at length of the economics of operations. But the system operators, who figure strongly in operation costs, concentrate on their own operating procedures. These have nothing to do with data divisions, account files, or the cost of processing each week's transactions. Meanwhile, the customers, the people to whom the data refer, understand none of this.

Thus, there is a need for a common language in which to express these concepts, because, for the system to work, all these individuals must somehow be in a state of agreement. Research shows that while user involvement is of variable value, lack of understanding of user needs is the best predictor of system failure. Central to this is the analyst, who must describe the system differently to each interest group.

In addition, the analyst runs up against each group seeing the information resource differently. As Figure 6-2 illustrates, the old tale of the Seven Blind Men of Benares

SUSAN REESE

FIGURE 6–2. A System Can Be Seen in Many Ways

applies equally as well to perceptions of systems. At least the blind men spoke the same language; imagine the confusion if each spoke a different tongue.

The programmer sees modules and code, lines of COBOL, and data bases. The operator sees forms and schedules. The system manager has concerns about lighting, air conditioning, machinery layout, wages. To the user, the system appears to be a talking machine, carefully and patronizingly guiding data entry. Customers think of the information resource as a befuddling and complex black box. It is unlikely that any of these people will be satisfied with a system definition until those elements they are concerned with have been fully described to them in concrete terms they can understand. That task falls to the analyst.

Finally, a "full" description may still be incomplete in the mind of each of the players. There will be suggestions for improvement. It will be necessary to express, argue, and counterargue intelligently. It is also unlikely that programmers, users, managers, system designers, and customers will argue on the same basis. After all, it is not the programmer's car payments that will be recorded, not the operator's COBOL procedure that will have to be debugged, and not the user's budget that will support development. Economic arguments and human factors, to cite just two considerations, are often at odds.

The user — a data entry clerk, for example — may know little about the functions of the machine, but will have an intuitive feel for when a procedure is "wrong." A programmer may argue that there is no way a certain function can be programmed given the tools available and will refuse to begin what she considers a fool's errand. The system manager may insist that "experience shows that this kind of thing won't work." The analyst is often called in to mediate and negotiate among inalterable demands (Figure 6-3).

ROLE	CONCERNS	BASIS FOR ARGUMENT
Programmer	Elegance Speed Coding Time	Clearly-Written Code, Easy to Follow and Maintain
System Designer	Efficiency Modularity Flexibility Clarity	No Loose Ends, Neat Package, Efficient Running, East to Change, Easy to Change, Easy to Predict Effects of Changes From
System Operator	Reliability Trainability	Easy to Understand, not Confusing, Little Chance for Error in Interpretation
System Manager	Cost Effectiveness Reliability Manageability	Good Return on Investment, will not Require Much Maintenance, Easy to Assign Work and Tasks to
User	Understandability "Friendliness", Usefulness	Knowing What Comes Next, Assistance (Help), Obviousness, Does the Job

FIGURE 6–3. The Basis for Communications Among Systems Professionals

In this welter of conflicting concerns and styles of argumentation, it helps to have a common basis for communication, something to agree to disagree about. Models offer that. They form the basis for presentations to management, operations staff, users, and implementers. A single, well-presented image can provide the basis for profitable discussion.

Of course, if the image is confusing or vague, argumentation is more than just difficult; people wonder exactly what it is that they are supposed to talk about. Fifteen years ago, the analyst's models, besides being confusing and difficult to understand, didn't tell the whole story. Since 1970, many new, sensible, and commonly understood diagramming techniques have been developed. They have tended to simplify and make the relatively rigid and specialized vocabulary of the information resource specialist more intuitive.

The **communication role** of models assists the analyst in bridging the gaps among the variety of players. The gaps of lack of common language, disagreements in point of view, and argumentation style and values can be narrowed through the selection and presentation of appropriate models.

The Documentation Role

Of course, just getting people to talk and agree is rarely enough. Diagrams, charts, and tables put ideas onto hard copy. Because they are hard copy, they tend to become official quickly, partly because of the value of the agreement necessary to draw them up.

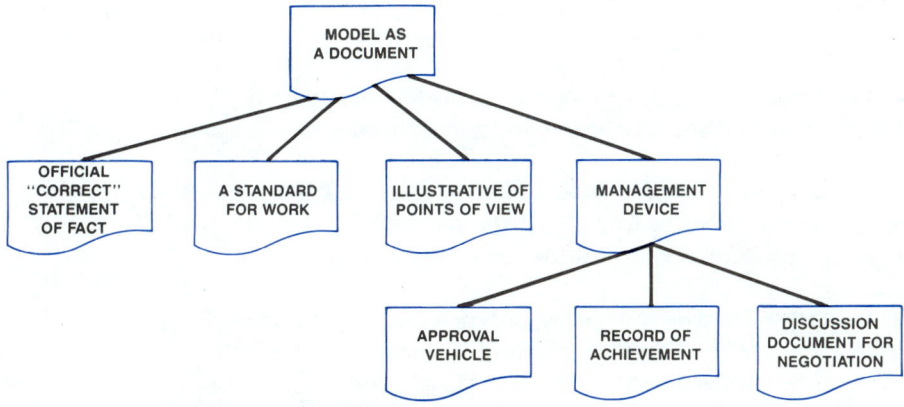

FIGURE 6–4. Documentation Roles of Models

More important is the effect that a particular diagram has on the design process. Because two correct models may have little in common, choosing a particular model over another probably sets a unique design course. Consider the following fable:

A potentate of a small but information-rich country had two daughters. One was fair of face and form, but stood only three feet six inches in glass slippers. The other was as ugly as a full-memory dump, but stood regally and was, as well, a swell orator. Together, they might have made one stunning princess.

The potentate, known as "Your Awesomeness" to his subjects and as "Daddy" to his daughers (named Beatrice and Bytilda, or Bit and Byte for short) had a problem marrying off his daughters, oddly endowed as they were. So he commissioned a Royal Portrait Painter to produce likenesses more flattering to his progeny.

The RPP labored mightily over several weeks and produced several different renderings of Bit and Byte. Some stretched Bit a bit and others took the bite out of

Byte. Eventually, Daddy picked one of each and sent them to the families of neighboring potentates.

In time, suitors arrived. Naturally, Your Awesomeness did not want the suitors to suit the daughters in the flesh. Instead, he had them serenade, sweet-talk, and eventually propose to paintings of Bit and Byte. Eventually two swains swooned so for the bogus biddies that they proposed. Needless to say, Queen Bit of Spreadsheet found herself married to a seven-foot-tall, wart-covered Frog Prince, while the Marchessa Byte of Database discovered her Prince Charming was a three-foot-tall, but gorgeous, gnome. Daddy was pleased, but found himself wishing that photography had been invented.

As you can see, almost any portrait would have sufficed, but the choice of a particular one set one particular monarchy down a particular marital alliance path. Deciding to accept one model as "official" may well determine an information resource that is unique.

By this logic, a particular modeling document dictates all subsequent design work. It is important that the document represent a correct **point of view**, that is, a set of standards through which an information resource is viewed. For instance, the point of view of system operation differs dramatically from that of system implementation. An easy-to-write, quick-and-dirty program might have operating instructions too complicated to be useful. A user-friendly program may be too difficult for a novice programmer to write.

The point of view is part of the general problem of the Uncertainty Principle. Remember, a system's description depends at least in part on the categories and tools available to the system observer. Of particular importance are these questions:

- How regularly was the system observed? Continuously, periodically, or occasionally?
- How were the observations made — coarsely or finely (for instance, number of documents/day versus number of documents of a particular type/day)?
- Was the observer also part of the the system? A user? A participant? An evaluator?

System observation time, or **time fabric**, is important. A continuous process needs continuous observation to be properly appreciated. Compare sports reports based on ten-second observations every half-hour (a sports news "spot") with continuous reports (a live broadcast). A periodic process need not be watched continuously; individual observations are appropriate, but trends may not be noticed if observation is relatively infrequent. An occasional process may be missed entirely if systematic but infrequent observation is used. Astronomers scan the heavens nightly for supernovae that may occur once every decade, and managers demand weekly reports looking for variances that may occur only annually.

An example of the importance of time fabric is the distinction that Gane and Sarson (1979) draw between *data flows* and *data stores*. One major difference is that a data store is relatively stationary. Observation of systems with a continuous sampling scheme provides different information than does occasional sampling. What looks like a data store under some circumstances may resemble a data flow under others. This distinction may not be important at early design stages; however, left in place, a data store will be implemented as a file or file segment. Data flows, on the other hand, are commonly thought of as records within files. Implementation of this distinction can mean a very different development budget.

In a similar way, measurement fineness, or *mesh*, is also important. Information content can be measured in bits, bytes, segments, or pages. Round-off errors which result in rounding small files to a page (4096 bytes) can seriously alter estimates of system loads.

An observer inside the system will see things differently than one on the outside. In particular, using a computer to measure the performance of a computer-based system makes the observers part of the system. The measuring device adds to the load on the system. But more important, it cannot take measurements of elegance, comfort, or psychological appropriateness. It merely measures what the system can do in its own terms. Documentation expressed in terms of the system's internal functions suffers from what anthropologists call the "inside-outside" problem. The document may say far more about the observer than about the system. Later, when the observer is not part of the system, the document becomes false or misleading.

A final role of documentation is assisting in project management. First, a document records who thought what and when and formalizes the results of agreements. Second, a document stands as a record of a particular analyst's achievements, just as a photo of Joe beside a twenty-pound fighting river trout is testimony of Joe's angling skills. Finally, a document provides something for others to sign, indicating that the work has progressed so far and a certain willingness, if not desire, to see it progress further.

Thus, the documentation role is threefold: It makes things official, it sets a course of action; and it provides milestones and records.

The Translation Role

Beyond the communication role of bringing people together and the documentation role of making things more or less official, models also serve to translate between concepts and the reality of an operating information resource. More particularly, models serve a *decision-support* role for analysts.

What sorts of decisions do models support? Figure 6-5 on the following page illustrates, through a data flow diagram, the logical design of part of a bill-posting system. Note that process 3 "Update customer status record" is central to bill posting. Assuming that the figure accurately depicts the activities of bill posting, we can subject this model to a number of tests.

For example:

- What happens if process 3 slows down?
- What happens if process 2 ("obtain customer status record") has several errors in it?
- What if a customer complains about the accuracy of a bill?

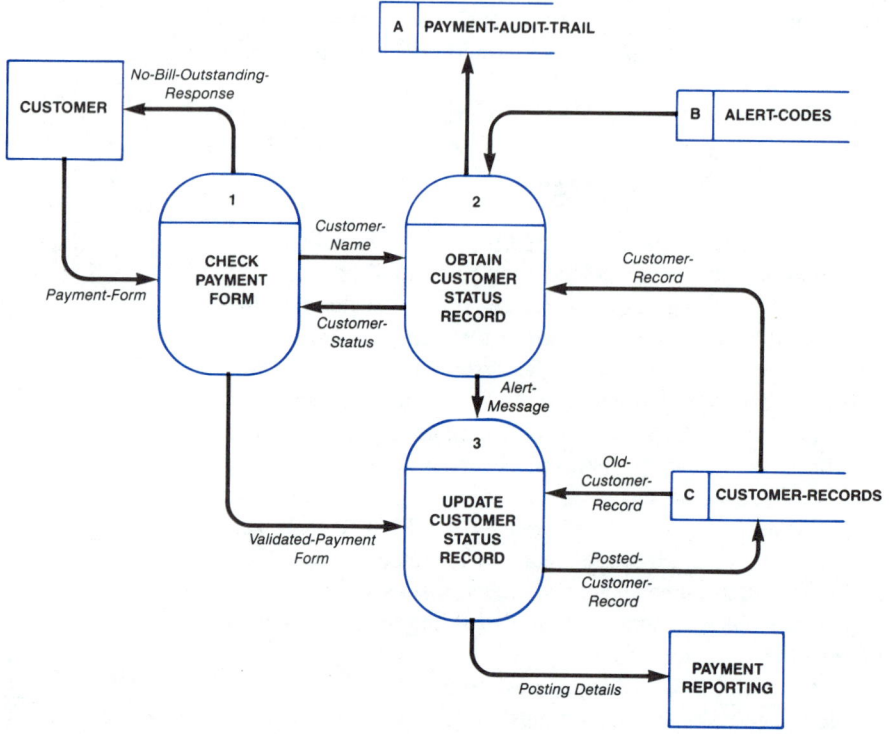

FIGURE 6–5. Part of a bill-posting system

Note the form of these questions: "What...if...?" These questions give voice to designers' dilemmas: "What if I change my design — what effect will that have on other parts of the system?" The answers to both the question and the dilemma are found in the models.

We can add information to Figure 6-5 about efficiency, flow, and reliability and derive conclusions about the efficiency and effectiveness of the system. For instance, we can see from Figure 6-7 (opposite) that by adding figures indicating flow rates, process 3 requires posting 400 bills daily, or 50 each hour. Since we know from our investigation data that one clerk, even with electronic aids, can post only about 20 bills per hour, we conclude that process 3 is a bottleneck under the best circumstances and that a redesign is necessary (given these constraints).

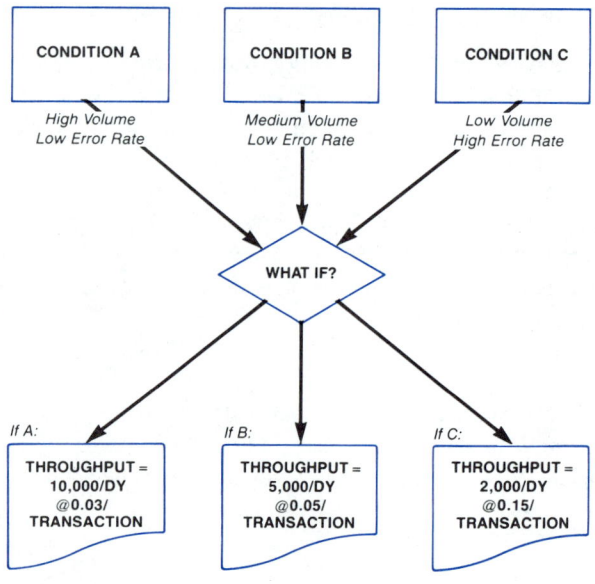

FIGURE 6–6. What-If in Modeling

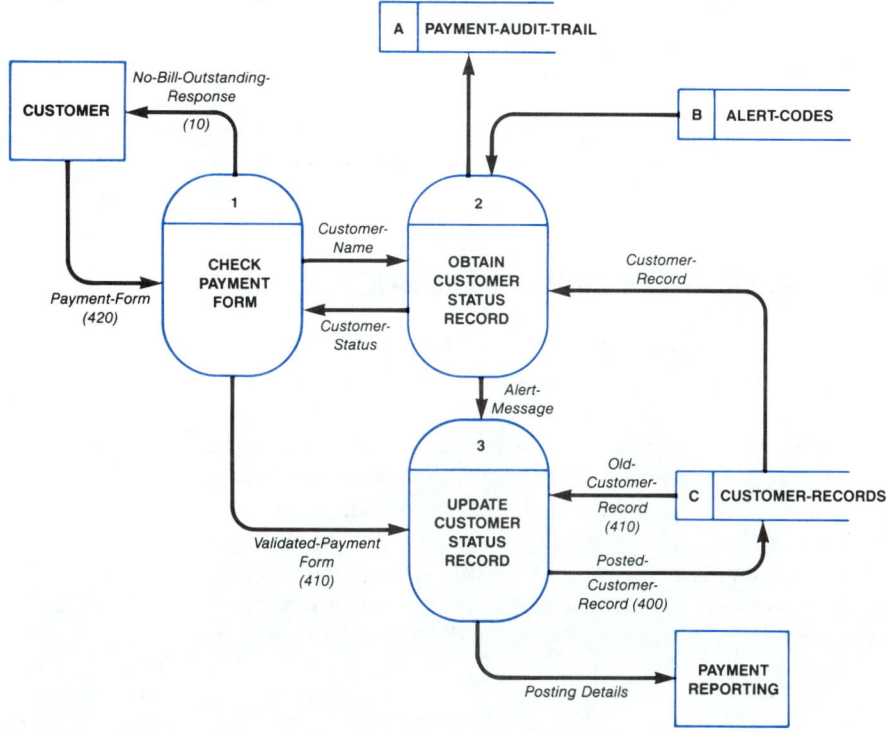

FIGURE 6–7. The Decision-Support Role of Models

In none of this did we require an actual system to draw the conclusion. Instead, we **constrained** the model with figures derived from the "what if" question.

Using models in the design process is a valuable, but insufficient, tactic. Ultimately we will have to test the implementation of our design against the real world. We do not actually have to build the system modeled in Figure 6-5, however, to answer these "what if" questions.

To use models as support for design decisions, several things are needed:

1. Analysts must be able to draw models correctly.
2. There should be some way to redraw models rapidly and accurately, probably using a computer.
3. Analysts should be willing to refine, redo, or discard models that are seen as flawed under the "what-if" regimen.
4. Analysts should be willing to accept criticism, and, just as important, they should be willing to listen to and respond to a wide variety of "what-if" questions.
5. The choice of models is critical; some are inappropriate for decision support because they are hard to understand or limited in what they represent.

6. Analysts need training in the use of models as decision-support tools; there are several new tools on the market, such as Excelerator™, which professionals use in design decision-making.

The three roles of models (communication, documentation, translation) support analysts in their task of producing logical designs for IRMaint. In the following section, the theory of simulation is discussed.

6.2   MODELING AND SIMULATION: SOME DEFINITIONS

Producing a logical system design from a large collection of user-experience data is a complex task. As the previous section discussed, models serve to assist this process, which we also term "modeling."

Modeling is a common activity. When repairing a lamp, for instance, one has to unscrew a number of components. Remembering the order in which these bits of metal and fiber came off is difficult. We write this information down, sometimes drawing pictures. The pictures, with instructions, are models of lamp construction. When an engineer builds a bridge, blueprints represent the parts of the structure and the ways they interrelate. In this sense, too, a computer program is also a model, as is a recipe. The instructions for bicycle assembly contain pictures, words, and a parts list. These are three kinds of models. But what is a model?

Definitions

We term the system being modeled the **base** system. The system that acts as the model is called the **model** system. For convenience, these terms are shortened to "base" and "model." The relationship between the model and the base is called a **simulation**. We say, "The model simulates the base."

A blueprint simulates a bridge, the base system. Bicycle assembly is simulated in the instructions. The relationship between the bridge and the blueprint is a simulation. Naturally, we cannot place the blueprint across a river and hope to drive on it. But depending on how well we simulate the bridge in the blueprint, we will find it useful for a number of purposes.

A formal definition is as follows (Figure 6-8):

M is a model of R if and only if:

1. M and R are both systems;
2. For every element x in R there is at most one corresponding element x' in M;
3. For every relationship p among elements in R there is at most one corresponding relationship p' in M holding among corresponding elements in M; and
4. For every set of elements [a', b', ...] related through a relationship p' in M, it is true that the corresponding set of elements [a, b, ...] in R is related by the relationship p in R corresponding to p' in M.

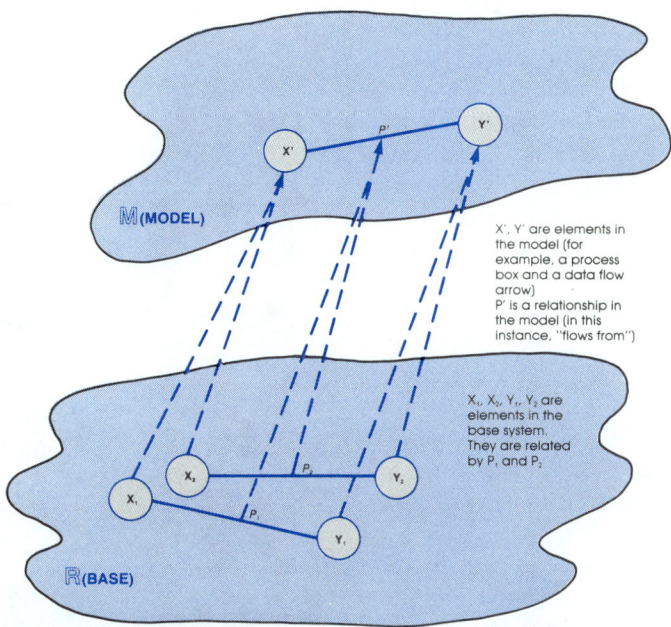

FIGURE 6-8. The Definition of "Model"

The first requirement simply ensures that both the model and the base are systems, with elements, relationships, and goals. The second requirement states mathematically that M has at most as many elements as the base; in that way it is "simpler." Likewise, the third requirement guarantees that the set of relationships is no larger and hence is "simpler." The model is no "busier" than the base.

It is the fourth requirement that makes models useful. Simply stated, it requires that everything that is true when speaking of the model is correspondingly true about the base. In other words, conclusions you draw using the model have some translation into true statements about the base. This requirement has an obvious connection to the decision-support function of models. We can examine the model, find out what is true about it, and translate those truths into truths about the base without having to build the base system.

SOME EXAMPLES

In Figure 6-8, several elements or relationships in the base correspond to individual elements or relationships in the model. We simplify the truth by requiring the model to say less about the base than the base does. This ensures that nothing that we conclude from the model as true will actually be false concerning the base. The model will not be misleading.

Figure 6-9 is a representation of the Mona Lisa painting, and thus of Mona Lisa herself (see the section "Transitivity," below). By our definition, the painting is a model of the woman: the painting and the woman are systems; the painting has fewer elements or visual aspects; the painting has fewer relationships among the elements (it is only two dimensional, for example); and statements we derive about relationships and elements in the painting have corresponding truths when translated back to the woman.

Aesthetic statements about the model are not truths about the model, but instead are truths about the relationship between the model and the observer, as are statements about the function or construction of the model. Obviously, the model is made of paint and canvas and the woman is not. These statements mention how the model is constructed; the canvas is not an element of the painting.

Let's examine some truths in the painting. (The [...] notation distinguishes elements and relationships in the model (the painting) from those of the woman). There is a [smile] [on] the [face] of the portrait. There is also a corresponding smile on the face of the woman. The [face] in the painting has a [nose] [on] it; the woman's face also has a nose. The [smile] in the painting is not a smile, it is a bit of paint, but it corresponds to a smile. Calling the woman who sits for the painting a "model" further confuses things; from our point of view, the term is incorrectly applied in that case.

FRENCH GOVERNMENT TOURIST OFFICE

FIGURE 6–9. A Representation of the Mona Lisa Painting

Now look at the equally interesting but less famous model in Figure 6-10 on the opposite page. My eldest son drew this at age 4. It hardly conforms to anything in the real world, although several years later, the artist still insists that it is a true representation of his father. There are several difficulties in referring to this drawing as a model of a person. For one thing, the [nose] is exactly [on-a-level-with] the [eyes] in the drawing. The modeled mouth (that is, the [mouth]) is below the [right-eye]. I assure you that these do not correspond to true statements about the base. Is Figure 6-10 is a model of me just because the artist says so?

The answer to that is complex. By our definition, Figure 6-10 is not a model of the intended base. It fails in some ways, but it succeeds in others. We usually say that models such as Figure 6-10 are **poorly conforming** models, or just "poor" models. Twentieth-century art has endured many controversies about conformity, and since this is not an art text, we'll not join that battle. We will merely say that systems analysts cannot afford the luxury of such controversy and should have their models conform to the bases as closely as possible.

It is obvious, however, that few models are perfect. There are always some errors. The simpler, more intuitive models adopted by systems analysts make it easier to spot errors and correct them than it was fifteen years ago.

FIGURE 6–10. A Systems Analyst?

Properties of Simulations

There are seven important properties of simulations: non-symmetry, reflexivity, transitivity, non-transferability, reduced complexity, non-partition, and irrelevance.

NON-SYMMETRY

Simulation is not a two-way street. It goes only in one direction. If A models B, then B cannot model A. (Why not?)

REFLEXIVITY

On the other hand, any system is a model of itself by the four clauses in the definition of a model.

TRANSITIVITY

If B is a model of C and A is a model of B, then A is a model of C. By extension, we can create chains of models. A drawing of a plastic model of a boat is also a model of the boat. Because models simplify, the result of a chain of models is increased simplicity (see "Reduced Complexity" on the following page). By this property, physical design specifications model the planned system (see Figure 6-11, also on the following page).

NON-TRANSFERABILITY

Two models of the same base are not necessarily equivalent, and the property of non-transferability almost ensures that they have no simulation relationship themselves. Stated simply, two models of the same base may not be comparable in a simple way. Two sets of blueprints for the same proposed building may not be comparable. Two systems analysts can each produce a strictly conforming data flow diagram for a proposed system, two diagrams that bear no relationship to each other. They may represent (simulate) different aspects of the planned system. Thus, it is often difficult to choose between models without other information.

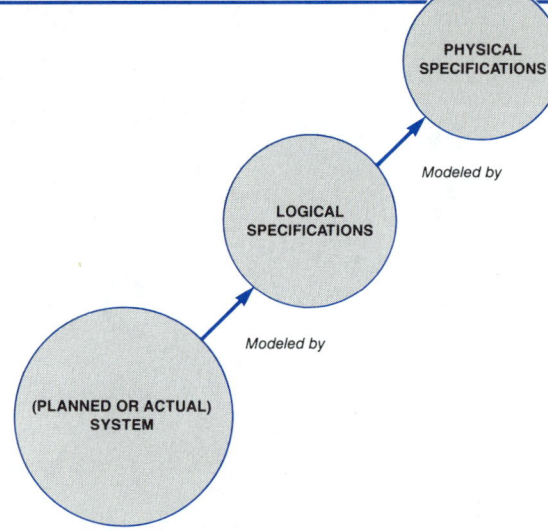

FIGURE 6–11. Specifications as Models

REDUCED COMPLEXITY

An obvious benefit of models is reduced complexity. The "at most one" phrases in clauses 2 and 3 of the definition ensure this. If A models B, then A is at most as complex as B. We say informally that "A is simpler than B." Simplification is achieved in one of two ways:

1. Grouping: similar elements or properties are merged. Figure 6-12 illustrates all the activities of bill posting with a single process box.
2. Elimination: irrelevant elements or properties are left out of the model. The data flow diagram in Figure 6-12 ignores the physical and temporal aspects of bill posting and models only the logical ones.

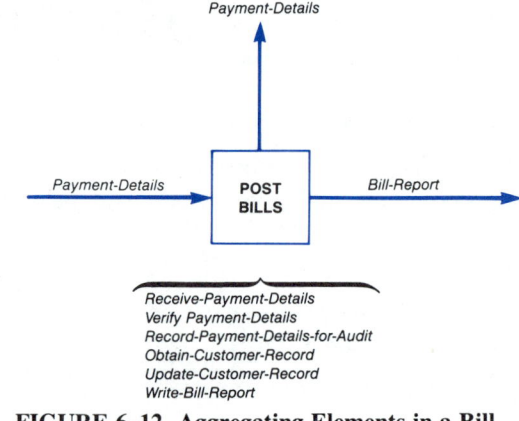

FIGURE 6–12. Aggregating Elements in a Bill Posting System

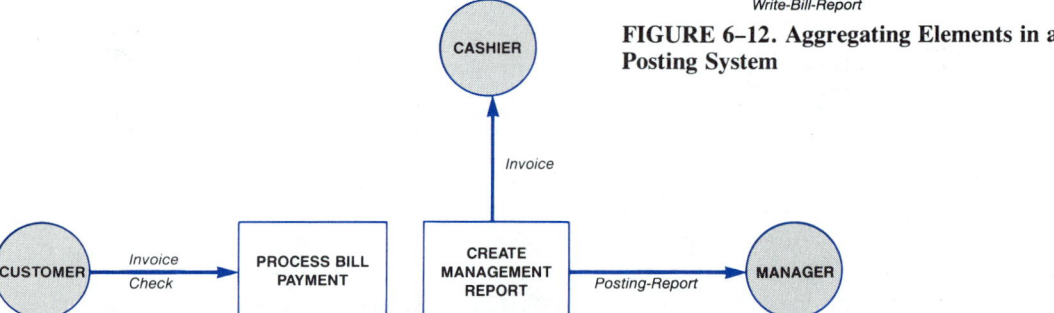

FIGURE 6–13. A Model Incorrectly Partitioned

NON-PARTITION

Because we cannot arbitrarily examine part of a system and hope to learn about the whole system's function, models share the same property. A model of part of a base system is not necessarily a model of the whole base, despite its automatic simplicity.

Figure 6-13 is a diagram of bill-posting activities in the form of a materials flow diagram. Note that only the start and end of the posting process are modeled. For purposes of this illustration, the operations that actually post the bill, reconciling it with invoices have been left out. The relationship between the incoming pieces of paper and the outgoing invoices is missing. Instead, it appears that the incoming paper is the outgoing paper. But this relationship is false in the proposed bill-posting system. Hence, Figure 6-13 is a non-conforming model (and probably useless, too).

How could this happen? Because we created a single system from unrelated parts. In doing so we related elements of the model (incoming sales slip, outgoing invoice) not related in a corresponding way in reality. By adding a new and invalid interface, we invalidate the whole model.

IRRELEVANCE

A kind of reversal of non-partition, the property of irrelevance states that any model of a base models the base plus irrelevant, unrelated elements. The diagram in Figure 6-14 models not only the bill-posting process but also shipping plus purchasing plus manufacturing. There is nothing true in the model in Figure 6-14 that does not have corresponding truths in the enlarged base. And because it will be difficult to "disprove" the model in this way, it is important to choose bases carefully. Having a model does not mean that the base itself is a unified, functional, or valuable system. Furthermore, since those reading the models tend to treat them as the actual base, it is important to educate

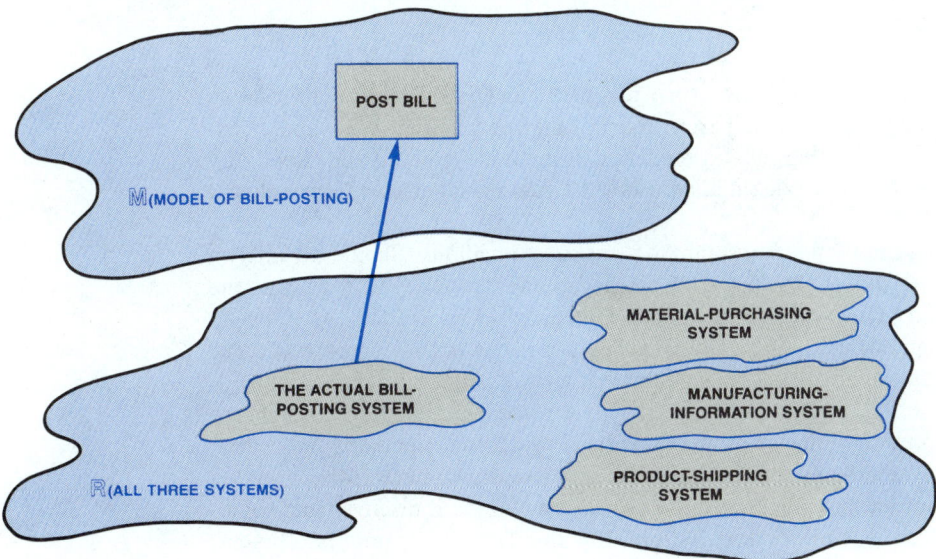

FIGURE 6–14. A Model of More than the Base System

those for whom the models are intended in the base system, too. The models can assist, but they are not everything.

These seven properties dictate the limits of usefulness of models. Models can be seriously flawed and still have value. They can be also perfect yet valueless. The next section examines the value of flawed models and introduces the important concepts of consistency, completeness, and validity.

The Value of Models

Three criteria determine the value of models for communication, documentation, and decision support: consistency, completeness, and validity.

CONSISTENCY

Readers do not like being confused. Analysts should not treat models as curiosities to be studied for hidden meanings or interesting ambiguities. Boxes representing bill posting should be labeled uniquely; drawing rules should be rigidly adhered to; words should unambiguously refer to single concepts. No model should result in a question such as "what does this symbol refer to?"

COMPLETENESS

The completeness criterion requires that important aspects of the base not be left out. Tables that ignore common or important situations, diagrams that leave out common relationships, texts that neglect important activities are incomplete. Because a model simplifies, incompleteness is relative, and fairly "complete" models may be difficult to understand and time consuming to read. Incomplete models may be more useful than complete ones in this regard.

VALIDITY

Validity is the major value criterion. An invalid model is one that conforms so poorly that incorrect implications result from reading the model. Inferences about system performance based on reading invalid models will themselves be invalid. Invalidity produces bugs, and bugs produce dissatisfied users, harried managers, and unemployed analysts.

The analyst is incidentally the most common source of invalidity. Should the analyst include elements in the model that have no corresponding "reality" in the base, the model becomes invalid. This results from poor or incorrect investigation. The analyst may decide that a relationship should exist when in fact it has not been observed. For example, an analyst, through laziness, may infer that a document is distributed according to some official, efficient plan when in fact it is routed circuitously through the organization. The model the analyst produces will show a direct, perhaps undesired, flow of information. When the system is constructed, workers who in the past depended on a delayed and somewhat random distribution of information will discover that someone else consistently receives information before they do. The new system is efficient, but maddening.

Another, though less likely, source of invalidity is poorly drawn models that do not conform to the rules. Such errors may result from lack of training, time pressures, or poor planning. Figure 6-15 humorously illustrates this source of invalidity.

A third source of invalidity is a poor investigation that collects inappropriate, confusing, or misleading data. When an analyst's questions are irrelevant, the resulting data are irrelevant; attempting to integrate the data would violate the principle of non-partition.

Confusing data, on the other hand, can be clarified through statistical analysis, unless the data items were collected in a confusing and inconsistent way (often the result of an investigation effort involving relatively untrained analysts).

TESTING FOR CONSISTENCY, COMPLETENESS AND VALIDITY

The following guide lists questions an analyst can use to determine how well a model meets criteria of consistency, completeness, and validity:

FIGURE 6–15. Poorly Drawn Models are Sources of Error

CONSISTENCY:

1. Is each element of the model well defined?
2. Do any two elements of the model seem to correspond to each other in any way? (Ambiguity?)
3. Does the model seem to be constructed according to the appropriate rules?

COMPLETENESS:

1. Is every relevant element of the base represented by an element in the model?
2. If not, are the deleted elements really irrelevant?
3. What is the value of the model without these elements?

VALIDITY:

1. Does each element of the model actually correspond to some element in the base?
2. Does each relationship in the model actually correspond to some relationship in the base?
3. What does the model seem to "say"? Are the inferences actually true about the base?

In Figure 6-16 on the following page, model A is inconsistent. It seems to model the basic DP cycle with a flow diagram, but it uses the wrong symbols and a graphic that is ambiguous (a line without an arrowhead). Model B is incomplete. It models the same DP cycle but leaves out an important aspect — output. Model C is invalid; it has a relationship (process output back to input) that is true only in specific cases, not generally.

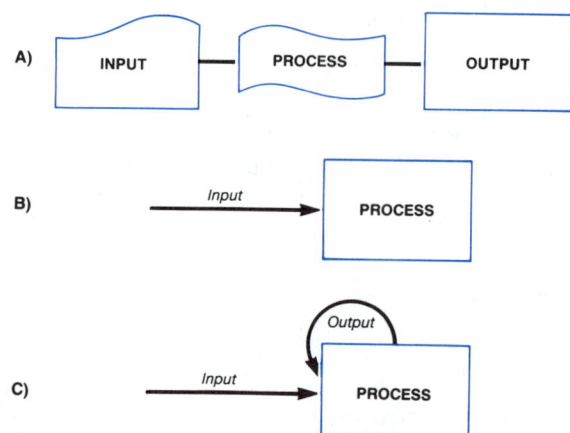

FIGURE 6–16. Three Incorrect Models

REPAIRING THE DAMAGE

Inconsistencies can be avoided through the use of glossaries and cross-reference dictionaries. The trend toward **data dictionaries** in IRMaint is a healthy movement in this regard. Automated systems of producing models, like Excelerator, also assist and draw symbols correctly.

Incompleteness is inherent in modeling. Strict attention to scope statements (what are we really investigating?) can help. The use of inverted dictionaries can show what is represented. This kind of attention can prevent embarrassing questions such as "Fine. Now, where is bill posting happening?"

Validity problems arise through improper, untutored, or poorly directed use of systems analysis techniques. Perhaps the greatest tool the analyst can employ is humility. During investigation, the analyst, no matter how brilliant, is a mirror, not a source of light. When the analyst interpolates or creates data during investigation, the resulting models are likely to be invalid, modeling the analyst, not the information resource.

Examples of Simulations

Chapter 13 will introduce over two dozen common modeling techniques employed by analysts; here we will consider four as typical of simulations.

The structured text in Figure 6-17 resembles a programming language as well as English. It simulates by eliminating unclear and confusing terms (such as adverbs and

```
GET PAYMENT-FORM
DO FIND-CUSTOMER-RECORD
DO VERIFY-CUSTOMER-BILL
DO UPDATE-CUSTOMER-RECORD
PUT POSTING-DETAILS
```

FIGURE 6–17. Structured Text

pronouns) and by aggregating similar base elements under common terms. The result, although dry and unpoetic, is to the point and quite clear.

The procedure for filling out the order form illustrated in Figure 6-18 is simpler than the actual process. The procedure does not tell how to hold a pencil or how fast to write — these are irrelevant. It aggregates all the boxes for descriptions under the term "Description." It uses only the term "box" rather than "box," "square," and "blank," eliminating a potential source of confusion. If some important aspect of form completion were eliminated — such as recording the customer name — the model would be incomplete. A line stating, "Shred form. Dispose of form," would obviously invalidate the model, since the base system would not do that.

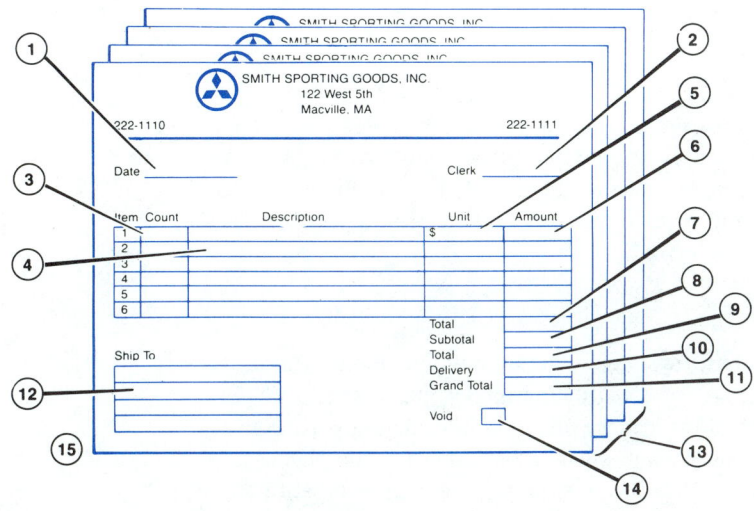

FIGURE 6–18. A Form-Filling Procedure

The flow diagram in Figure 6-19 uses arrows to indicate the flow of control between processes. Each box depicts a single process. The arrows mean "after this process is finished, go to the next process." This model can be further simplified by aggregating boxes 2 and 3 into a box labeled "stamp all documents." If a process were left out, the model would be incomplete. The order of stamping the documents may be unimportant; if that is so, the model is somewhat non-conforming, because the order is explicitly stated in the model, but not in the base.

In Figure 6-20 on the following page, the decision table depicts nothing in motion. Here we model a decision's "logic" or "sense." The top part of the table indicates conditions to test. Only one of them can be true at any given moment. The bottom half indicates actions. The table is read like this:

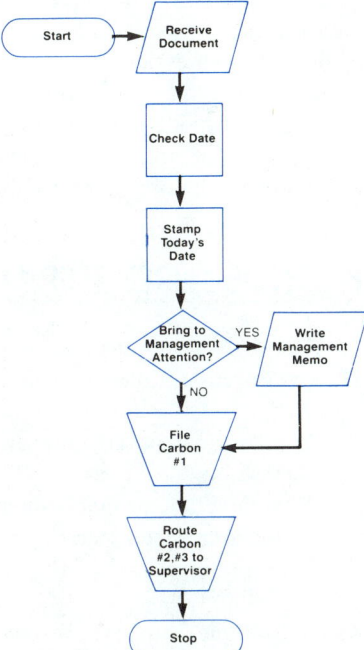

FIGURE 6–19. A Flow Diagram

1. If a particular condition across the top is true...
2. then move down the column until an X is located...
3. and read off the recommended action at the left of the row with an X in it.

	BLUE-CHIP			NON-BLUE-CHIP		
	High Rate of Return	Medium Rate of Return	Low Rate of Return	High Rate of Return	Medium Rate of Return	Low Rate of Return
BUY NOW				X		
OBTAIN MORE INFORMATION		X	X			
PUT OFF DECISION					X	
REJECT	X					X

FIGURE 6–20. A Decision Table

A single decision table such as this one cannot describe very complex decisions. The danger in making and using decision tables is that some important conditions may be left out. An error in placing the Xs in the table invalidates the model. In fact, Figure 6-20, which is supposed to model the decision to invest in a particular stock in a portfolio, is invalid for sane investors. Can you spot the error?

A Review

Models help us express and test base systems without building them. Because models are systems themselves and the rules of constructing models make them simpler than the base systems, an economy of effort can be achieved. Models are naturally incomplete expressions of their bases, but analysts dictate the kind of incompleteness, limiting it to removal of irrelevant aspects. Because invalidity is the major threat to the usefulness of models, the training of analysts stresses good data collection and analysis techniques. The following section discusses the modeling process in some detail and elaborates on data-handling techniques.

6.3   THE THREE-STEP MODELING PROCESS

The three-step modeling process is illustrated in Figure 6-21 on the opposite page. The three steps are the following:

1. Abstracting: reducing and arranging a large amount of data in a systematic fashion
2. Analyzing: further reducing information to make it more understandable, with the aim of summarizing the data
3. Representing: depicting the data as a model

As an example, suppose we have collected a large volume of data concerning users' needs for on-line abstracting of texts such as reports and memoranda. The data consists of transcripts of interviews and some tables counting the number of documents, the

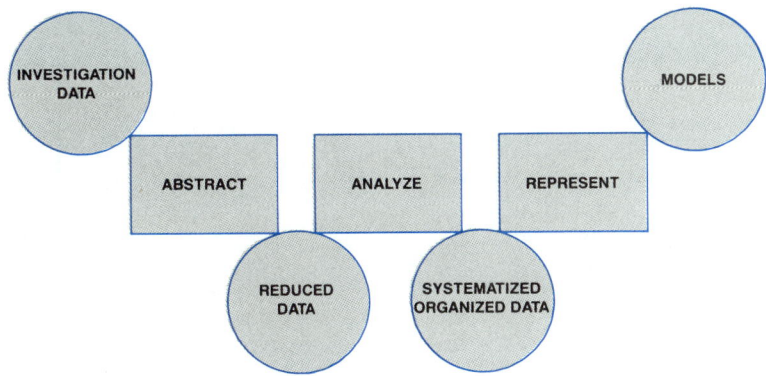

FIGURE 6–21. The Three Step Modeling Process

existing turn-around, and the prevalence of error. The modeling process proceeds as follows:

First, we **abstract** the collected data by summarizing the interviews, perhaps boiling them down to key points, main expressions of needs, and major concerns of users and managers who read the abstracts.

Next, we **analyze** the data. This analysis may produce a statistical summary indicating general trends such as average number of documents needed each month, average turn-around time delivered and desired, and average satisfaction levels. We can also analyze by producing examples or case studies of specific events involving the current abstracting system. Each type of analysis assists us in understanding the general trends in the data and creates specific examples for us to consider.

Finally, we **represent** our analyzed abstracted data in some tangible form. These forms are usually visual, almost always pictorial or schematic in format. But we may also produce a table, a narrative that outlines a series of procedures to be followed, or even a motion picture that demonstrates the abstracting process.

Abstracting: Choosing the Important Data

There are three aspects of abstraction: choosing the appropriate data from the available data, reducing the data in a consistent and reliable fashion, and maintaining the data in a usable, accessible form.

Data selection is important only if the appropriate data have been collected. One cannot make a reasoned choice from poor, insufficient, or irrelevant data. Ensuring that enough of the right kind of data has been collected is important in the planning stages of logical design. As Figure 6-22 on the following page illustrates, data selection is determined prior to data collection, to ensure that the right questions, such as the following, are asked:

- What is to be modeled? What kind of system is it?
- What is the purpose of modeling? What will the models be used for?
- What modeling tools are available? Which diagrams, tables, and texts will be produced?

- What kind of model is to be produced?
- What analytic tools are available? Can we perform statistical analyses? Are we trying to derive cases or examples?

The nature of the base system is important because relatively simple systems should not require a great deal of modeling. The principle of **parsimony** applies here. A simple system, such as a manual order-filling system in a small restaurant, should not require a great deal of data collection to have its essence captured. On the other hand, a complex, semi-automated, semi-finished-product inventory system that follows products through production stages in a large factory may well require volumes of data to be understood. Because such systems can exhibit complex behavior, a large number of observations will be necessary. For some very complex systems, such as the human brain, almost all activity appears unsystematic, and one must resort to basic research techniques (particularly experiments) that are not commonly employed in systems analysis.

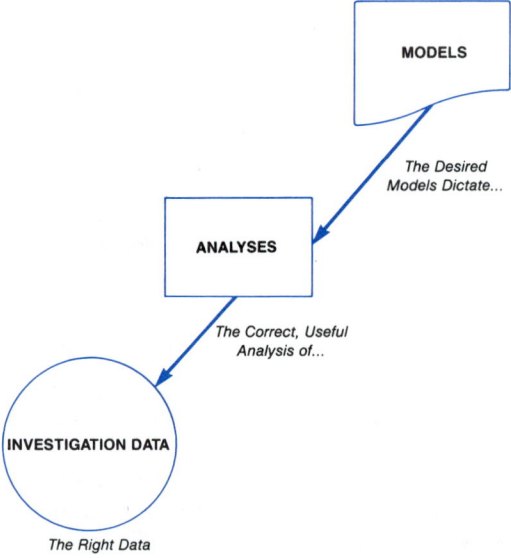

FIGURE 6–22. How to Select Data

If the purpose of modeling is only for communication, it is likely that the only data necessary relate to "typical" cases. On the other hand, typical cases rarely cause problems; it is the atypical case that costs additional effort in information resource maintenance. Hence, design and evaluation require a great deal more data and data in more detail.

Where modeling tools are quite detailed in nature, obviously the appropriate data to choose relates to the details of the operation. Where only gross representations are to be made, details become unimportant and may be abstracted out. This is the case, for instance, in constructing high-level structure charts. In other cases, specific aspects of system functioning are unimportant to logical design. Data flow diagrams, for instance, ignore time considerations because at the level of logical design, we are concerned only with logical dependencies among processes (what is necessary for what) rather than temporal dependencies (what preceeds what). Constructing a data flow diagram requires this sort of "necessity" data. Building system flowcharts, however, requires the temporal, timing information, so that the correct order or sequence of activities is modeled.

Finally, good investigation principles dictate that we not remove data we intend to analyze or attempt to analyze inappropriate data. Suppose we have collected information on screen-format preferences for a management information system pertaining to inventory from seventeen managers. We cannot perform a numerical analysis aimed at producing statements about the "average" preferred screen format—you cannot "average" seventeen descriptions. Rankings of preference for specific, known screen formats are better data for this purpose.

Data Reduction

Having chosen the appropriate data, we should reduce it to manageable proportions in any of four ways (Figure 6-23):

- cataloging
- categorizing
- characterizing through statistics
- exemplifying through case studies

Suppose our investigation involved the inventory system's screen formats, and we have collected the following data in the form of comments from four managers:

Manager A: I like to have the entire screen arranged in rows and columns: products in the rows and product data in each column.

Manager B: I don't care about seeing lists of products. I merely want to be able to see data on a particular product displayed so I can read it, you know, spaced out properly.

Manager C: I don't care about seeing lists of products. What I need desperately is to be able to see what the discrepancies are between on-hand and depletion rate, between expected order date and receiving date, and so forth.

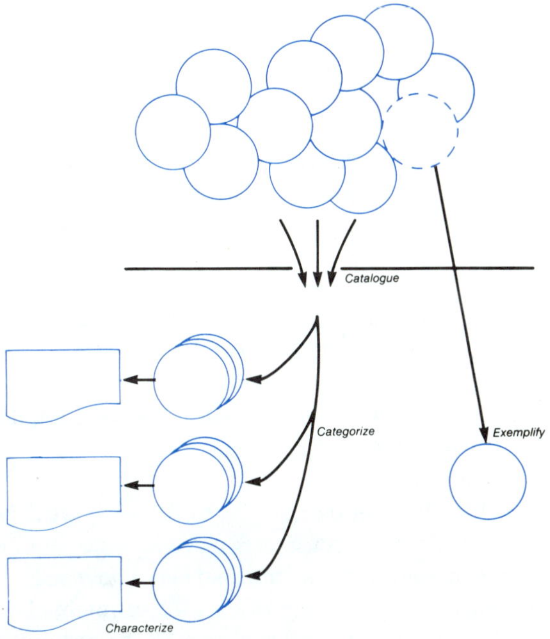

FIGURE 6–23. Data Reduction

Manager D: Who cares about the products themselves? In this department we really care only about gross overall space requirements and materials flow. I couldn't care less about a widget, but I need to know how many free square feet I've got in the warehouse.

Here are four items of data, each different from the others. We can create a **catalog** of preferences by listing the four types of request and describing each:

A: product by attribute table
B: attributes for a particular product retrieval
C: status exceptions for attributes
D: aggregate space and flow attributes

In describing the data this way, we discard most of the comments and concentrate on a list of factors that seem to distinguish the responses.

Categorizing goes a step beyond cataloging. It provides a count of the number of data elements in the catalog entry for each different data item. If four managers responded the same as manager A, three as manager B, six as manager C and none as manager D, the data would be reduced to this:

A:5
B:4
C:7
D:1

If most of the responses appeared similar or if trends appeared in the responses such that we could characterize most of them as "X" or "X, given Y," then we would perform rudimentary statistical or **aggregational** judgment. We might, for instance, reduce the above data to a statement such as this:

X(rows and columns) 5/17 (29%)
Y(retrieval by product code) 4/17 (24%)
Z(detailed analyses) 8/17 (47%)

This means that half the responses favored detailed analyses, with the rest evenly divided between displays of rows and columns and retrieval by product code. The "trend" is more apparent.

Finally, a further step is to derive **typical cases** as representative of the data. One way would be to consider a manager who requires a broad overview, perhaps requiring some summary data, and another who needs to know specific things about specific products. While *neither* of these cases might actually be any certain manager's situation, each would represent a summary of the responses:

P: a manager who needs to gather a broad overview of the products in inventory, including summary statistics about flow, capacity, and storage requirements
Q: a manager who needs to know specific things about specific known products or who needs to detect anomalies in the inventory such as long order intervals, overly high capacities, and so forth

Each of these reduced sets of data "captures" the essence of the original data in some sense. The choice of what level to reduce the data to depends on the requirements of the data analysis and representation stages.

Maintaining Abstracted Data

A few words are in order about maintaining abstracted data. First, data reduction does not imply that data are thrown away. On the contrary, data should be carefully cross-indexed and filed for future use. Interview transcripts are useful sources of quotations for the final report and proof that the interviews were conducted. It is also possible that on rereading data subsequent to model construction, the analyst can gain additional insights into the experience of users.

Second, data should be protected against corruption. Because they represent abstracted experiences, they are less likely to be considered private information (Chapter 11 discusses the need for security for interview data). But because abstracted data are uninterpreted, they may be misinterpreted by others. While data do not necessarily identify individuals, they may identify teams, units, or departments. What may appear after some abstraction to be a trend might be, on analysis, only a statistical fluke. It is important that abstracted data not be treated like mathematical conclusions.

6.4 ANALYZING INVESTIGATION DATA

This section examines ways of *analyzing* data have been collected and abstracted. Because merely reducing the data does not help us understand and characterize them, we resort to analysis techniques to make data more "palatable."

There are two kinds of analysis relevant here. The first we term "aggregational" because we add numbers together to look for clusters, trends, and characteristic values. The second kind is called "case," because we are concerned with finding common or unusual cases in the data. An example, or case, resulting from this kind of analysis may not actually appear in the data, but may be constructed from several cases. Whereas aggregational analysis attempts to characterize all the data — and provides estimates of how well the characterization works, case analysis strives to create cases for discussion, that is, typical cases or unusual examples.

Aggregational Analysis

In **aggregational analysis**, we put sets of data together. The set of observations below can easily be characterized "in the aggregate" as "just about always 3":

3, 3, 3, 3, 3, 4, 3, 3, 4, 3, 3, 3, 3, 3, 3

While statistics and the underlying mathematical principles add precision and reliability and undoubedly aid in the systematization of analysis, one need not always use powerful statistical techniques in making aggregational statements.

On the other hand, statistics are quite valuable in situations like the following:

1. Characterizing observations that appear scattered among many values (such as the set 1, 4, 6, 7, 7, 8, 12, 13, 13, 13)
2. Infering the cause of some phenomenon (for instance, stating, "Adding a screen driver contributed significantly to reducing human resource needs for maintenance reprogramming.")
3. Making statements about relationships between values (such as "Response speed seems to be inversely related to the number of users logged on at any given moment.")
4. Demonstrating that small differences between groups are actually meaningful (as in asserting the "Throughput of jobs was greater in January than in December." or "Throughput increased throughout the first year of the trial system.")

In each of these cases, it is the ability to perform mathematical operations on the data in a consistent fashion that enables one to make these statements with confidence. We can make similar statements without statistics, but the precision and systematic nature of the computation provide us with a measurable degree of confidence. Without statistics, our confidence depends on the credibility of the person making the statement, our own intuition, and a host of uncontrollable factors. Thus, the same statement may be believable one moment and incredible the next. Statistics, on the other hand, are consistent, precise, and accurate.

How we select statistics depends on:

1. the type of analysis to be performed (parametric versus non-parametric; see below)
2. the goal of the analysis (description versus inference of causes)
3. the "behavior" of the data
4. the type of data
5. the knowledge and experience of those who have to understand the results of the analysis

TYPE OF ANALYSIS

In **parametric** statistics, data values can be added together as numbers, while **non-parametric** statistical analysis cannot make that assumption. The distinction boils down to this: with parametric statistics, one is concerned with values alone, while in non-parametric statistics, one is concerned with frequencies of values (how often particular values occur).

Consider this simple example. A program invokes five subroutines the following number of times:

Subroutine	A	B	C	D	E
Invocations	1	3	100	10	4

A non-parametric statistical analysis orders the subroutines in decreasing frequency of invocation: C, D, E, B, A. A parametric statistic would be the average value of invocations across programs: 120/5, or 24. Parametric statistics result in a single "parameter" or characterization of all the data: 24 invocations per subroutine. Non-parametric statistics do not produce the single number; instead they produce tables of frequencies or rank orders.

THE GOALS OF ANALYSIS

There are two general goals of statistical analysis: description and explanation (see Figure 6-24). Common statistics merely describe the whole set with a single parameter or table and include: (1) the **mean**, or average value, (2) the **mode**, or most common value, (3) the **median**, or middle value, (4) the **frequency distribution** or list of frequencies for each possible measured value, and (5) the **variance** or measure of how well the mean characterizes the entire set of data. These statistics are diagrammed in Figures 6-25 and 6-26.

Use	Shows up as...	In
REDUCING SCATTER	MEANS, MEDIANS, MODES, FREQUENCY DISTRIBUTIONS	DESCRIPTIVE STATISTICS
SHOWING RELATIONSHIPS	CORRELATIONS, CHI-SQUARES (GOODNESS-OF-FIT), REGRESSION ANALYSIS	DESCRIPTIVE STATISTICS
DEMONSTRATING DIFFERENCES	T-TESTS, ANOVA, MULTIPLE ANOVA	INFERENTIAL STATISTICS
MAKING INFERENCES (WHAT CAUSES WHAT, WHAT IF)	ANOVA, MULTIPLE ANOVA, MULTIPLE REGRESSION	INFERENTIAL STATISTICS
SIMPLIFYING, CHARACTERIZING	FACTOR ANALYSIS, CLUSTER ANALYSIS	DESCRIPTIVE STATISTICS

FIGURE 6–24. Goals of Analysis

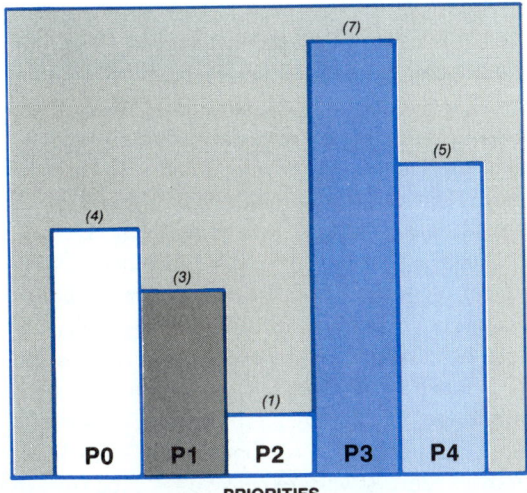

FIGURE 6–25. One Way to Describe Data

Explanation is the goal of **inferential statistics**. We utilize statistics to infer the cause of relationships between or among groups of data items. The most powerful inferential techniques can be used to infer causes, to discover why something has happened. These techniques attempt to show that observed differences between groups are inherent, differences actually existing between the groups.

In a previous example, such an inference is made concerning the addition of a screen driver. Adding the driver significantly lowered human resource needs. In another example, it was inferred that the difference between two months' throughput was a real, rather than a chance, event.

Of course, inferential statistics can go only so far. In each of the above examples we do not know what really caused the difference: what about January made the throughput higher? What about the screen driver decreased manpower needs? To find out the answer to those questions, we need to do more analysis.

Common statistics used to make inferences include the following:

1. **T-tests**: Do two groups have different mean values?
2. **ANalysis Of VAriance** (ANOVA): Are there systematic and real differences among several groups in their mean values?
3. **Chi-square**: Do the observed frequencies in several groups differ from what would be expected by chance?
4. **Regression analysis**: How does one set of observations relate to one or more other sets of observations?

JOB PRIORITY FREQUENCY DISTRIBUTION

08/17/86   10:00–11:00

PRIORITY	COUNT	PERCENT
0	4	20
1	3	15
2	1	5
3	7	35
4	5	25
	20	100

MEAN = 2.3 (average)

MODE = 3 (most popular)

MEDIAN = 3.2 (half exceed 3.2)

VARIANCE = 2.0 (spread*)

*approx. 2/3 of all values are within 2.3 ± 2
19/20 of the values are within 2.3 ± 4
99/100 of the values are within 2.3 ± 6

FIGURE 6–26. Four Descriptive Statistics

TYPE OF DATA

The choice of statistics also depends upon the kind of data being analyzed. **Nominal** data consist of names that are not arithmetically comparable. Values such as "subroutine A" and "subroutine B" are nominal values. They cannot be compared in terms of their arithmetic values. One could not say that "subroutine A" is twice as much as "subroutine B" or even that the first is "more" than the second. For nominal data, we can use only non-parametric statistics, simply because there is no way to add values together.

Ordinal data goes a step beyond nominal data. They are values that can be compared with one another in terms of degree, such as "more" or "less." In other words, the values can be placed in an *order*. Typical of ordinal data are values for attitudes. "Like very much," "like somewhat," "dislike somewhat," and "dislike very much" lie along an ordinal scale because each value can be compared with the others in terms such as "more" or "less". But we cannot say how much more or less. Other commonly used ordinal data are frequency-of-use data ("continuously," "often," "sometimes," "rarely," and "never"), value scales ("indispensible," "important," "of some value," "of little value," and "of no value"), and estimates of expense ("very expensive," "expensive," "inexpensive").

NOMINAL (NAMES) CATEGORIES

Type of transaction	Count
BILLING	483
PAYMENT	404
ORDER-ENTRY	763 (***MODE)
SHIPPING/RECEIVING	90

ORDINAL (ORDERED) CATEGORIES

Order Lead-Times by Period	Count
SHORT (lt. 2 days)	288
MEDIUM (2 days to one week)	142 (***MEDIAN IS ONE WEEK)
LONG (1 week to one month)	330
INDEFINITE (gt. one month)	3
	763

INTERVAL (NUMERIC) CATEGORIES

Order Response-Times	Count
0 days beyond promised	653
1 day beyond promised	100 (***MEAN IS 0.2 DAYS)
2 days beyond promised	4
3 days beyond promised	3
4 days beyond promised	2
5 days beyond promised	1

FIGURE 6–27. Three Types of Data

When we can put a value on the difference, we move from ordinal to **interval** data. The term "interval" is a bit uninformative, because the important quality of interval data is that the values express "how much." The "interval" referred to is a unit value, the quality missing from ordinal scales. Not only can we say that x is greater than y but we can tell just how much greater: x minus y.

For instance, an estimate of magnitude of expense can be expressed in dollar terms. These terms *are* arithmetically comparable, with a unit of one dollar. If manager A states that a new system will cost $20,000 and manager B estimates the same system's cost at $25,000, we can express the difference between the two estimates as $5,000. Had we merely asked for a magnitude, we would not have been able to perform this computation.

In fact, interval data are very much like numerical data. The properties of numbers are the properties of interval data: we can add, subtract, multiply, and divide the values.

Common interval scales include costs (in dollars), time to complete a project (in days), frequency of use (in accesses per day), run time (in seconds), file size (in bytes), operational speed (in MIPs). Figure 6-27 illustrates the differences among nominal, ordinal, and interval data scales.

It is also common, but inaccurate, to treat ordinal data as though they were interval by translating successively greater values as successive numbers. Although this tactic is technically incorrect and, in some cases, misleading, it sometimes can be used to make estimates. The example below illustrates that tactic.

Suppose twenty five managers are asked their estimates of the value of a specific spreadsheet application. Their responses are:

Value rating	No. of managers	Assigned value	Aggregated values
Very valuable	1	5	1 x 5 = 5
Valuable	4	4	4 x 4 = 16
Of some value	7	3	7 x 3 = 21
Of little value	9	2	9 x 2 = 18
Of no value	4	1	1 x 4 = 4
Total	25		64

The average value (64/25, or 2.6) lies somewhere between "of little value" and "of some value." Thus, our assumption that relative ranks can be treated as interval values of some unit leads to the conclusion that the managers, *in the aggregate*, are not wildly enthusiastic about the value of this application.

We can also use **non-parametric** approaches to this data and arrive at a similar, but more reliable, conclusion. If we line up the twenty-five responses in order

5, 4, 4, 4, 4, 3, 3, 3, 3, 3, 3, 3, 2, 2, 2, 2, 2, 2, 2, 2, 2, 1, 1, 1, 1

we notice that the middle value is 2 (there are twelve greater than this "middle 2" and twelve less). This *median* value is interpreted as saying that as many managers find the application of little or no value as find some value or more.

THE BEHAVIOR OF DATA

Thus far, we have mentioned the type of analysis, the goals of analysis, and the type of data being analyzed as determining the type of statistics to use. An additional factor is the "behavior" of the data. How can data "behave"?

By "behavior" is meant the **distribution** of the values found. Are most of the data of one value? Are the data distributed more or less equally across all values? Do they seem to bunch up in the middle of the range of possible values or are there several apparent concentrations?

Figure 6-28 illustrates a graph called a **frequency distribution**. The data for this graph are estimates of the number of hours that could be saved daily by using a certain spreadsheet application:

3, 3, 3, 3, 3, 4, 3, 3, 4, 4, 3, 3, 3, 3, 2, 2, 2.

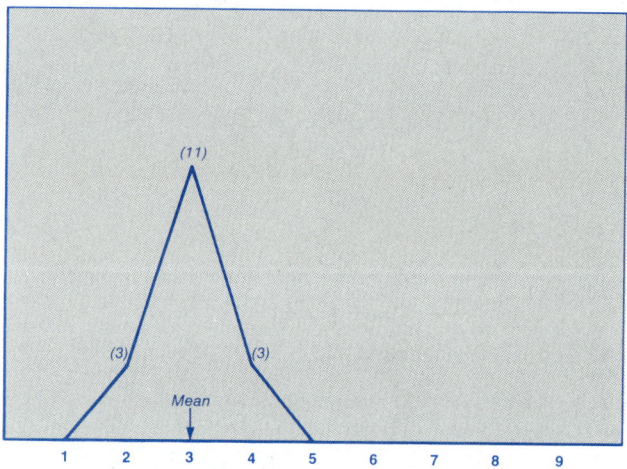

FIGURE 6–28. A Frequency Distribution

Most of these seventeen observations (eleven, to be exact) have a value of 3. Others cluster closely about it. Another set of data, graphed in Figure 6-29, exhibits a different kind of behavior in response to questions about lost worker-days to training:

1, 4, 6, 7, 7, 7, 8, 12, 13, 13, 13, 13, 13, 15, 17.

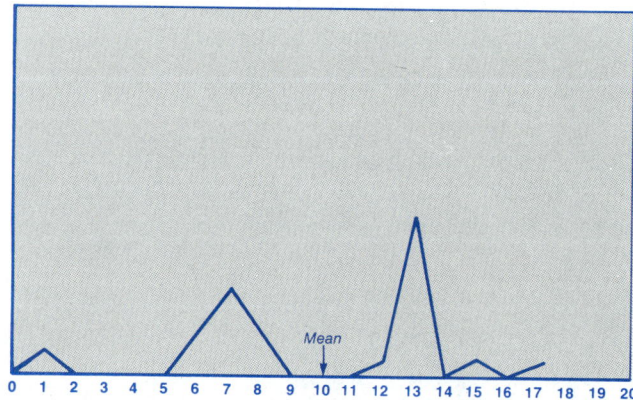

FIGURE 6–29. Clustering of Data

In this set, there appear to be two clusters, one about 7, the other about 13. In technical jargon, the first set exhibits low **variance**, the second set relatively higher variance. (Of what value is the variance?)

Note that the arithmetic average of the first set (3.00) is close to the cluster around three hours. But in the second set, the average (9.93) is quite different from either cluster value of seven and thirteen worker-days. Low-variance data are more accurately characterized by the average than high-variance data; that is, where variance is high, parametric statistics are less valuable.

In other words, we require relatively well-behaved data to use parametric statistics. To say "something" about the data, the data have to "hold still and be photographed", so to speak. If the values are "all over the place," a single value does not describe them well.

THE AUDIENCE

Our final consideration in choosing statistics is one of communication with those who have to follow our discussion. We must limit ourselves to what our audience can understand which sometimes rules out any statistical discussion whatsoever.

Therefore, our choice of statistics — and even our choice of the presentation of results — is limited by circumstance. A pie-graph (Figure 6-30) oftens tells more than any statistic. Frequency distributions, such as in Figure 6-29, may be more productive than a regression analysis. Finally, statistical inferences may be counterintuitive and cause divisive arguments. Inferring that system downtime may be due to the failure of a module that has never failed before may be statistically correct; however, this argument may not appeal to the intuition of system managers, so other evidence may have to be collected and presented in a convincing way. Statistics can present sophisticated arguments for subtle effects, but people may not be able to understand either.

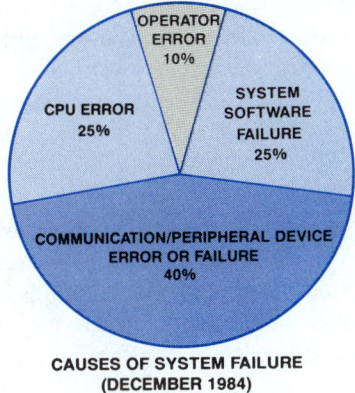

CAUSES OF SYSTEM FAILURE
(DECEMBER 1984)

FIGURE 6–30. A Pie Graph

Case Analysis

The other major kind of analysis is the derivation of examples or cases. As previously mentioned, examples can relate to either typical or extraordinary cases. Where the emphasis in design is on the ordinary and expected, an example that captures essence and stimulates thought is valuable. This kind of analysis makes statements such as those beginning with, "The manager around here needs...." Aggregational analysis does not make statements of this sort since it concerns not individuals but the group as a whole. It makes statements that begin, "The *average* manger around here needs...." There may not be, of course, an average manager.

Other cases may emphasize unexpected events or **critical incidents** in the investigation. In that case we make statements such as:

> On occasion we may run up against a manager who needs something different. One such case is that of manager X, who needs....

The term "case" is critical here. The *case study method*, employed in this text and in many business curricula, relies on deriving cases that are instructional and interesting. In computer-based information resource maintenance, where the ordinary is even more routine than in, say, social systems, it is the *critical* cases that can spell the difference between failure and success.

Discovering just those instances which lead to system failure satisfies the goals of investigation in which most of the design is predetermined. Consider what might have happened if analysts ignored the warning of this vice president of marketing:

> I really don't care how you present sales data in your reports. But you've simply got to redline outrageous reports and values that don't follow trends. Yeah, I'd guess that 99-percent of the time our sales are within 5-percent of forecasts. But one time a turkey in Tennessee sent in a summary which claimed he'd sold eleven gross of widgets in January. Based on that claim, our market analyst predicted a 10-percent increase in widget sales, so we tooled up for more widgets. Needless to say, our average widget sales were about one gross, which was exactly what he'd sold, not one-one, you see? Now if a report redlined 1000-percent sales performance, I could check up on it instead of finding out about mistakes after we'd invested fifteen thousand dollars in production machinery.

Using case studies requires a great deal of sensitivity, tact, and writing ability. A case study touted as typical may, in fact, fail to occur if the data are widely variable. A critical incident may never occur again. A poorly described case may be unreadable or illogical. Finally, if a particular critical incident is recognizable, heads may roll — managers and supervisors do not like to find out about expensive foul-ups a year later in a systems analyst's report.

Cases serve well for two kinds of readers. Upper-level management is quite unconcerned with technical details and needs a quick overview, which a case can provide. Technical management, on the other hand, sweats over the details and worries about extraordinary events. They need to understand the user's predicament before it occurs. In each situation, the selection of the appropriate case is an important step in building understanding.

Below is an example of a typical case:

Typically a manager in the marketing division handles from ten to twenty products. All phases of marketing, from research through test trials of a product to maintaining marketing data for analysis are the responsiblility of each product-group manager. Each day the marketing manager must sift through six to ten detailed reports on product performance. Reading these takes two-and-a-half hours on the average. A number of decisions have to be made. Typical decisions concern advertising campaign management, sales personnel deployment, customer credit advancement, track down data, finding salespeople, and making presentations to upper management. Once a week an entire day is spent in a divisional sales strategy meeting. At this meeting, all reports must be available in summary and tabular form; all data presented must be verified since it will not be checked higher up. Typically the marketing division manager complains about incorrect actual sales data, his inability to locate sales staff in the field, a lack of cooperation from billing, shipping, and manufacturing areas, and, most important, a lack of time to do anything. Decisions that should be made in a considered fashion are often forced in minutes during meetings.

And now, the extraordinary case:

One manager notes what happens in the divisional sales strategy meeting when his sales data are incomplete. One week, owing to a mail strike, he had no sales data from the Eastern region for his products. In this case, he improvised by hand-calculating a trend analysis based on last year's data. He had an estimate of how well his products *might* have done. Since he had no previous data for newly introduced products, he used a product-use-substitution scheme, interpolating sales from estimated sales of similar products. To his surprise, his data were not only accepted in the meeting, but later proved to be dead on. Furthermore, after the meeting, he was able to assign one of his analysts to the task of computerizing this estimation scheme. The result was an *ad-hoc* program that has proved quite useful in subsequent strategy meetings over the past year.

In each case, little statistical data is presented and the use of numbers in a broad range makes argument less likely. The language employed is descriptive, and claims are strong and without qualification. The important use of cases is in discussion leading to decision making. The precision of aggregational analysis is foregone in return for the intuitive value of cases. Where the data are not clear-cut and unequivocal, cases facilitate early design discussions.

Our section on analysis has presented an overview of issues relating to the analysis of investigation data preparatory to creating models. Two types of analysis, aggregational and case, have been discussed. Each is appropriate for certain situations. Where precision is important and attainable, aggregating the data is valuable. Where quick, clear indicators of typical or extraordinary events are important, case analysis serves well. In either type, the statistics and examples make concrete the words and numbers collected in the data. Model construction makes the statistics and examples visual, as discussed in the next sections.

6.5   MODELING TOOLS

After the analysis of the data is complete, the analyst should build representations of the analyzed data for use in design. In selecting modeling tools, the analyst is guided by tradition, the task at hand, and the potential readers and listeners. This section discusses the alternatives in modeling tools and a procedure for selecting them.

Classifying Models

Modeling tools can be characterized in these ways:

- physical form or format
- codes used in the representation
- attributes represented in the model
- use of time in the model
- items represented in the model
- certain aspects of the use and interpretation of the model

The physical form of the model is often limited to two-dimensional paper drawings, tables, and texts. Sometimes it is helpful to build a three-dimensional physical model (especially when discussing layouts), but this is usually left until physical design. Most of the models we will discuss are two-dimensional "paper and pencil" models.

An example is the book you are now reading. It is more or less two-dimensional; presenting other forms would be difficult. An alternative to the paper-and-pencil model is the computer simulation. Such simulations often include two- and three-dimensional representations and may extend to moving images, action graphics, and models that can be manipulated in real time. Our goal and method are much more modest.

Considering only paper-and-pencil models now, there is considerable variation in the kind of code used in the model. By this term we mean the set of symbols that make up the drawn models. Codes can be any of the following:

- language or language-like text
- tables of numbers or words
- graphs
- networks of boxes or circles connected by lines
- pictures or layouts

Examples of each of these are in Figure 6-31 on the opposite page.

In many cases a good idea or concept can be represented equally as well by one form of code as by another. The choice of the form of code depends as much on the familiarity of those viewing them as anything else.

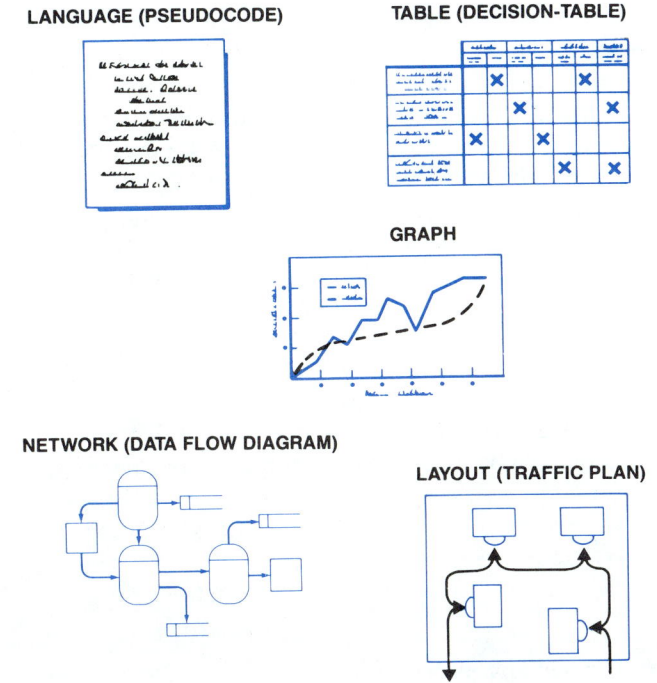

FIGURE 6–31. Codes Used in Models

Each model is limited by its codes and is further bound by what attributes and items it can represent. Language, for instance, can represent anything "expressible." In fact, language is often too wordy and metaphorical; that is why structured English and pseudocode were developed. Other models may represent only certain aspects of the base.

Typically analysts are concerned with representing the following: **content**, **flow**, and **structure**. That is, analysts concentrate on representing the content of the base (for instance, a process, a report, or a physical action), the flow of something within that content (information, control, materials), and the structure holding the elements of the content together (authority structure, distribution in space or over time, logical dependencies). Furthermore, given the aspect to be represented, some models are specialized for particular items or subsystems. A system flowchart, for instance, depicts the flow of information and control among system processes during operations. A structure chart represents the structure of a problem in terms of the relationships among elements of the solution (problem A breaks down into problem B, problem C, and problem D).

Models can be classified into how they represent time. Some represent changes over time (**dynamic**), while others view a snapshot (**static**), and still others ignore time altogether (**asynchronic**). A system flowchart is dynamic; it shows how data moves over time through a set of operations. Each arrow reads "and then...." A data flow diagram, on the other hand, is asynchronic. Rather than meaning "and then....," the arrows indicate logical necessity: "to have this data flow, such-and-such a process must have been started." In decision trees, decisions at single points in time are examined. Inexperienced users may think that the tree is dynamic as they read it from left to right; in fact, this is only the custom of reading and has nothing to do with interpreting the decision tree.

In addition to these dimensions for classifying models, these are attributes that concern how the models are drawn and read. Some sprawl over an entire sheet, while others are rigidly drawn in one direction. Contrast a data flow diagram and a decision table in this dimension: the first can have symbols placed anywhere; the second, only in a rigid box format. Another attribute is **precedence**. Some models are read in a predefined fashion, while others can be read in any order. Decision trees and tables are rigid in that sense, as is an English-language description of a process. Reading the model in some fashion other than the convention can lead to confusion. Data flow diagrams make sense only in the direction of the arrows, as do most flow diagrams.

Finally, some models have ways of highlighting specific aspects. Emphasis can be indicated through color, line width, placement on the page, bullets and arrows, and so forth. For instance, although decision tables are relatively limited in what they can represent, the list of the subsequent actions can be used to represent the relative importance or value of various choices.

Figure 6-32 lists a variety of models in terms of several of the dimensions discussed. Examples of each of these are found in Figures 6-33 through 6-37.

Evaluating Modeling Tools

The selection of modeling tools is guided by the characteristics of the tools and the needs of the situation. Four important criteria are: completeness of the representation, the ease of modifying the design, the ease with which the model will be understood, and the value of the model to physical designers and implementers.

Of course, one cannot always choose specific modeling tools. Shop traditions and standards as well as one's own skills have to be considered. Nonetheless, the analyst must consider *all* of the foregoing discussion when beginning the modeling exercise. Otherwise, the model may be incomplete, misunderstood, or misused.

An Overview of Models

For purposes of discussion, we will consider only three classes of models along the dimension of the aspect of the base represented: flows, structures, and processes.

FLOWCHARTS

Flowcharts model the flow of materials, control, or data over time. If flowcharts involve time at all, they are dynamic; otherwise, they are asynchronic. Flowcharts are the workhorse of systems analysis, because analysts have a primary concern with the flow of information among processes and across time. The major flowcharts used by systems analysts include:

FORMS

   Drawings
   Text
   Physical Models

CODES

   Language
   Table
   Graph
   Box-and-Line Network
   Layout/Pictorial

ATTRIBUTES

Depicts $\left\{\begin{array}{l}\text{Flow}\\\text{Content}\\\text{Structure}\end{array}\right\}$ of $\left\{\begin{array}{l}\text{Materials}\\\text{Control}\\\text{Data}\end{array}\right.$

USE OF TIME

   Static (a snapshot in time)
   Dynamic (demonstrates changes over time)
   Asynchronic (effects of time ignored)

OTHER

   "Sprawl"
   "Fussiness"
   Style of Reading (L-R, Top down, random)
   Use of Highlighting
   Availability of sub-modeling for sub-systems

FIGURE 6–32. Classifying Models

- system flowcharts
- Warnier-Orr diagrams
- data flow diagrams
- materials flow diagrams
- document flow diagrams

- layout flowcharts (traffic patterns)
- process (manufacturing) flowcharts
- Nassi-Schneiderman charts
- precedence (scheduling) charts
- Gantt charts

Figure 6-33 characterizes flowcharts.

FLOW CHART	CODE	REPRESENTS	ATTRIBUTES	TIME
System Flowcharts	Box & Line	Operations	Control Data	Dynamic
Program Flowcharts	Box & Line	Procedures	Control	Dynamic
Data Flow Diagram	Box & Line	Data Flow Processes	Data	Asynchronic
Material Flow Diagram	Table	Operations	Material	Dynamic
Document Flow Diagram	Table or Box & Line	Operations	Documents	Dynamic
Layout	Pictorial	Traffic People Furniture Electrical Machinery	Data Control	Static or Dynamic
Process Flowcharts	Table	Processes	Control	Dynamic
Nassi-Schneiderman Chart	Boxes	Procedure	Control	Dynamic
Precedence Charts	Box & Line	Operations	Control	Dynamic
Gantt Chart	Table	Activities Operations	Control	Dynamic
Warnier-Orr Diagram	Table	Process	Data, Control	Asynchronic

FIGURE 6–33. Common Types of Flowcharts

STRUCTURE CHARTS

By their nature, **structure charts** are static. They represent "snapshots" in time; hence, they have no way of representing time. Instead they show how things "go together." Whereas flowcharts relate to processes, structure charts relate to construction. A flowchart can be likened to a movie, a structure chart to a photograph.

Structure charts depict the contribution of elements of the chart to other elements in a graphical way. In a sense, structure charts are photographs of how complex systems are put together subsystem by subsystem. For instance, a module structure chart depicts the way in which a given module (say, A in Figure 6-34) contributes to the performance of a higher-level module (say, AA). Structure charts tend to depict necessary relationships (what is necessary for what); for that reason, they are "logical." Whereas flow diagrams are the meat-and-potatoes of the classical systems analyst, structure charts are the fine herbs of the modern logical system designer. The technique described in this book, the structural-functional

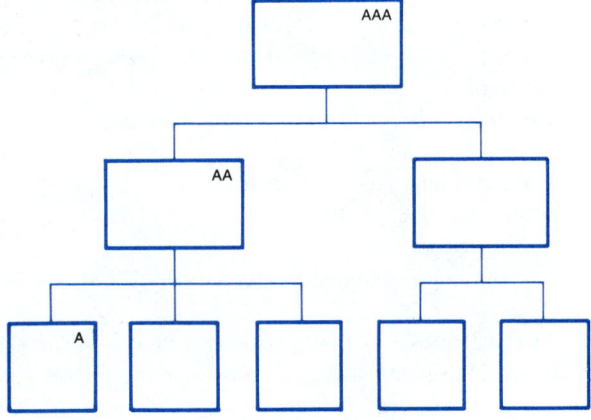

FIGURE 6–34. A Structure

method, is essentially an outgrowth of the interplay between the dynamism of the data flow diagram and the logical expressiveness of the structure chart.

Common structure charts available to the systems analyst include:

- module structure charts
- decision trees
- decision tables
- HIPO diagrams
- labor distribution charts
- organization charts
- tables of contents
- data dictionaries and their entries
- DIADs (refer to Gane and Sarson, 1979, pg. 147)

Structure charts are categorized in Figure 6-35.

STRUCTURE CHART MODEL	CODE	REPRESENTS	ATTRIBUTES
Module Structure Chart	Box & Line	Subordination	Logical Necessity
Decision Tree	Box & Line	A Decision	Structure of Decision
Decision Table	Table	A Decision	Structure of Decision
HIPO Diagrams	Table	A Process	Structure, Input & Output
Labor Distribution Chart	Table	Labor Assignment	People versus Tasks
Organization Chart	Box & Line	Subordination	Reporting Authority
Table of Contents	Outline	Subordination	Semantic Contribution
Data Dictionary	Outline	Subordination	Semantic Contribution
DIAD	Box & Line	Data Reference	Access Among Data Items

FIGURE 6–35. Common Types of Structure Charts

PROCESS LOGIC

The term "**process logic**" refers to the way in which a process operates. The term "how to operate" would do as well, but "process logic" has gained some currency from structured system design. Because it is logical (that is, structural) but refers to basic hardware operations, process logic is put in this class.

Many process logic models appear in language or language-like modes, such as structured English or pseudocode. Since the basic model for a process is Subject + Verb + Object, process logic models tend to be dynamic (static descriptions are possible, but uncommon).

Common process representations include:

- natural language or "description"
- structured English
- pseudocode
- programming languages (especially BASIC, Pascal, and COBOL)

The list above is in decreasing order of expressiveness and increasing order of precision. Interestingly, the more precise the process logic representation, the less work is necessary to convert it into software. Modern structural-functional techniques make it difficult sometimes to separate logical and physical design.

Process representations are categorized in Figure 6-36.

PROCESS REPRESENTATION	CODE	USEFUL FOR	ATTRIBUTES
Natural Language	English	Anyone	Full English Language
Structured English	English Subset	Designers	Poetic, Imprecise Words are Removed
Pseudocode	Nouns and Keywords (Verbs)	Programmers	Terse, Precise, Automatable
Programming Languages	Varies	Coders	Standardized, Relates to Implementation, Highly Testable

FIGURE 6–36. Common Types of Process Representation

Selecting Modeling Tools

There are five criteria for selecting modeling tools:

1. completeness
2. understandability and readability
3. valuable products of the model
4. existence of a test of adequacy
5. existence of a test of validity

Several models contribute *valuable by-products*. Because of their nature, dynamic models can be followed and manually simulated through desk-checking. Others can be neatly segmented for closer examination (for instance, Nassi-Schneiderman charts and structure charts).

PERT, CPM, and Gantt charts are useful in project management, while data flow diagrams can be used to determine bottlenecks. Examination of specific morphs can yield a lot of knowledge about the system that is not directly expressed as data flows.

Some models are more easily validated than others. A model represents the base, but only the base. There is nothing true of the model that is not true of the base. Validity tests are quite difficult to make, however. The simpler the model, the easier the test. Several techniques exist for *program* testing that may in the future assist logical designers. Data flow diagrams can be systematically checked for validity with respect to logical data requirements. A logical model, however, cannot be checked with physical tests.

Selection of a specific modeling tool is guided by the following considerations:

1. Characteristics of the base: is time important? Are we going to model processes? What kinds of structures are to be represented?
2. Characteristics of the tools: what do they represent easily? How easy are they to draw and test?
3. Tests for completeness, adequacy, and validity
4. Pragmatic considerations concerning readability and understandability.

Many shops have specific standards for modeling. Some traditional shops maintain that the essential element of a well-designed system is the system flowchart augmented by program flowcharts. "Structured" shops (that is, shops in which at least lip service is paid to structured design techniques) require data flow diagrams, structure charts, HIPO diagrams, and data dictionaries. Most project management techniques require Gantt and PERT charts. A project "book" will have a table of contents. Organization charts are still considered primary analytic devices for new system development.

In other situations, the choice of diagrams and charts may be left up to the analyst, especially where strong user involvement is necessary, such as in prototype development and user-generated software. In these cases, model types are not prescribed in advance and careful consideration of the choice of models is necessary.

6.6 PRODUCING SYSTEM MODELS

There are two methods for producing paper-and-pencil models. The *manual* technique uses templates to produce standard symbols for line-and-box models. Models that resemble forms are produced by filling out pre-printed diagrams.

Many modeling techniques have no standard symbols are drawn by hand with pen and ink. Those that are essentially textual in nature are written or typed. These include pseudo-code, data dictionaries, HIPO diagrams, and structured language. Diagrams whose physical size increases rapidly with an increase in the complexity of the base require careful planning to find sufficient space. And forms are available with pre-drawn symbols that can be linked in standard ways.

The latter forms are similar in aim to **automated charting**. Most automated flowcharts, which have been in existence for over twenty years, actually work in reverse. That is, they take written code and produce flowcharts that should have been drawn previously. Of course, program flowcharts are not part of logical design *per se*. Other automated chartmakers have been integrated into packages. For instance, CPM diagrams are an inherent part of the Super Project Manager, available for microcomputers now. Intech's Excelerator is capable of integrating a number of charting techniques with a complete data dictionary and a word processor.

Automated tools are most valuable in producing two-dimensional flow diagrams, although the static bar-charting types (such as Gantt charts) are easier to implement. Most flow diagrams suffer from "spaghetti-ism," the tendency to become a mass of twisted lines — a large number of boxes connected by a large number of lines is quite difficult to plan manually. Most such charting efforts become multi-pass activities, with later passes cleaning up and simplifying the results of earlier ones. The computer can locate appropriate locations for boxes, simplify complex paths, and note simplifications. Figures 6-37 and 6-38 on the opposite page, illustrate a data flow diagram before and after simplification.

Other automated tools are available to maintain models that grow during logical design. Complex and multi-level flow diagrams such as system flowcharts, data flow diagrams, and Nassi-Schneiderman charts require numerous revisions, each having enormous implications, not only for the layout of symbols but for the logical design itself. It is simply not humanly possible to check all data-flow connections between all processes

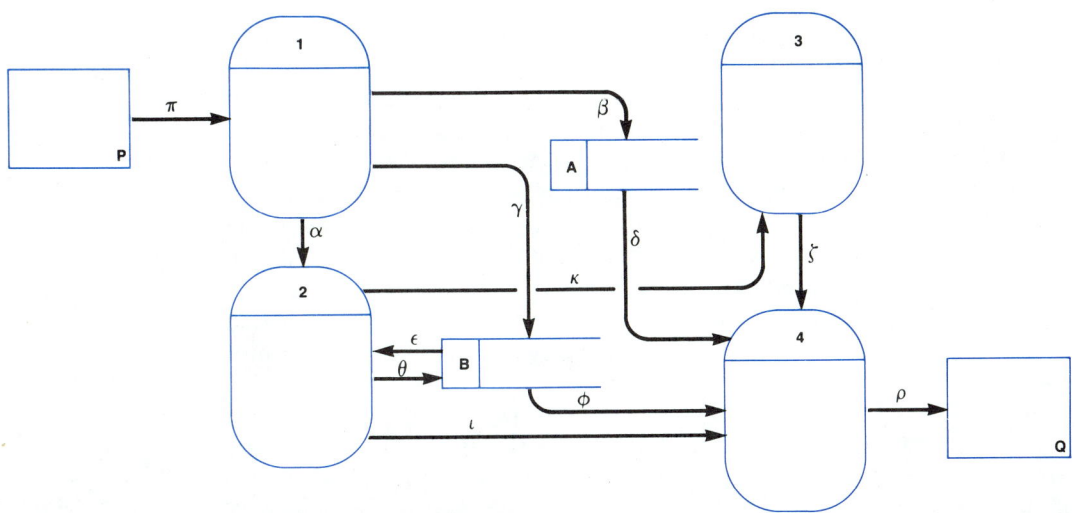

FIGURE 6–37. A Complex DFD

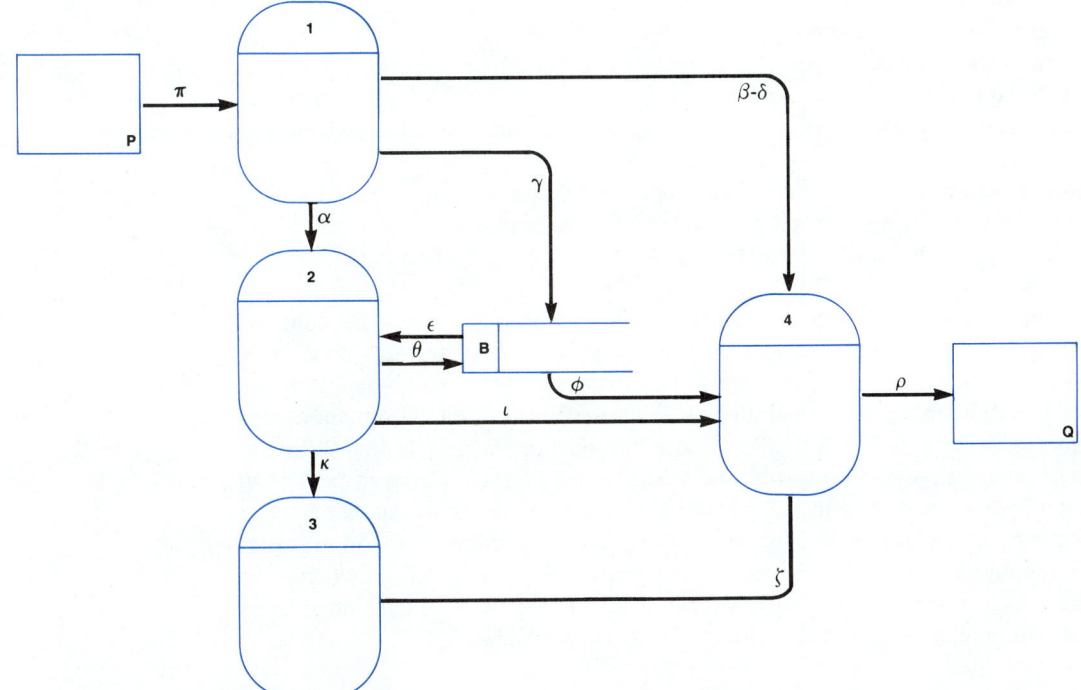

FIGURE 6–38. A Simplified DFD

in a data flow diagram in any real system without assistance of some sort. In fact, the standard arsenal of structured logical design includes an automated data dictionary, computer-assisted data flow diagrammers, computer-based HIPO charts, and automatic inclusion of decision charts and pseudocode. The computer maintains control over all links through the data dictionary. Some automated systems for design are built around this concept; most full-feature database development systems also provide these facilities.

Another valuable aspect of automated model production is that the resulting models are more easily manipulated and analyzed within the computer itself. The previously mentioned Super Project Manager utilizes CPM diagrams and computes the critical path for various situations, as well as maintains associated resource lists for each task. A Gantt chart can be processed into a PERT diagram with a little designer intervention, but without redrawing; a critical path can then be computed. Labor distribution charts can be derived by computation from Gantt charts.

Graphics tools assist in the development and display of the models. Such tools can be developed easily from off-the-shelf microcomputer graphics packages and can provide these facilities:

- automatic **highlighting** of critical paths, bottlenecks, illogical or incorrect drawings
- automatic subsetting and graph analysis producing **subgraphs** (that is, graphs of partitionable sub-systems: see Chapter 1 for a review of this concept)
- **word-processing** interfaces to provide labeling and detail
- **color graphics** for highlighting or increasing readability
- automatic **zoom** facilities for dynamically focusing on or stepping back from areas of the model
- standard **graphics interfaces** such as NAPLPS, to provide machine-readable formats for previously created models;
- **communication** facilities for distribution through a computer network
- **automated camera work**, including slide or film production
- inclusion of other graphic information or a data dictionary on-line

The word-processing interface can be made quite flexible, allowing for production of multiple versions of the model at various degrees of complexity and in various type faces or graphic formats, including color. Special direct interfaces to high-quality printing facilities greatly enhance the readability and impressiveness of the presentation.

One final advantage to an automated model production facility is the ability to interface mathematical programming with the model to test it. Such testing at the *logical* level is currently unsupported by theories. This is not the case, of course, at the physical or programming specification level, for which there exists a number of theories (structured programming, and Jackson design, for examples). As systems analysis begins to utilize its automated systems more often, we can expect to see theories developed here, theories that will help analysts test their logical designs for validity.

6.7   CONCLUSION

Producing the logical system design requires designing specifications that conform to the needs expressed in the investigation report. *System models* bridge the gap between the mass of data gathered in the investigation and the relatively compact and limited specifications. System models include charts, tables, diagrams, and structured text that express the collected data in restricted and systematic forms. In addition, they are the basis for creating the logical system specificaion and therefore must be compatible with the specification methods.

Modeling is the process of making a simplified version of the data. Models are limited by their **consistency**, **completeness**, and **validity**. They are useful because they aid in *communication*, provide a basis for *more systematic analysis*, can be used directly in the *design process*, and can be *more easily evaluated* than a mass of data. Models satisfy three needs in the design process: (1) the need to *communicate among individuals of diverse backgrounds and interests* during the development process, (2) the need to *translate vague and abstract terms* into more concrete and visualizable formats, and (3) the need to *move from what exists now* to what the design must accomplish.

Modeling consists of three distinct activities: **abstracting**, **analyzing**, and **representing**. Abstracting is the selection of pertinent data from the mass collected. Analysis makes succinct statements about the abstracted data, statements that evaluate, prioritize, and relate elements of the data to one another. Representation creates concrete forms (such as diagrams or charts) from the analyzed and abstracted data. These forms are the models themselves.

Modeling can be done manually, or it can be computerized (to an extent). There are over two dozen kinds of charts, graphs, tables, and structured text commonly used by systems analysts. These models can be grouped by appearance, content, and use. Choosing modeling tools is an important aspect in systems analysis work; recognizing the variety from which to choose is crucial.

The next chapter discusses the process of producing new systems or system elements. It outlines the principles of logical design.

In terms of a structured approach to systems analysis training, Chapter 6 has introduced the second step in the task of producing logical system designs. The creating of a system model follows the gathering of investigation data. Chapter 7 completes the process by introducing the interpretation phase, which results in logical systems specifications, the embodiment of the logical system design.

6.9 CASE: IN CASE OF FIRE

Harry Brighton works with an implementation team on a management information system for sales managers in the Warren Widget Company. The history of the company's development is checkered. Most of the original team has left WWC; only Harry is still around. Documentation is spotty. Harry's task for the next several weeks is to develop a training plan for sales managers who have to use the system, which is intended to supply the managers with a greater level of knowledge about the sales activities in each district. Some production information is also available, and there is a cute package to make sales forecasts with special commands. During development, several sales managers were involved in the design, as well as the VP of marketing.

All Harry has to go on are system flowcharts, some program flowcharts, mockups of screens, and a list of the commands that the system accepts. In addition, he has access to the data from the preliminary and detailed investigations, including several high-level data flow diagrams of the proposed system. And that is about it. Other documentation such as program listings and lists of bugs noted by the programmers will not be helpful. None of the managers has had extensive experience with computers, and Harry doubts that training will be easy. Besides, he has been told that training must take place within a three-week period prior to the release of the system and is limited to two hours a day, no more than two days a week. The training manuals must be complete and ready for the trainers. Harry is in a panic.

Questions

1. What is the root cause of the problem here? What *should* have been done already?
2. Given this situation, what kinds of models should Harry present in his training manual? Discuss the logic of his choices.
3. What can Harry *capitalize* on? What information will help Harry choose training materials for this particular audience?
4. In the training sessions, what kinds of models can Harry use that may not be adequate in a training *manual*?
5. Harry is participating after design is complete. Suppose he had had a chance to work during the design phase, with an eye toward constructing a training manual. (This is partly the answer to Question 1.) What kind of influence would Harry have had in the design phase? Which major role or roles of models would have helped Harry then? How would Harry's needs actually have *aided* system design then?

DISCUSSION QUESTIONS

1. Consider a proposed system for processing payroll records as a base system. Suppose there exist both a data flow diagram and a program flowchart as models of the proposed system. Suppose further that programmers produce (1) a COBOL program that processes payroll records and (2) a report generator package that produces reports. The COBOL program is based on the data flow diagram, and the report generator on the program flowchart. Discuss (1) which systems model other systems, and (2) what each of the seven properties of models can tell us about this situation.
2. A proposal exists to develop an automated order-entry system for a petroleum products firm. Analyst S has been asked to prepare some audio-visuals for a meeting at which the proposal will be introduced. Attending the meeting will be the division VP, two sales managers, the head of the order-entry department, the Finance VP, and the director of information systems. Discuss which models might best suit this audience and the process by which analyst S might select appropriate models to show.

3. What are the trade-offs among consistency, completeness, and validity? Which trait is easiest to verify? Which trait is most difficult? What procedures followed by the analyst carry the greatest threat to validity and completeness? What steps can the analyst take to improve validity and completeness?

4. Given that producing a system is a complex activity, do you consider the communication role of modeling one that is likely to be satisfactorily carried out under typical circumstances? In what situations will communication prove difficult? How can models improve communication? When will this be easiest?

5. Why is the point of view so crucial in system design? What are the costs of an "incorrect" point of view? What steps can be taken to ensure the "correct" point of view? Are there situations in which there might be no single correct point of view?

6. How can automated modeling techniques help the analyst handle problems in physical placement, size, form, and vocabulary for particular modeling tools?

7. In the three-step modeling process, what are the responsibilities of the following individuals: systems analyst? project librarian? manager of systems analysis? programmer? user?

8. Outline situations in which each of the four data reduction techniques (cataloging, categorizing, statistical characterization, case study exemplification) are just adequate.

9. When can the physical form of a model be critical? Describe a situation in which pencil-and-paper modeling is insufficient.

10. In Salvatore Die and Tool's manufacturing division, seventeen different major parts are used to put together nine products. Each product goes through six to nine manufacturing stations. At each station, the product is tested. If it passes the test, it is sent along to the next station. If it fails, it goes to a recycle station where a technician performs disassembly (if required) and fills out a form indicating where the product failed the test, what the apparent source of the difficulty is, and where the (perhaps disassembled) product is being rerouted. The product and its form are sent to the appropriate station or, if the recycle technician so judges, to a waste bin. When a product is tested at a station, a product tag is updated with the test results, the time and date, and the operator's initials. When the final station is passed, the product tag and any failure forms are removed and sent to production control for analysis. Each day, a production summary is written detailing and summarizing production volume, station volume, and testing results.

    Based on the description above, discuss the relative advantages and limitations of each of these modeling tools: data flow diagram; process flow diagram; materials flow diagram; decision table; organization chart; and the above natural language description. Refer to the five attributes of models introduced in the text.

11. Since is it possible to "program" decision tables and decision trees, in what sense can we say that each is equivalent to program flowcharts and BASIC programs? In what sense can we say that they are NOT equivalent? What are the advantages of each of these to a programmer who must write a COBOL program?

12. Consider Pascal, COBOL, Fortran, and BASIC. Imagine that a given program flowchart is used as a base system to "model" it as a program in each of these four languages. Study the specifications of these languages and discuss whether they are "equivalent" in terms of what they can model. What are the trade-offs among them in terms of completeness, consistency, and validity?

13. Since structure charts are asynchronic (Chapter 13) looks at symbology for sequencing in structure charts), is it true that they have no value in representing dynamic systems? What value do they have?

DESIGN EXERCISES
Using dBASE III

1. Start up the HAIR system as you did in Chapter 1, entering a control key of 546372819 and examine the first menu. Note the **options** available: ADD, UPDATE, QUERY, and MAINTAIN. Select the ADD option and add yourself to the CLIENTS file when the FILE SELECT menu comes up. Note the appearance of screen as you add data to this file. Note also that you are not really adding data to a file; you are answering questions posed to you by a program which THEN adds the data. After you've added a record, get out of HAIR by responding with ENTER to each request.

Here are the commands that dBase III uses to add and delete records:

DISPLAY RECORD RECNO( )	Displays the record you just entered. "RECNO( )" refers to the record number of the currently-examined record
DISPLAY ALL	Displays ALL the CLIENTS records
DELETE	Erases the record you just entered
DISPLAY ALL	Note that the record you deleted has an asterisk (*) after the record number. The record is **marked for deletion** when you ask to have the file PACKed. Until then, the record is still there.
RECALL ALL	Records marked for deletion are "unmarked"
DISPLAY ALL	Note that the record is unmarked
DELETE	Delete the record again and. . .
PACK	Note what happens now. . .
DISPLAY ALL	The record is GONE (for good!)

DELETE and RECALL use the same format as DISPLAY. When you APPEND records at the end of the file, you will have to reSORT the file if you want to put the current contents in order.

Now, as an exercise, hire yourself as a stylist (USE STYLISTS), put yourself back in as a client (USE CLIENTS), and delete product 1500 from PRODUCTS. Then, find out who has the same hair color as you do and which other stylists work the same days as you do (if any).

DESIGN EXERCISES
Using dBASE III *continued*

2. The text mentions three major activities in modeling: **abstract**, **analyze**, **represent**, and briefly introduces you to three classes of models: **flow, structure**, and **process**. Explore the HAIR system (DO NOT DELETE ANY RECORDS!), abstracting and analyzing your experience with the menus. Then, represent

1) how data flows among processes (as you determine them),
2) the structure of the menus, and
3) the processing of entering data.

You may select any models from the lists provided in the figures. It is best to work in pairs for this exercise.

CHAPTER 7

INTERPRETING: DESIGNING SYSTEMS AND THEIR ELEMENTS

OBJECTIVES

1. To distinguish logical and physical design in terms of goals, techniques, and products
2. To use the laws of information in a model-based logical design exercise
3. To employ the principles of input, output, process, data, and boundary design in a logical design exercise
4. To list the major decisions based on logical designs and discuss the criteria on which they are made
5. To describe the major links between logical and physical design, including input forms, output reports, data base, security, control, and processor design

INTERPRETING: DESIGNING SYSTEMS AND THEIR ELEMENTS

7

CHAPTER

7.0. WHAT DOES IT ALL MEAN, ANYWAY?

"What are you doing?" Abdullah Jackson demanded.

"Programming. What do you think?" replied Evelyn Donne, a programmer/analyst working for Abdullah, senior systems analyst. Evelyn had been employed by St. Cloud Industries' data processing division for only a week, but fresh out of Bigtown College with a degree in software engineering, she was eager to impress her boss. Now she was very confused. Hadn't she been hired to program?

"Listen, Eve. I'm sorry I yelled at you, but you can't program from those specifications."

"Why not? You gave them to me to program and that's what I'm doing."

"Eve, those are *logical* specifications. They tell only what functions need to be designed and programmed. They don't have anything you can program directly from. Look, [shuffles through some papers] here is some pseudocode, right?"

"Right." Evelyn saw nothing she couldn't program from.

"OK, what does 'DO Verify-Account-Code' say to you?"

"Well, I was going to ask you about that one. What is that, looking for a check digit or something? And what does the account code look like? Is it six or seven digits long? And the other thing is, what kind of message do I print out if the code is wrong, anyway?"

"Eve, that's why we call these logical specs. Those are questions that you will have to answer before you start programming. Look here. If you program in a six-digit code, then we're limited to only 99,999 account codes with one check digit. We've got to do some *physical* design work *before* you can program anything. It's based on these specs, but you can't code these subroutines until those questions are answered."

Evelyn cannot be faulted for wanting to make an impression, but she is only halfway through the design exercise and programming is not called for yet. Beginning with the analyst's *investigation* of a system, the analyst produces working models of the system which are scrutinized for enhancement, repair, development, or regeneration opportunities. Interpretation of models results in logical system specifications, describing in detail the functions that are to be carried out.

The **design exercise** is this interpretation. Chapter 7 looks at this in three ways. First, there is the contrast between logical design and physical design, corresponding to the difference between saying *what* you want and *how* you are going to get it. Next, model-based design capitalizes on the behavior of information under environmental constraints to subject the models to testing for the need for change. Finally, component-based design looks at five specific aspects of information flow and concentrates on designing inputs, outputs, processes, and data structures and reinforcing system boundaries.

7.1   THE DESIGN EXERCISE

Evelyn and Abdullah's misunderstanding is really over the definition of logical design and its products. Logical design produces a set of specifications, detailing which functions are to be built. These specifications are independent of the real world in the sense that they do not indicate how the functions are to be built or how they will actually work. We call them **implementation-independent** specifications.

The IRMaint design exercise is a two-stage affair. First, logical design details *what* is to be built. Then physical design reflects on those specifications in the real, operating environment and shows *how* they should be built and operated. The principles and goals of design that guide both phases are quite general, whether applied to an information system or to living room furniture.

There are two ways of proceeding with any design exercise. There is a **model-based approach**. Here, the designer produces models that are subject to a variety of tests and investigations. The other way to design is to utilize a **component-based approach**. Here, a single *general* model is used, and elements are designed around prototypical examples. In IRMaint, the model employed is derived from data processing: input, process, and output. This is augmented by the data component and boundary design. As you will see, the component-based approach can be employed with models as a kind of accounting system. Have all the inputs, outputs and processes been considered? Has the data been adequately specified? What boundary considerations have we missed?

Logical versus Physical Design

How do analysts distinguish **logical** and **physical** design so that they will know when the logical design is complete and physical design takes over? Why is this distinction so important?

Logical design comes before physical design simply because taking down and rebuilding physical systems that do not work is expensive. A physical design specifies how the components of an information resource actually perform in their environment. Because the environment is turbulent, it will usually change between the initial conception of the system and system installation. Therefore, decisions that commit to particular physical elements and relationships are postponed as long as possible.

Consider the following examples:

"Very early in our design we decided to go with dBaseIII™ as a programming tool. This decision necessitated using dBaseIII files with their limitations and the particular way in which dBaseIII writes reports. For example, we couldn't have any reports that ranked tactics by their average prognosis.

This necessitated our writing these programs in BASIC, which took a great deal of time and introduced another series of control files."

"We eliminated most theft-control schemes as unlikely or expensive. Because the firm couldn't afford to employ a clerk in the parts room, we decided that the most effective control would be *post-hoc* weekly reports on inventory. Of course, this wouldn't eliminate theft, but it would make it more likely that missing parts would be noted as missing."

"Most operators already know the control and destination codes for the manifests by shipper. We designed the system so that they could enter these directly on the forms. Only after we designed and printed the forms did we realize that the operators we'd spoken to were twenty-year veterans of the shipping department; new clerks had to look the codes up in a book. So we had to put an extra module in to allow them to look up the codes on-line. In effect, we had to build a dictionary — and a whole system to create and maintain the dictionary — after the system was released."

Here are three examples of the pitfalls of making physical design decisions during the logical design phase. In the first case, a decision to use a particular database manager severely limited logical design. The second example shows what happens to an inventory when arbitrary decisions on economic feasibility are made prior to looking at the logical possibilities. Finally, premature considerations about the power of the operators made it necessary to reprogram and enlarge an existing system.

Logical design is limited to implementation-independent decisions. We sense the limits of logical design when decisions require knowledge of the operating environment of the system being designed. In each of the above examples, a physical design decision was made: database and report structure; frequency of checking inventory and reporting on variances; and the availability of certain codes in the minds of operators. Logical design makes no particular assumptions about the operating environment, postponing the following considerations:

- timing and sequencing of operations
- data formats and layouts
- report and form layouts
- identity of the agent (person or equipment) performing particular processes
- implementation concerns, such as the programming language to be used
- life-cycle concerns: the payback period, the economic viability, the political visibility or exposure of the project; or the development expertise.

This eliminates a great deal of worry. But is the resulting design useful if these important factors are ignored?

Yes and no. The design will prove useful only if the physical designers **interpret** the design into a working physical form later, using the available tools. Naturally there is no guarantee at the early stages that any particular physical design will work for the best.

Consider process 2 (Register Part Movement) in Figure 7-1, a DFD for the parts control example. It seems that every time a service person removes a part from inventory to put into a vehicle being repaired, someone (probably a clerk) tracks the destination of that part at the source (the inventory shelves).

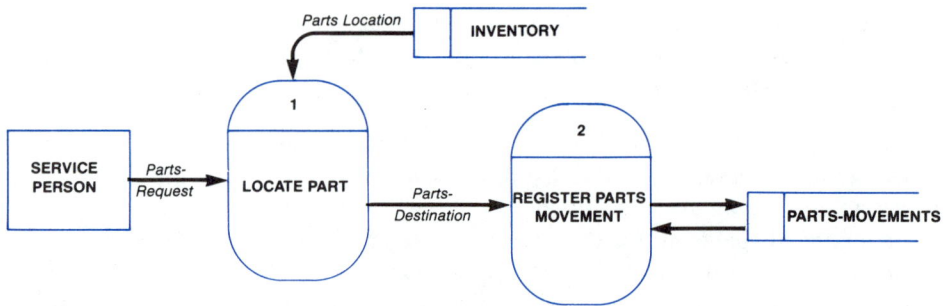

FIGURE 7–1. An Example of Overzealous Interpretation

But this is an illusion. Our minds automatically desire a concrete example of the function. In fact, there are a number of other ways this could be done. The service people could do it themselves. They could fill out parts-usage stubs as part of the job order and file them in a shoebox when they take parts from inventory. Or they could simply hand in the stubs to the manager, who will check them for reasonableness before preparing the final bill. Simply stated, there are a variety of physical implementations for the simple logical process of recording the movement of a part. These decisions should be made later, based on knowledge of the actual operating environment.

Principles and Goals of Design

This section outlines eight design principles common to all design exercises and important to logical design: parsimony, simplicity, structure, black-box, top-down, transportability, transparency, and conviviality.

Parsimony literally means "stinginess," but economy is a better term. A parsimonious design is one that is as small as possible. This could mean restricting the number of data elements, processes, and structure chart modules. Note that processes 1.5 through 1.8 in Figure 7-2 merely rearrange and total certain data elements *without the use of additional input and without producing additional output*. The design could be more parsimonious by merging these four processes.

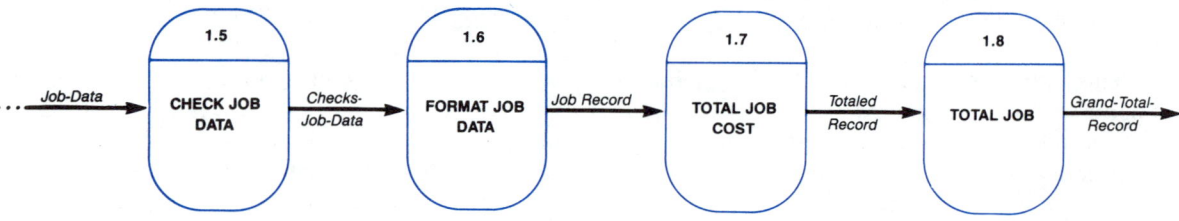

FIGURE 7–2. An Unparsimonious Design

Simplicity requires simple, rather than complex, expressions in the design. Usually the simplest design solution is the best. The chair illustrated in Figure 7-3 solves the problem of poor posture through a simple design that aligns the back, shoulders, and arms into a position appropriate to desk work. Other solutions are usually more complex, requiring headrests, back adjustments, cushions, and footrests.

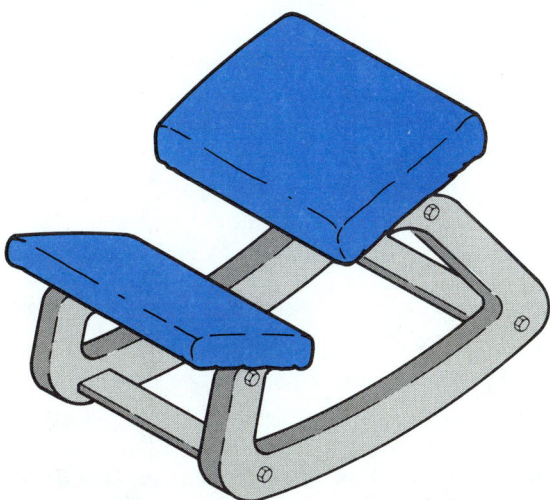

FIGURE 7–3. A Simple Design

Structure principles are well known but rarely voiced. They include dividing systems into functional subsystems that cohere and contain a limited number of elements. For example, a module labeled "Prepare Invoice" should prepare an invoice as its result, contain between two and nine processes, and do nothing unrelated to preparing invoices (such as preparing letters of credit). Figure 7-4 illustrates a module that violates this principle.

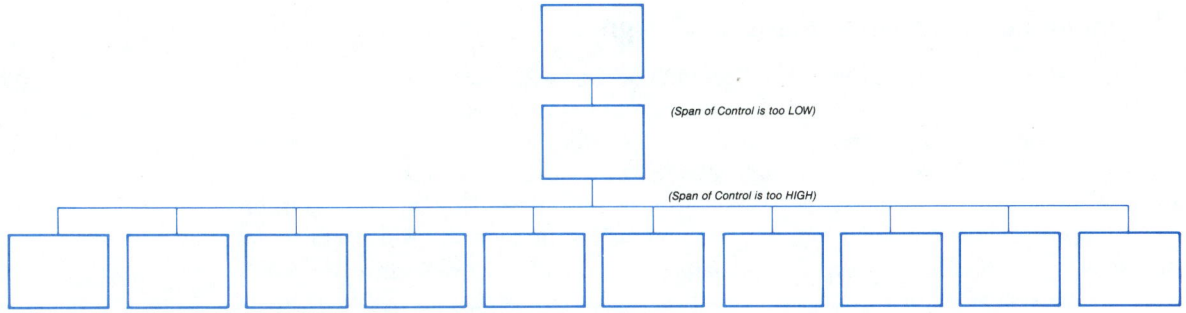

FIGURE 7–4. An Example of Poor Structuring

Black-box design means treating each information subsystem within a given design solely on the basis of its inputs and outputs. Nothing is assumed about its detailed workings, that could result in suboptimal designs. If we knew that the billing department also prepares shipping invoices, we would not design additional shipping invoice preparation into an order-management system. But we need designs that are impervious to changes in modules, since because they are *not* part of the design, they may change on a whim.

Top-down is the basic approach to structured systems analysis and is described in Chapter 4 in detail.

Transportability means that no assumptions about the operating environment go into the design. Thus, the system can be built to work in a variety of places. The use of pseudocode (Chapter 9) ultimately produces transportable software, which can be run on many different systems. The price of transportability is additional translation of the design according to the particular needs of the environment. These needs are not located deep within the design, where they are difficult to locate and change. Figure 7-5 illustrates the specifications of a screen that makes few references to the particular display device or form used. (See Chapter 13 for this kind of specification.)

Transparency means that operating procedures make no extra demands on the operator beyond those "apparent" from the system. The chair in figure 7-3 is not transparent, because it is hard to judge how to sit in it. A small manual is required for its proper use.

Much has been written recently about "user friendliness." A system can be "user friendly" without being the least bit friendly, and "friendly" systems can, in fact, be quite unusable. The better term is **conviviality**, which means a kind of joy or comfort in use. A convivial system is a pleasure to use. Menus, for example, are often seen as "user friendly" because they make selection of functions easier than typing in their names. However, many menus are poorly designed and hardly convivial. In fact, unless a system has an "expert" mode, menus often get in the way of experienced users.

These eight principles are good design guidelines to apply to logical designs. But they do not tell us what to design, only how to judge the design exercise. The next section looks at two ways to perform design: model-based and component-based.

Screen = Status-Line
Data (1–20)
Request-Line
[Menu-Choice]

FIGURE 7–5.
A "Transportable"
Data Design

Techniques of Information Resource Design

There are two major approaches to design. The first approach is building a model and testing it. A system is prototyped as a model and the model is successively refined through a number of tests. The use of models in *decision-support systems* also follows this approach. The other approach is **cumulative** rather than **holistic** in the sense that the design is constructed in chunks, rather than from a refined whole. This "component-based" approach is commonly employed in construction projects for which individual components have been established through experience. A house consists of a kitchen, a living room, a number of bedrooms, and so on. Each component can be worked on (relatively) independently of the others.

Model-based design is inherently top-down, since increments are based on the functioning of the entire model. Component-based design is not necessarily a bottom-up approach. The menu of components selected depends on accumulated expertise with "standard" systems. In IRMaint, the components are derived from a model of information processing that includes input, output, and processes, enlarged by including data and boundary design.

IRMaint employs both approaches. We construct and modify models generally, but we are guided in our model construction by the need to consider the five important components. Some texts on systems analysis rely on a long list of components (which may also include procedures, data communication, security, disaster recovery, database, and screen design as separate components). The huge variety of systems under development these days negates the applicability of most of any list of components to any particular design. The use of alternate development strategies such as prototyping and end-user-developed software means that many components come pre-packaged or built-in. Decision-support systems may not use *any* of the traditional components. Instead, we will use the five components mentioned as a kind of auditing procedure for logical designs, to make sure we have considered these five general areas.

7.2 MODEL-BASED LOGICAL DESIGN

Model-based logical design works like this:

1. The analyst investigates the current system and accumulates data on its current functions.
2. The analyst constructs an initial set of logical models of the current system.
3. These models are examined systematically for bottlenecks, logical contradictions, violations of principles of good design, and ineffectiveness.
4. The models are refined in light of these evaluations and subjected to futher examination.
5. When no further logical improvements can be uncovered, the models are then considered complete.

In carrying out this process, the analyst employs both **formal** as well as **practical** skills. Formal skills include an understanding of system dynamics, the laws of information, and how to construct models. Practical skills include knowledge of specific kinds of systems: inventory systems, accounts receivable systems, and query systems. In other words, the analyst has to know about systems in general and then interpret this knowledge through experience with the specific kind of system to be designed.

Using Information Laws

The information laws introduced in Chapter 1 are the guiding lights of logical design. Based on these laws, the analyst examines the logical models (we will consider structure charts and data flow diagrams for now), subjecting them to a series of tests:

1. **Law of Conservation of Information**: can each of the outputs from a given process be generated merely from the sum of the inputs to the process? Otherwise, the process is "intelligent" (see Figure 7-6).

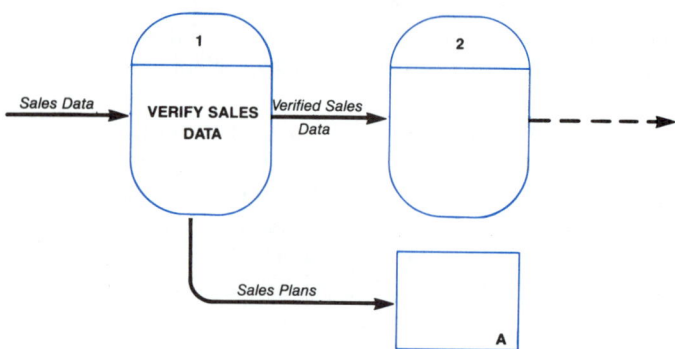

FIGURE 7–6. An "Intelligent" Process

2. **Law of Utilization of Information**: is each of the inputs to a given process necessary to generate at least one of the outputs from the process? If not, the input is unnecessary (Figure 7-7).

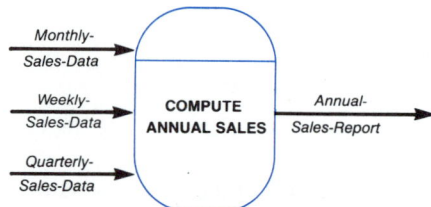

FIGURE 7–7. Some of These Inputs Are Unnecessary

3. **Law of Logical Data Flow**: is it necessary to assume that all of a process's outputs are available (complete) before all the inputs have been looked at? If so, these outputs do *not* depend on the inputs. You may be dealing with an intelligent process or an inefficient one that has to process unnecessary information (see Figure 7-8).

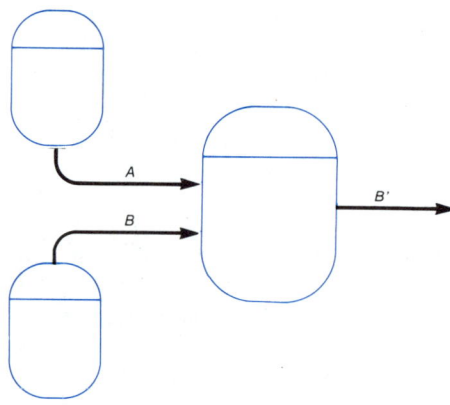

FIGURE 7–8. Violating the Law of Logical Data Flow (B' is Available Before A is Complete)

The principles of design also apply directly to models. **Parsimony** considers morphs like Figure 7-9 an extravagance, they should be included with previous morphs. **Simplicity** dictates that the complex design for inventory update in Figure 7-10 should be replaced by the simple one in Figure 7-11 on the following page. **Structural** principles limit DFDs to a small number of processes that are, in fact, inherently related to one another. **Transportability** means that we terminate logical design on a level at which detailed factual information about the specific operating environment is necessary.

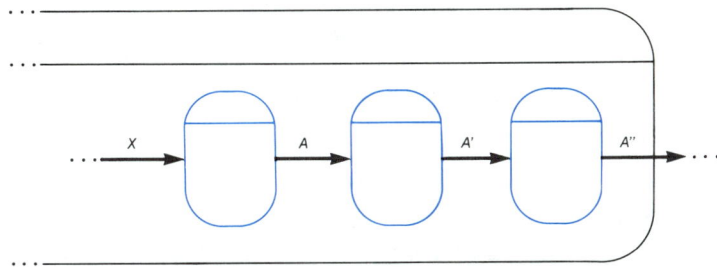

FIGURE 7–9. These Four Processes Are Actually only One Process

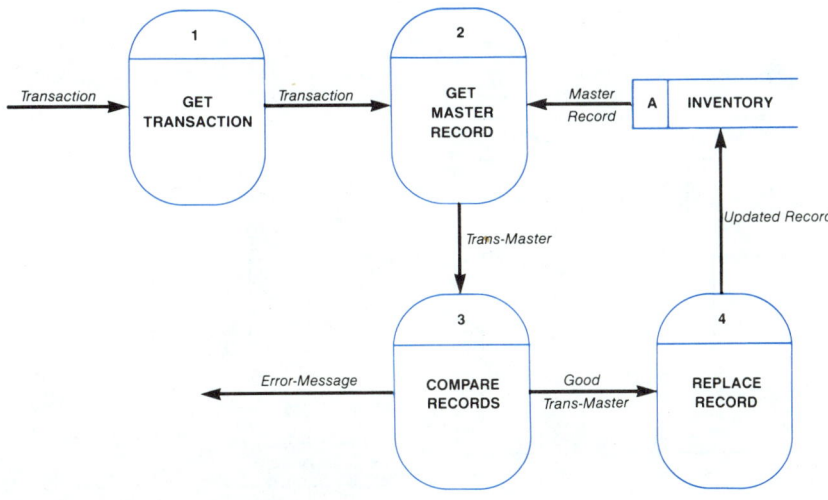

FIGURE 7–10. A Design with Complexities

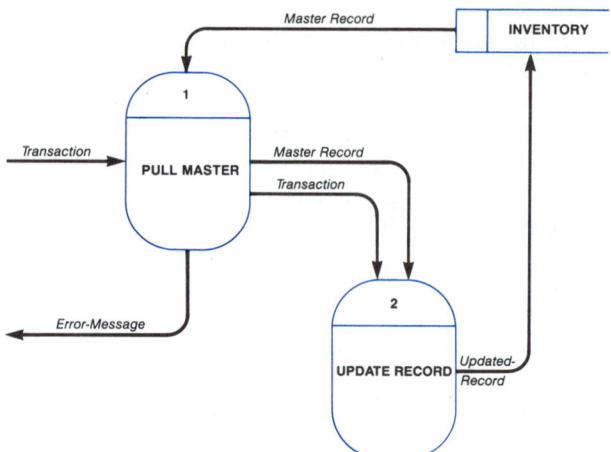

FIGURE 7–11. A Simpler Design for Figure 7–10

For practice, examine the DFD in Figure 7-12. Can you find seven examples of poor design? As a hint, consider the three laws of information flow and the principles of parsimony, simplicity, structure, and transportability.

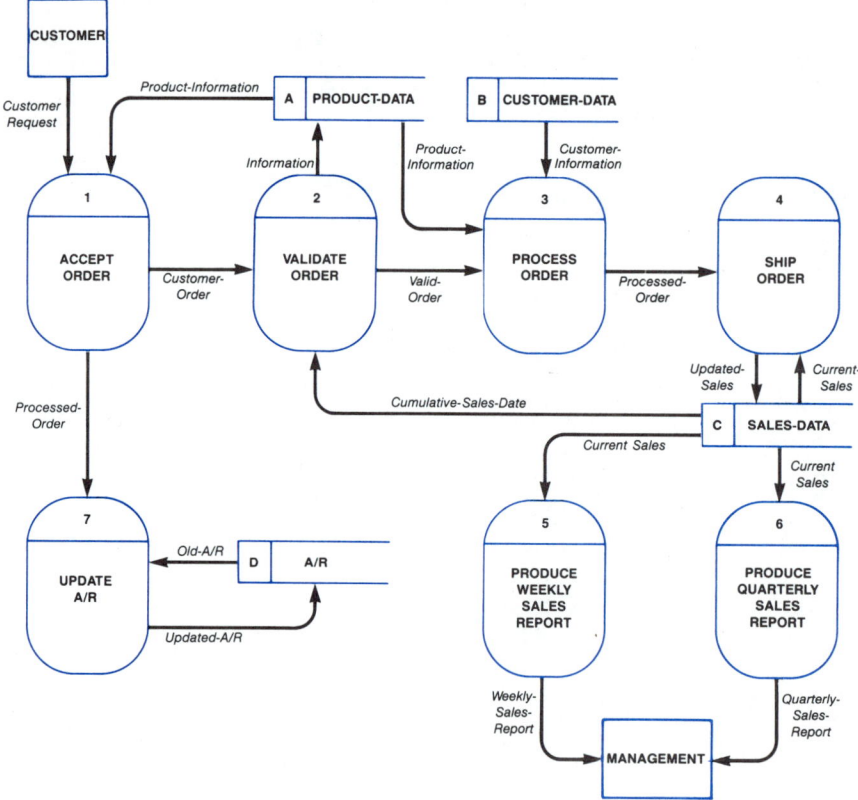

FIGURE 7–12. Find Seven Errors in this DFD

Interpreting Models to Designs

How do we create a design out of a model when a model is only part of the design? The model in Figure 7-13 is obviously supposed to be a model of an inventory system, but it surely does not *look* like an inventory system. What is missing?

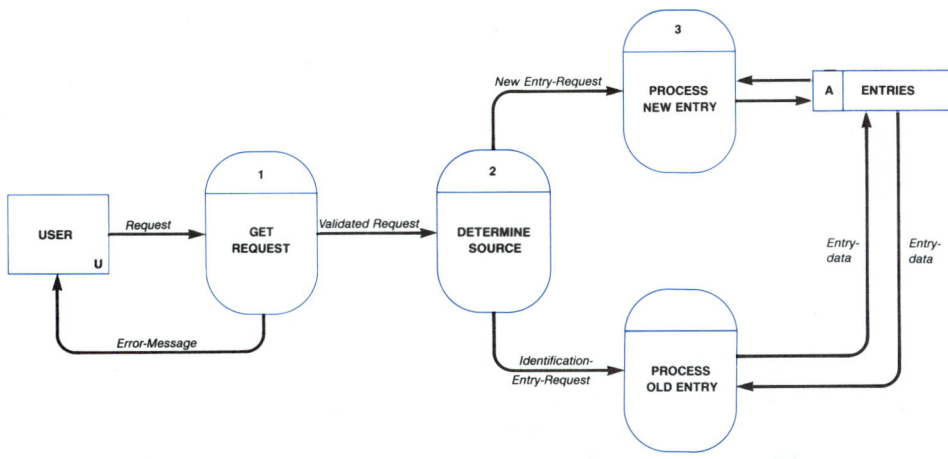

FIGURE 7–13. An Inventory System?

First, the terms used do not reflect the common idea of an inventory system. "Process New Entry" does not have the same ring as "Add Item Name." It is too generic and ambiguous. What is an entry? How do we know it is "new?"

Second, the system described does not differ from any other transaction-oriented system. (The generic morph is shown in Figure 7-14.) What makes an inventory system

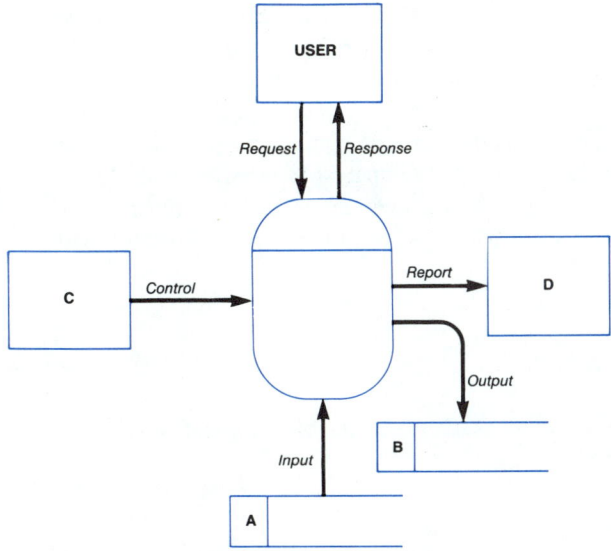

FIGURE 7–14. A Transaction-Processing Morph

an *inventory* system? The peculiar function of such a system is the tracking of goods from a central store. "Get transaction," "Process Transaction," and "Write Transaction Report" do not apply specifically to this tracking any more than they apply to a library circulation system. Most analysts know that there are specific kinds of transactions that distinguish an inventory system:

1. a new name entered into or removed from the list of items in the central store
2. an item (but not a name) removed from the central store (a sale)
3. an item (but not a name) added to the central store (stocking)
4. a report requested on inventory activity
5. an exceptional activity performed if a request cannot be satisfied (for example, a back order or receipt of the wrong item for stocking)

Logical design requires this sort of knowledge to subject the logical design to a second set of tests. These tests revolve about questions such as:

- Does the system handle each of the kinds of transactions that this type of system is supposed to handle?
- Can the system recover from the kinds of errors that such systems are likely to make (such as losing track of current stock levels)?
- Can the system counter the kinds of actions that its environments are known to make (such as overshipments)?

None of these questions requires detailed knowledge of the *actual* environment, however. Rather, they refer to *generic* situations. Overshipments are a way of life with inventory systems. However, the specific way in which overly large shipments arrive — size, frequency, day of the week — are deliberately ignored in logical design. No doubt it will affect physical design if such overshipments occur on a daily, rather than a yearly, basis. A complex procedure to correct overshipments is not feasible if it is used only once a year. The logical design merely specifies that such a procedure exists. The cost of not having one can be calculated later.

Therefore, the analyst's knowledge of specific systems comes into play during the second phase of model-based logical design. When the analyst is satisfied that (1) the design is logically sound according to design principles and (2) the design is functionally sound according to the analyst's experience with systems of this type, model-based logical design is complete.

7.3 COMPONENT-BASED LOGICAL DESIGN

Component-based logical design examines five aspects of a logical design for completeness (Figure 7-15 on the following page):

1. Input: are all inputs present, available, and accurate?
2. Output: are all desired outputs actually created when they are needed?

3. Data: are all data items defined and accounted for in a meaningful way?
4. Process: do the processes work, effectively producing the outputs from the inputs?
5. Boundary: is the system boundary secure from infiltration, loss of data, and error?

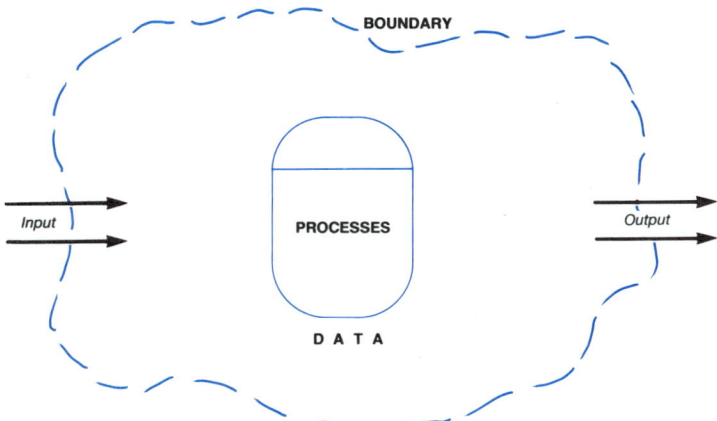

FIGURE 7–15. Component-Based Logical Design

Component-based logical design differs from the second phase of model-based design. There, the analyst looks at the type of system being designed to see that all *functional* components are present. Our version of component-based design looks at the overall design as a system and poses questions of an accounting nature: Do things add up? Is everything defined? Are all elements promised actually present?

Input Design

Input design concentrates on the content, but not the layout, of forms, screens, and other input matter, asking whether the items are available and accurate when needed. Input design questions include:

1. Is every input *effectively* available? That is, can the item be made available when needed for computation? If not, an operator may be tempted to make one up.
2. Is every input *accurately* available? If not, what kinds of inaccuracies usually appear?
3. Is every input available in the *structure* required? Sometimes a summary is needed rather than specific data, or only a sample is required. In other instances, the data needs to be made available by month, sales location, or type. Sometimes the data have to be presented in a sorted order by date, customer number, or customer name.

Input design is guided by the same principles that guide all design. Here, we pay special attention to parsimony (do not require more inputs than are absolutely necessary), simplicity (simple inputs are better than complex ones), and black-box design (we need not know precisely how the inputs look, or which codes or values actually appear). The result of good input design at logical stages is a list of inputs.

Output Design

Output design is more important than input design for the logical specifications, because our design technique is essentially top-down, or *output-based*. Output-based design requires looking first at the outputs required, and then working backward to the necessary inputs.

Outputs generally consist of the following "types" of data:

- reports
- data stores
- flags and indicators
- graphical displays
- tables
- completed forms

REPORTS

A text **report** often follows a structure like this:

- Title
- Summary
- Table of contents
- Body (introduction,…,conclusion)
- Supporting materials

The processing necessary to produce a report will (1) be identified by a title, (2) include a summary, (3) be indexed in some fashion, and (4) include a number of other, cross-referenced materials.

The body of a report is normally divided into chapters or sections (each identified by a title or header), and then into pages. This analysis may seem trivial, but it is a common failing to produce elegant and informative text in poorly organized volumes. Most modern word processors produce the required formats. Many also gather footnotes together and aid in indexing.

DATA STORES

Data stores (often called files) are merely collections of records. Reports are represented in DFDs as data stores, but the term is used here to mean simply a file, such as is often found in a file folder. Data store design is discussed below.

FLAGS AND INDICATORS

Flags and **indicators** are data flows that reflect the current status of a system during operation. A flag is actually data *about* data and is used to make decisions within processes about processing options. Was the previous operation successful? Does the data item pass a reasonableness test? Should the process send an error message back to the user?

A flag is simply a yes/no value, while an indicator is a value chosen from a fixed range:

- Data-OK: Is the data format correct? (flag)
- Valid-receipt: Was the receipt logged and calculated correctly? (flag)
- Invoice-found: Was the invoice located in the data store? (flag)
- Transaction-type: Which type (Add, Sale, New, Delete, Report) of transaction is this one? (indicator)
- Receivable-age: Which 30-day period does this receivable fall into, from 1 to 13 months? (indicator)

GRAPHICAL DISPLAYS

Graphical displays contain data that has to be interpreted by special output machinery to make sense. As display technology advances, the list of available displays grows. Line, full-color, three-dimensional, plotter, and musical graphics are only some of the more commonly available display technologies. Speech can also be included here.

TABLES

Tables are a particular problem in output, primarily because there are so many ways to produce them. In describing tables, however, we can isolate a number of factors that are of concern to the logical designer:

- table title
- data source (often identified in the title)
- date, place, and format of the table
- table margins
- column and row values (for two-dimensional tables)
- legends
- interpretation notes

A table can be stored as data; in Figure 7-16, the title and elements of date, place, and source are used as unique identifiers. Using sophisticated word processors, tables are often easy to integrate with text. Therefore, analysts may assume that formatting considerations can be ignored during logical design. Even the dimensionality of the table may be of no concern to the analyst, although five-dimensional tables (sales broken down by type, by item, by region, by year, and by month, for example) are beyond human comprehension.

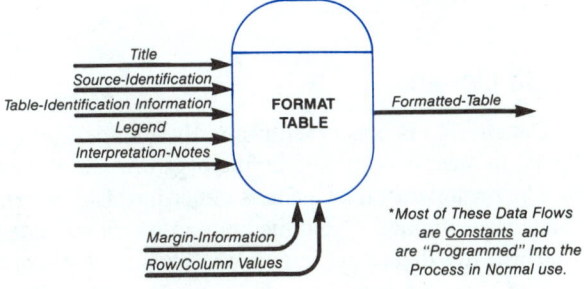

FIGURE 7-16. A Table

COMPLETED FORMS

A common output format is the **completed form**. A user may be asked to complete a blank form (Figure 7-17). In other instances, semi-completed forms are sent from

Name _____
Age _____
Occupation _____

Request Type
(Check One)

□A   □C   □E
□B   □D   □F

Date of Request

_____ / _____ /19 _____

FIGURE 7–17. A Blank Form

person to person. This will be indicated in a DFD, as in Figure 7-18.

Output design is *not* concerned with formatting, codes, and specific physical appearances. As illustrated above, analysts are concerned only with the type of output "document" and the data structures that make them up. The next section discusses data design in some detail.

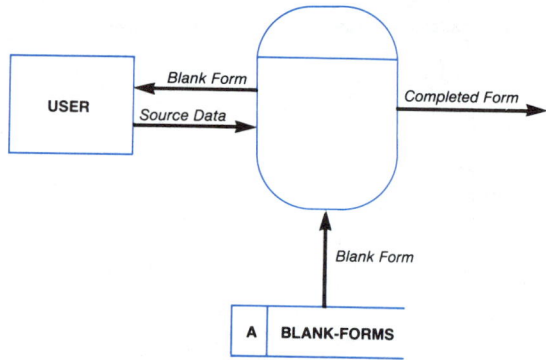

FIGURE 7–18. Morph for Form Completion

Data Design

Data design is concerned primarily with designing access to data within data stores. This is, in turn, conditioned on the organization of data within the store.

The **organization** of a file is either intrinsic or extrinsic. An **instrinsic** organization is one that is implied by the medium on which the data are stored. Thus, a book's access is intrinsically **direct** and a tape's is intrinsically **sequential**. Direct-access files can be accessed more or less "directly" on a record-by-record basis given the **address** of a particular record. On the other hand, sequential access files must be accessed in a fixed order: the twelfth record can be accessed only after the eleventh is read (see Figures 7-19 and 7-20).

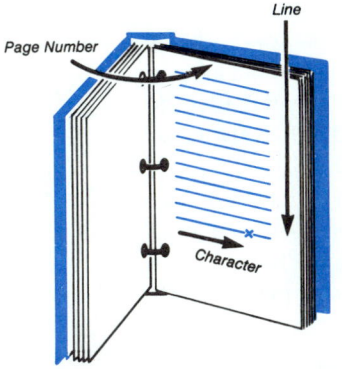

FIGURE 7–19. Direct Intrinsic Organization

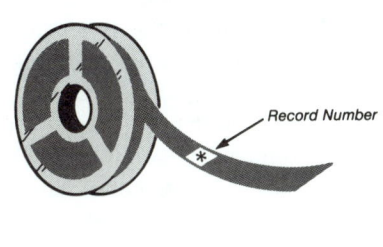

FIGURE 7–20. Sequential Intrinsic Organization

When records are accessed according to an algorithm or another process independent of the storage medium, its organization is **extrinsic**. Popular extrinsic organizations include random, indexed, tree-structured, indexed sequential, and linked list (Figures 7-21 through 7-25).

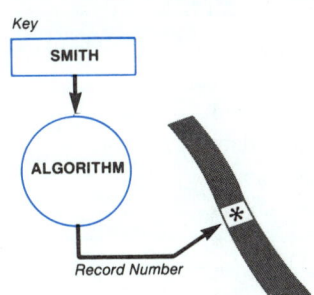

FIGURE 7–21. "Random" Access

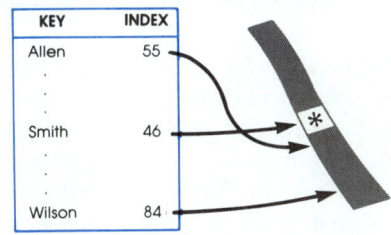

FIGURE 7–22. Indexed Organization

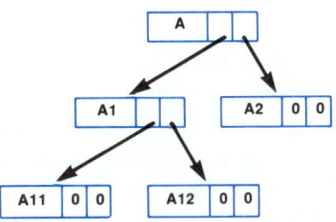

FIGURE 7–23. Tree Organization

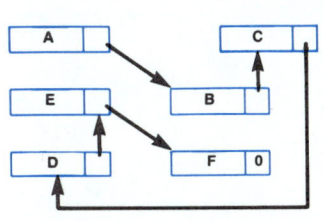

FIGURE 7–24. A Linked List

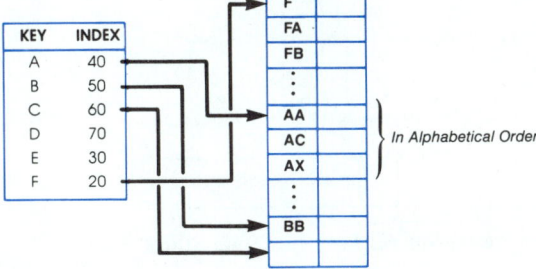

FIGURE 7–25. Indexed Sequential Organization

A **random** organization really has *no* organization. Records are located through an **algorithm** that computes the addresses of records from some key value and then uses that direct intrinsic address.

An **indexed** organization provides a **look-up table** for each data item key, enabling items can be stored anywhere, provided their addresses can be found later in the table.

An **indexed sequentially-organized** file is an indexed set of sequential files. It combines the features of the indexed file (random organization and relatively rapid access) with those of the sequential file (efficient use of storage media).

Each data item in a **linked list** has data embedded within it that can be interpreted as an address for random access to the next item of data. The linked list is an efficient user of storage devices, but housekeeping is time-consuming. A **tree structure** (Figure 7-26) is really a complex linked list, with each item having the address of one or more "next" items stored within itself.

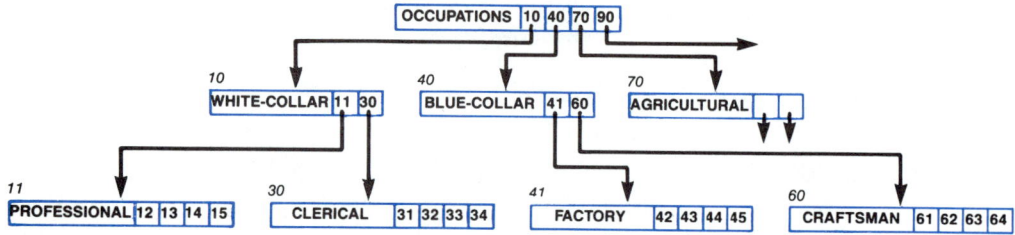

FIGURE 7–26. A Closer Look at a Tree (Numbers Indicate Direct Addresses of Records)

Data organizations are usually determined independent of the logical design. **Keyed** organizations (random, indexed, and indexed sequential), however, require that data elements and structures have **unique identifiers** or given non-unique identifiers, some way of solving the problem of which item to deliver. For data flow diagrams, this is often indicated as shown in Figure 7-27. Here, a key is represented as a data flow into a data store. The output from the data store depends on a computation based, in turn, on this key.

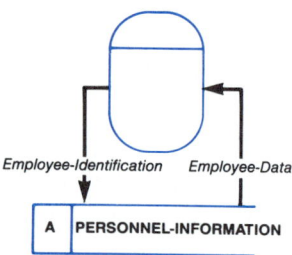

FIGURE 7–27. Using a Key to Access Data in a Data Store

Data design is concerned with determining how identifiers relate to structures of items, which relate to the items themselves. **Entity-relationship** charts (Chapter 13) spell out these relationships and assist in deriving data designs, given logical relationships.

In most organizations, for example, individuals have more than one skill. Several individuals work for one individual (called a supervisor or manager), who in turn works for other individuals. A job (identified by a job title) is defined in terms of the skills required for the job, while a position is defined as a job title plus the name of the person who holds the job (the incumbent). A department consists of a manager and a number of workers. These relationships determine a data design, as shown in Figure 7-28.

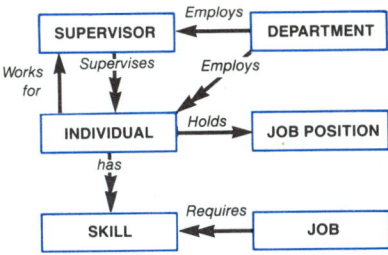

FIGURE 7–28. A Data Design

A person can hold only one position, but many persons can hold the same position. Position identifiers, not individual names, are unique keys for positions. Because a supervisor is unique to a department, the department may be located uniquely by supervisor name or department identifier. Skills belong to individuals, but not uniquely. An individual's skill profile may be found through his name or built up through an exhaustive list of skills. Given a skill, we can compute the names of all workers having those skills, and through these workers, the names of their managers, and hence departments that require particular skills. A department's skill profile can be computed by going through all the workers, checking their managers' names (and therefore checking to see if they are in the required department) and then computing a worker-by-worker skill list.

The principle of storing only information unique to an entity and computing other, related information based on those unique identifiers is the basis of **database management**. The data base itself contains only bare-bones information. All other information is computed through processes. In effect, "intelligence" is taken out of the data store. However, what we really accomplish here is even more interesting than that.

By removing the need to store the responses to all possible queries, we avoid the problem of anticipating all queries. Instead, the query results in a user-activated computation to retrieve specific information. Intelligence is therefore moved completely out of the system, making the system far more controllable and flexible. The data stores become "very dumb," because they are logically constructed without regard to detailed knowledge of the nature of queries that users may wish to apply later. Instead, processes "compute" the queries. This is the basis of what has been termed "**fourth-generation languages**" (4GLs), which allow users to "talk to a data base." Note, however, the need to build processes to interpret user needs during operation (Figure 7-29 on the following page). The advantage is that the *analyst* need not "compute" all possible queries before the system is designed.

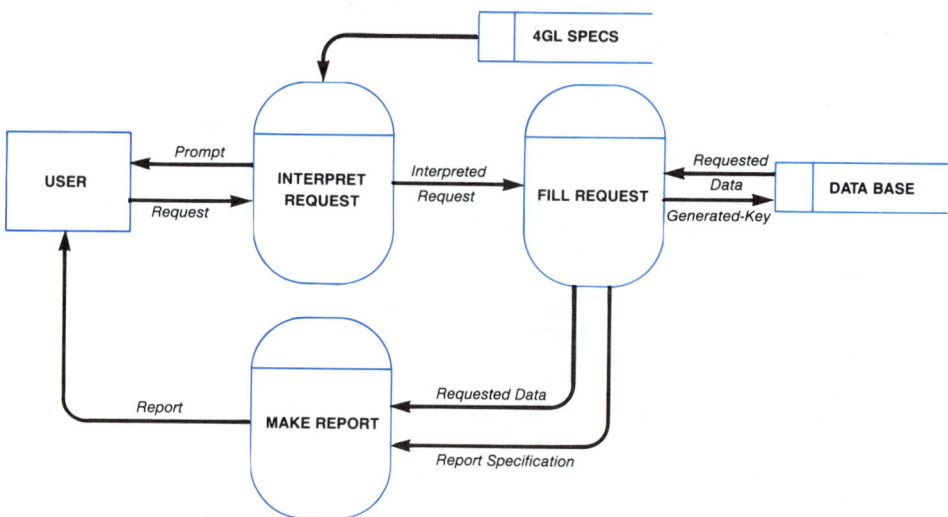

FIGURE 7–29. How a Fourth Generation Language (4GL) Can Work

At the beginning of this section, we noted that data design is concerned with designing access to data within data stores. This is partly determined by the organization of the data within those stores. Another consideration is that of **logical dependency** — designing data structures so that their elements somehow cohere logically and functionally.

This aspect of design is commonly termed **normalization**, the reduction of a data description to a "normal" or standard description. While normalization is a complex, often mathematical concept that goes beyond the scope of this text, a single example will illustrate how normalization works.

Consider a telephone list for a student organization. The list contains names, addresses, phone numbers, honorific (Ms., Mr.), year, department, office held (if any), and indication of interest in any of ten activities. We can say that with regard to a single student,

> student-information = = honorific + name + address + phone number
> + year + department + [office] + act-1 + act-2 + ... + ... + act-10.

(= = means "is defined as" and [...] indicates an option.)

Now, let's relate this design to logical data structures.

First, consider how you would find an individual student's data. Since there may be two John Smiths, the name is not unique. Since there may be several students residing together, the addresses and phone numbers are not unique either. In fact, there is no simple, unique way to find a particular John Smith's information. However, student ID numbers are unique. Therefore, we can add student-ID to the information and have a unique identifier for each record.

We indicate the fact that student-ID is the unique identifier by underlining it, as follows:

student-information = = student-ID + honorific + name + address + phone-number + year + department + [office] + act-1 + act-2 + ... + ... + act-10.

The next concern is that students may hold several offices over several years and we would like to save some of that history. Rather than provide room for an unknown number of offices, we simply create a new data store that contains records of students, years, and offices, removing office from the student-information and putting it in the new data store called office-holders:

student-information = = student-ID + honorific + name + address + phone-number + year + department + act-1 + act-2 + ... + act-10.

office-holders = = student-ID + year-held + office-held

The unique identifier for office-holders is year *plus* office, since in each year, only one student holds an office. (Let's ignore impeachment for now).

A third confusing aspect of the student information file is the appearance of ten activity interests. If we wanted to expand the list to twelve interests, we would have to redesign the file completely. It is far better to create a file that relates students to these activities, which might appear like this:

student-information = = student-ID + honorific + name + address + phone-number + year + department

office-holders = = student-ID + year-held + office-held

activities = = student-ID + activity-name

Here, the unique identifier is in fact the *entire* data item! If we want to find a particular student's activities, we will scan the list for all activities carrying the student's ID.

A final concern is yearly updates. Not every student passes enough courses to advance one year academically, so our data design requires that we query each student annually year to determine which year (freshman, sophomore, junior, senior) the student is in. The "year" in student-information does not depend on the student, but rather on some outside source of information. We would do better to eliminate year and replace it with some aspect unique to a student such as year-of-matriculation:

student-information = = student-ID + honorific + name + address + phone-number + matric-year + department

office-holders = = student-ID + year-held + office-held

activities = = student-ID + activity-name

Our data design is now complete and "normalized." Each record of data can be uniquely accessed by a key. Information on each student depends solely on the student, and if it changes, the entire set of files need not be redesigned. Certainly it will require some effort to compile a list of all activities for a student, but that is not really what we wanted. The activity list is useful to contact students to volunteer to help with activities. And while it may be interesting to know which offices John Smith held at Bigtown U., it is far more important to find out who was corresponding secretary in 1984, should an important letter go astray. Our **data design** now meets our **information access requirements**.

Be aware that our design method was a bit backward. We began with a list of "what we know about students;" however, most analysts begin with a list of "what we need to find out about students," beginning with the access requirements (from output design), working back to the list of specific data items required.

Process Design

The data items required consist of stored and computed data. **Process design** is concerned with laying down those computations necessary to meet output requirements. There are several aspects:

- determining what needs to be computed from static data that does not change and needs not be computed
- understanding the computational requirements (It makes no sense to compute something 400,000 times if it *never* changes.)
- determining whether the output can be computed effectively from the input in a finite amount of time
- designing layers of interpretation between users and data so that users can easily specify what they want

The first consideration has already been covered. It is easier to compute lists of activity preferences for a single student than to put up with the bother of redesigning the data base each time another activity appears. On the other hand, and as a second consideration, when data change infrequently, it may not really be worth recomputing, especially if demand is high. For example, if *resumés* of students holding office in the organization become important, then it may be worthwhile to re-sort the office-holding file by student-ID so that *resumés* of individual students can be produced quickly.

The third consideration is difficult to determine during the logical design period. We could easily define "selling-price" as the 200,000,000-th digit of the decimal expansion of *pi*, but the computation is just too hard! While some queries may expect a six-way breakdown of data, this table might contain thousands of entries, far more than anyone could successfully read and understand. Given the amount of data, 97 percent of the cells in the table might be empty, with half the remaining 3 percent merely estimates. The output cannot *effectively* be computed from the input.

The fourth consideration is one of the user-interface language. How will users actually specify what they want? Will they select from a menu? Will they have to use abbreviations? If so, which ones? Can they stop complex computations in mid-byte and look at some sort of meaningful partial results? Can they effectively create and store their own procedures, ask for their own abbreviations, design their own language?

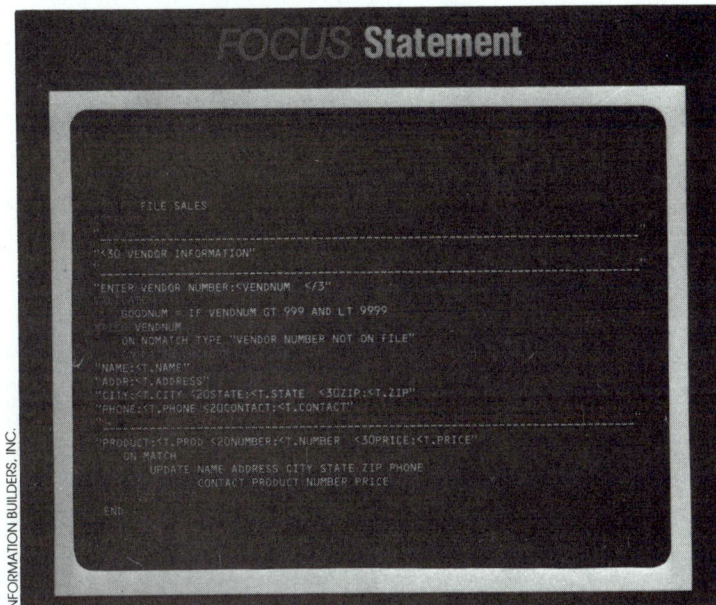

Focus™ is a 4GL

These questions border on physical design, but they are also logical questions. Is it possible for a user to request a computation that will result in an error? If so, what will the system do with the results? If the user gets "stuck," can the system proceed anyway? What will the user have "control" over? What will the system assume as defaults?

A major concern in process design is how to handle error conditions and security at the system boundary. These topics are discussed in the next section.

Boundary Design

Chapter 1 introduced the concept of "morph," or form. Many morphs are concerned with managing the **system boundary**, points of entrance or exit of data. Of primary concern in managing the boundary is system defense in two senses:

1. "Bad" data should not be allowed to enter the system
2. "Good" data should not be allowed to leave the system except to intended sinks.

Concern with **data integrity** begins with the quality of data brought into the system. Here we will be concerned with the following (examples are in parentheses):

1. range checks (ID numbers are all between 11111 and 99999.)
2. reasonableness checks (Withdrawals should be less than $1,000,000.)
3. format checks (Name is last-name, first-name, initial.)
4. presence checks (First name is present.)

5. batch checks (The total of all transactions equals the batch total previously computed.)
6. error checks (The check digit computed from the account code matches the last digit in the account code.)
7. size checks (The number of transactions entered matches the size estimate entered at the start of the session.)
8. access code checks (The user password was located in the password file under the user's ID number.)
9. access checks (This user is validated to access the accounting data.)
10. authority checks (This user is authorized to change the accounting data.)

Considerations 8 through 10 are not data checks, of course, but validity testing of the *source* of the data. Just as we are concerned that the data presented are error-free, so, too, are we anxious that the source of the data be pre-validated as an unlikely source of error. We are concerned here with **security** as much as data integrity.

On the output side of the system boundary, our concerns turn to access of data rather than errors, since the other components of design ensure accuracy:

1. Destination verification: is this the sink to which data are destined?
2. Audit: do we have a record of which data are going where?
3. Backup: if data are to leave the system, are copies readily available or computable?
4. Proliferation: if data are to be copied, how can we ensure that the data leaving the system do not re-enter it later after they are obsolete?

These are often difficult questions. Destination verification is often handled as part of a sign-on procedure; files of data cannot be moved out of the system without this verification and corresponding auditing of the copy or removal. System backup is usually routinely performed on a daily or weekly basis. The problem of proliferation is more difficult, since printed copies of data might not be dated, and updating of voluminous listings is a non-trivial task many users avoid. The use of a query-based system may only enlarge this problem.

A final boundary concern is the survival of the boundary itself. **Disaster planning** is still an art in most large organizations; smaller ones just hope disasters will not happen. The increase in the use of hard-disk-equipped microcomputers without a concommitant increase in availability of high-security file management systems leaves many users exposed to theft, fire, sabotage, or accidental loss of all their data. Reasonable boundary design includes provisions for frequent and routine backups and recoveries, graceful system take-down, and paper backup procedures.

7.4 DESIGN DECISIONS

System designs are the result of the interpretation of models with an eye toward their subsequent use in making decisions about system implementation.

Project Decisions

Logical system specifications are the major document on which go/no-go decisions are made following logical design. Specifications include recommendations for file design, boundary management (including user interfaces), and processing that physical designers use to estimate time and dollars for development effort. For example, if complex reports are necessary and a report generator is not available, it may be necessary to purchase one, increasing project costs and possibly delaying project completion dates.

Implementation Decisions

In the previous example, the need for complex reports dictated either the purchase of a report generator or laborious hand-coding of specific reports. This example shows how implementation decisions are forced by logical design. A complex data design may necessitate building a database manager. The need for *ad-hoc* queries may require construction of a query interface. Boundary requirements may mean utilizing a programming language that makes data validation easier, especially if unreliable sources are to be tapped. If security is going to be a problem, then hardware to encrypt transmitted data may be needed. These decisions reflect the dependence of physical design on logical design.

Installation and Operational Decisions

Logical specifications seem far removed from operational ones, but there is still a connection here. Most logical designs indirectly suggest sequence and timing, ultimately reflected in physical designs. Validation of input comes before computation; batches must be complete before they are processed; data tapes have to be identified and readied for mounting before the end-of-year run is begun. These come indirectly from the logical specifications. While a database administrator may rightly control the data element names and types used and available for operation through user queries, it is the logical specifications that show the relationship between query and response; query types unforeseen during logical design may not be computable during operation.

Decisions about Users

Even more removed are personnel and training considerations, yet even here there is a link. The functional structure of a system is often reflected in the high-level menus users see. The "menu-ability" of a system may depend on the care the logical designer takes. A reasonable functional design is easy for users to understand — after all, it is derived from the users' views of their information needs. An unnecessarily complex or technical design may result in an unconvivial system that is unintuitive or non-transparent.

User training and manuals reflect the functional design. Most poor manuals are based on physical designers' ideas of how systems or computers work. Uncomfortable training sessions oppress potential users with an overly-technical viewpoint. Users can take care of themselves if the apparent architecture looks familiar and is intuitive. Otherwise, users may simply refuse to try to understand, learn, or apply.

Approaches to Decision Making

The basic design principles already outlined also guide our decision making. Logical specifications go only so far before physical design takes over. Most important are principles of structure, simplicity, conviviality, and transparency.

Logical designs should be structured, and the structure should reflect functionality. The user should feel confident about what to do next from examining a list of functions. This implies that menu-driven systems are superior. While this is not always so (sometimes menus are annoying), as a *design* approach, asking "Will the user be able to select what to do next from a menu of choices?" is a good decision-making criterion. If the answer is "no," then the design must be reworked.

Consider the set of possibilities for processing a bank-loan application. At each moment in time, the user must know what to do next or at least know how to find out what to do next. The logical structure (Figure 7-30) should mirror the entire list of options available to the user at this point. They should be mutually exclusive (different) and exhaustive (all present). It falls to the physical designer to choose the words to make the distinctions effective.

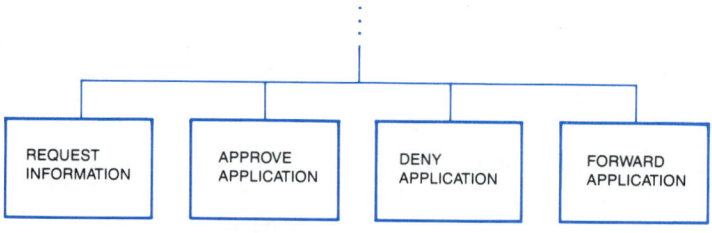

FIGURE 7–30. A Set of "Logical" Alternatives

We have already discussed simplicity, transparency, and conviviality in some detail. Again, the questions to be asked here suggest logical design decisions:

1. Is the purpose and function of the system at each point simple enough for the user to understand what is required to make the system meet its purpose? If not, there will surely be errors in operation.
2. To make this decision, do we (the logical designers) have to know something important about how the operational environment will work? If so, then we know that this is a decision for the physical designer.
3. Will the user feel comfortable with the function the system is performing?

A relatively simple, well-structured, transparent, and convivial system will exact fewer training hours, demand a smaller overhead in terms of catastrophic user-caused problems, need a smaller implementation team (in general), and meet with less resistance from users than one not meeting these criteria.

This approach to decision making lessens the burden on the physical designer, too. Because the physical designer bows to the will of operational considerations, the more the system meets the above-mentioned criteria, the less the physical designer has to design "around" the logical design to make it work. Figure 7-31 illustrates what happens when logical designers stumble in their task. The logical structure shows dependencies that can be resolved only at the time the system is actually operated; but since they must be resolved before then, the analyst will have to guess at the most likely situation. The distinct possibility that "likely" situations may not be likely in a turbulent environment makes the situation a gamble.

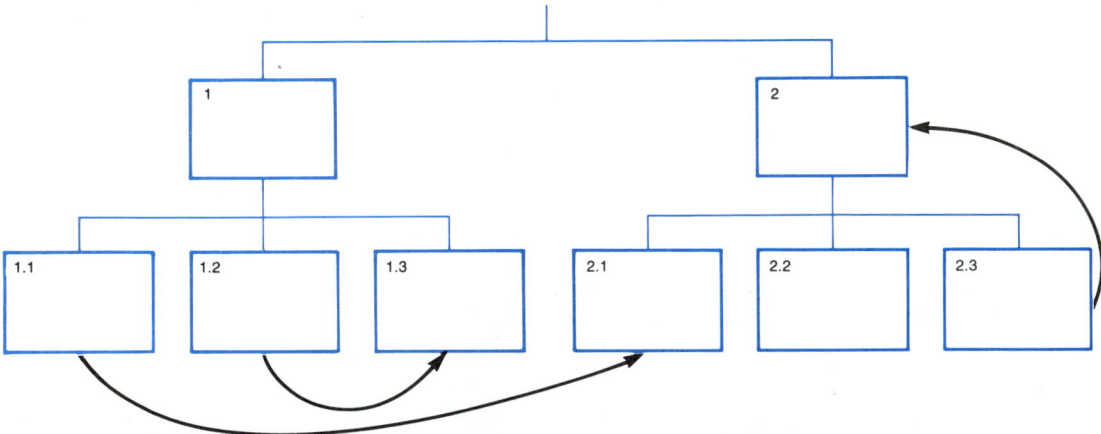

FIGURE 7–31. What Happens When Logical Design "Depends" upon Physical Design Considerations

7.5 LINKS TO PHYSICAL DESIGN

Physical design is to logical design as a blueprint is to a sketch. The blueprint shows *how* something will be put together and how it will work when connected. The sketch shows how various functions link up to meet the customer's needs. Five areas of physical design come directly from "sketches" the logical designer provides: input forms and screens, output reports, data bases, security and control procedures, and processors.

Input Form Design

Logical design specifies the sources of information for each process; data design shows the logical structure of that information. At each point where a data flow cuts the system boundary, the physical designer has to create a form or format.

Data come into a system on paper, a **form**, and through *smart* and *dumb* **terminals**. Where relatively *intelligent* terminals are employed, a **screen** simulates the paper form and the user fills in the blanks. With lower orders of intelligence, systems engage in conversations with their users on a line-by-line basis. In all three cases, the logical structure of the data should dictate the physical format.

Form and screen design are complex undertakings that are beyond the scope of this text. Principles of structure, simplicity, transparency, and conviviality guide the designer. Screens and form segments should logically "cohere" to meet structure requirements. Transparency dictates that completing the form or screen should be obvious. Simple codes are better than complex abbreviations. In a computerized system, the computer should assist in case of an error in range, type, reasonableness, and so on. Manual systems should have built-in checks, check digits, check sums, and so forth. The burden in form and screen completion should be low; this means that redundancy is a drawback unless errors are likely (and high levels of redundancy, such as requiring data to be copied several times, often lead to these errors). Figures 7-32 and 7-33 illustrate the relationship between the logical and the physical design of screens and forms.

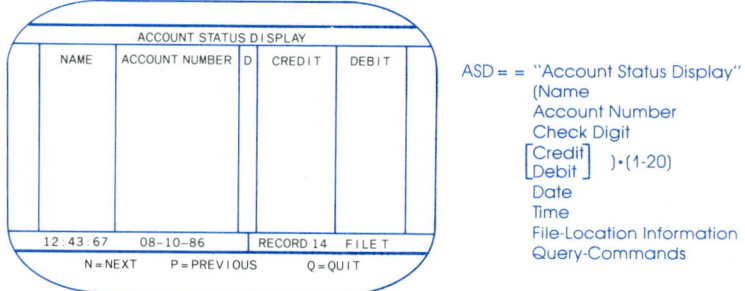

FIGURE 7–32. SCRFENS: Physical and Logical Designs

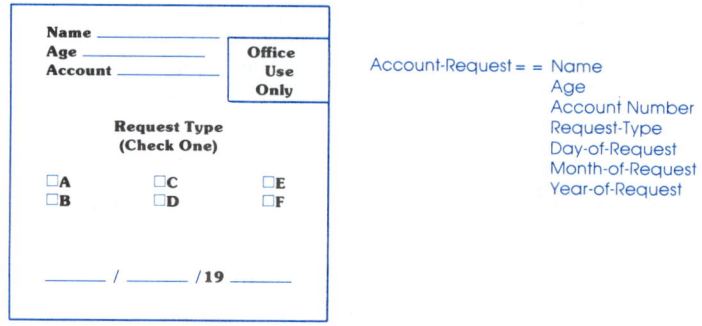

FIGURE 7–33. FORMS: Physical and Logical Designs

Dumb terminals actually "interview" those entering the data. The principles of interviewing help the physical designer create a **script** that is simple, structured, transparent, and convivial.

Report Design

Creating a report is simple; creating an informative and useful report is difficult. The logical specifications for a report are often sketchy, and, because they do not refer to the report medium (paper, screens, cards, etc.), fitting the data to the shoes of reality often requires mental shoehorning.

Consider a report that is to present sales figures broken down by region and product for each month of the year. Figure 7-34 (opposite) shows the logical specification of the report. The problem is fitting the report to a standard 132-column printout. If a standard report generator is available, the flexibility the physical designer has may be limited to a specific table format with standard page and column-heading formats.

```
Sales-Report = Month-Report (12)
Month-Report = Month + Region-Report (8)
Region-Report = Region + Product-Report (64)
Product-Report = Product-Code +
                 Unit-Sales +
                 Gross Sales
```

FIGURE 7–34. A Complete Logical Design Which Is Independent of Physical Media

On the other hand, if the report has to be hand-coded, the physical designer may decide to alter the logical requirements to fit the physical page. Twelve months plus a yearly total plus a 25-character product description may require 155 columns of information for the 132-column page. The need to paginate (avoid printing across page separators, number pages, and group functional data on separate pages) limits what can go on a printed page vertically.

Reports intended for screens may be even further inhibited by the common 25-line-by-80-character displays. The 25-line limit on a display introduces a requirement to scroll from screen to screen, and the logical design must be augmented with additional user communication and control functions ("Press C to continue, Q to quit, N for the next page, P for the previous page"). Standard screen managers can ease the design workload or make it impossible to produce functional screens for the output required.

Where more exotic output media are concerned, logical designs may ultimately prove completely infeasible. Cards sent to customers for reply may not contain enough room for the data to be printed. Mailing labels are small. Multi-part forms require special printers and raise project costs. These are important and difficult physical design decisions that need not concern the logical designer. On the other hand, experience may inform the logical designer about these considerations. In that event, the strict rules of logical design for transportability *might* be bent to avoid arguments, disappointment, or failure. There is no science here beyond the art of management.

Database Design

Because database design is properly a field of its own, no superficial commentary can suffice. Here, we will concentrate on the central roles of the database management system and the database administrator.

We have already hinted strongly that many physical design decisions are moot in a database environment. But what *is* a database environment?

Figure 7-35 on the following page, illustrates the components often found in a database environment. A **database manager**, usually a computer program, controls all access to data within application programs. This frees programmers from the necessity of defining data types and structures within programs. It also gives the database administrator a tool to enforce good database design principles.

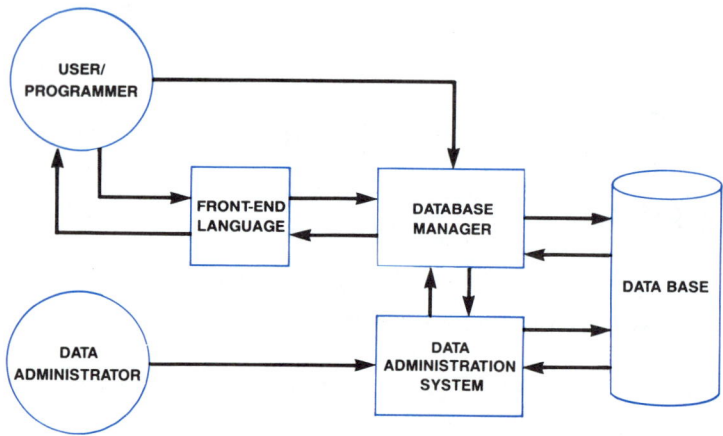

FIGURE 7–35. The Database Environment's Components

The **database administrator** is the data "czar" in the organization. This person controls definition of files, records, and fields by reviewing these definitions with analysts and programmers and amending the **corporate database** definition. The database administrator alone can change that structure. Because databases are forced to be normalized through administrative procedures, redundancy is reduced and the need to update programs with each physical system change is often eliminated.

The database environment also makes it possible for users to access data directly, without programmer intervention. Database management is the key to fourth generation languages (4GLs). Because users need not be concerned with either how data are defined or how or where they are stored, they can query the database in relatively natural ways, for example:

DISPLAY FIELDS EMPLOYEE-SALARY, EMPLOYEE-CLASS, EMPLOYEE-ID FOR EMPLOYEE-AGE > 50.

TABLE EMPLOYEE-SALARIES BY EMPLOYEE-CLASS AND EMPLOYEE-AGE.

COMPARE EMPLOYEE-SALARY BY EMPLOYEE-CLASS AND EMPLOYEE-AGE RANGES ARE (20/30/40/50/60).

A fancy database manager will allow users to save their queries (or the results) for reuse, editing, inclusion in a report, use in a word processor, or distribution to others through an electronic mail network. **Report generators** are important in this environment. They handle the bridge between the logical requirements (which are expressed in the three concerns mentioned in the section "Approaches to Decision Making), and the physical limitations of the system.

While database environments are almost invariably at least partly computerized, the concept of **data control** enforced by a data administrator can be used even in manual environments. One aspect of this, **forms management**, is an older discipline concerned with supervision of the content and format of a firm's forms. Forms often become outmoded or fall into disuse and need to be revised or discarded. A forms management group can maintain a smaller, better-designed repertoire of forms by applying data control and good design principles to form creation, deployment, and distribution.

A final aspect of database design includes the design or specification of the physical machinery to store and access data. We have already mentioned data organization, but this is independent of the physical data requirements implied by the mechanisms. Tapes are inherently sequential and updates generally run from tape-to-tape, rather than on a given tape. Disks and diskettes are inherently direct access. Information stored in books is inherently accessed in an indexed fashion (as is this book).

Most media have size and speed limitations. A diskette may hold only 360 KBytes or it may hold 1.2 M Bytes; a tape can hold between 10 and 60 million characters. Most business-employed microcomputers have two diskette drives, although hard disks storing between 20 and 40 million characters are common; in some cases these disks are **dismountable**. As technology progresses, the variety and range of capacities and speeds of storage devices will only increase. In a database environment, the physical designer does not have to know the optimum way to store data on Tuesday's devices given Monday's configuration.

Where database management techniques and systems are not employed, it falls to the physical designer to lay out files and estimate their sizes. As with the automatically managed systems, the physical layouts of files depend strongly on the kinds of access needed, which are determined through the logical specifications. For example, if student-information were defined as before, each record on a student would have to contain the information listed. In addition, an index would have to be created that matched a unique student-ID with an actual record address. If the student-ID could be converted through a computation into a record address, there would be no need for a separate index file. What are the trade-offs here? Does logical design depend at all on this decision?

Security and Control Design

As data flow into, through, and out of a system, boundary considerations become physical design problems:

- developing and maintaining a list of passwords and access codes and keeping these secure
- developing, maintaining, and reporting on a set of audit files to track activity on sensitive or critical files
- developing, maintaining, and testing disaster recovery plans, both automated and manual
- making data checking independent of fixed values guessed at before operation is begun
- developing easy-to-use, foolproof algorithms for maintaining the integrity of files
- developing and enforcing procedures (such as encryption) to ensure that data cannot be used if it is seen or obtained by those who should not see or have it.

Each of the problems listed above is actually the goal of a system that supports the system being designed. Thus the physical designer is concerned with a password-data system, an audit-data system, a disaster-recovery system, a data-about-data system, a data-integrity-data system, and an encryption system. These *may* be designed independent of any target system (for example, many database management systems allow the data administrator to specify range checks on input data and easily modify these through an on-line terminal), but in many cases the specific physical requirements of the system may necessitate a unique security and control system. For many complex, secure systems, it is common to find that the support systems compete heavily for resources. Among these resources are the processors, those elements that actually do the computation.

Processor Design

Processors can be thought of as computers — which they often are, although many are software or manual procedures. A processor is the physical equivalent of a logical process. Each logical process should be a processor when physical design is complete.

It may seem odd to consider hardware, software, and procedures to be in the same class, but each has the function of transforming data. In fact, the theory of computation, on which most software is constructed, states that each stored program actually simulates a hard-wired (*hardware*) device which has a specific function. The computer on which these stored programs work is called, not surprisingly, a "**stored program computer**." There is good historical reason for lumping hardware and software together.

On another level, hardware and software share a spectrum of facility, too. Encryption can be introduced either as a (tedious) manual or mechanical procedure, a (relatively more efficient) program, or a (very efficient) chip. The process may be logically identical for the three cases, but the ultimate processors differ tremendously. A human encryptor may be able to process one character per second; a program, perhaps 10,000 characters per second; and a specialized chip, maybe 1,000,000 characters per second. The decision as to which kind of processor to build depends on technology, money and time; the decision to *have* such a processor is made during logical design.

Processor design often involves the purchase of computing equipment such as microcomputers, front-end processors for telecommunications, minicomputers, mainframes, and automated record-management systems. Again, these decisions depend on technology, money, and time and go beyond logical design. Because logical design does not specify how often, how many, or how quickly information is transformed, estimating machinery needs cannot be based solely on logical design. Consider these situations:

1. A manager wonders if a microcomputer would help him better perform statistical analysis on sales data
2. Users want to know how rapidly they can expect to get back their daily update logs after they perform the updates
3. The director of MIS needs to know whether the existing mainframe can accommodate yet another on-line application without seriously degrading response time to the existing terminals

4. The warehouse is worried that an increase in paperwork to cut down on pilferage or untraced damage of goods would slow everything down to a crawl
5. The controller is afraid that a daily work-order summary report may tempt shop managers to overreact to short-term work load increases and decrease the overall efficiency of the plant.

These concerns about frequency, volume, and speed are based on considerations that the logical designers cannot anticipate accurately; they are, however, very real considerations during physical design. What if they make implementation impractical?

There are two alternatives. The first is to cancel the IRMaint activity. In that event, most of the effort to date will be lost, although the experience gained can be used in future similar proposals. Expenditures on logical design, however, are not high, perhaps 10 to 15 percent of the project budget. There is loss, but it is not as high as it would be if physical design decisions were made early, committed to, and then found to be impractical during implementation or installation.

The other alternative is to rework the logical design. Here, there are several alternatives:

1. Restrict the investigation boundary to those areas judged practical for implementation.
2. Redesign the logic of the system so that implementation will be more practical.
3. Recheck the logical design for complexities, functional loopholes, and poor structuring that might require a complex or unworkable physical design.

The first alternative merely looks for a "smaller," cheaper, and more easily-implemented design based on new knowledge of the implementation environment. It is a common solution to the problem. The second alternative is normally infeasible since logical design does not take implementation requirements into consideration in the first place. Nevertheless, in the real (as opposed to the theoretical) world of IRMaint, where teams rather than individuals work on projects, it is possible to *try* to work a logical design toward physical implementation.

The third approach assumes that the logical specifications are either inaccurate or incomplete, that the physical designers have had to "fill in" the sketches in an impractical way. Security and speed considerations may make some functions impractical in the real world. Reworking the specifications would reduce, for example, the need for security by reducing the number of functions; this would cut down on processing time as well.

7.6 CONCLUSION

This chapter discussed the third of the three major activities of the analyst: interpreting. Analysts interpret the models through their experiences to derive logical system specifications. These specifications are the basis for further design decisions.

The design exercise for IRMaint is a two-part operation. The first is called **logical design**, and the second, **physical design**. The principles of design include **parsimony**, **top down**, **conviviality**, **simplicity**, **structure**, **black-box design**, **transportability**, and **transparency**.

A design exercise is simultaneously **model-based** and **component-based**. Model-based design strives to construct a set of specifications based on system models, which are subject to critical scrutiny. Component-based design revolves around the major components of a data flow diagram, including **input**, **output**, **process logic**, **data structures**, and **system boundaries**.

The logical system specifications contribute to important design decisions that affect IRMaint work in the future. These include decisions on **project continuation**, **implementation into physical form**, **installation** and **operation**, and **hiring and training the users**.

The logical specifications are linked to physical design in a more direct way. Input design ultimately results in **input forms**; output design, in **report formats**; data design, in **data base creation**; boundary design, in **security and control mechanisms**; and process design, in the selection of **processors and hardware**. These, in turn, specify the five major physical components of any information resource: hardware, software, data, people, and procedures.

This chapter introduced the reader to the output side of the IRMaint effort: logical specifications. Chapter 10 continues with a detailed look at the content of those specifications. Chapter 14 examines the question of economic feasibility and exposes a number of ways of asking the question, "Is it worth doing?"

Chapters 8 through 10 discuss the *processes* of systems analysis. Chapter 8 looks closely at producing the preliminary investigation report. Chapter 9 focuses on procedure writing and techniques for creating models of those procedures. Chapter 10 details how to write and present logical system specifications.

7.7 CASE: LOGICALLY, THERE'S SOMETHING WRONG

Bigtown Lignite is a major producer of coal-based chemicals. For over three decades, Bigtown Lignite has been a leader in the application of high technology to production facilities, but they are lagging in the use of technology in the office. Since becoming director of administrative services, Sarah Cohen has been pushing to introduce computers into the front office, not only for accounting, but for general filing, correspondence, and routine administrative work such as reimbursement for travel expenses, and parking assignments.

Sarah contacted Harry Broadfoot, MIS director, about putting computers into the secretarial and clerical offices. Harry said that he, too, would like to see computer use extended to accounting, purchasing, and production, but that such a project would have to be approved by B.L.'s president. A pilot project was recommended. Within a few weeks, analysts and programmers were working away, implementing electronic-filing, word-processing, spreadsheeting, and forms-completion programs for the administrative assistants in the executive offices. Then came the bad news.

Candice Kowolski noticed it first, because she transcribes the travel requisitions. "This system you've described doesn't have any way to check on the values. The way it is now, I can tell if $500 is too much for a day in a hotel or if the person put in too little for the car rental." Edna Michaelson also saw a problem: "Listen, if I'm typing in a new parking assignment, I've got to see exactly where that space is going — and whom it's going next to. There are some very tender egos working here!" Gerry Marsdale spoke up with this comment: "Did you ever stop to think about how we're going to type correspondence without being able to look at past correspondence and compare it? Imagine the hassles if I forget to acknowledge a response in the next letter and say 'your letter of the 18th'!" And Mary McDougal pointed out a real problem for her: "I can't type. I mean, I can't type well. Are you going to send me to typing school?" Viona Kates was aghast: "You mean that anyone who gets on this system can work on anybody's correspondence? Well, you just try putting that one past the 'General' (the president). On the other hand, what if one of us is sick? I mean, just how *will* it work?"

Sarah continued hearing complaints, trying to sift the real problems from the others. As far as she could tell, most of the resistance came from secretaries used to working with well-understood paper forms. Resistance was one thing, but what about those other problems with terminals, typing, and checking?

Questions

1. What about this situation is tied up with the difference between logical and physical design? What obviously has not been done here or done poorly?
2. Which specific components have been isolated in these five complaints? Identify each and discuss the specific kind of consideration that has been ignored or poorly done.
3. Refer to the design principles outlined in this chapter. Which seem to have been violated here? What steps might have been taken to make sure that the problems that seem imminent will not occur?
4. What is wrong with the basic design involving the four subsystems of correspondence, forms completion, electronic filing, and spreadsheeting? Use a structure chart and try to complete it downward. What about extending it upward? What *would* be above these four subsystems in the structure chart?
5. Consider what you know now about how secretaries work with correspondence. What physical design considerations might make a given logical design unworkable? Are any of these operating in Bigtown Lignite's situation?
6. Before Sarah Cohen "signs on the dotted line" and accepts any logical design, what questions should she ask of the analyst? Compose such a list. Which are the most critical? How can she go about separating her *operational* concerns from her *functional* ones? Can the analyst help her do that?

DISCUSSION QUESTIONS

1. A consulting systems analyst is approached by a medium-size business that wishes to computerize its billing, accounts receivable, and accounts payable functions. What are the first things the analyst should do? What should the analyst *not* do?

2. Prepare a paragraph of guidelines for a novice analyst to follow to determine when design is "getting physical." Can you actually sum this up in one statement?

3. What are the *drawbacks* of implementation-independent design?

4. You have been asked by a friend to help design a procedure for cleaning house. What would each of the design principles mentioned in the section "Principles and Goals of Design" advise you in this particular case?

5. What are the relative advantages of model-based and component-based design? Would you ever use one to the exclusion of the other? Given your design for housecleaning in question 4, outline a model-based approach and a component-based approach. What do you have to know for a component-based approach that is far more apparent for designing an information resource?

6. You have been asked to design a system for maintaining the grades of students in large undergraduate classes. Discuss the specific components of this system, as listed the section "Component-Based Logical Design." List several design concerns for each component.

7. Continue with the design in question 6. List the required text reports, data stores, graphical displays, tables, and completed forms. Is this approach helpful? What are the dangers in this approach?

8. Data for the grading system include the following:

   • Student name
   • Student ID
   • Course name
   • Course number
   • Instructor name
   • Weights for each component of the grade
   • Evaluations on each component for each student
   • Total evaluation for course for each student
   • Evaluation-to-grade table for each course

   What sort of organization might be appropriate for this data base? (Do not immediately consider using a computer!)

9. Consider the grading system data for question 8. Design the data base for the system based on the principles of normalization mentioned in section "Data Design."

10. Grades are important and sensitive. What boundary considerations would you put forward for the grading system? Be specific as to the kinds of checks you would institute. Look to the output as well as the input side.

11. What decisions have you made that might influence later development in this project? What decisions concern structure, simplicity, transparency, and conviviality? What would be the costs of having to redo the design at these later stages?

12. Try your hand at physical design. Assume that the system has to run on a microcomputer system that uses a 25-line-by-80-character monochrome screen, an 80-character letter-quality (very slow, about 15 characters per second) printer, with data stored on floppy

diskettes, one per course per instructor. Design the following, being careful to listen to yourself make decisions as you design:

1. input forms for data collection (paper backup is required)
2. input screens for data capture
3. grade reports by course
4. grade reports by student within a course (based on queries from the keyboard)
5. grade reports by component within a course (based on a keyboard query)
6. security procedures for user verification, data validation, and backup and recovery.

13. If you have access to a microcomputer with a database manager, complete the physical design started in question 12 by laying out the database specifications.
14. In the section "Data Design," the text points out that it is desirable to remove "the need to store the responses to all possible queries." Why? Why is it valuable to have "dumb" data stores and remove intelligence from the system?
15. The section "Database Design" ends with a number of questions about the student information system previously defined. What are the tradeoffs in indexing a file as opposed to designing a computation that generates record addresses directly? What happens if logical design is made to depend on physical design in this case?

DESIGN EXERCISES
Using dBASE III

1 The files in HAIR are not just a collection of records randomly organized. As you discovered in Chapter 5, in most HAIR system files, each record in most of the files has a unique identifier. These identifiers are called "keys" and are used through index files. dBASE III makes it easy to create and use index files. To create an index file for the STYLISTS file, first USE it and then enter these commands:

INDEX ON STYLNAME TO STYLISTS	Creates an index in *alphabetical order* into a file called STYLISTS.NDX
SET INDEX TO STYLISTS	Tells dBASE III that when it tries to match records (see below), the file STYLISTS.NDX contains the keys. There can be any number of index files, but you can use only one at a time.

Now, if you list the file, it will come out sorted. (Try it.) There is a special command to find records using the index, rather than searching through all the records. Try this:

FIND "FRANK"	The record keyed to the STYLNAME "FRANK" is located, but not displayed. You have to do that.
FIND "MARY"	Mary does not work here...

DESIGN EXERCISES
Using dBASE III *continued*

We know that there is no sytlist with the name MARY (see note below) because we end up at the end of the file. If you enter this

    ? EOF()   Asking "Are we at end of file?"

dBASE responds with

    .T.   Meaning "Yes, it is true."

One thing to be aware of is how dBASE matches. When you bring dBASE up, it matches by prefix. This means that entering FIND "MARY" is met by the first record beginning with the letters M, A, R, and Y in that order. The following ALL match "MARY": "MARY," "MARYANN," "MARYLEE," and "MARYMARYMARY." HAIR normally runs with pre-fix matching. You can switch to exact matching, so that "MARY" will match only "MARY" and not "MARYLEE   " with this command:

    SET EXACT ON   Use OFF to restore prefix matching.

You can index a file on any field or set of fields. You can use a set of fields by indicating a concatenation or stringing together of the fields by a plus (+) sign. For example, you can index the client file like this:

    USE CLIENTS   Use the clients file
    INDEX ON NAME + PHONE   Keys are name followed by telephone number
      TO CLIENTS

    SET INDEX TO CLIENTS   Tells dBASE which index to use
    FIND "HUFF, KELLY   " + "558-1654"
                               The key is a 20-character name followed by an eight-
                               character phone number

Now find out the favorite stylists of the third and fifth clients (in alphabetical order). Note that by convention, HAIR names its index files with first names identical to the files being indexed. It is not required, but it is easy to remember.

DESIGN EXERCISES
Using dBASE III *continued*

2 Several other HAIR files should be indexed, according to the table below:

FILE	KEY1	KEY2	INDEX FILE
CLIENTS	NAME	PHONE	CLIENTS.NDX
STYLISTS	STYLNAME		STYLISTS.NDX
SKILLS	SKILL		SKILLS.NDX
SESSIONS	RCPTNO		SESSIONS.NDX
SERVICES	ACTIVITY		SERVICES.NDX
PRODUCTS	ITEMCODE		PRODUCTS.NDX
SUPPLIER	SUPPNAME		SUPPLIER.NDX

The files STSKILLS, RECEIPTS, and ACTSKILL are not indexed. Note that the index files have the same name as the files themselves. Although this is not necessary, it is convenient for our example. Create the index file by following this pattern:

USE xxxx	Examine the file xxxx
INDEX ON field1 [+ field2] TO xxxx	Tell dBASE III what the index field(s) is (are). Any number of index fields can be used so long as they match the index file's data (created by INDEX ON...)

Index these files knowing that in the HAIR system, index files have the same "first name" (xxxx) as data files (data files end in .DBF and index files end in .NDX). If you try to write over an existing file, dBASE normally requests verification and you normally respond with a Y (for "yes").

3 The design of this system can be examined, as the text explains, in two ways: through models or through components. Using the available information on file indexing and structure, the models you have already built, and the appearance and demands of the menus, describe the logical design. Is all information that is entered pertinent? Is all the information available from sources (clients and sales) or does it have to be "made up"?

4 Now look at the components. List
(1) the inputs required
(2) the outputs apparent from menus
(3) boundary conditions that you have detected.
Data has already been explored in some detail and process will be explored in Chapter 9. Given your role-plays, what (if anything) seems to be missing from these lists?

CHAPTER 8

PRODUCING PRELIMINARY INVESTIGATION REPORTS

OBJECTIVES

1. To describe the function of the Preliminary Investigation Report (PIR)
2. To produce a problem statement, scope statement, and project summary for a sample investigation
3. To distinguish the components of a needs analysis including organizational, information, and procedural needs
4. To be able to produce an outline for a preliminary investigation report
5. To be able to make a preliminary investigation presentation, including a recommendation for action based on data collected

PRODUCING PRELIMINARY INVESTIGATION REPORTS

8

CHAPTER

8.0 INTRODUCTION

"Officer, I tell you I've never seen such a frightening sight in my whole life. I mean there we were, eleven people all dying and there was no one who could save us!"

"Now calm down, madam. Just the facts. What exactly happenend?"

"Well you're going to find this really hard to believe. No one, I mean *no one*, had prepared us for this massacre. You can't imagine what it was like. It was beyond human comprehension!"

"Madam, just the facts, please."

"OK. Well, let me try to remember. Yes, we all got together at two in the conference room. Larry Hegeldinger, the consultant from Primex Data, was going to make a presentation to us. We're a steering committee, you know."

"Steering committee?"

"That's a group of people who oversee a project. You see, this project's going to take about two years and there are eleven of us from finance, personnel, and business services who read the reports and listen to the presentations. We picked Primex to do the logical design work based on a presentation over four months ago and today at two, Larry Hegeldinger was supposed to present the results of his preliminary investigation to us for our approval."

"Beg pardon, Madam. *We* are doing the investigation, not some consulting company."

"No, Officer West, a preliminary investigation on our organizational, informational, and procedural needs in the area of business services — that's typing, mailing, accounting, finance, publicity, and so forth. Larry's company gathered some data on our needs and was supposed to define the new information resource for us logically. We were to listen to his presentation and then either approve going ahead, modify his suggestions, or reject the project and cancel it."

"OK, so what happened?"

"Well, Larry began by outlining Primex's view of the problem. He said that we needed to computerize our offices and go heavily into office automation. He spent a while telling us what office automation has to offer, and that's when it happenend...."

"What, precisely, happened?"

"Nothing. Not a thing. You see, we had all died from listening to him tell us about office automation. I mean short-haul modems, local area networks, 2400-baud communication, store and forward messaging, and so forth. We were all *dead* in the water! We didn't understand a thing, nothing. Instead of defining the problem in *our* terms, telling us how big a project it might be, and detailing our needs to us in our own terms, he tried to teach us the technology and snow us with *his* deep knowledge of electronics. I don't think he spoke with a single manager here, never mind a secretary or a mail clerk. He just rambled on about bus structure, mailboxes, virtual communities and computer conferencing until we were incoherent. That's when I ran out and told Ms. Madigan that Larry was killing us softly with his voice...."

Officer West might swim his way past this logjam, but poor Larry is going to be sent down the river. It's obvious that neither Larry nor Primex has met the goals of a preliminary investigation. Not only did Larry kill his audience, but he failed to complete a *preliminary investigation*, substituting what was obviously a sales talk for a planning meeting.

This chapter looks at the **preliminary investigation** (Figure 8-1), the second major activity in the **system development life cycle**, following the initial contact with the client and preceding the detailed investigation. While, in many ways, preliminary and detailed investigations are similar, their goals are quite distinct, as the following sections detail.

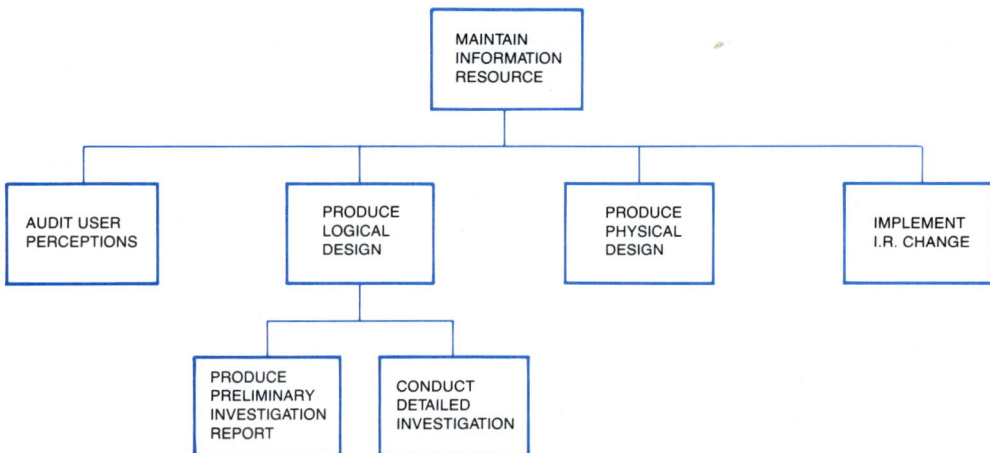

FIGURE 8–1. The Preliminary Investigation in the System Development Life Cycle

Chapter 5 contrasted the preliminary investigation with the feasibility study. The differences lay in (1) the emphasis on defining the problem or opportunity as opposed to proving the existence of one or more feasible solutions and (2) the top-down, goal-oriented, and analytic nature of the investigation contrasted with the bottom-up, product-oriented, synthetic nature of the feasibility study.

The result of the preliminary investigation is the **preliminary investigation report** (PIR), which requires determining system goals, determining the client's needs, defining the nature and scope of the problem or opportunity, and writing the report (see Figure 8-2). The necessary tools are basic investigation techniques and structural-functional logical design. The PIR serves three purposes: documentation of the assumptions, process, and results of the preliminary investigation; discussion of these points with clients; and agreement on the next steps. The report itself should cover (1) the problem statement, or what the investigation was about (This may be an opportunity or a true problem or obstacle); (2) the scope of the project to be undertaken to achieve the goals of solving the problem; (3) the limitations that have been discovered and assumptions that underlie the findings; (4) the needs which have been discovered through analysis of the problem; (5) recommendations for action, typically involving a more detailed investigation and logical design, or cancellation of the project at this stage; and (6) supporting documentation in the form of models, excerpts from interviews, project management forms and reports.

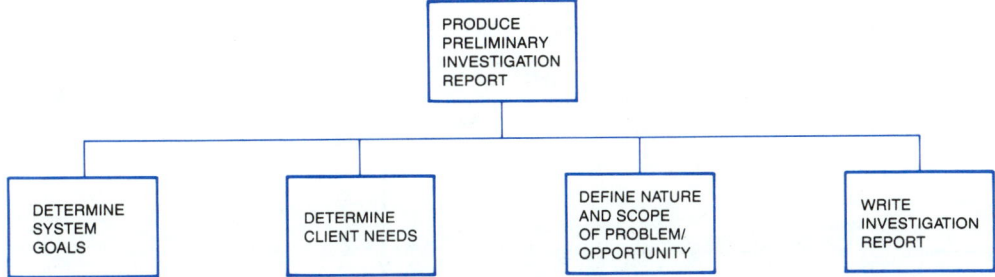

FIGURE 8–2. Goal Analysis of Preliminary Investigation Report Production

The needs analysis is crucial to the PIR. It concentrates on three separate areas. First, it looks at the needs of the client organization. These needs are derived directly from organizational goals. While many of these needs may not involve information resources, it is just as important to understand the goals of the client organization. Second, the needs analysis lists, categorizes, and prioritizes *information needs* and uses of the client organization. While a detailed look at the use of information is reserved for the detailed investigation, a functional analysis of information use at a high level is necessary at this stage. A discussion of the *procedural needs* of the client organization in its operational environment forms the third side of the needs triangle (Figure 8-3). Here, a discussion of how the client organization can use information to meet its goals brings the PIR to its "logical" conclusion.

The PIR also details the methodology of the preliminary investigation itself. For the sake of credibility, analysts should clarify just *how* they worked to draw their conclusions, including a discussion of data collection and analysis techniques and a candid exposure to the limitations of the data and conclusions. Clients need to know what risks they are taking when they base $1 million worth of work on a $40,000 preliminary investigation.

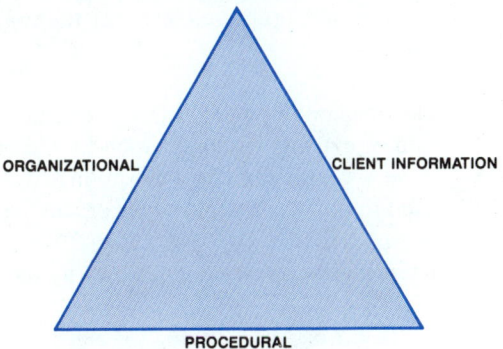

FIGURE 8–3. The Needs Triangle

The following sections provide the details of this introduction. We end this chapter with guidelines for making oral presentations. Most PIRs are presented in person and accompanied by at least a question-and-answer session.

8.1   WHAT THE PIR DOES

The preliminary investigation report serves three general purposes: documentation, discussion, and agreement. For each purpose, the PIR completes a contractual relationship between the analyst and the client to report on analysis activities. Thus, the report *documents* what the analyst has actually done, provides the basis for *discussion* of the analyst's findings, and gives the analyst and client a set of statements and recommendations about which to reach *agreement*. Figure 8-4 illustrates these relationships.

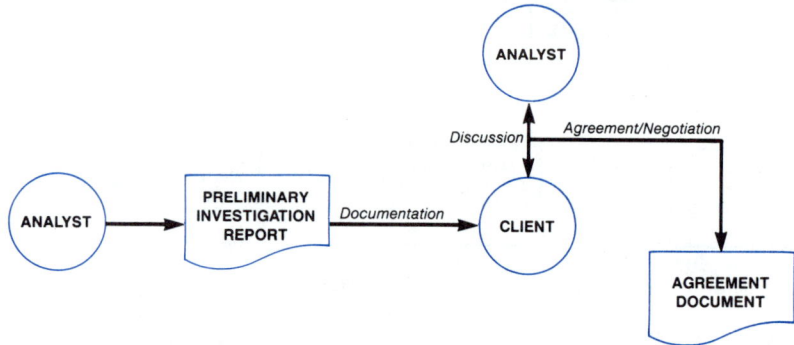

FIGURE 8–4. The Roles of the Preliminary Investigation Report

Documentation

The following excerpt from a PIR illustrates its **documentation** purposes.

> Our goals in performing the investigation were essentially derived from a need to understand the scope and function of the existing order-entry system. By this we mean (1) which people have some influence on how orders are entered and processed and (2) the major functions which comprise those processes.

> Our methodology consisted of one-hour prescheduled interviews; examination of existing documentation on the order-entry system, including a study done two years ago by a consulting firm; and some examination of the forms used in order entry and order processing.

The interviews were conducted as follows:

1. Oct 9. Ms. Van Horten, VP Sales
2. Oct 11. Mr. Yves Matson, Dept. Head, Order Entry

3. Oct 12. Mr. Willy O'Callaghan, Clerk, Order Entry
4. Oct 16. Ms. Tammy Fletcher, Purchasing Director, Elbow Enterprises (Customer)

Note that the analyst dispassionately lays down precisely what she did and the order in which it was done. In this case, she interviewed two managers, one clerk, and a customer and collected existing system procedure documentation. Her purpose in doing this is clearly spelled out in the first paragraph: to understand the scope and function of the existing order-entry system.

The reasons for documenting the process are these:

1. to present the basis on which data were gathered to increase credibility of the conclusions drawn
2. to dispell any fears that conclusions are drawn on any other basis
3. to illustrate the tentative nature of the conclusions (to show that they do, in fact, depend on the particular data collected)
4. to educate the client about the kinds of techniques analysts use to gather data

Discussion

The purpose of any document is to foster further action (unless the document is intended for the archives) on the **informed consent** of those agreeing to the action. Therefore, it is important that documents like the PIR provide ample opportunity for **discussion** among their readers.

Examine this excerpt:

Each interview was conducted in the following manner. First we contacted managers who referred us to their subordinates. We made it clear that managers should contact their employees before interviewing in order to explain who would be interviewing them and why cooperation was important. Our questions were taken from our major goals (previously mentioned) and were pretested on other analysts for clarity and wording.

One major conclusion is that the error rate increase as seen by the order-entry clerks is not as high as that reported by the sales managers. Because of the recent change in automating warehousing and because sales managers have recently been given responsibility for tracking these errors down, there is a higher *perception* of error. In fact, examination of correspondence concerning errors shows that there has been a pretty constant rate of error — about one in sixty orders has some kind of mistake. In view of these findings, we cannot conclude that there has been a *rise* in error rates.

On the other hand, we are not yet sure why errors do occur, since insufficient data were collected to shed light on this problem. We *have* isolated four possible sources of errors.

It is clear that controlling errors is important. However, making too much of errors seems to have damaged morale somewhat. Clerks are clearly upset by the large number of errors they seem to make in filling out the forms. Yet as we have stated, there is no reason to believe that error rates are increasing. In fact, in view of the recent jump in business, one might have expected error rates to rise and they have not.

Our recommendation is that we continue the detailed investigation, emphasizing not only error rates, but looking at processing speed and volume, also.

In this excerpt the PIR promotes discussion in the following ways:

1. **Terms of reference** are clearly defined. "Our goals in performing the investigation were..." "Each interview was conducted in the following manner..." "The following specific documents were examined..."
2. **Technical terms** are either avoided or defined.
3. **Conclusions** are stated as based on specific evidence using constructions such as "Because..." or "In view of..."
4. Areas of doubt are clearly spelled out, as in "We are not sure of why..." or "Insufficient data were collected to shed light on..."
5. All **sources of data** are clearly described, especially where conclusions are drawn based on analyst experience rather than actual data collected
6. **Conclusions** are all logically based on clearly-stated assumptions. Emotional appeals are avoided, although emotion may be a significant player in the data (as in "Clerks are clearly upset by the large number of errors they seem to make in filling out the forms")
7. The report is organized in this order: assumptions, data, conclusions, recommendations.

Agreement

Because the PIR is to be used as the basis for further action or inaction (if that is recommended), **agreement** becomes important, since the client is asked to follow the recommendations laid out by the analyst.

There are several kinds of recommendations an analyst can present:

1. Discontinue efforts at this time.
2. Postpone any work until a specific time or specific event.
3. Continue the work with redirected aims.
4. Continue the work with the original aims.
5. Enhance the effort by speeding it up, enlarging the scope, or raising the budget.

Usually the PIR will *define* the aims more clearly rather than redirect efforts, since it would be illogical to have collected information in other areas. Reasons for making other recommendations are:

1. Work may be *halted* if the existing information resource is actually performing well; it may be that the original stimulus for the PIR was spurious or reports were incorrect.
2. Work may proceed *later* if the technology required will not be available for some time, if funding will be delayed, if needed personnel will not be available until a later date, if the problems to be addressed are not yet serious enough to warrant repairs or improvements, or if IRU work schedules dictate a later continuation date.

    Work may be *expanded* (more staff, more dollars, wider scope) if the preliminary investigation has uncovered a greater need for work, a need for more extensive repairs, a greater expectation discovered among workers for enhancements than predicted, enhanced technological opportunities, or further management directives to widen the scope.

These have to be agreed on, which implies negotiation among the parties (the client, the analyst, the IRU) for continued work. Agreement is *not part of the PIR*; it follows the PIR. But the report should set the basis for agreement by spelling out the advantages and disadvantages of feasible courses of action. Analysts need to be clear in their presentation of alternatives, whatever they may be.

The following illustrates part of an analyst's recommendations. "If the current situation continues...," "An alternative is to wait until...," and "It is likely that additional work...." all spell out the possibilities.

> We see three possible alternatives for future work. First, we see that the status quo can remain. That is, the existing manual system can be operated as is. If the current situation continues, however, a great deal of work will have to be done to improve morale in the Order Entry department to counter the attitude that they cause a lot of errors, which is, in fact, not really true. In addition, procedures need to be written down and made available as part of training, even if the current situation remains.

> An alternative is to wait until errors show a measured rise (which seems inevitable as the volume increases, given the policy of keeping staff levels as low as possible). We can propose a number of objective, reliable measures of error rate which can be used to signal the need for some improvement. In addition we can prepare the documentation necessary for this approach.

> However, neither approach is forward-looking or proactive in the sense of anticipating and countering problems before they occur. It is likely that additional work is necessary detailing system functions to locate all sources of error (i.e., to validate the list provided previously) in order to redesign procedures to catch these errors in the earliest possible stages of data processing. This work can be done within one month, well before any real pressing error situation will arise and well before seasonal peaks in order processing.

The following is an appropriate scope statement:

> The study will examine clerical functions (order taking and forms handling) in the order-taking, billing, and warehousing departments, concentrating on manual subsystems as they now exist, with some projections made to the mid-range future (2-4 years). We will be speaking with clerks in these departments, some sales managers, and a few customers.

The **limitations** of a preliminary investigation generally involve permissions, time, money, and staff. For example, the analyst may have permission to interview only one clerk or interview only on Fridays. He may not have permission to speak with customers. Time and expense limitations are obvious. Staffing limitations are generally placed by the IRU.

Other limitations refer to the outcomes expected from the *whole* IRMaint exercise. Consider the system we have used as an example. Management may be quite wary of further computerization and may make that a limitation. They may restrict recommendations to areas of proceduralization rather than working environment or staffing. There may be stringent limits on total expenditure or time to implement. Existing computer equipment may be pre-specified as required or there may be instructions not to disturb existing manual systems.

The following illustrates some typical limitations on the preliminary investigation and subsequent IRMaint work:

> The preliminary investigation was limited to fifteen working days and the interviewing to Fridays only because of the pressure of work on the clerks. Also, because of illness, only one analyst could be spared for the work in this period. The clerical supervisor was also interviewed, as requested by the order department manager. Because the billing department catches most of the errors, we had hoped that several billing clerks could be interviewed; however, this proved impossible in this time period, as vacations severely limited billing clerks' availability. Only the billing supervisor was interviewed.

> The project budget is tentatively set at $4,500, and no computerization is foreseen. Existing manual systems are to be overhauled with special attention paid to keeping personnel needs at or below present levels. Emphasis is to be placed on streamlining processing of orders and catching order errors while the customer is still on the telephone.

These stringent limitations may ultimately cripple any development work (the detailed data will shed light on that), but they must be clearly spelled out.

The Needs Document

Based on goal analysis, the **needs document** clarifies organizational, informational, and procedural needs. In other words, the document spells out what the organization's goals are, what role information plays in meeting those goals, and what procedures need to be followed to achieve the goals using the information. A *detailed* needs analysis comes from the detailed investigation that follows, but the PIR begins this analysis.

It is important that the needs document distinguish among **perceptions**, **opinions**, and **requirements**. In fact, some systems development methodologies call this phase *requirements* analysis to clarify that distinction. People may perceive one need in quite concrete terms ("I need more help around here," "What we need is a computer," and "I need to get rid of these errors"), but these perceptions may have little to do with their actual job requirements. For instance, the feeling that staffing needs to be increased may arise more out of a psychological need to have a crowd around than from a true analysis of productivity. What the third statement means is that the manager feels beset by the problems that errors entail — if no one cared about the errors, then those problems would go away and there would be no need.

Opinions are the bane of analysts. Everyone is, of course, entitled to an opinion. But what academics call "untutored opinions" (statements of fact arrived at without any study or thought) are useful only in determining the psychological climate of an organization. *Attitudes* are statements of evaluation, such as "I don't like having so many people working on simple projects." *Beliefs* are statements of truth taken on faith, such as "We have to support management in everything they do." These further complicate needs analysis.

The purpose of the needs document is to focus attention on the structure of the functions that have to be performed for the organization to meet its goals. In other words, a clear identification of what *must* be done, uncontaminated by what people say they *want*, is necessary. *Wish lists are important* and have a role in needs analysis; however, they are *not* needs. A wish list represents one's psychological state modified by his knowledge of solutions to those needs, further mediated by pressures in the environment (see Figure 8-6). *They are not system needs!* Justifying individual psychological needs with the cold, hard facts of system needs is difficult. Analysts should not ignore these psychological needs, but they do not form the core of the needs document.

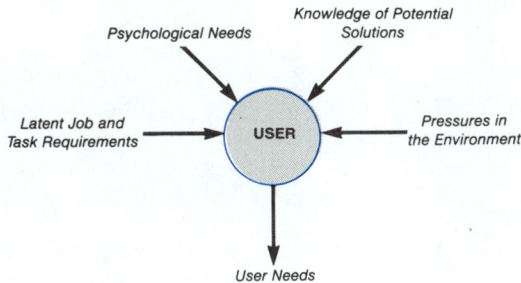

FIGURE 8–6. Where User Needs Come From

Recommendations

In a feasibility study, **recommendations** bear on the feasibility of preliminary system designs. A PIR makes recommendations *only* as to whether further study is necessary. In other words, the recommendations made by the PIR refer only to the work of the analyst in completing logical design. The decision of readers of the PIR is: "We go on,"

"We stop here," "We wait a while," or "We'll go do something else." *In no case does the PIR recommend a specific system design*, since the design can be based only on data not yet gathered in a detailed analysis.

Supporting Documentation

Supporting documents consist of the following classes:

- memoranda that initiated the work
- lists of individuals interviewed
- lists of documentation examined
- interview schedules or informal lists of questions asked during the interviews
- samples of forms, reports, and input documents collected (a subset of the collection to be made in a detailed investigation)
- high-level data flow diagrams
- organization charts
- high-level structure charts of system functions

Summary

The contents of a PIR are best illustrated by a sample table of contents (Figure 8-7). Remember that the use of a PIR presupposes that its major goal is informed consent by the client either to proceed or to halt. Therefore, the contents should inform and motivate consent. Because the detailed analysis may be performed by those not involved in the preliminary investigation, there is also a need to transfer information, usually in the form of supporting documentation.

1. BACKGROUND	**6. SUPPORTING DOCUMENTATION**
2. PROBLEM STATEMENT	6.1 Memoranda
3. SCOPE AND LIMITATIONS STATEMENT	6.2 Individuals Contacted
3.1 Scope (Organizational, Functional, Type, Time, Labor Class Involved)	6.3 Interview Schedules
	6.4 Documents Examined
3.2 Known Limitations	6.5 Samples of Forms, Reports, Documents Collected
3.3 Derived Limitations	6.6 High-Level DFD
3.4 Trade-Offs	6.7 Organization Charts
4. NEEDS DOCUMENT	6.8 High Level Structure Charts
5. RECOMMENDATION	

FIGURE 8–7. PIR Table of Contents

The following sections look at the needs analysis, planning, and data collection in some detail. Investigation techniques are covered in greater detail in Chapters 11 and 12.

8.3   PLANNING A PRELIMINARY INVESTIGATION

This section provides a guide to planning and executing a preliminary investigation. The next section discusses the three aspects of representing needs, which are central to the content of the PIR. Following that, we will look at ways of collecting data during the preliminary investigation and what should be paid special attention. The last section provides pointers for making an oral presentation of the preliminary investigation findings.

Defining the System Boundary

In the order-entry example we have been looking at, is the system boundary clearly defined? A major aspect of planning the preliminary investigation is determining what will be studied and what will not.

The **study boundary** may differ from both **design** and **implementation boundaries** (see Figure 8-8). The study boundary is, by definition, the largest (compare with

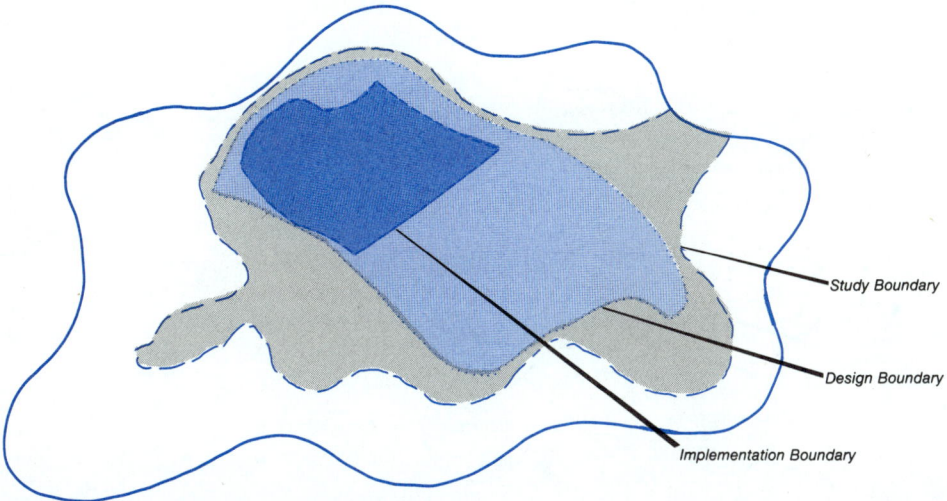

Study Boundary

Design Boundary

Implementation Boundary

FIGURE 8–8. Study, Design and Implementation Boundaries

Figure 5-4) and may not necessarily be a single system — in fact, it rarely will be. Since we are concerned with studying the interfaces through which the target system interacts with its environment, we must study some of the environment. This may entail studying the entire organization or a significant proportion of its environment. Consider order-entry as a "system." Do we not also have to be concerned with the billing "system," the warehousing "system," and the sales "system?" Do we risk ignoring the customers who are in the environment? What other elements of the order-entry system's environment should we consider?

To determine the study boundary:

1. Include, initially, all formal organizational departments that are part of the client organization (in this case, just order entry).
2. Add ingestor (input) elements in adjacent systems (billing and warehousing interfaces).
3. Add other elements from the larger environment that represent consumable resources (customers).
4. Add control and sense elements that treat the selected departments as subsystems.
5. Repeat step 4 until you are high enough in the management chain to build a single "tree" (see Figure 8-9).
6. This forms the initial cut at the study boundary; eliminate subsystems that are inaccessible, politically off limits, or otherwise denied to the study.

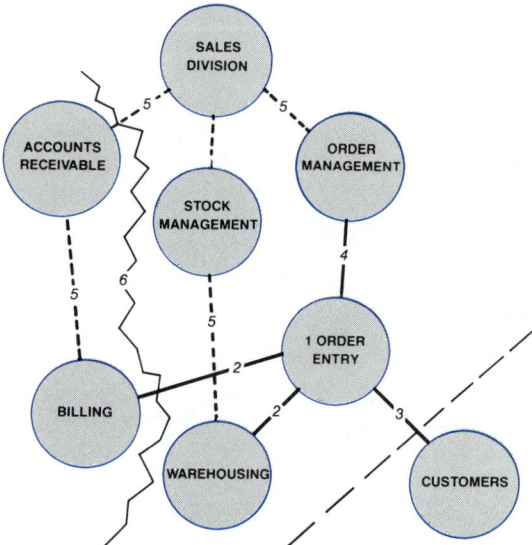

FIGURE 8–9. Setting the Study Boundary (Steps 1-6)

The **design boundary** is determined by looking at organizational and informational needs. The study boundary contains not only a single system, but a network of systems and elements. The preliminary investigation should point out where the system actually *is* and provide enough data to restrict the design boundaries to profitable areas. **Implementation** boundaries are determined by physical design to further restrict work to subsystems in which work is feasible.

Choosing Respondents

The preliminary investigation usually collects data by interviewing individuals and observing some activities. Looking in the right place is the most critical aspect here. The key to finding the right place is in the definition of the study boundary. Respondents for interviews should be chosen either because they are representative (see Chapter 6) or because they know something of value to the investigation and can talk about it.

The following guidelines provide some help:

1. The larger the study boundary, the more individuals you will need to interview.
2. Select management at the highest levels; you will need their permission to interview at lower levels. These interviews need not be detailed, but management has to be informed about what you are doing.
3. Within departments, interview at least one person at every level of responsibility.
4. Every adjacent system's interface with your target deparment(s) is staffed; at least some data on this interface is needed, although a complex series of detailed interviews is not required. If detailed written data are not available on these interfaces, interview a person in each interface.
5. Environmental elements, such as customers, would have to be extensively and systematically sampled; the preliminary investigation is not the place for that unless you have been asked to do so.

Thus you will need to interview several managers, workers at all levels, and interface personnel. For a compact study boundary, this could be as few as two persons; in complex systems you may be interviewing ten or more. Remember that the preliminary investigation is exactly that — *preliminary*. Save a detailed series of interviews for later; now, you are only trying to define the system and its needs.

Asking Questions

The following topics should be covered in each preliminary investigation interview:

1. **Introduction**: who is the analyst and what does the IRU expect to accomplish from the preliminary investigation? What techniques will be used? What is the time frame of the preliminary investigation? (These are obviously questions that respondents will have of the analyst.)
2. **Orientation**: who runs each department? Who has which responsibilities? What does the department do? Who are the staff? Where do they work? How can they be contacted?
3. **Mission**: what is the organization, department, or individual supposed to accomplish?
4. **Impression**: what is the impression that the interviewee has of the problem or opportunity? How crucial are the issues around the problem or opportunity?
5. **Restriction**: what can and cannot be examined? What are some political considerations that the analyst is unaware of? What else ought to be looked at other than the problem or opportunity?
6. **Permission**: who can be interviewed when about what?

The following excerpt illustrates the beginning of a preliminary investigation interview containing several of these kinds of questions. We will look at interviewing in more detail in Chapter 11.

(Introduction to an interview with Millie Van Horten, VP Sales):

SA: Good afternoon, Ms. Van Horten. I'm Ella Paderesky, a systems analyst with the Information Resource Management group, which you contacted two weeks ago about a problem with order-entry errors.

MVH: Pleased to meet you. Yes, I did phone over there. We've got some sort of problem with errors on our forms and I thought you could help us.

SA: Well, we'll try. First, let me tell you how we work and what we try to do.

>   (Spends a while explaining how the investigation will be
>   conducted and what the results might be.)

SA: Now, let me ask you a few questions about the Sales division. Ms. Van Horten, what are the goals of the Sales division?

>   (Some discussion of these goals)

OK, I think I understand *what* the division does. Now, who are your immediate subordinates and what responsibilities do they have?

>   (Ms. Van Horten responds, giving the analyst an orientation
>   to the sales division and its personnel.)

SA: Good, now which of these individuals would you recommend that I speak with about the order-entry problem?

MVH: Well, most important would be Yves Matson, since the problem is in his department. I think you'd best speak to one of our sales managers, if they're not too busy and, oh yes, while I think about it, here's the number of Tammy Fletcher over in Elbow Enterprises. She's really the most steamed up about this; I know she'd be glad to speak with you about what effect these errors are having on her.

SA: OK, now one last question. Is there anything we should *not* be looking at, any areas which are off-limits here?

MVH: No, none I can think of except for the fact that my sales managers are really busy now. If you want to speak to any of them, let me know first.

SA: One more thing. Do you think we should be looking *outside* the order-entry area, say into billing or accounting?

MVH: No, I don't, not now. Let me think about that. From what you've told me, your preliminary work begins here in any event. I'll get back to you by next Monday, after I talk with Zack Philo over in accounting. That's a bit tricky, since he handles billing and accounts receivable. I'll let you know later.

Collecting Data

The preliminary investigation collects data in other ways, too. Foremost among the data collected are memos and manuals that relate to information, forms, sample reports, and input documents. Previous investigation reports also should be examined.

This may make for a mountain of data, much of which will not be critically examined at this time. Printed matter should not be collected at random, however. After all, a preliminary investigation is not a fishing expedition, and analysts are not police detectives looking for "clues." Analysts should plan to collect data that seems relevant, but it is hard to know that in advance. Several guidelines can limit the collection, however:

1. Previous studies, where not confidential, must be collected, read, digested, and understood.
2. Memoranda, especially those concerning known problems, should be collected.
3. A list of manuals should be prepared for later collection.
4. Only those forms, reports, or input documents that are mentioned as being troublesome or problematic or that illustrate an opportunity should be collected.
5. Confidential documents or those that present a privacy problem are best not collected at this stage. Otherwise, they will be part of the permanent documentation for a project that might be cancelled in the early stages. Of course, the collection of these documents may be unavoidable for certain kinds of systems (police, military, top-secret industrial, for example).
6. Information collected from one department might not be releasable to another department. That is, your client may not have permission to see some things you have collected. Since this is awkward, ask first.

Writing and Presenting the Report

Allow time for writing, typing, editing, printing, and distributing the report, which should be presented orally. It is not always necessary that the analyst who did the preliminary investigation be the one who makes the presentation; however, a great deal of preparation will be necessary to train the understudy.

The Plan

Figure 8-10 illustrates a preliminary investigation plan. One week is allocated to planning. The data collection phase will involve four interviews and is scheduled to take one week, also. The needs analysis will take two weeks, and it will take a week to write and type the report. With company mail being what it is, the oral report cannot be scheduled until a week after the report is printed. Therefore, six weeks will be needed to complete the preliminary investigation. (Chapters 15, 16, and 17 illustrate the principles of project management and contain more detail on time and expense planning.) Note that

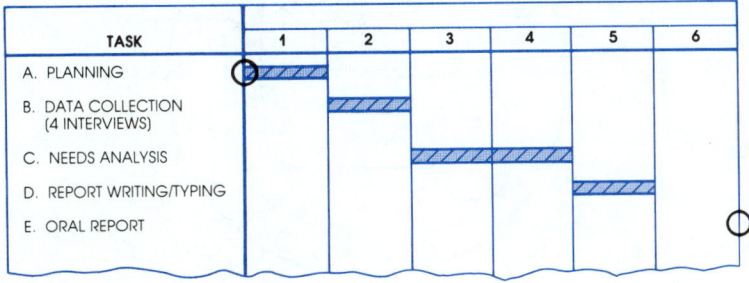

FIGURE 8–10. A Preliminary Investigation Plan

this analyst plans to work alone, full-time on the project, meaning six person-weeks or 30 person-days at $240 per day, bringing the labor cost to $7200 for this small study. There might be good reasons for the analyst's manager to lengthen the calendar time while decreasing the labor expenditure (neither planning nor preparing an oral presentation is going to be a full-time effort, and waiting for typing is "free"), but this is covered in Chapter 15.

8.4 PRELIMINARY NEEDS ANALYSIS

The needs analysis forms the basis for argument and agreement in the PIR. The task of the analyst is to *analyze* the expressed needs, values, opinions and ideas, evaluating them and discovering the actual, demonstrable, and, one hopes, satisfiable needs.

Needs analysis in the preliminary investigation is taken from structural-functional design. We begin with the major goals of the client, equating these with needs. These needs are then decomposed into subneeds, sub-subneeds, and so forth. The unique aspect of the preliminary investigation needs analysis is that three distinct areas of needs are examined: organizational, informational, and procedural.

Organizational needs are the goals we have just mentioned. However, decomposition is aimed at *organizational* activities such as project definition, human resource allocation, financial limitations, and so forth. In other words, organizational needs are analyzed to determine the organizational context of the proposed work, its assumptions, and its limitations.

Informational needs are analyzed to construct a series of high-level data flow diagrams of the information flow, accomplishing the subgoals concerning that information. Figure 8-11 illustrates such a set of DFDs. Note that analysis is at a very high level, with each DFD process referring to a subsystem seen from the point of view of the information resource.

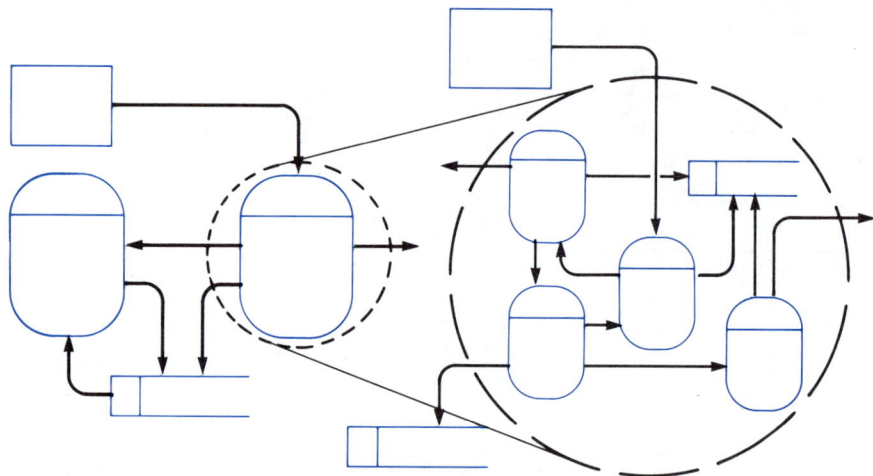

FIGURE 8–11. A High-Level DFD and an Explosion of One of the Processes

Procedural needs are operational requirements that detail the physical assumptions and limitations of the possible solutions. Procedural needs are not crucial to logical design, but they will be important during physical design, installation, training, and maintenance. Procedural needs are analyzed until a very high-level *system flowchart* can be drawn, illustrating how the system is expected to operate. This is not a diagram of the solution system, however, but a tentative and very schematic diagram of the way the system is expected to work, especially with regard to other systems.

The following sections provide some detail on analyzing organizational, informational, and procedural needs.

Organizational Needs

In the very first interview about the order-entry error situation, the following questions were asked of Milly Van Horten, VP, Sales:

SA: Ms. Van Horten, what are the goals of the Sales division?

MVH: First, to increase our market share. Second, to provide quick response to requests for our products. And of course, to respond to the needs of the market.

SA: Are those goals currently being met?

MVH: The first? Yes. We've increased our share to 11 percent from 8 percent in only one year. Response to our customers…well, you know about the problems there. And market research? We've got the best in the business — we're usually ahead of the market.

SA: Now, about response to customers. Can you tell me in your own words what you think the problem is?

MVH: Sure. We've been getting a lot of flack from our sales managers. They've been hearing from their best customers that our orders are getting messed up somewhere. You know how important good relations with customers are no matter how good the product, so they're all upset.

SA: Have you seen any evidence of the problem?

MVH: No, just what I've heard from my people. But they know what they're talking about.

SA: I'd like to speak to one of them, as well as people in the order entry unit. Can you refer me to the appropriate people?

Note that the interview begins with *organizational* goals and proceeds to analyze these from Millie Van Horten's point of view. Completing the sale by delivering the goods is one of the goals. Apparently, the goal is not always being met. A lot of work will have to be done now to find out where the roadblock lies. The analyst is beginning organizational needs analysis.

The next step is to prepare an organizational chart that indicates functional responsibility for each area. In this case, the sales division has three departments: marketing, order processing, and market research, each headed by a department head (Figure 8-12 on the following page). Marketing makes the sales, Order Processing completes the paperwork and prepares orders for shipping, and Market Research investigates marketing opportunities.

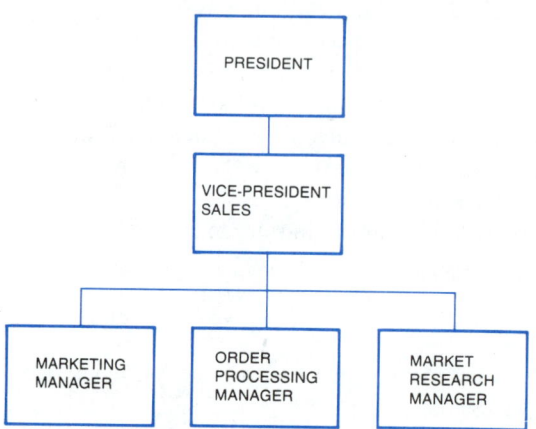

FIGURE 8–12. An Organizational Chart

Of course, organizational needs may not always correspond so neatly to actual functional divisions. In these cases, the analyst must be alert to crossed responsibility lines and may need to speak to several people to determine organizational needs at lower levels.

For instance, although order processing does orders, not all the processing goes on there. Here's a segment of the interview with Yves Matson, head of the department:

SA: Mr. Matson, what are the goals of this department?

YM: We process orders that sales personnel have stimulated for us.

SA: What does that require of your people?

YM: Well, we take phone orders, fill out the paperwork, and send picking and packing slips to the warehouse and completed sales forms to billing.

SA: So there are three subgoals, so to speak: taking a phone order, completing paperwork, and sending forms out?

YM: Right.

SA: I'm not clear about the paperwork — does that mean filling out these three forms? [He points to the forms on the desk]

YM: Yes, but we don't really do all of the paperwork, either. For example, if an item's not in stock, we call the sales manager to alert a sales type about a customer's problem. We also check on each customer — if their credit's no good, we won't ship. So Billing and Accounting get involved, too, although that's usually routine.

SA: So some of your functions are actually farmed out to other departments or people?

YM: Uh-huh.

SA: Do you know of any problems...?

Here the analyst is trying to determine if the goal of processing sales to completion is actually met in the order processing department. She determines that not all the work is actually done in order processing, but that sales managers, billing clerks, and accountants also get involved.

In an attempt to isolate the "problem," the analyst will indicate on the structure chart that some parts are inaccessible — within accounting, for example — restricting the system boundaries. She may later determine that the errors actually arise in accounting or sales management, *not* in order entry.

Informational Needs

Another aspect of the needs analysis is to draw a high-level DFD and a structure chart of the use of the information resource in meeting organizational needs.

After interviewing Yves Matson, the analyst is referred to Willy O'Callaghan, an order-entry clerk. Yves has assured the analyst that Willy is both informed (he has worked in the department for two years, about the average) and well spoken. Here is part of the interview:

SA: Now what do you do after you take the call from the switchboard?

WO: I take down the information about the sale from the customer.

SA: And that information goes where?

WO: Well, it doesn't go anywhere. You see, first I've got to do an in-house credit check, so I put the guy on hold and call over to accounting to see if he's paid up or what. If he's not, I transfer the call over there. If he is, then I get back to him.

SA: Do you ever lose anybody in the phone system?

WO: Oh, sure, but they call right back, or I call them, because the first thing I do is get their name and number. Anyway, I tell them the order will be processed in six days, blah, blah, blah, you know how it is, and then I hang up.

SA: And then...

WO: You want to know what I write down in each box?

SA: No, not at this moment. But what do you do with each form you fill out?

WO: Oh, yeah. I send this blue one here to accounting, the red copy goes to billing, the yellow one I send off to the warehouse, and the grey one I keep here in our files.

SA: OK, now let's talk about what happens to those copies....

The analyst is beginning *informational analysis* (sometimes called "data analysis"), looking at data flows at a relatively high level. Figure 8-13 on the following page, shows what the analyst has learned so far from this interview. After interviewing and collecting data, the analyst can then prepare a series of high-level DFDs. They illustrate related external entities in the environment, the approximate system boundary, and at least one process for each system subgoal. It is not necessary at this time to uncover individual transactions, although surely that will be the topic of later interviews.

The goals of information needs analysis are to understand how the information resource is used to further organizational goals, to position the system boundary, and to detail the interfaces of the system with the environment. The design boundary will probably be smaller than the study boundary; if it is not, the study boundary might have to be enlarged. System interfaces need to be detailed now to set the limits for detailed investigation later on. Finally, the information resource should be described in terms of information exchanges to begin the logical design. The first task in logical design will be to indicate *where* the problem lies (that is, in which subsystem) and to redesign that subsystem.

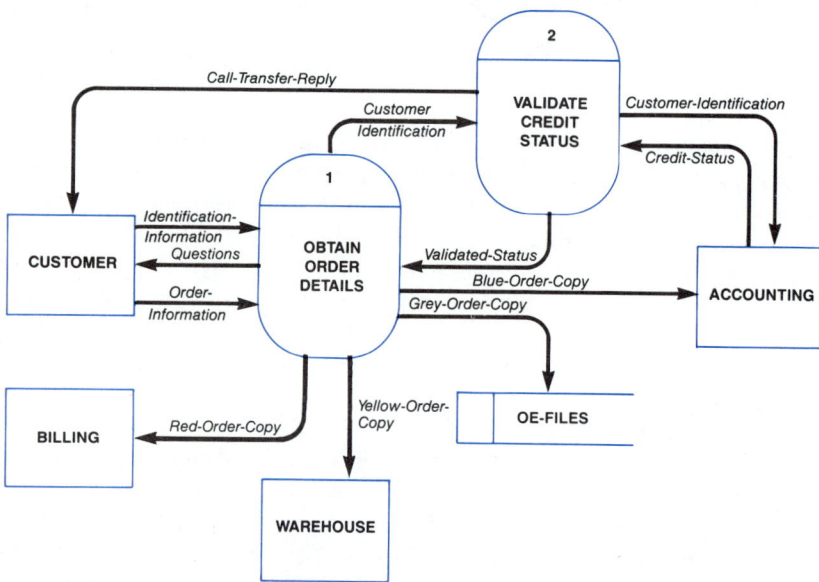

FIGURE 8–13. The High-Level DFD Derived from the Interview

Procedural Needs: Operational Requirements

The preliminary investigation needs analysis will not be complete without an understanding of the operational constraints of the system under investigation. These constraints are important mostly in physical design, but they need to be available in decision making at early stages.

Operational requirements may appear in any of the following areas among others:

- speed
- volume
- reliability, accuracy, precision
- redundancy
- cost
- physical layout, access to materials
- staffing
- security

A process may be expected to take no longer than A hours per transaction, of which there will be B daily, with no more than one transaction error in every C transactions. Transactions have to be backed up on media D and the cost per transaction must not exceed E dollars. All work has to be done in area F, staffed by no more than G people and with security at an *H* level at all times. Limitations *A* through *H* may not affect logical design, but there is no reason why physical designers should have to go back and get this data. More procedural needs will also come out in the detailed investigation.

Summary

Organizational needs stem from the goals of the organization. These needs are met in a variety of ways, some using the information resources of the firm. This in turn dictates needs to receive, process, and transmit information. Finally, the organization may place constraints on how information is to be used in terms of cost, time, precision, and so on.

8.5 COLLECTING DATA FROM RESPONDENTS

Chapters 11 and 12 will provide a primer on interviewing and observing. We have already examined some issues regarding asking questions and collecting data and discussed how to select respondents for interviewing in the preliminary investigation. This section is concerned with three issues in data collection that present unique problems in the preliminary investigation:

- the kind of data that CAN be collected from respondents
- how to avoid contaminating the data
- how to draw conclusions from the data

Limitations on the Data Collected

Because a preliminary investigation *is* preliminary and because it may be the first exposure many individuals have to the IRU and its techniques, a number of unique limitations arise (see Figure 8-14).

1. Psychological limitations
   1.1 Uncertainty
   1.2 Fear
   1.3 Nervousness
   1.4 Inarticulateness
2. Poor problem definition
3. Incomplete preparation of the respondent
4. Changing investigation goals
5. Management resistance to change
6. Contamination by user attitudes, opinions, emotions, and expectations

FIGURE 8–14. Limitations on Preliminary Investigation Data Quality

First, some respondents may be unsure, unsettled, nervous, and even frightened. Their responses may be guarded, evasive, impulsive, or diversionary. If so, these may not be the right individuals to interview. Someone called "articulate" may also be quite shy with strangers, technical people, or professionals. Others may be very aggressive.

These concerns make it difficult to collect information in the first place, to trust the dependability of the information gathered, and possibly to process the data at all. Although all interviewing is subject to these limitations, during these *first* interviews, these effects may be enhanced.

Second, the analyst may not be sure of what is being investigated. Thus, questions are vague, at a higher conceptual level, and have to be followed by probes. Respondents may become overly sensitive to the tentative nature of the interview and respond in ways they feel will either keep you away or bring you back, depending on how they feel. On the other hand, they may simply not know what you are talking about.

Third, preparation of respondents may be incomplete or totally lacking. You might be interviewing someone the boss collared that morning. Workers may be loathe to respond to questions when they do not know what their supervisors expect them to say.

Next, it is harder to be systematic in the early stages, when the goals of the investigation are changing. The analyst may not be sure about the people and their work, and may not be secure in feeling "welcome" or even "allowed." Furthermore, because only a few people will be interviewed, it will be difficult to cross-check responses.

The fifth limitation is that analysts have to contend with *management resistance* to computers, computerization, systems analysis, and the world of technology. That resistance can be overwhelming, and in these early stages, it is often fatally damaging to the project because permissions may be denied.

Finally, the data in the early stages are necessarily more contaminated by opinion, attitude, emotion, folklore, expectations, and beliefs. We will discuss this more in the next section.

Overcoming fear or shyness is a general problem. An analyst will press managers to make sure they recommend individuals who are articulate and knowledgeable. The tentative nature of the investigation content should not interfere with the analyst's technique; preparing an interview agenda before each interview is most helpful.

The surprise that respondents exhibit when faced with on-the-spot interviews can be reduced by insisting that managers prepare the employees. In any event, the analyst will try to put employees at ease as much as possible. The lack of a systematic process and the uniqueness of each interview could challenge most television talk-show interviewers, but analysts have to make each interview a unique experience and must remain flexible.

The final major limitation, management resistance, may occur where a department has already had a bad experience, or where a department is unfamiliar with technology and has preconceived negative feelings. It is hard for an analyst to stand up under a withering attack by a vice president. Letters from IRU managers may not be a sufficient shield against executive wrath. One large consulting firm matches the level of the interviewer with that of the interviewee: a senior partner in the firm interviews high-level executives, juniors interview middle management, and consultants interview everyone else.

Keeping the Data "Clean"

Data collected are often contaminated. Some sources of contamination have already been mentioned: fear, reticence, diversion, opinion, attitude, belief. Others include:

1. Data can be lost if not maintained systematically. A good filing system is imperative.
2. Analysts cannot "fill in the gaps" where questions were not asked or were not understood.
3. "Facts" need to be verified. Two people saying something does not necessarily make it true, but it is hard to base a million dollars worth of development on the testimony of one individual.
4. Opinions, attitudes, beliefs, mythology, and folklore may be repeated in any group because of group norms, which operate to keep the group together. Sam and Sarah may say that errors come from the warehouse because they are protecting each other. Tilly, Tommy, and Terry may each swear that cut-off customers always call back because they believe it is the customer's responsibility, even if half the customers are never reconnected. Ellen, Eileen, and Irene may say that they never make errors on the forms, because they do not want to admit to making them. Getting to know — and making allowances for — group norms is an important aspect of the preliminary interviews.
5. Other contamination creeps in when old data (from previous studies) are used or when interviewing takes a long time and situations change (as might happen when a major system is being installed while another is being investigated or when staffing, particularly supervisory staffing, changes).

Clean data are important for more than ethical reasons. It is difficult and dangerous to draw conclusions from contaminated data. The next section shows why.

Drawing Sound and Understandable Conclusions

Here is another part of the transcript with Willy O'Callaghan:

SA: How were you trained in how to use these forms?
WO: Well, I never had any training. The boss just sat me down with St. James (Rhea St. James, another clerk) and told her to show me.
SA: And was that sufficient?
WO: Oh, sure. I made a few mistakes for awhile, you know, but she helped me out. Everybody here looks out for everybody because we know how important those forms are.

Now, is the analyst justified in stating that clerks receive only spotty, unsystematic on-the-job training from helpful coworkers?

No. Willy joined the staff two years ago. Perhaps there is a new training program. Rhea may be better or worse than others in training coworkers. Willy may be a fast study or a dullard. Just because Willy thinks the training was sufficient does not mean everyone does. What about individuals who left? Did they leave because they didn't know what they were doing? Willy says everyone sticks together, but maybe he is friendly and everyone looks out for him. What about those who left? It would be difficult to draw general conclusions about individuals' training based on Willy's few comments.

Another problem with the conclusion is that it is vague. What is "spotty" and "unsystematic" to the analyst may be a coherent training program to Rhea St. James. The "only" used in that statement biases the reader to expect that a "larger" training program would be better. Finally, the term "on-the-job-training" implies that knowledgeable people train others. Not only did the analyst not find out about Rhea's qualifications (Was she a good worker two years ago? Did she know her stuff?), but one might think that there was a "program" working here to provide training. In fact, all Rhea did was show Willy how to fill out the forms and later respond to his questions. She did not provide any supervision, exercises, or remediation. In short, Willy was never trained at all.

Understandable conclusions based on data resemble these: "The problem occurs in department A," "Workers know that they make errors and this affects morale," "Management finds the level of errors in order processing unacceptable," and "At least one in sixty orders has a detected error; undetected errors seem to run about the same rate according to customer complaints."

Summary

Special considerations in gathering data from respondents arise in the preliminary investigation, including limitations of the reliability and content of the answers provided, the effects of the questions asked, and the confidence the analyst can place on the value of the answer. There are several special sources of data contamination that the analyst collects, only some of which can be controlled. Finally, it is critical that each conclusion be supported by data and phrased in an understandable way.

The final section of this chapter discusses important factors in making oral presentations of preliminary investigation results.

8.6   PRESENTING THE REPORT

Three aspects of oral presentations are of specific importance: the need to discuss implications in person; preparing a first-impressions "show"; and obtaining approval at the first step.

The Need for an Oral Presentation

Because the preliminary investigation is both preliminary and first, it is important to get individuals involved in managing the project together in person. This increases credibility, allows instant clarification of confusion, may provide for a higher level of trust in later sessions, and shows that the IRU has sufficient faith in its own workers.

Credibility is increased by the ability to elaborate on conclusions and provide additional evidence if asked. False impressions and confusion can be worked out on the spot.

Trust is often higher in face-to-face meetings if only because individuals get a lot of information from others' faces, non-verbal "data" that is not apparent in reports and the stilted style often adopted by writers. Trust also depends on how much trust the IRU apparently has placed in the analyst. A professional presentation may enhance an already well-done piece of work. On the other hand, a poor presentation may ruin a good impression. Both, obviously, have to be well done.

Preparing the Presentation

This presentation may be the first time all the important players (client, analyst, IRU management, interested observers, executives) get together, creating first impressions that will be hard to counter later. Research in impression formation shows that first impressions are very important. Here are some general guidelines for preparing the presentation:

1. Prepare the room. Creature comfort is important. Tired, vexed, uncomfortable people seldom agree to anything beyond leaving.
2. Use visuals and do them professionally. A number of microcomputers do beautiful graphics overhead-ready. Too much talk puts people to sleep. But *do not read the transparencies*, either. Visuals should not be overcrowded, wordy, or sloppy.
3. Publish an agenda and stick to it.
4. Prepare a structured meeting with breaks. It is during the breaks that much of the informal coalition-formation necessary for later agreement takes place.
5. Introduce everyone to everyone else.
6. Watch posture, eye contact, gestures, nervous habits, and tone of voice.
7. Make extra copies of the report available, along with pencils, pens, and notepads.
8. Prepare the lighting and sound systems; make sure the microcphones and projectors work.

Obtaining Approval for the Next Steps

The reason for the meeting is to obtain approval of recommendations. Because this is the first such meeting, it sets a pattern for future meetings. Managing the group's decision making is critical.

Monitoring the group is especially critical now because limitations, impressions, and expectations that arise early can be countered later only with great effort. It is important to reach a confident and widely applauded decision.

The presentation is a passive event from the audience's perspective. They all will have read or will read the report, but for now they are there to listen. Making a decision, on the other hand, requires active participation. Moving the group toward consensus is important and shifts the mode from information reception to action.

Moving toward consensus can be done by putting the agreement on the agenda (for example, in Figure 8-15 there is an item labled "Next Steps"). If discussion seems to drift, the analyst can remind the group that "after we've clarified these points, we should consider what our next steps are." A form that allows the user to "sign off" on the receipt of the report can be stapled to the agenda to serve as a reminder.

Decision makers reach consensus. It is not usually enough that an unpopular decision be made (although, of course, this may be necessary);

AGENDA

1. Introduction of the staff
2. Problem definition
3. High-level view of the situations
4. Recommendations
5. Next steps: action items

FIGURE 8–15. Preliminary Investigation Report Presentation Agenda

action requires motivation among all parties. Consensus is best built in several steps:

1. Make sure everyone knows what is being agreed to.
2. List all points of concern or disagreement and handle them one by one.
3. If they all cannot be settled, and if the situation is so justified, argue that they need not all be settled.
4. Watch out for interpersonal conflicts that you simply cannot resolve.
5. Present a consensus statement that refers to unsettled concerns and lays out a schedule for meeting them.
6. Ask for a vote as a last resort.

No one will agree to a blank check. Action items have to be spelled out clearly. Of course, the analyst's advantage here is that recommendations themselves will be action items.

Action items are statements in the form "X will do Y by (or during) time or date Z." Matching Xs, Ys, and Zs is hard and many analysts prefer that managers determine them later. Nonetheless, the report will refer to next steps to be done by the IRU and the client, and these form the basis for the action items.

Agreement is then reached and the meeting is adjourned. The following transcript shows how consensus is reached around the order-entry preliminary investigation:

SA: OK, now I think we've clarified most of the points you've brought up. Are there any which you still feel vague or concerned about?
Client: Yes, there's the question of....

(Some time passes in discussing and listing these concerns)

SA: I think there are only two concerns we have not adequately addressed, the question of timing and the feeling in management that streamlining the order-entry system without looking also at billing would be a duplication of effort. Is that about it?

(Some discussion here, but general head nodding and murmuring of agreement)

SA: Thanks, now I'll turn the meeting over to Al Chambers, who will discuss the action items.
AC: Thanks. Now as we see it, there are several things to be done. We'll enlarge the size of the project and do another look at billing now, if that's what you want. It will delay the project, but that's OK with me if you want to pay the price.

(Lengthy discussion about this)

AC: Hold on, I think we're drifting a bit. You've all agreed that we have to do the work, but unless we agree on a tentative time table, we'll get nowhere. Let's do these action items one by one and later put some times on them, OK?

(Some more discussion of timing, then agreement on the budget)

AC: OK, I'll get our analysts working on billing, too. I'll get a memo to you by Friday with a tentative schedule. You'll have to meet with your people in billing to ready them for interviews. I'll get copies of this report over to them as soon as you've made contact and gotten their cooperation. Does that about cover it?

Maybe all meetings do not go as smoothly as this, but they do not have to be contained warfare, either. Remembering that people are people helps. Agendas help. Having a good report helps. And remembering the reason for meeting — to agree — helps, too.

8.7 CONCLUSION

The preliminary investigation report is the first formal document based on data collected by the analyst. It defines the problem to be investigated in a **problem statement**. It identifies the scope, bounds, and opportunities for the investigation in the **scope statement** and **project summary**. The latter document forms the basis of the **contractual aspects** of the project, laying the foundation for agreement between the information resources unit and the client organization staff. In this document the responsibilities and rights of each organization are spelled out, along with identification information necessary to begin the work. Typically the project summary and scope statement are formally "approved" before work may begin.

Next, the analyst investigates the needs of the client organization. Three sorts of needs are examined: **organizational needs**, **informational needs**, and **procedural needs**. This set is derived from the basic function of an information system, as laid out in Chapter 1. The organizational needs are the goals of the organization as seen from the perspective of the information resource: how can the information resource contribute to the organization's goals? Structural-functional design plays a key role here. Second, analysts detail needs by determining which reports, statements, and documents workers need to meet organizational needs. The analyst outlines procedural needs by explaining how the current operations function and where improvements may be noted.

The preliminary investigation consists of the following steps: (1) initial contact, (2) needs analysis, (3) redrafting of scope, boundaries, and opportunities documents, (4) determination of the next steps, and (5) the project plan. The **project plan** is a rough outline of the timetable and cost of the next steps (see Chapter 14 for details on costing and forecasting).

The PIR is completely data-based in that all conclusions are drawn from data collected in the investigation as interpreted by the analyst. Data collected are limited in a variety of ways. These include fear (of automation, of change, of the unknown), political pressure, ignorance of the possibilities, and analyst bias. Techniques for keeping data as valid as possible include training of respondents, honesty in preparing respondents, careful choice of respondents, empathy techniques, and self-awareness. Sound and understandable conclusions depend on the analyst's skills, experience, and awareness.

Although many PIRs are simply handed over to the client and management, a presentation may be necessary. Where a **formal presentation** is conducted, a certain protocol should be followed, including inviting the right personnel, distributing the PIR in advance, focusing on the next steps, and obtaining consensus on the advisability of proceeding. The PIR and its presentation should include supporting models, especially high-level DFDs, a general structure chart detailing how information resource activities contribute to client organization goals, and definitions of terms.

Chapter 9 looks at procedure writing and chart drawing, following Chapter 6 in the use of models in representation. Chapter 10 treats the presentation of logical specifications in much the same way as this chapter has looked at the preliminary investigation report, concentrating on specifics, with more detail on writing and presentation techniques.

Chapter 8 discussed investigation techniques, but it is in Chapters 11 and 12 that details on the tools of investigation (interviewing, questionnaires, observation, simulation, diaries, and other indirect observation techniques, as well as data analysis) are discussed.

8.8   THE CASE OF THE THROWN RIDER

Ralph Gunther is the general manager of Wilson's Western Wear, a "purveyor of the best in western outfits for cowboy and tourist alike" in Weston. Wilson's has been in business for eighty-four years, the last seven of them as a wholly-owned but locally-operated subsidiary of Duds International, a clothing chain. Duds operates out of Bigtown, almost eight hundred miles away, but Ralph has had few problems to call them about.

This month, however, Ralph has just about had it. Wilson's, as do many similar firms, uses an open floor, with packaged clothing laid out in bins and some of the better garments on hangers. Dressing rooms are provided. A large area is set aside for saddles, boots, and other leather accessories, but these are handled as a separate division; leather items have to be rung up before they are brought out of the area. This is to correct a major shoplifting problem Wilson's had a few years ago.

Ralph's current problem is that the end-of-day totals rung up on the registers for the leather area do not match the batch totals derived manually from sales slips. The totals in the general merchandise area usually do match, but on days when there is a large difference in the leather totals, Ralph has noticed there are some differences in the general merchandise totals, too.

Ralph does not know whether the problem is with his clerks, the computerized cash registers he had installed six months ago, or the way batch totals are being computed overnight by his assistant bookkeeper. He has phoned Duds International headquarters to request some help. It was at Duds' insistence that the computerized cash registers were installed, although Ralph fought for, and retained, some manual controls (such as the batch totals on sales slips, which are still written up manually). Now he regrets having two parallel systems. It was never like this in the rodeo — just him and a horse. But this computer has thrown him.

Questions

1. Duds responded by sending out a systems analyst who has had some experience with the computerized cash registers installed in Wilson's. Outline the contents of a preliminary investigation report that the analyst might present to Ralph Gunther.

2. Consider what the analyst might hope to find out in a preliminary investigation. The problem statement as originally given to the analyst (from phone calls from Gunther) was this:

"Daily batch totals don't agree with the computer's."

Fill out this problem statement to be more informative based on what you suspect should be looked at.

3. The scope, as originally suspected, is only the computer system. What arguments might you make (if you had already performed the preliminary investigation) to enlarge the study boundaries?

4. Gunther apparently is not familiar with computers or very comfortable with them. What might an anlyst say by way of introduction to put him more at ease in a first interview?

5. Given the limitations as you see them, list two needs in each of the areas of organization, information, and procedures that would come from interviews with Gunther.

6. Gunther encourages the analyst to speak with Ramon Garcia, who is the sales manager for the store. What questions are appropriate for Garcia about the way sales are made and documented?

7. What are two recommendations you would definitely *not* make to Gunther at this time? Why?

Discussion Questions

1. Examine the table of contents below for a PIR submitted to an engineering group. Which (if any) of the functions do you feel the PIR stands a chance of meeting?

   1.0 Executive summary
   2.0 Detailed findings
   3.0 Next steps

2. The section "Planning a Preliminary Investigation" discussed needs analysis and introduced three classes of needs: organizational, informational, procedural. How do these three classes mutually interact? Which logically depend on others? Provide an example that shows how they are interrelated.

3. Refer to the material in Chapter 1 on system boundaries and review the questions asked in the section "Defining the System Boundary." In the order-entry example, is the system boundary clearly defined? What arguments can an analyst use to extend study boundaries? Why should the study boundary be *larger* than the ultimate design boundary? On what basis will the design and implementation boundaries be made even smaller? Provide an example in the case discussed in this chapter.

4. Examine the organizational chart in Figure 8-16. As an analyst, you have been called in to find out what kind of investigation you should perform. Customers have been reporting that the amounts of goods they receive do not match the amounts they have ordered, although the invoice amounts do match. Whom would you interview in the investigation and in which order?

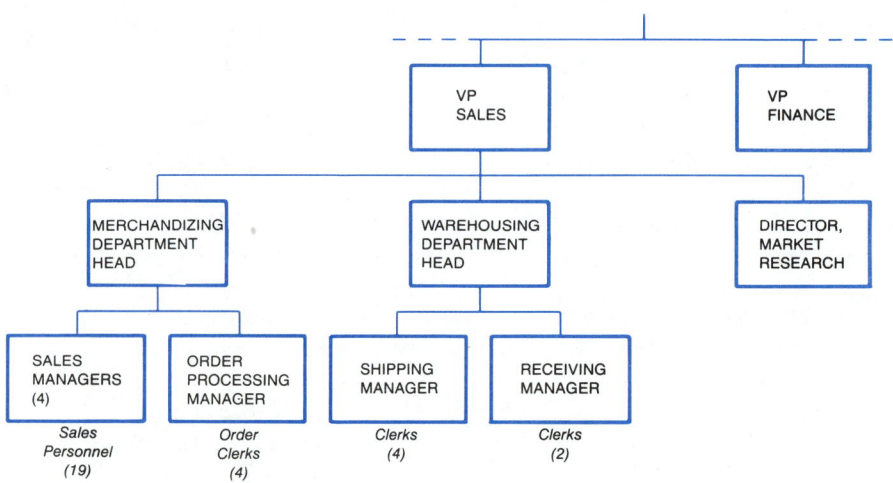

FIGURE 8–16. The Sales Organization

5. Here's an interview between an analyst and the sales manager in question 4. Critique the kinds of questions the analyst is asking. This is the first time they have met.

    SA: So, your organization is having problems?

    SM: Well, no, we like to think we're doing a good job here. Anyway, what about these problems?

    SA: Well, everyone knows that your guys are writing down the wrong numbers and that has led to some errors in orders.

    SM: No, it hasn't. Anyway, what are you here to find out?

    SA: I'll tell you. We think the errors are at the source. I want to interview all your salespeople starting tomorrow morning to get to the bottom of this. Now, can you give me a list of all salespeople who are in town?

6. This chapter discussed six limitations on the value or usefulness of data analysts collect that are particularly important in the preliminary investigation. Which do you think are most important? Which are the most difficult to overcome? Whose responsibility is it to see that these limitations do not arise?

7. John has interviewed the VP Sales, two salespeople, and a clerk who processes the orders. On the basis of this he makes the following conclusion:

    If errors *are* occurring in processing orders, it is not in the sales department.

    Criticize this statement in terms of soundness and clarity. On what bases would such a conclusion have to rest? In what different senses could this kind of conclusion be interpreted?

|DESIGN EXERCISES|
Using dBASE III

1. So far, we have looked at existing files and added and deleted a few records. Now it is time to **create** files, enter data and edit the results. Bring dBase III up again. Enter the following at the prompt:

> CREATE IDEAS   Create a new file called
> IDEAS.DBF

At this point, a screen appears with prompts to enter the names, types, and lengths of fields. Press the F1 (**HELP**) key. Press it again. The HELP key provides additional information which is **toggled** (turned on and off) by repeated pressings. You may want to leave HELP on while you enter file creation information.

Define four fields. The first field, called TOPIC, will save a topic, and will be 15 characters long. The next field, called PRIORITY, will be a number between 0 and 999 with an importance judgment. This importance judgment will be a three-digit numerical field with NO decimal places. A third field, called IDEA, will contain up to 160 characters of idea on the topic of TOPIC. If you give each IDEA under a TOPIC a unique PRIORITY, what should the key field(s) for IDEAS be? Finally, a fourth field called FILE will be 8 characters long and represent a file that concerns this topic. It can be left blank or filled in later.

When you've defined these fields, press ENTER when asked for a fifth. You'll be asked to **confirm** with ENTER that you have completed defining fields and dBase III will ask you if you want to enter records now. Reply N for no.

Now, go back over your role-play and the lists you created in 7.2.2 and APPEND records with your ideas, organized by TOPIC (You will have to USE IDEAS to be able to APPEND at this point). When you run out of ideas for IDEAS, press ENTER without entering a TOPIC. Now use DISPLAY to list ALL the TOPICs. How many different topics did you enter? Select the third topic and DISPLAY ALL the IDEAs for it. How would you get a list of all ideas for all topics that made sense? (Hint: try SORT.)

2. What if you want to change something? LOCATE or DISPLAY the record you wish to change, and type the following:

> EDIT   Change a record

DESIGN EXERCISES
Using dBASE III *continued*

and dBase III brings the entire record up. Press HELP to get some guidance on what keys to press. dBase III uses some WordStar™ key conventions ($^$ means to hold the Ctrl (Control) key down while you touch the following key):

$^$A	Go back to previous word
$^$F	Go to next word
Ins	Toggles insertion and type-over modes
Del	Deletes the letter to the left of the cursor
Right and Left arrows	Repositions the cursor by one character
PgUp	Go back to the previous record (if there is one)
PgDn	Go ahead to the next record (if there is one)
Up and Down arrows	Move up a field and down one field.
$^$End	Quit editing

CHAPTER **9**

WRITING PROCEDURES AND SYSTEM CHARTING

OBJECTIVES

1. To distinguish between the variety of charts that document system activities
2. To list the three basic structures of a procedure and to draw inferences from these concerning the limitations of procedure-writing
3. To be able to create a pseudocode procedure for a given sample exercise
4. To be able to draw operational conclusions from a procedure model
5. To describe how to produce and maintain users' guides, operators' guides, and technical reference manuals for operational systems

WRITING PROCEDURES AND SYSTEM CHARTING

9

CHAPTER

9.0 INTRODUCTION

"Refer to Figure 19A for the following instructions," Ted read to his roommate Sal. "Locate the end unit assembly, part 14B' — I think that's the little red piece over there with the squiggle on it — 'and insert this gently into the body unit' — that's the one we built this morning, Sal."

"Yeah, but which end goes in where? And weren't there two parts to this body unit?"

"Well, the picture's not too clear. I think it goes behind the door there, except I don't see any door on this body unit; maybe there were two parts to it, let's see...."

"OK, listen, let's move on to the next page. I see a picture there of the latch assembly and I know we've got that over here, it's that blue thing."

"Yeah, Sal. It says here, 'To assemble the latch unit, grasp the latch assembly in your right hand, and holding both the number 14 cotter pin and the #8-14 latch screw, attach the latch assembly to the latch unit, just behind the point where you attached the end unit to the body unit. Then....'"

"Uh, Ted. Just a minute. We don't have any end unit here, remember?"

"Oh, well, Sal, we just gotta have *something* here to put together. I mean this diagram's all full of pointy things and little squiggles and all we've got is a bunch of half-assembled parts here. I've read these instructions to you twice and we're almost nowhere! What do we do *now*?"

Ted and Sal aren't the only ones who have been frustrated by do-it-yourself, step-by-step guides to assembly. Ordinary clerical workers, computer system operators, and highly paid executives are similarly confused by badly written, poorly diagrammed, and improper instructions.

This chapter is concerned with writing a *procedure*, which is the act or method of carrying out some action. The word "procedure" is derived from the same Latin root as the word "process," namely *procedere* — to go forward. A procedure, then, is a method by which one goes ahead and does something. Does it sound like Ted and Sal are getting ahead?

Ted and Sal are simply carrying out instructions from a printed page with diagrams and words. The procedures that accompany an information resource activity are often far more complex, if only because the tasks generally involve paper, displays, filling out forms, getting approvals, and keying in commands or responding to menus. It is often hard to visualize the finished product and to get some kind of feedback on how well you're doing. In fact, computers are famous for responding with "SYNTAX ERROR" or "FATAL ERROR B312X" — uninformative messages that support the myth that programmers and analysts cannot speak to human beings.

This chapter looks at structured procedure writing. We begin with the theory of structured programming and its implications for writing procedures. Then we will turn to a primer on pseudocode, a kind of half-English, half-computer-programming-language tool for writing procedures. The next section looks at the use of operational models in the design of systems. Analysts use operational models to find bottlenecks, slowdowns, and loopholes in processes. Finally, we will turn our attention to writing manuals: users' guides (which inform a user about what to do when), operators' guides (which similarly inform system operators of system activities), and technical manuals (which assist those who have to maintain the system after implementation).

The chapter ends with a discussion of procedures as programs for people (are people programmable?), probes what is not "proceduralizable" (what you cannot hope to program), and discusses the limits of knowledge in systems analysis.

We will define **procedure** as follows:

> A procedure is a *sequence* of *steps* taken to *convert* one situation into another *over time*.

This definition has three parts. First, a procedure consists of **discrete steps** or stages, sometimes called *instructions*. Second, these instructions are **sequenced** with one following all previous ones and preceding all the subsequent ones. In other words, individual steps can be numbered. Finally, the steps must be **active**, converting one thing into another or changing some situation. More important, each step must be **demonstrable**; we must be able to show that each step has a **measurable** *effect* and, even more important, there should be a **noticeable** *result* of the cumulative activities of these steps.

Figure 9-1 shows the procedure that sets up the entry of a chapter of this book in WordStar™. Note that there are nine discrete steps, in order, and that each step has a noticeable result. The result of these steps is a page setup for entering a new chapter.

1. Examine list of file names beginning "CHn" where n is the number of the chapter to be entered.
2. If such a file exists, perform the EXAMINE CHAPTER procedure.
3. Press D
4. Respond with the name of the chapter outline file: CHn.OUT and press the Enter key.
5. When the message NEW FILE appears and the command menu is drawn at the top of the screen enter CONTROL-V. "INSERT" will go off on line 1.
6. Then enter CONTROL-O followed by S, followed by 1. Spacing is set to 1.
7. Enter CONTROL-O followed by J. Justification is set off.
8. Enter CONTROL-J followed by H followed by 2, Help menus disappear.
9. Enter .HE   CHAPTER n

FIGURE 9–1. A Procedure for Using WordStar

Figure 9-2 shows something that is *not* a procedure. There are steps, but it is not clear in which order the steps are to be performed. There also seem to be several steps without any demonstrable effect(s), and overall it is not clear that the end result matches the predicted result.

1. Turn on the computer or leave it running
2. Notice the number of disk drives.
3. Don't be concerned if the red light is on. It will go off.
4. Select an appropriate drive for the diskette.
5. Insert the diskette the right way.
6. Ignore error messages at this time, but remember them for later.
7. The machinery will work for a while.
8. You'll notice that a list of files will come up.
9. Don't do anything now.

FIGURE 9–2. A Non-Procedure

In many ways a procedure is like a tightly-coupled system. Each step is an element, and the steps are related so that each one depends in some way on one or more previous steps. The goal of the procedure can be equated with the goal of a system, so we could define a procedure as a system of instructions in which relationships are restricted to a kind of "serial dependence" of the sort just described.

The difficulty with procedures is not that they are complex, but that we have to *specify* or write them. What seems clear to us in mental images suddenly looks like a bowl of wet, steaming spaghetti, hard to grasp and too hot to handle. Remember the first time you had to help someone program in BASIC or Pascal?

There are several reasons why it is hard to describe procedures. We often lack a good, complete, **intuitive language**. In giving traffic directions, there is a certain language you use:

Go about three traffic lights north on Highway 17 until you see a fast-food chicken outlet on your left. At that intersection turn right and then make an immediate left about fifty yards down the road. Look for a red house with a yellow fence around it just a short drive down the first gravel road on your right. You can't miss it.

Each kind of procedure seems to have its own language and mastering that language is almost as hard as mastering the procedures themselves. In fact, most learning situations involve a period of time learning the instruction language. We call that language a **metalanguage**, or language about language. The metalanguage of systems analysis is the terminology of General Systems Theory (GST), as introduced in Chapter 1. Most "chapter ones" introduce terminology and attempt to bridge the language canyon.

The second problem with procedure specification is **clarity**. Being clear means avoiding words with special, private meanings. For instance, the phrase "just a short drive" in the directions given above may be only half as long for the resident as for the visitor. Being clear also means keeping the sequencing obvious and the references apparent: which part that we assembled in which step do we use now?

A third problem is one of **feedback**. The city I live in is notorious in a special way. Neighborhoods usually have only one or two exits, and most streets in a neighborhood have similar names (such as Dalford Road, Dalford Hill, Dalton Court, Dalton Mews, Dalton Drive). It is easy to find the neighborhoods but once you are in the neighborhood, try to find a street address! I always assist visitors with feedback information: "You know you've gone too far if you hit 53rd Street....I live on top of the hill....It's a white bungalow....If you run into a T intersection, you've gone too far...." While not an active part of the instructions, feedback helps the performer of the procedure to know that the procedure has been performed correctly so far.

Specificity, clarity, and feedback are the major criteria for a good procedure language. Not all "languages" resemble written or spoken language. We can specify procedures in a variety of ways (see Figure 9-3):

- written in English or in a special jargon
- written in "structured" English, with unclear language removed (see below)
- written in pseudocode, a very highly restricted kind of structured English;
- composed in a programming language such as COBOL, BASIC, or Fortran
- detailed in a graphical language such as decision tables, decision trees, or Warnier-Orr diagrams
- demonstrated in a film or videotape or a sequence of images as in cartoons or videodisk.

1. **COMMON PARLANCE** — Common Language of Users
2. **DECISION TREE**
3. **DECISION TABLE**
4. **STRUCTURED PARLANCE**
5. **PSEUDOCODE** — Almost A Programming Language
6. **PROGRAMMING LANGUAGE**
7. **FLOWCHART**
8. **MOVIES**

FIGURE 9–3. Varieties of Process Logic Representation

A tutorial on pseudocode shows up later in this chapter, and some graphical "languages" are discussed in Chapter 13. Generally, IRMaint procedures are composed in pseudocode, augmented by a variety of drawings, screen mock-ups, forms, and report outlines, although recent advances in graphics and videodisk-computer hybrids make it possible to present procedure specifications in enhanced visual modes.

Structured English is preferred by many as a first approximation to a restricted but clear specification language for procedures, especially procedures that are handed to people rather than implemented in a programming language. An example of the evolution of a procedure from an idea to English to structured English is given in Figure 9-4 (opposite). Note that in this passage, adverbs of time and quality disappear, prepositions are fine-tuned, verbs become active, "is" disappears, and references by pronoun or adjective (this, that) are replaced with specific descriptions.

While structured English *is* a good vehicle for presenting procedures to people, we will use **pseudocode** for three reasons. First, by having only one standard, everyone will be able to read procedures during the logical design stages — "padding" can be added later to make the procedure more palatable to human tastes. Second, structured English is appropriate only for English, not for French, German, or Swahili, whereas psuedocode is easily translated among natural languages. Finally, pseudocode sits precisely between natural language (or, as it should be termed, "common parlance," the common style of speaking) and programming languages. In fact, pseudocode is easily translated into structured BASIC, as you will see later in this chapter.

To dial a number that you've already called, the procedure should be natural, easy to use and obvious. It also can use only the buttons on the instrument and the switch hook.

When you want to redial a number, push the switch hook down twice rapidly and when you have a dial tone, press the number sign, located on the lower right-hand corner of the set.

REDIALING:
1. Depress switch hook twice, within two seconds.
2. Press # key.

FIGURE 9–4. Idea to English to Structured English

9.1 STRUCTURED PROCEDURE WRITING

Writing a procedure resembles writing a story. There are three important aspects: the "plot" (the structure of the activities to be performed), the "action" (a description of each activity), and the "motive" (the goal of the procedure). Each must be taken into account. To make production of the procedure's text as consistent as possible, analysts have adopted a style of procedure writing modeled after the style of programming, called "structured programming."

Structured programming is actually a theory of the construction of computer programs for what is termed the **von Neuman processor**. Based on a theory of computation put forward by John von Neuman in the 1930s, the original electronic digital computers, called "von Neuman processors," shared a number of common traits:

1. The processor could do one thing at a time.
2. The processor performed a stored program or list of actions one at a time.
3. The processor had addressable memory units that could contain either data or program steps.
4. Program "stepping" was normally performed beginning at one specified memory address in ascending numerical sequence unless altered by an action that explicitly directed sequencing to another location.

Von Neuman processors are not limited in terms of the actions they can perform, but traditionally they have performed arithmetic, comparisons, conditional branches, or alterations in sequence based on comparisons, input, output, and halt. A reader familiar with BASIC will recognize that language's heritage in early theories of computation in the LET, IF, GOTO, READ, PRINT, and STOP statements.

In the early 1960s, researchers showed that programs composed to run on von Neuman processors also shared common traits, by virtue of the definition of the processor. In a classic paper (Bohm and Jacobini, 1964), the concept of "structured" programs was developed as a way of describing all programs. The following section describes this theory.

Three Basic Structures

The thrust of Bohm and Jacobini's idea is that each program written for a von Neuman processor can be expressed equivalently as a structure of the following three structures:

1. sequence ("AND THEN")
2. alternation ("OR ELSE")
3. repetition ("WHILE")

Sequence (Figure 9-5) is a "structure" of two activities in order. The order is given by the arrow joining the two boxes: A AND THEN B. When describing a sequence, we

FIGURE 9–5. The "SEQUENCE" Structure

may or may not use the "AND THEN" connective. The normal directions of reading (left to right, top to bottom) signals AND THEN:

> Separate the three copies. Send the PINK (top) copy to BILLING. Send the BLUE (middle) copy to the WAREHOUSE. Retain the YELLOW (third) copy for your records.

Although the *sequence* structure mentions only pairs, the theory allows us to drop the outer layers of boxes for convenience. This four-step sequence is diagrammed as in Figure 9-6.

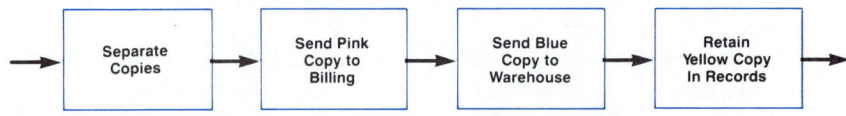

FIGURE 9–6. An Example of the a Complex SEQUENCE structure

Alternation (Figure 9-7) is sometimes called "IF-THEN-ELSE" because that is the way we read it. IF a condition holds, THEN do one structure, ELSE do another. There are two alternatives: one is used if the condition is TRUE, the other if the condition is FALSE:

IF the student has graduated within the last five years,
THEN access the record in the recent-grad file.
ELSE access the record in the archives file.

The clause following IF is called the "ante-cedent," or condition, clause; the one following THEN, the "consequent" clause; and the one fol-lowing ELSE, the "alternative" clause. It is not necessary to have both a consequent and an alter-native. Consider this:

IF the student is late for the examination
THEN fill out the examination-tardiness form.

There is no need to do anything if the ante-cendent is FALSE.

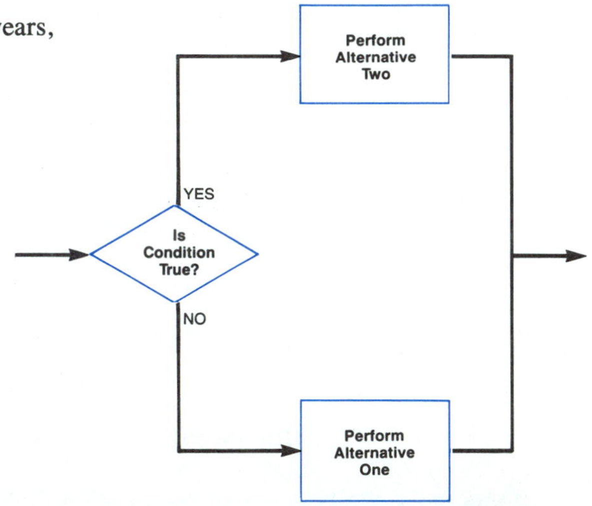

The **iteration** structure specifies that a struc-ture is to be repeated (Figure 9-8 on the following page). Sometimes called "DO WHILE," the use of this structure is illustrated below:

FIGURE 9–7. The IF-THEN-ELSE Structure

DO WHILE there are any more students on line
    Accept exam-admission slip
    Check student-number with student-id-card
    IF the numbers match
     THEN admit the student
     ELSE send the student to the registrar
    END-IF
    Send the student to the next available desk
END-DO

For ease of reading, we indent each new clause, thus indicating which activities belong in which structure. However, there is still the possibility of confusion. The activity "send the student to the next available desk" could easily be lined up to look like part of the alternative structure in the IF-THEN-ELSE structure, although it would not make sense. It would confuse. We get around this potential confusion by bracketing clauses with END-IF and END-DO.

The DO-WHILE structure allows us to **repeat** activities as long as some condition is true (in this case, as long as "there are any more students on line"). A variation of this is called DO-UNTIL. This structure repeats until some condition *becomes* true. (What is the equivalent DO-UNTIL form for the procedure above?)

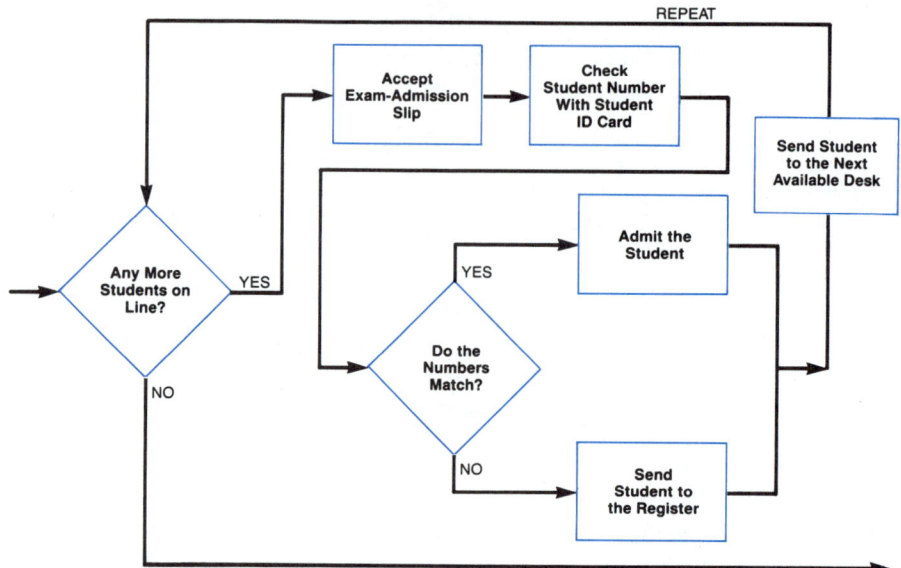

FIGURE 9–8. The DO-WHILE (REPEAT) Structure

If this appears to be a way of organizing flowcharts, that is an apt perception. The theory states that all programs intended to control a von Neuman processor can be converted into one or more equivalent structured programs. (What is another version of the student exam admission "program" above?) But the theory can be used to say a lot more about both programming and procedure writing.

The Assumptions in These Structures

Figure 9-9 lists the assumptions and limitations of the structured approach. First, a structured specification of a procedure is limited to describing sequential processes that begin at a certain time and end at a certain time, termed **linear, sequential** processes. Some processes are neither linear nor sequential. Consider a student studying for an exam or a manager trying to think up alternatives for a decision.

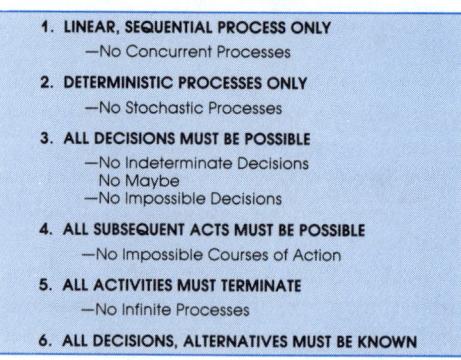

FIGURE 9–9. Limitations of Structured Program Approach

Second, we can describe only **deterministic** processes. Each arrow signifies that an activity is definitely determined by previous actions and previous actions only. We cannot say, "Maybe this activity is performed and maybe not." There are many important non-deterministic, random processes in business and management.

Third, the IF-THEN-ELSE structure requires that all antecedent clauses be either true or false and demonstrably so. In other words, anything described following IF must be **testable** for truth in some period of time. We cannot effectively diagram unresolvable alternation structures such as "IF the barber of Seville shaves only those who do not shave themselves and he shaves himself THEN ..." since the antecedent clause is logically ambiguous (or so said Russell in his *Principia Mathematica*).

The fourth assumption is that all **antecedent conditions** have either consequent or alternative actions that can be taken. In other words, there are no "impossible" courses of action. Boxes cannot be left empty.

Another assumption is that the von Neuman processor has to quit at some time and say, "Here is your result." The implication of this is that every box in the structured specification diagram has precisely one exit or termination.

Finally, the major assumption is that the structure and nature of the activities do not change over time. We know, in advance, which decisions have to be made, which structures have to be repeated, which activities have to be accomplished *at the time the procedure is designed*. There is no way to specify a procedure that changes itself somehow.

These limitations arise due to the nature of the von Neuman processor and the purpose of the theory (which is to describe programs intended for the von Neuman processor). The implications are many and are presented in the next section.

The Implications of These Assumptions

Figure 9-10 on the following page, illustrates how the theory of structured programming applies to computer programs. Some subset of human activity is to be modeled as structured programs in a flowcharting process. When these programs are implemented in code (coded in a language), they control some subset of computer activities. If we wish to apply the theory of structured programming to the specification of procedures for people as well as computers, we should be aware of the limitations that these modeling processes create. While in most cases the limitations are not crucial, we *are* limited in what we can attempt to model.

The following procedures cannot be effectively modeled or expressed as structured programs (except through tricks of language) for the reasons given:

1. coming up with a solution to a complex problem (non-deterministic)
2. trying to understand why someone is acting a certain way (non-linear)
3. understanding arguments someone is making (nonsequential)
4. making a complex business decision (lots of "what if" or "maybe" decision points)
5. carrying out a dubious decision (some impossible actions)

6. carrying out a complex business, sporting, or military maneuver (the structure is not known in advance)
7. running an operating system (infinite process with much concurrency)

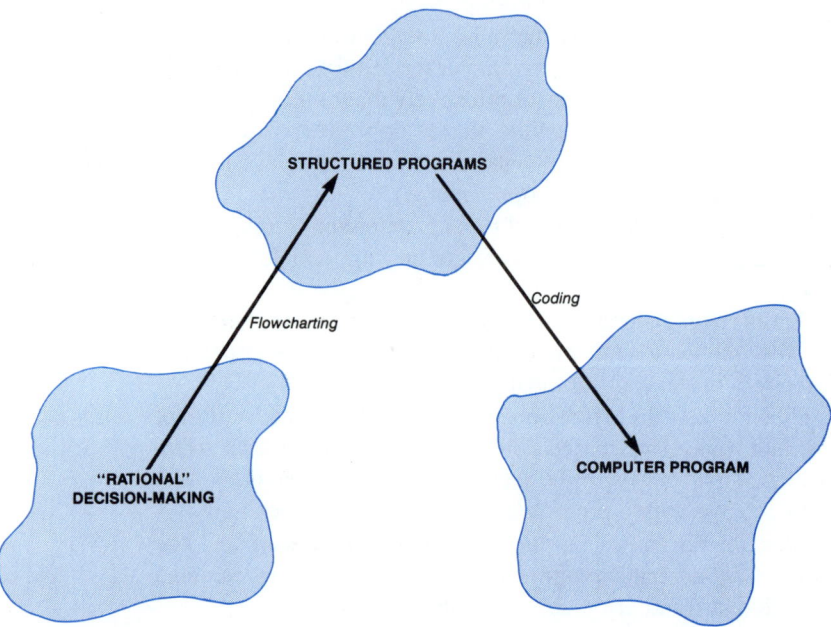

FIGURE 9–10. Applying the Theory of Structured Programming to Programming

A quick analysis of this list shows that many nonroutine aspects of business and most creative or artistic behaviors are difficult to handle using the language of the theory of structured programming (perhaps we need a "non-Neuman processor?"). In other words, we use structured procedure specifications at our own risk.

On the other hand, our purpose *is* to **proceduralize** aspects of managment and business. When doing this, we should pay attention to those procedures that are profitably proceduralized, and the discussion above provides tests to the analyst for what not to proceduralize. The following section expands on this idea.

Writing Procedures for Real Situations

A real situation, unlike a textbook scenario, is full of indeterminacy, concurrency, dead ends, and lack of design. In a real situation, people often do not know or do not wish to know all the facts, have to do many things at once, are sometimes stymied, and often do not have instructions for some conditions and are left to their own "discretion." In the face of this relatively pessimistic statement, what is the analyst to do?

First, merely recognizing that not everything can be specified step by step is a qualitative leap for many analysts and managers. The human mind is capable of accepting a great deal more fuzziness than most computers; experienced workers often muddle their way out of problems on the flimsiest of data.

Second, the theory of structured programming *does* provide a loophole in that the activities can describe almost anything. The following is an acceptable "structured" procedure:

```
IF a meteor is falling on your head
  THEN call Superman for help
    IF Superman arrives
      THEN breathe a sigh of relief
      ELSE breathe your last!
    ENDIF
ENDIF
```

The third thing we can learn from the limitations is that in writing procedures, often we will be stumped. *If a procedure cannot be written, then more research needs to be done.* The theory keeps us honest (if we use its language honestly).

Finally, we can introduce **concurrency**, **indeterminacy**, and **flexibility** to design by expanding the language of structured programming. Because a business organization is demonstrably *not* a von Neuman processor, we can add terms to indicate concurrency ("MEANWHILE"), indeterminacy ("SOMETIMES"), and flexibility in design ("REFER TO X"). Consider the following procedure:

```
Find out if the dam will hold for another hour.
REFER TO manuals for new procedures to be available 10-22-89
MEANWHILE begin emergency procedures
  Phone the regulatory authorities
    IF the regulatory authorities are available
      THEN pass the buck to them
      ELSE SOMETIMES phone the EPA
        SOMETIMES phone the local authorities
    ENDIF
END-MEANWHILE
```

The style of procedure writing we have developed so far in this chapter is called "pseudocode," a programming language that is not intended for any computer. The following sections present a tutorial on pseudocode for the sensitive procedure writer.

9.2 PSEUDOCODE

Pseudocode is a programming-language-independent language with which to specify procedures that are linear, sequential, deterministic, and predefined. Because it is independent of the specifics of any programming language, analysts can use it to specify the logical structure of procedures without limiting description in the early stages to "what can be done in COBOL" or "what Cosmos 944 BASIC can do." It is therefore somewhat **machine-independent**, too. We say "somewhat" here in light of the previous lengthy discussion of the assumptions underlying the theory of structured programming on which pseudocode is based.

A Definition of Pseudocode

Pseudocode is not meant for computers; it is a way of expressing structured procedures using the following language:

1. (nothing): expresses sequence in normal typographical order
2. IF-THEN-ELSE: alternation
3. DO-WHILE (or DO-UNTIL): repetition
4. ENDIF and ENDDO: brackets
5. RETURN: end a procedure before the last line and return to the procedure that invoked this one
6. CANCEL: stop a procedure (RETURN from a main procedure is equivalent to CANCEL)
7. DO: perform a procedure

Some writers alter the definitions slightly. For example, William Davis (1982) uses the construction IF *a* THEN *p* SO *x* ELSE *q* SO *y*, where *a* is some condition, *p* and *q* describe the possible situations, and *x* and *y* specify consequent and alternative activities (for example, IF it is March 31 THEN it is the end of the fiscal year SO DO End-of-year ELSE it is normal end-of-month SO DO End-of-month). Some writers specify the word END alone to replace END-IF and END-DO. The words CANCEL and RETURN are not in the original vocabulary of the theory but have been added to make this pseudocode consistent with languages like BASIC and dBase III.

An Example: Putting on Your Shoes

DO WHILE there are any feet left
  Pick up a shoe
  Grasp right shoelace in right hand
  Grasp left shoelace in left hand
  Pull laces tight with hands and side foot
  IF you are right-handed
    THEN move right lace over left lace and grasp
      both laces with right hand, letting left
      hand drop
      Bend right lace around and under the left
      lace, pulling it through from above by the
      left hand

    ELSE move left lace over right lace and grasp
      both laces with left hand, letting right
      hand drop
      Bend left lace around and under the right
      lace, pulling it through from above by the
      right hand
    ENDIF
    Pull the laces tight.
    IF you are right-handed
     THEN use left thumb and right hand to make a loop
      in the left [now to the right] lace.
      Move the loop over the right [now the
      left] lace and grasp both with the right
      hand thumb and forefinger.
      Using the right thumb and forefinger, push
      the right lace over the bottom of the loop,
      loop the right lace over and through the space
      between the right lace and the loop.
      Pull the loop through with left thumb and
      forefinger and pull the laces tight.
     ELSE use right thumb and left hand to make a loop
      in the right [now to the left] lace.
      Move the loop over the left [now the
      right] lace and grasp both with the left
      hand thumb and forefinger
      Using the left thumb and forefinger, push
      the left lace over the bottom of the loop,
      loop the left lace over and through the space
      between the left lace and the loop.
      Pull the loop through with right thumb and
      forefinger and pull the laces tight.
    ENDIF
  ENDDO

Expanding Pseudocode — User Objects

    Pseudocode can be expanded to make it look more like a programming language
(see the next section). We can certainly simplify the procedure above by using subproce-
dures (that is, subroutines) that are written identically for left- and right-handed people
by saying something like "DO make-loop WITH hand" where hand is either "right" or
"left." We then can define the procedure "make-loop" to have a parameter called
"hand," which is substituted for the actual value of the parameter as in "DO make-loop
WITH 'right'" or "DO make-loop WITH 'left.'" This change simplifies procedure writing through **parametrization**.

Another way to expand pseudocode is by the addition of **user objects**. These are reserved words or phrases that have specific meanings as defined by a dictionary, for example:

DO-WHILE NOT EOF( ).
IF FIRST-TIME THEN...
DO procedure 10 TIMES.
WAIT

These objects introduce (1) the **logical operator** NOT, which inverts a condition, (2) the **reserved condition** EOF( ), which has obvious meaning, (3) the **processing condition** FIRST-TIME, which signals that this is the first time this activity has been performed in this particular procedure, (4) the DO-WHILE, which is done exactly ten times, and (5) the "programmed" **pause** to wait for synchronization.

Pseudocode and Programming Languages

Derived from the language of the theory of structured programming, pseudocode expresses the plot of procedures, with the text providing the action. The resemblance of pseudocode to languages like BASIC and some forms of COBOL is intentional, to make implementation of designed procedures easier and quicker to verify.

In fact, a language such as that used by dBASE III is very close to the pseudocode developed here. dBASE III is the language used to code procedures to be processed by the database manager in dBASE III. An example of a dBASE III program is given in Figure 9-11. A similar program written in BASIC is found in Figure 9-12. The pseudocode for these examples is found in Figure 9-13.

```
dBASE III™ CODE

NAME = SPACE (20)
@ 2,0 SAY "What is your
   name?";
   GET NAME
READ
ID = 0
@ 3,0 SAY "What is your ID
   number?";
   GET ID
READ
IF ID < 100
   @ 4,0 SAY "SORRY, you are
   not registered."
   WAIT
   RETURN
ELSE
   DO REGISTER
   DO ENROLL
ENDIF
```

```
BASIC

10 LOCATE 2, 1
20 INPUT "What is your name?";
   NAME$
30 INPUT "What is your ID number";
   ID
40 IF ID < 100 THEN
   PRINT "Sorry, you are not
   registered"
   PRINT "Press any key to
   continue...";:J$ = INPUT$(1)
   STOP
   ELSE GOSUB 1000 : GOSUB 2000
50 STOP
. . .
1000 REM ROUTINE TO REGISTER
. . .
2000 REM ROUTINE TO ENROLL
```

```
PSEUDOCODE

GET NAME
GET ID-NUMBER
IF ID-NUMBER < 100
   PUT "Sorry, you cannot
   register"
   STOP
ELSE
   DO REGISTER
   DO ENROLL
ENDIF
```

FIGURE 9–11. The Process in dBASE III

FIGURE 9–12. The Process in BASIC

FIGURE 9–13. The Process in Pseudocode

Analysts are cautioned that when they write procedures in pseudocode they are *not* writing programs; therefore, they should fight the tendency to become overly telegraphic and mnemonic. Pseudocode is really a technology for thought as much as a documentation technique: it helps the analyst puzzle out what the procedure means as much as what the actions actually are.

9.3 OPERATIONAL MODELS

So far in this chapter, we have spoken about procedures as though writing them constituted mere documentation of design, as though they were dry, lifeless specifications. In fact, procedures — and their models — can be used to *analyze* designs in specific ways. This section discusses the use of operational models (models of procedures) in finding bottlenecks, slowdowns, loopholes, and so forth in a proposed design.

Analyzing Procedural Models

Examine Figure 9-14, a traffic-flow diagram illustrating trips starting at D. What does it say to you? Apparently it indicates which path seems to be the most popular, but it also hints at some problems. Can you see them?

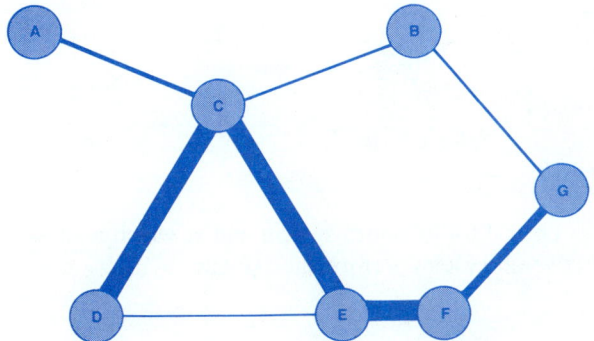

Thickness Indicates Approximate Traffic Volume

FIGURE 9–14. A Traffic Flow Diagram

First, the most popular route is not the most direct. It is a longer distance, but seems to be preferred. Second, the most direct route is the least popular; the shortest distance between two points obviously is not a straight line. Third, even though this shortest route is part of a path leading elsewhere, it is the least popular. Of course, if you also knew that this segment of highway is not paved, you might understand why it is avoided.

Operational models, models that show how something works over time, are useful in determining conditions such as those mentioned above.

What *is* an operational model? It is a model that follows the order of activities, one that can be described in terms of time and resource requirements needed to produce a known, predictable product.

A system flowchart is an operational model. Data flow diagrams seem not to be, but we often use them where strict time order is not important for discussion. A set of construction directions is an operational model, whereas an architectural drawing is not.

Directions to get to the corner store constitute an operational model, whereas a map of the city streets does not.

Finding Bottlenecks, Slowdowns, Loopholes, Error Generators, Losses, Misunderstandings, and Omissions

Operational models can be examined for the following kinds of problems or potential problems:

- **bottleneck**: point where work flows together and at which the ability of a subsystem to process work fast enough does not meet volume and time demands (Figure 9-15)

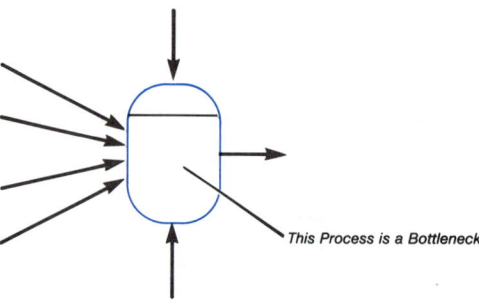

FIGURE 9–15. An Information Bottleneck

- **slowdown**: a subsystem that is critical to the entire system and at which a slowdown in processing will affect overall system performance (Figure 9-16)

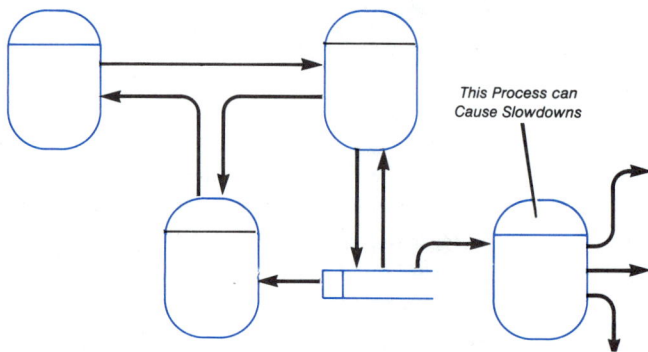

FIGURE 9–16. A Slow-Down

- **loophole**: a condition that is not accounted for, allowing an activity to proceed when it should not
- **omission**: a condition that is not accounted for, disallowing an activity when it really should go ahead
- **error generator**: a source of errors, brought about by a situation for which the probability of error is high

- **loss**: a "crack" in the process that allows resources to escape or become unaccountable
- **misunderstanding**: a subsystem whose operation is not completely understood or whose description or operation is unreliable
- **superfluity**: a subsystem that is never used because the condition it is designed to handle cannot occur
- **redundancy**: a subsystem that repeats the work done by another subsystem

Bottlenecks and slowdowns are **structural deficiencies**; they are problems with the structure of activities in the procedure. A bottleneck common to most students in the registration procedure is the usual long line to file completed rosters. A slowdown occurs if a professor is unavailable to approve entrance to a course designated "Instructor approval required." Until the professor becomes available, no approvals can occur.

Loopholes and omissions are **logical deficiencies** in the design of the procedure. Some conditions that occur have not been properly accounted for; a loophole lets something happen that should not and an omission prevents processing when it should not. A registration loophole is one that limits ordinary registration in a particular section of a course, but does not similarly limit audits. A class could fill up with eighty auditing students. An omission situation would occur when a student is wrongly refused registration because of an apparently incorrect registration form. For instance, the student may have two majors and have listed the "wrong" major in the single space allowed and is therefore denied entrance to a section marked "majors only."

Error generators and losses are evidence of **efficiency and effectiveness problems**. An error generator may be a process that is complex or difficult to perform. Students sometimes find the instructions for class registration confusing and forms are often poorly designed, thereby increasing the chances of error. Where a single individual handles registration processing, the chances for error rise with the volume of students registering. Loss situations can occur, for example, when multipart forms come apart ("Where is Ms. Jones's yellow copy?") or when tear slips are accidentally torn off.

The final three problems are **design and documentation difficulties**. If a process is not understood, then its description will be "nonprocedural" in the sense that the structure or content of the process cannot be specified in advance. One symptom is the well-known look of frustration that people exhibit when they don't know what to do. Superfluous processes do not harm anything, but they waste time in design and construction. There is no reason to design a procedure and train people to process a form that will not be used. However, there is usually no reason to build in redundancy. Why do something twice if once will suffice?

Sometimes, though, redundancy is a way of checking for error generators. If a process is error-prone, then another process can check against the error. Consider a machine that scores optically-scanned tests. It randomly makes one error in every 1000 tests. Suppose the tests are rescored with another machine, which also makes one error on every 1000 tests. If we compare the test scores, we will notice differences in scores on just under two tests in 1000. Redundancy helps us catch the errors. As an exercise, how could we use a third machine to eliminate the need for a manual check on differences in scores?

We can use system flowcharts and DFDs to check for bottlenecks and slowdowns in the detailed investigation. Loopholes and omissions are found through logical analysis of all possible conditions in examining decision tools such as decision trees, decision tables, and Warnier-Orr diagrams. Error generators and loss situations can be discovered by examination of procedures for clarity, complexity, human factors considerations such as readability, sequencing, speed, and the layout of forms and screens, as well as the physical integrity of forms and the existence of control tactics such as batch totals, audit listings, and history records (or audit trails). Superfluity is difficult to check for. It is hard to know that something will *never* occur, but certainly the analyst can sample situations and try to get an accurate figure for the probability that certain events will occur. Redundant processes that are not concerned with error control (batch totals are "redundant," but that are useful for testing the validity of data) may be hidden; a systematic approach to examining procedures will assist in finding them.

Misunderstanding or an unclear or incomplete procedure is easily determined from system documentation. If the analyst cannot complete a diagram, if arrows go off into the wilderness, or if boxes are essentially unlabeled, the procedure is subject to misunderstanding.

Improving Models through Research

We can improve the model of a process by using three techniques: prototyping (Chapters 4 and 16), simulation, and protocol analysis (Chapter 12). Since they are described in other chapters, only a few words are offered here on their use in procedure writing and charting.

Prototyping is a way of improving a particular model of a system or procedure at the time it is built. A model can be "mentally" prototyped in the same way. In fact, the "what-if" facility built into most spreadsheet processors provides precisely this capability. Consider the decision tree in Figure 9-17. Suppose we listen to a conversation

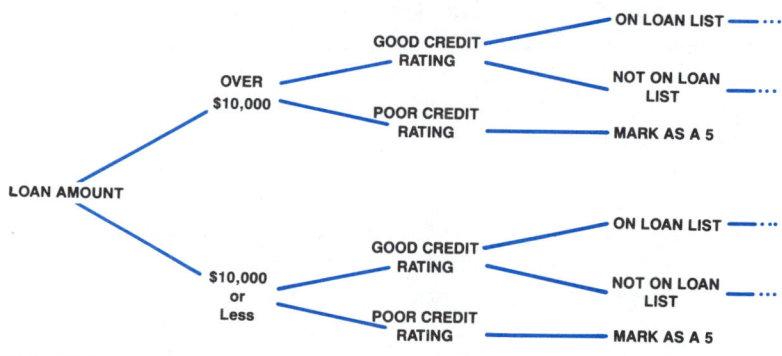

FIGURE 9–17. Bank Loan Decision Tree

between the analyst who drew up the model and the clerk who performs the processing of the form in question:

SA: Let's look at how you decide how to classify this applicant.

CL: OK. I see you've got things laid out neatly there on that chart.

SA: Yes, it's called a decision tree.

(Explains how to read the decision tree.)

Now I think from what you told me last week that the most important thing is the amount of the loan, whether it's less than ten thousand dollars or greater than ten thousand?

CL: Yes, although sometimes we get requests for half a million and then I take them to Mrs. Warren.

SA: OK, I'll add that over here. Do you have an upper limit?

CL: No, not really, at least not here. Maybe other branches have it but not us.

SA: Now, the next thing you consider is the credit rating.

CL: Yes, and that really is the thing that slows down these applications, because I've got to call about every one, even the $500 ones.

SA: Uh huh. Does it really not matter for the small ones?

CL: Not really. We've never refused a small loan to a depositor, in fact most just use their StarCards to get those small loans.

SA: So someone who didn't use the StarCard and was a depositor with a need for $500 would wait as long as a stranger who wanted half a million?

CL: I guess so.

SA: OK, next, if they fail the credit test you mark them as a five. Otherwise you look at the loan list?

CL: That's right. We can't lend money we haven't got.

SA: Shouldn't that be the first thing? Can't we make that test first and send people away if you haven't got money?

CL: I guess so, but we expect the loan officers to have already told them about our loan list. I guess we really don't know if the customer knows, but we have to go ahead and process them this way.

The analyst is discovering bottlenecks, loopholes, and misunderstandings that can be improved.

SA: What if we change the procedure, put the loan list check first?

CL: Well, then we'd get back to the loan officers quickly, I guess, and make sure the customer knows.

SA: Now, suppose you just skip the credit rating for small loans to depositors, say those less than the StarCard limit?

CL: Well, even if we did it that way, I'd have to know their outstanding balances on accounts.

SA: We could put that up quickly on your terminal.

CL: Well, if we did it that way, then, sure, I guess we could speed up approval for small loans.

SA: And they are what percentage of all applications?

CL: Maybe twenty, thirty percent?

The analyst and the clerk are prototyping a procedure. While it is not clear yet that the new procedure is going to be accepted by management, the analyst is finding out what it is possible to try with the existing procedure. Nothing suggested so far is so outrageous that the clerk cannot understand or accept it.

Simulation goes one step beyond mental prototyping. By building a "model" system, the analyst can observe actors carrying out a procedure and note bottlenecks, slowdowns, and so on. In the previous example, for instance, the analyst could have the clerk attempt the procedure in a number of different ways using dummy, mocked-up forms. To speed up his observation, the analyst would simulate such processes as credit checking (otherwise the overall simulation would take days instead of minutes).

Protocol analysis is a combination of observation and interviewing and can be combined with simulation. The analyst not only observes the subject performing a procedure but also engages the subject in conversation about the procedure and its progress. In this way logical problems, the perception of errors or loss, and difficulties with misunderstanding are quickly cleared up. Let's listen in on the analyst and the clerk:

> *SA*: Here's the form. Follow this new procedure we talked about and let me know what you're thinking as you follow it.
>
> *CL*: OK. So here's the name, Mr. French, and he wants $600 for home improvements. Now I turn to the terminal — you want me to do that now?
>
> *SA*: Yes.
>
> *CL*: OK, so I key in the name here and I see he's got $1100 on deposit and only $240 outstanding against his StarCard, so I mark in a two and...then what do I do?
>
> *SA*: You send the form back to the credit manager...just put it in this box...and now?
>
> *CL*: I guess I pick up another form?
>
> *SA*: From where?
>
> *CL*: No, I don't. I see. If it's a two, then I take it to the credit manager, after all why should French wait until I've processed everyone else's forms?
>
> *SA*: Good. Then you do another?
>
> *CL*: You know, it would make more sense if I took all the day's forms, took out the small loans and did them first, wouldn't it?

The situation is moving rapidly from an I-design-you-try to a we-both-try-to-design situation. Using protocol analysis, both analyst and subject quickly understand the implications of changes to procedures.

The key to all procedures is documentation. In the next section, we turn our attention to the writing of manuals. Normally they are written during the physical design stage when the physical procedures are well known. Later we will examine the implications of manual writing to logical design.

9.4 WRITING MANUALS

A procedure is no good without documentation. Procedures that will be implemented in software usually are written in pseudocode, with documentation of the procedure intended for software designers, coders, and maintainers. Procedures that people have to follow are documented in two ways. First, those intended for user-operators are put into an operations manual that describes step by step what must be done. Those meant for user-consumers have a different emphasis. They are written in terms of the consumers' goals: produce a report, retrieve an item, enter some data. Documentation does not illuminate how the system functions so much as how the user has to think about the problem to get it solved by the system.

This section looks at these kinds of documentation. It first discusses the general function of documenting procedures. Next it discusses **users' guides**, handbooks for people who want to use an information resource to meet a need. The third topic will be **operators' guides**, which are technically-oriented documents that instruct an operator of a piece of software-controlled equipment or an operating system. Finally, we will turn our attention to **technical manuals**, which detail the internal operation of the software and hardware and the overall design of the system.

Documenting Procedures

We have been emphasizing the modeling role of procedures, but do not forget that they eventually will be carried out. This implies several documentation needs:

1. Procedures have to be written so that the ultimate performer can understand them to carry them out.
2. *How* and *when* to employ a procedure must be documented as well as the form and content of the procedure itself.
3. Procedures may have to be taught if they are not habitual or obvious.
4. Updating software is easier than updating human beings' ideas about how to employ a procedure.
5. Manual procedures have to be audited and maintained just as do the software procedures.

Computers do not ponder the meaningfulness of what they do. "What's the point?" is not a question that computers naturally ask. But people will perform poorly-justified and seemingly irrational procedures grudgingly, haltingly, tentatively, and often ineffectively. People need to understand more than the words, the sequence, the content. They have to feel comfortable and confident that what they are doing at each step contributes to the whole. As work by Hackman and Oldham (1976) on the Job Diagnostic Survey has shown, job satisfaction depends on feeling integrated with the work, understanding the meaningfulness of what one is doing, and getting feedback on the value of the work.

Good procedures make their intent known. It is not obvious that there are two different procedures in Figures 9-18 through 9-20 and that each procedure requires a different set of keystrokes. The need to document procedures whose use is non-obvious often leads to a redesign of the procedure itself.

Another aspect of documenting *how* and *when* is to use advertising. Analysts may often assume that effective systems and well-designed procedures will be used because they are good and well designed. In fact, users cannot employ a procedure they are unaware of. "Undocumented features" are the treasure troves of hackers, but they surely do not help users who haven't the time to try all possible keystroke combinations — or put up with the results. Often a company newsletter or IRU pamphlet or update serves the public information purpose. Users who need to hear about a new, more efficient procedure have to be approached effectively — yelling at them will not do.

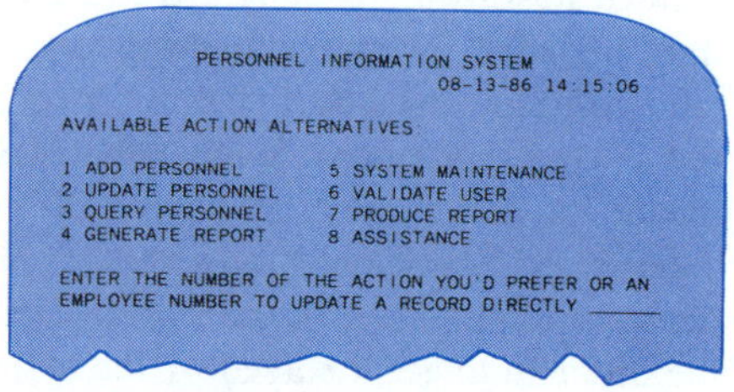

FIGURE 9–18. Sample Retrieval Screen (A)

Complex procedures are difficult to learn from documentation of use, and thus tutorial documentation needs to be written. Increasingly, tutorial software accompanies complex packages. Many of these include enhanced **assistant** modes. For example, the popular dBase III has an ASSIST mode, which prompts for every keystroke. The writing of **tutorial packages**, computerized and not, is now big business. Most analysts have little experience in doing this and require the assistance of professional educators.

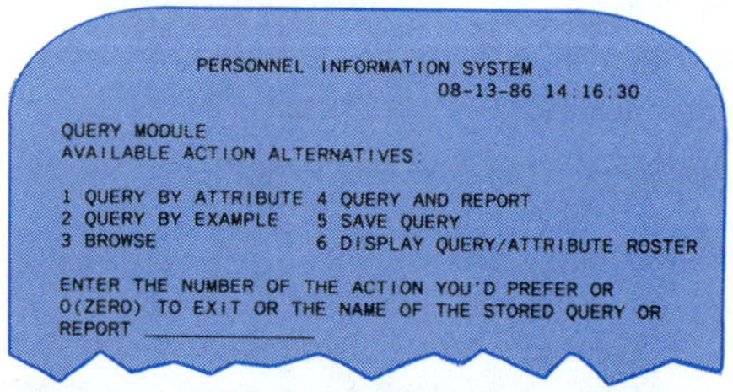

FIGURE 9–19. Sample Retrieval Screen (B)

One cannot merely plug a new procedure in users' ears and hope they will use it judiciously and correctly. A procedure change may have enormous implications that may result in effects ranging from flawless processing to a fatal shutdown. Where procedures are to be changed, public relations, education, and technical assistance are paramount.

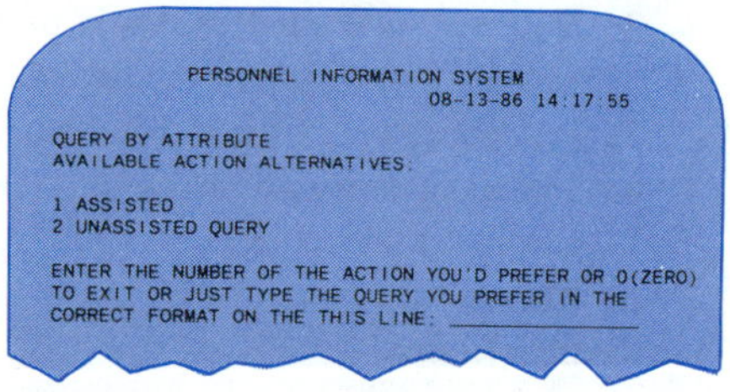

Procedure changes should be part of a **planned maintenance** program. While a lot of attention is given to maintaining procedures implemented in software, there is little literature on maintenance of human procedures. Analysts often incorrectly assume that users will let them know about procedures that need to be maintained and how to change them. But the performance of human beings is so complex that even observing a poorly-performed procedure may be not be illuminating.

FIGURE 9–20. Sample Retrieval Screen (C)

The famous studies by behavioral scientists from Yale of the Hawthorne relay assembly unit at Western Electric (Roethlisberger and Dickson, 1966) illustrate this. The conclusions of this study are still being debated, but the observations are valid. In an attempt to discover ways to improve productivity in the unit, the scientists altered lighting levels. Each time they raised the lighting levels, productivity increased. But when they started decreasing the levels, productivity still increased. In fact, productivity seemed highest at very low lighting levels. Scientists later attributed productivity increases to the effect on the women in the unit of being observed and paid attention to.

Later, re-analysis of the data tentatively concluded that the increases came from better group cohesion rather than individual feelings. In each analysis, however, the complexity of human behavior shines through. Currently there is no science of procedure maintenance. While many talk about "user friendliness," this concept is more hype than science. Analysts must understand not only the psychology of the individual user but the sociology of the context of work, and this "science" is not well developed.

The following sections look at three kinds of documentation of procedures: user guides, operator guides, and technical manuals. These manuals are produced much later in IRMaint, but logical considerations can have great implications for the content of the manuals.

The Users' Guide

The **users' guide** informs the user of how and when to employ a system (thought of now as a set of procedures accessing machines, software, data, and people). Therefore, the users' guide has to be written from the user's perspective, beginning with the task a user thinks is to be performed.

The following examples illustrate this point succinctly:

A: To update records in a patient's file, the user should depress function key 1, then tell the computer which keyname to look for. If it seems to go into an infinite loop, interrupt the computer.

B: To change a patient's record, press the CHANGE key. When the prompt line asks for "PATIENT NAME," type in the name of the patient, last name first without punctuation. If there is no response within thirty seconds, press the BREAK key.

The first is written in **cyberbabble** and does not address the user's task of changing patient data. The second example is written in the language of the user. Of course, certain technical terms (BREAK key, prompt line) may have to be defined, but these will appear in a glossary at the start of the guide.

The users' guide normally will be structured according to the goals a user needs to achieve rather than a technical person's need to segment system activity in some peculiar way. For instance, access to data may be done either sporadically or organized into batches. The analyst may be tempted to conceive of user activity only in one or the other sense, whereas the user may need instruction on both modes.

The users' guide may be augmented with a **reference manual**. The reference manual is a lexicon, organized so that a user who has a particular need can look up that need.

Here is a guide to the contents of users' guides:

1. a functional overview of the system's purpose and products
2. definitions of important, necessary technical terms
3. details of use of each function
4. efficiency considerations that the user can control
5. error conditions and their effects and how to recover from them; a list of error messages

6. security, privacy, and access threats (if required)
7. user responsibilities for data integrity
8. management considerations

Appendix A contains some examples from the PHYSICALC™ Users' Guide. PHYSICALC is a microcomputer-based package for physician billing through a provincial health agency in Alberta, Canada. The manual is intended for both the purchaser of the package (the physician) as well as the office staff who operate the system.

The Operators' Guide

Whereas the users' guide is written on the basis of users' problems or goals, the **operators' guide** stresses operational activities in interaction with machinery or operating system software. Generally, operators are concerned with providing service without consuming the results of that service themselves. They are shielded from the results of the application and usually observe only the hardware. Messages they handle are from lower-level operating system software and refer to the operational status of programs rather than to their products.

For instance, an operator may be informed that a specific program has begun, that it requires 240 Kbytes of memory, that it has been allocated one entire disk unit, and that it will utilize specific networking capabilities. Another message may illuminate the status of a program that has crashed with an E06 condition. A third message may request the operator to mount a disk pack or a tape.

These messages do not refer to what the users want the information resource to do for them, but what they need to get the programs to work. Therefore, operators' guides are written from the operations viewpoint.

Operators are concerned with **continuity**, **efficiency**, and **reliability**. A system should function with as little downtime as possible. It should produce results as quickly as possible, and it should not have to be restarted because it has made errors. Operators want to know about the status of systems, the hardware, and operating system software. Is the system running? If so, is it running up to par? If so, is it working right? If not, what needs to be fixed, replace, or removed?

Operational procedures are as important as others, and perhaps easier to write because most analysts are familiar with operators and their skills. However, the technology changes so quickly that it is not always possible to anticipate the hardware's buttons and lights. Operational procedures can be composed in a logical functional sense early in the design work ("System operation requires a start-up procedure that consists of powering-up, checking system status, and invoking the main subsystem.") and the specific physical actions put in later.

Appendix B details a section of the PHYSICALC Operators' Guide.

The Technical Manual

The programmers, analysts, and coders who follow logical design need to be informed about how the system works to accomplish the users' goals. This is documented in the technical manuals, of which there may be many:

- **technical reference manual**: documentation of the system design and all functions
- **programmer's guide**: details of programming conventions and special programming considerations
- **installation manual**: how to install the system, with emphasis on software generation, machine diagnostics, and connection of boxes
- **technical overview**: general design of software and data, often including the logical design
- **operational specifications**: benchmarks, efficiency considerations, error conditions, and recovery from the operator's viewpoint.

Technical manuals are based on the physical design of the system, which depends on the logical design. Most technical manuals are composed long after logical design is completed. On the other hand, the structure of the procedures is dictated early in the logical design, so it should be possible to begin writing some of the technical background early in the design exercise.

Appendix C presents a portion of the PHYSICALC Technical Manual. This manual is intended to assist programmers who want to make changes to the code. PHYSICALC was written in BASIC for the IBM PC and the technical manual is organized in terms consistent with that language.

9.5. PROCEDURES ARE PROGRAMS FOR PEOPLE

This chapter has concentrated on procedures as programs that people carry out. This final section examines two philosophical questions concerning procedures. The first is whether or not people can be programmed. The second looks at the limits of what can be made into procedures.

Are People Programmable?

If you give people procedures to follow, can you expect them to carry out those procedures as though they were machines? Obviously not, as this chapter has stressed. People require **motivation, rationalization**, a picture of the **whole**, and a feeling of **integration**. They make mistakes, they comment on what they are doing, they think of ways of improving their own work or the quality of the work situation — things machines are not supposed to do.

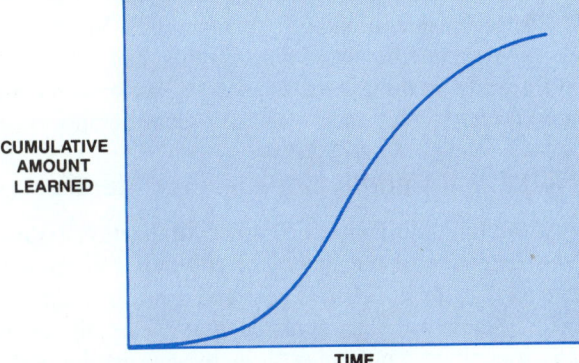

CUMULATIVE AMOUNT LEARNED

TIME

FIGURE 9–21. A Learning Curve

Even more crucial is the learning time. Figure 9-21 illustrates the **learning curve**. In the early stages of learning, an individual has little to relate additional experience to in order to derive a concept. Until that critical mass is reached, growth in learning is slow. Later, growth is explosive as each additional experience is similar enough to past ones to be integrated. After a while, no really "new" situations are encountered and learning plateaus. The learner is now maximally competent — until something changes (as it always will in a turbulent environment).

The time to learn is rarely factored into a system design. An elegant procedure may be unlearnable. Even more problematic is the fact that people may learn the wrong things. Consider a worker who, in puzzling how to write a report, stumbles on an old report and styles his on that one. If the report appears acceptable, the worker may continue to use that format, although it may not be optimal or even valuable. *Unlearning* this writing style may prove difficult for the worker.

Other workers may learn how to "get around" a system to make it easier to use. "It works" is what users tell consultants when asked why they use a particularly silly procedure to do something. That it didn't really work or that it was not really appropriate in any but special circumstances wasn't an early consideration during learning.

Other workers may decide to change a procedure because they feel it ought to be changed. If these workers then teach this new procedure to other workers, it becomes folklore — unofficial "software for the brain" that analysts will have to discover painfully during their investigations. Because work is a social situation, this kind of cultural diffusion is often as important as official training.

Finally, people just don't like being told what to do. In a controversial book called *Robopaths*, Yablonski (1972, pp. 6-18) discussed a pathological approach to procedures that some people adopt, slavishly seeking out procedures to follow. While workers may feel comfortable with a good procedure and may in fact attempt to proceduralize all their work (often to the disgruntlement of clients and customers who need something just a little bit different), few of them love being told what to do all the time. In famous experiments by Milgram (1974, pp. 13-26), his associates got apparently normal, middle-class Americans to "kill" others (the events were staged and no one was harmed) to demonstrate the lure of authority, but that lure is unreliable. Workers seldom feel comfortable with every procedure just because they are told to like it.

These considerations can be augmented by ethical, moral, and religious arguments, too, which would be out of place in this text. People can be indoctrinated, brainwashed, and forced to do things, but as analysts, we must remain aware of the kernel of humanity that will resist — for good reasons — our attempts to get people to behave like machines.

What You Cannot Hope to Proceduralize

If we limit our concept of proceduralization to those procedures that can be couched in the language of the theory of structured programming, the set of proceduralizable behaviors is quite small. The trend toward building decision-support systems (DSS) began because scholars realized that the really interesting and difficult problems in management were not going to be treated merely with information (which was the goal of the MIS movement). Instead, computers could assist decision makers without actually making decisions.

Keen and Scott Morton (1978, pg. 1) defined a DSS as a computer system that would assist decision makers without replacing them. This definition really asserts that computers can do some things well and people do others well, so why not have a felicitous mixture instead of trying to proceduralize management decision making?

Over the past ten years, the results have been disappointing. The jury is still out, but the trend has been to discard the original caveat and attempt, through the use of artificial intelligence approaches, to capture decision making using another model of procedures.

This model is called the **heuristic model**. Basically, a heuristic model is a learning system rather than a merely cybernetic one. The current DSS approach is to build what are termed *expert systems* that attempt to learn about a phenomenon from an expert and then turn that knowledge back on a decision-making problem.

In fact, such approaches have proven remarkably effective in well-defined decision-making areas such as chess playing. Other applications, however, especially in medical areas, have proven productive but not yet highly reliable. Where the rules are complex, changeable, poorly shared among experts, or inexact, this approach is still not very effective. In fact, we still have doctors, lawyers, engineers, and professional managers. What we no longer have is the humility that the original DSS thinkers brought to their task. **Knowledge** may be structural in nature and **expert knowledge** may be capturable, but **wisdom** so far eludes the trap of the written procedure.

Perhaps the clue lies in a comment by Gregory Bateson, famous for his work in communications and systems approaches, especially for therapy. He developed the concept, mentioned in Chapter 1, of **deutero-learning**, learning to learn. When the environment is merely disturbed, it is possible still to learn all the probabilities using a learning system. But where there is a sufficient degree of threatening turbulence, learning is not enough. We have to change how we learn, change our conception, and only then can we reprogram, that is, rewrite the procedures. So far we just have not approached this ability with computers or without them.

9.6   CONCLUSION

Chapter 6 detailed the function of modeling in systems analysis; this chapter looked specifically at **procedural models** that document how a system is to function. Chapter 14 will look at other diagrams that illustrate a system's logical, structural, and operational aspects; this chapter focuses on writing procedures for people to follow.

A procedure is a "program for people." Procedures are written to instruct people as to what to do, when, and with what, depending on the goals of the system and the relationship between the **agent** of an activity and the **results** of that activity. A procedure details activities to be performed with specific tools to accomplish specific goals at specific times.

There are three essential structures of procedures: AND-THEN, IF-THEN-ELSE, and DO-WHILE. These are taken from the **theory of structured programming**. The first is also called **sequence** because it specifies that two operations are to be performed in sequence. The second is also called **alternation** because only one of two alternatives is performed, based on some determinable condition. The third can be called **iteration** because an activity is performed iteratively until some condition is met. Sometimes the third structure is termed "DO-UNTIL" because the operation is performed as long as some condition is met. These three structures correspond to common programming structures found in most languages.

These structures have hidden assumptions. The first is that all processes are **linear** and **dependent**, meaning that a process depends on the results of previous processes and cannot begin until all prior processes are completed. Second, processes are **deterministic**, that is, the outcome of a process is completely determined by the input and the design of the process. Third, all decisions (IF-THEN-ELSE conditions) are possible — there are no "maybes" possible, no impossible decisions. Fourth, if an alternative is chosen, it must be possible at the time it is chosen. Fifth, every procedure has a single exit. Sixth, a procedure implies that all alternatives and the structure of the procedure are knowable in advance and that once specified cannot be altered.

These limitations make it difficult to create some procedures. In particular, office procedures are **concurrent**, **nondeterministic**, **indeterminate**, extremely **turbulent**, often **not repeatable**, and particularly difficult to plan — at least in some circumstances. It is, in fact, the major task of the information resource to collect and process data to remove concurrency, increase determinacy, reduce turbulence, increase repeatability, and facilitate planning.

Guidelines for procedure writing include locating concurrency, indeterminacy, stochastic elements, turbulence, unrepeatability, and planning problems before writing procedures. Prototyping may help reduce problems in procedure writing, as might simulation and protocol analysis, discussed in more detail in Chapters 16 and 12, respectively.

Procedural models are often written in **pseudocode**, a kind of simplified English that addresses the objective aspects of an operation. Pseudocode expresses the three basic structures of procedures and allows introduction of **user objects** into the procedure specification. Once procedural models are produced in pseudocode, the analyst can examine the model for **operational problems**, such as **bottlenecks**, **slowdowns**, **misunderstandings**, **loopholes**, **omissions**, **error sources**, **losses**, **superfluities**, and **redundancies**.

Written procedures are placed into three kinds of documentation. The first, the **users' guide**, details how a user sees, approaches, and uses the information resource. The second, the **operators' guide** details how the system is to operate. Since the system includes the user, the operators' guide must take user procedures into account. Finally, the **technical manual** details system functioning from an internal viewpoint, effectively treating both operator and user (who may be the same in small systems) as input and output devices. Technical manuals treat procedures as programs that these devices are expected to follow.

9.7 CASE: A CASE OF EYE-IN-THE-SKY PLANNING

Digital Eye-in-the-Sky, a firm specializing in digital photogrammetry, uses high-technology scanning equipment to convert aerial scale phototgraphs to digital records. Its customers include city planners, public transit systems, geologists, public safety departments, farmers, and others. The firm has been in existence for only five years, and last year it did over $2 million in business, with hopes pinned on doing over $4.5 million this year with an expanded staff of 27. Its equipment, run by skilled technicians, can locate trucks, trees, and even people. Digital Eye's services are especially prized by ski operators who peruse snowpack analyses, looking out for advance warning of avalanches.

Ken Johnson and Marie Wright have operated Digital Eye since it began. Over the years, the customer mix has changed, moving more toward the public sector and away from private enterprise. The number of jobs, the scale, and the geographic scope have changed, and marketing emphasis has moved to meet this trend. Sources of maps have expanded, too, to include satellite photos to augment, and even replace, lower-level airplane-originated material.

Ken's equipment changes fast. "If we don't constantly change to keep up with the technology, we're dead. We're the best and fastest in the area — our customers pay for quick response and the ski operators can't wait 'til summer — and we can do that only if we constantly stay abreast of the technology. It's expensive, but we do it."

Another expense is training the technicians who operate the software-controlled equipment. Digital Eye buys special photogrammetry systems, but Marie is in charge of a small group of information resource people who have to integrate the systems into the existing information resource, train the operators, and rewrite billing programs. Tania Coolidge does the marketing, keeping the pressure on Marie.

The latest piece of equipment will be on-line in a week. Marie is busy with last-minute testing of the software. Now she is turning her attention to writing procedures for the operators.

Questions

1. What are the criteria that Marie will be seeking for procedures she will write? What will she be looking for in training operators to carry them out?
2. What distinguishes user and operator documentation? Who is the real user of the equipment as opposed to the operator?
3. One aspect of procedure writing is sequencing and tracking of work for billing purposes. At Digital Eye, there are three shifts, but accounting works only during the day. Can you think of some procedures that need to be written for this sequencing and tracking of work?
4. Digital Eye is going to grow by over 125 percent this year alone. What effect is that growth going to have on the need for procedures? Assume that staff will be distributed this way by end of year:
   Management: 3, including marketing
   Sales: 1 full-time
   Production: 15 operators, 5 per shift
   Technical: 4, including 1 supervisor
   Clerical: 3 (marketing, billing, correspondence)
   Accounting: 1, plus shared use of clerical staff
5. The firm owns a minicomputer on which it does its accounting, billing, payroll, and some marketing; it uses word processing extensively for correspondence and production of manuals. There are no plans to tie photogrammetry work into the mini yet. The equipment they own is from several vendors; operators are trained on each piece as it arrives. Plans are to change the equipment yearly if necessary.

What problems do you foresee in attempting to link this equipment with the minicomputer from the point of view of procedure writing?

6. What difficulties do you see that Marie has in training operators on procedures? What, if anything, can be done?

Discussion Questions

1. Sal and Ted are having problems carrying out their assembly tasks. This chapter introduced three criteria for procedure writing. Analyze Sal and Ted's difficulties in terms of these criteria and propose changes that might have made their work easier.

2. Consider a relatively simple task such as obtaining a driver's license or registering an automobile in your jurisdiction. Write a structured procedure in pseudocode for this task, using the language introduced in the text.

3. Read the procedure below and convert it into a structured procedure. Discuss the relative merits of each method of expressing the procedure in terms of communication with other analysts, users, and programmers (you can use structured English):

   I'd say the best way to rent an apartment in this town is to look in the classified ads each day and complile a list of those in your area in the right price range with the right amenities. Then start phoning the list and ask about availability. I mean if they're not available, no sense going there, right? Then take this smaller list and go see the apartments. Make sure you check out those amenities, too, because they may be advertised but not available or not working or they may be miserable. Then you've got to choose, and I don't know how you do that, but make sure you can get the one you want at least two days before the end of the month because moving's a pain.

   What changes did you make and why?

4. Examine the procedure written below and discuss the changes you feel would be necessary to make it more efficient and effective, removing operational problems:

   IF it is within five days of the end of the month
     THEN IF it is within five days of the end of the year
       THEN DO End-of-year procedure.
     END-IF
   END-IF
   Accumulate sales figures for the month
   Produce weekly sales totals
   Produce monthly sales totals
   DO Produce sales totals for Department 1
   DO-WHILE there are any more departments
     DO Produce sales totals for next department
     Accumulate monthly sales totals
     IF sales-returns exceeds sales for department
     THEN Write Emergency Report
     ELSE Accumulate sales-returns totals
   END-DO

5. At the end of the section "Three Basic Structures," a procedure is described using the

DO-WHILE structure. The DO-UNTIL structure was mentioned as an alternative to the DO-WHILE. Rewrite this procedure using the DO-WHILE structure. What changes have to be made? Are the two exactly equivalent?

6. In the section "Finding Bottlenecks, Slowdowns...," redundancy is discussed as a "cure" for error generators. The question, "How could we use a third machine to eliminate the need for a manual check on differences in scores?" is posed at the end of the discussion on optically-scanned tests. Can you answer the question? Is redundancy always the best approach to error control? Can you think of situations in which redundancy is merely a waste of time?

7. The interview in the section "Improving Models Through Research," (right after the "Bottleneck" section) between the analyst and the loans clerk forms the basis for a prototyping exercise. Can you complete the analysis attempted on this interview in terms of locating efficiency, effectiveness, design, documentation, structural, and logical problems with the procedures the clerk follows?

8. Why are manuals so important if procedures are well written? What can the analyst contribute to users' manuals in the early stages of design? What limits the possibilities for completely self-documenting procedures?

9. This chapter ends on an apparently pessimistic note on the limits of proceduralization. People are painted as error-prone, hard-to-control free agents in performing procedures, and there seems to be little chance of proceduralizing management decision making to lighten this load. Is the message, in fact, pessimistic, though? What "hope" do we have in the fact that people do resist proceduralization? What evidence is there that mechanization (the use of mechanisms to leverage work) and automation (the routinization of aspects of work) are harmful and should be limited? What role should the analyst play in making the determination of what should be proceduralized? What consititutes valid training or valuable experience in doing this? Can these decisions be left to technical people?

⌐DESIGN EXERCISES
Using dBASE III

1. dBASE III has a built-in programming language that follows the conventions of structured programming very closely. As you TYPE the .PRG files, you will see the major programming language elements:

IF x	If x is true, then...
a	do these lines.
ELSE	Otherwise...
b	do these lines until...
ENDIF	here.

[The part between ELSE and b inclusive is optional; the IF and ENDIF are required.]

DESIGN EXERCISES
Using dBASE III *continued*

DO WHILE *y*	Repeat the lines labeled *c*
c	so long as *y* is true.
ENDDO	
DO CASE	Do one block of lines depending on
CASE *i*	which case is true at this time.
d	If none is true, then do the
CASE *j*	block after the word OTHERWISE.
e	The block beginning with OTHERWISE is optional. If OTHERWISE does not appear and none of the cases is true, then nothing is done.
OTHERWISE	
f	
ENDCASE	

There are also statements that look a lot like BASIC:

I = 1	Put the value 1 into I
J = "ABCDEFGHIJ"	Put the value "ABCDEFGHIJ" into J
K = STYLNAME	Put the stylist's name into K
ITEMCODE = "2030"	Change the itemcode to "2030"
PHONE = AREACODE	
+ "-555-6666"	Concatenate -555-6666 to the value of the string in AREACODE
X = .T.	Give X the logical value *true*
Y = .F.	Give Y the logical value *false*
? "[" + PHONE + "]"	Display the value of PHONE in square brackets
ACCEPT "?:" TO	
NEWNAME	Display the prompt ?: and have the user enter the NEWNAME

There are many statements for controlling the screen, such as GET, and READ. These will be covered in the exercises for Chapter 10. Statements which determine the environment include SET COLOR TO, SET TALK ON/OFF, SET DEFAULT TO, among others. You can explore all these statements by reading what dBASE III says about them using the HELP command. Here is an example:

HELP READ	Get advice on the READ statement.

Programs are created by the MODIFY command. Try this:

MODIFY COMMAND	
MYOWN	Create a program called MYOWN

DESIGN EXERCISES
Using dBASE III *continued*

The modify command provides a full-screen editing facility to create a program. Use the following commands in addition to those mentioned in Exercise 8 (^ means hold the Ctrl key down while pressing the next key):

^N	insert a line below the current one
^Y	delete the line at the cursor (CAREFUL!)
^T	delete the next word
^W	save the file when you are finished
^Q	quit and do not save the changes

dBASE III always makes a backup when you say ^W; its name ends in ".BAK" and you must use a number of DOS commands to rename files if you want to edit the old (.BAK) version. Be careful when making changes.

On the disk is a procedure called MONDAYS.PRG that goes through the STYLISTS file and counts the number of stylists who can work each kind of shift on MONDAY. The following codes are used:

M: Morning
A: Afternoon
E: Evening
D: Day (Morning + Afternoon)
N: Night (Afternoon + Evening)

D counts for each of morning and afternoon; N counts for afternoon and evening; a blank indicates that the stylist does not work that day. Unfortunately, MONDAYS does not work well. DO this program, testing and then modifying MONDAYS so that it works correctly. How many stylists work on Monday afternoon?

2. You can use MODIFY to write procedures that use dBASE III commands to go through files, produce reports, and interact with users. If you want to go through a file item by item, use this sequence:

LOCATE FOR *a*	Locate the first record that meets criterion *a*
DISPLAY	(or other computations)
CONTINUE	Find the next match

DESIGN EXERCISES
Using dBASE III *continued*

You can test for not finding a match by using this IF statement:

IF EOF()   If you are end of file then...

Using CONTINUE and EOF(), write a program named BUSINESS that goes through the SESSIONS file and prints out the NAMEs and DATEs for a given STYLNAME. For inputting the name, model your statements on those in MONDAYS. What information is there on work done by Frank?

CHAPTER 10

PRESENTING LOGICAL SYSTEM SPECIFICATIONS

OBJECTIVES

1. To list and describe the four functions of logical systems specifications
2. To describe the scope of system specifications in the areas of system operations, management, implementation, and evaluation
3. To distinguish the uses of logical system specifications by managers, designers, analysts, project managers, and system implementers and to outline their specific uses
4. To list the seven criteria for acceptable logical system specifications
5. To evaluate a given set of specifications against these criteria
6. To draw up and present logical system specifications that meet these criteria
7. To describe the function of approval of specifications at various levels of design work
8. To distinguish among three kinds of design reviews for approval purposes

PRESENTING LOGICAL SYSTEM SPECIFICATIONS

10 CHAPTER

10.0 ANYBODY WANNA BUY A HANDSOME BRIDGE?

"Now I know all of you are here because you *have* to be here, so there's going to be no bull about our agenda. I mean, after all, everybody *knows* we're a few weeks behind schedule, but, what the heck, you know the trouble we've had.

"Now, first I want to direct your attention to the tables we've included. There are forty-three of them, each a masterpiece of graphic art. The captions and explanations are good enough by themselves, so I'm not going to review them. What you see here is a summary I wrote after talking to our junior analysts, you know, the ones who asked all those questions. Well you gotta hand it to those youngsters, they sure do have a lot of energy. So I took all of their data and boiled it down to three sentences. You see those at the bottom of page three.

"Now that's not to say anybody really thinks that the exploration system is really just those three sentences, but I think they read well and are true. You'll notice that we've dispensed with all those funny charts and graphs you usually see in these documents; they're hard to read and, frankly, none of you can understand them, so I just boiled it down for you.

"If you look at pages sixteen and seventeen, you'll see the meat of the system design. It's built around front-ending several packages we can buy and putting some code together in Fortran for the rest. If you guys don't know Fortran, you can learn it; I've got some examples in there. The tables show what tapes have to be mounted when and that's going to be useful for you operators. How well the system will work is another matter, but we don't have to worry about that now.

"Over here, on page twenty-four, is a table showing the specific modules that have to be coded. We're firm on that, otherwise it can't be done. We've purposely left out the details on several of these, however, because we all know what a listing is and where reports go, don't we. It's all built around these packages that we can pick up cheap with the hardware we're proposing. These micros are really selling well; our competitors at Golden Sunset Oil bought fifteen last week. What they'll do with them, I don't know, but these packages are supposed to work for us.

"The other eighty-five pages of the specs tell about the system and how it's going to work, based, again, on those three sentences. I know that's the right design because, well, I've been in this business a long time and that's how these things always work.

"As a favor to you guys, I've left out all the usual stuff on installation, training, maintenance, evaluation, human factors, organization, and personnel. I've never seen anybody predict those right and besides, it's not part of coding and testing. So you've really just got the bare bones there.

"Now if you'll just look at these sixty-three slides I've prepared, we can go over the details..."

Here is a good example of how not to present logical system design specifications at a design review. Previous chapters have spelled out how to produce the final products of the investigation and representation activities of logical system design. This chapter discusses how to produce the specifications, which signify the end of interpretation. It is here that the analyst brings more than just skills into play. Knowing what things mean and how to take advantage of knowledge is just as important as the brute force of physical and mental agility. This is how the analyst supplies information and uncovers or applies it.

We begin the chapter discussing the role specifications play in achieving IRM goals. The middle part covers the criteria for good, useful logical system specifications and presents a guide to reading and composing them. The end of the chapter looks at how to conduct meetings leading to approval for project work.

Chapter 10 concludes the section on the products of systems analysis. The next three chapters discuss the tools used to collect organize and interpret data into specifications. Readers are urged to review the goals of logical system design discussed in Chapters 4 and 7.

10.1 SYSTEM SPECIFICATIONS

MSA The Software Company

Specifications convey the analyst's ideas to others.

The creation of logical system specifications is the culminating activity in producing the logical system design (Figure 10-1 on the following page). In a very real sense, specifications *are* the system design. In terms of the analyst's basic activities, specification is the result of the interpretation of available information. The analyst translates, then adds that information to what has been collected and represented. The analyst consciously adds to the pool of collected data. The result reflects, then, more than the data uncovered; it reflects something of the analyst, too.

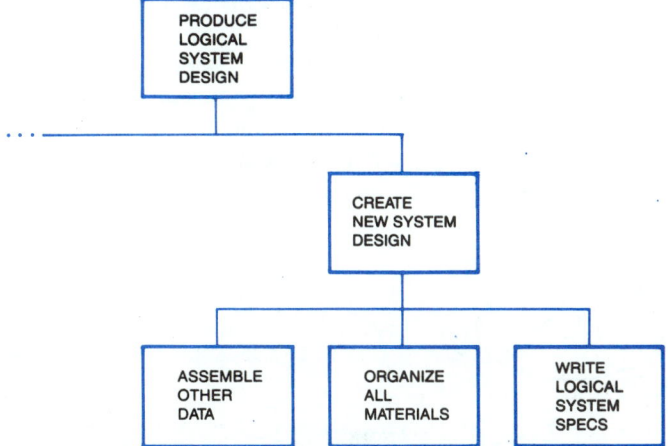

FIGURE 10–1. Goal Analysis: Creating Logical System Specifications

Interpretation is the expression of ideas, needs, and concerns in tangible form. The word comes from the Latin meaning "negotiator," the go-between, the person who irons out disagreements and cements contracts between people. The analyst has precisely this role: justifying seemingly incompatible data, merging seemingly inconsistent views, showing how impossible dreams can be reached in a logical fashion. Thus, the primary purpose in creating logical system specifications is *to interpret the collected data into a meaningful and valuable form* (see Figure 10-2). In particular, this interpretation refers to the *proposed system design* in terms of its logical structure and, often, its operation.

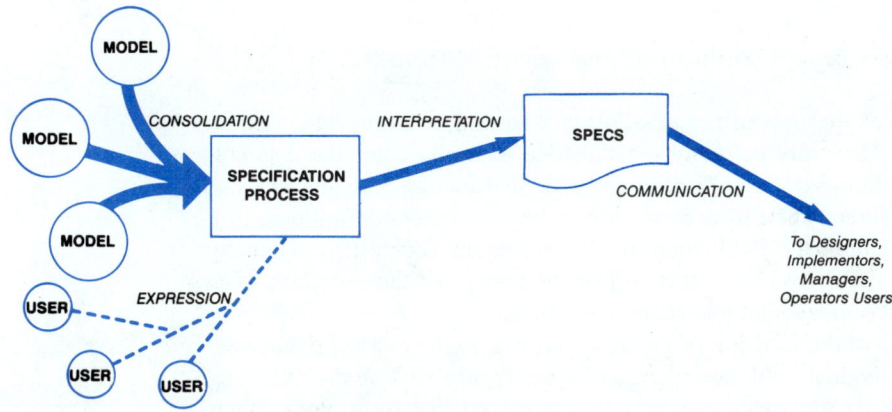

FIGURE 10–2. The Roles of Logical Specifications

The components of the specifications, which incorporate ideas from several minds and the needs of many groups, take the form of the preliminary investigation report, external information on hardware and software performance of the proposed elements, predilections and biases of analysts, consideration of what can be done (feasibility), and constraints placed on both development work and operating information resources. The analyst alloys these basic materials together and forges a design, stronger than any of the constituents.

A second function of the specifications is *to consolidate and justify seemingly conflicting streams of thought* (Figure 10-3). Contrast a supervisor's needs to reduce error rates and transit time on transaction processing with workers' needs for autonomy, discretion, and interesting work. Certainly there is a conflict here, not only of management and worker but also of the conception of the work process, its goals, and its products.

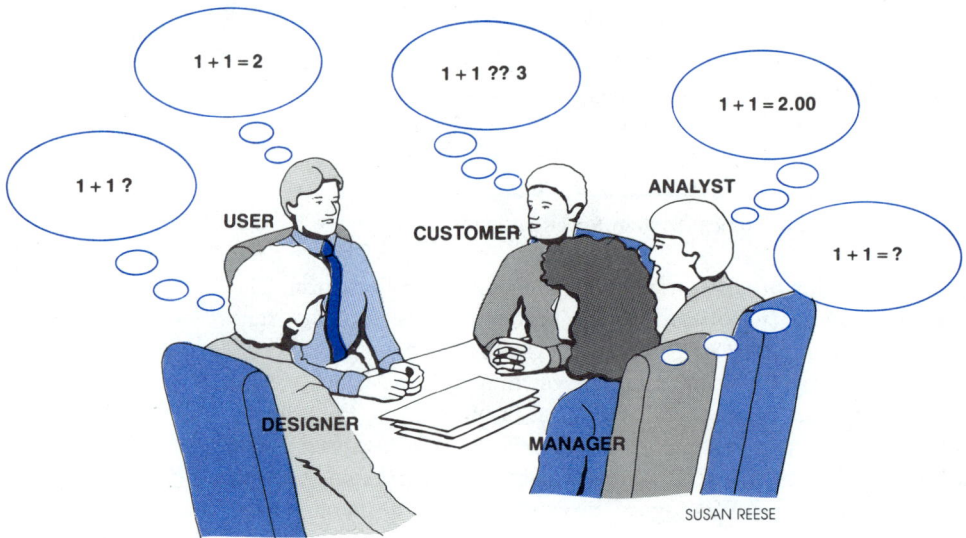

FIGURE 10–3. Specifications Bring Together Conflicting Schools of Thought

Consider the fears some express about the possibility of making work trivial, routine, mechanical, and boring. These are certainly incompatible with and color the data collected about potential computerization. Finally, consider that members of the analysis team may themselves disagree. Their disagreement may be over simple definitions, or it may extend to the feasibility of the IRMaintenance work overall. Given the complexity of many modern systems, such disagreements will be the norm, not the exception. The analyst must somehow consolidate and justify these conflicts.

Specifications also serve the function of *giving expression to the craft of the analyst* as well as the needs of individuals and the organization (see Figure 10-4 on the following page). That is, specifications are statements written by analysts that form part of their portfolios. An analyst who works because systems analysis is intriguing also works to produce high-quality, appreciated products. There is a high involvement of ego in designing systems. For instance, one analyst may be fond of on-line data-entry systems and may take criticism personally and react accordingly. Logical system specifications are expressions of the competences, insecurities, and dreams of analysts; managers and users cannot forget this.

Specifications give strong voice to the organization's stated and unstated goals: the appearance of being state-of-the-design-art, of being on top of or leading technology, of having an aura of efficiency and a high degree of cost-consciousness, or of being a "great place to work." Good analysts will not lose sight of these goals.

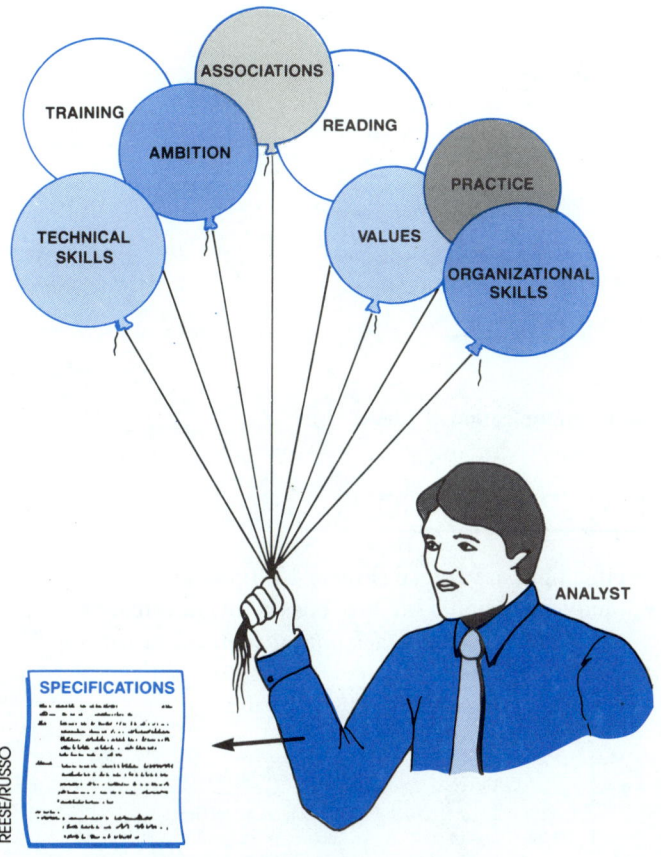

FIGURE 10–4. The Components of the Analyst's Expression

Specifications communicate all of these needs and goals to those who actually assemble, operate, and pay for the system (Figure 10-5). They serve as channels of communication between users and system designers, organizations and system operators, and analysts and programmers. They also chronicle the analysts' work: what they did, thought, got paid to do. Practically speaking, they pinpoint responsibility for deviations from expected performance. In stronger terms, they provide *a standard against which to assess system performance* and therefore act as a measure of the validity of design, coding, installation, and evaluation work.

In summary, logical system specifications interpret the data, consolidate thoughts and opinions, express the needs of the analysts and others in the organization, and communicate these expressions to others who have IRMaintenance responsibilities. The next section spells out in more detail the value of specifications to those who create and use them.

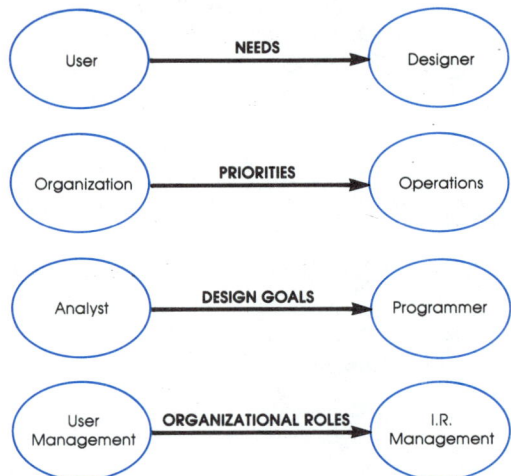

FIGURE 10–5. Specifications Assist the Communication of Ideas

10.2 THE VALUE OF SYSTEM SPECIFICATIONS

There are four areas in which specifications are crucial (Figure 10-6): system implementation, operation, management, and evaluation. All are phases of Information Resource Maintenance and Management. *Unlike system models, the specifications carry over into all other aspects of providing an information resource.*

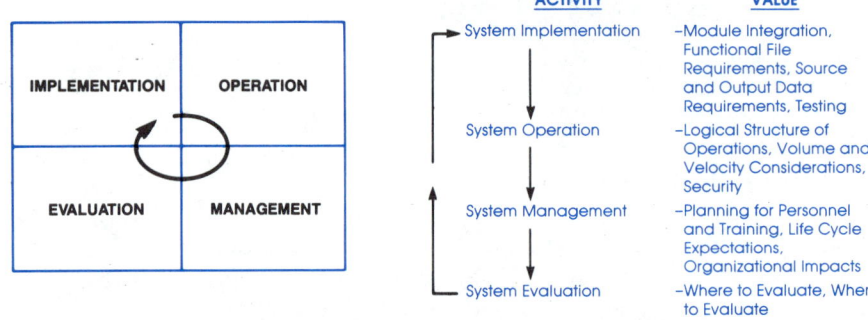

FIGURE 10–6. The Value of Specifications

System implementation must live up to the expectations set forth in the logical specifications (Figure 10-7 on the following page). These expectations spell out how modules are related to one another (at approximately one module per logical function); how modules interface with one another in terms of data; in what order modules should be implemented; where source data originally comes from; where output data (reports, documents, and forms) must ultimately go; file, data base, and record-management functional requirements; and ultimately a coherent schedule for installation that preserves logical functions. In a truly structured system implementation effort, testing is also performed in an order dictated by the logical system specifications, and test data can be derived from logical system specifications.

___ How modules relate (invoke, utilize) one another

___ How modules interface (input/output standards)

___ Order in which modules are implemented

___ Order in which coded modules are tested

___ Qualities of the source data

___ Destinations of the reports

___ File requirements (how many, where, contents)

___ Database requirements (i.e., what is needed to set up a database)

___ Records management requirements (type of access to type of data)

___ Schedule for implementation, installing, training

FIGURE 10–7. The Use of Specifications in Implementation

Logical specifications indicate *how elements of the system depend on one another*, whereas physical specifications, produced in subsequent phases, show how programs are to be coded, how files and forms are to be laid out, and how information is to be handled. Physical design follows from the logical system specifications.

After the system is implemented and installed, **system operation** then depends on both logical and physical specifications, although the latter are certainly more important to operators. However, certain aspects of system operations are spelled out in the logical specifications (see Figure 10-8). For instance, input-output relationships are of a logical nature: process x needs a and b as input to produce y and z as output. You cannot have the monthly sales report unless raw sales data are available. Smith will have to wait for her summary report until the tape librarian logs in this month's data. Jones cannot complete his spreadsheet until his assistant finishes his analysis of certain errors that continually creep into the accounts.

___ Input/output relationships

___ Volume considerations

___ Frequency considerations

___ Maintenance decisions

___ Operations philosophy

FIGURE 10–8. The Use of Specifications During Operation

Certain operations, too, cannot logically begin until others are finished. A tape cannot be dismounted until all the data on it have been read. Some volume and frequency considerations are reflected in logical dependencies: if a given input record produces n output records and if each transaction consists of m such records per day, then logically there will be $m \times n$ output records each day. Large files require large indexes; larger files should be approached through a database philosophy. Small files may not ever be necessary; major maintenance efforts on small files do not make sense. These and other *opera-*

tional considerations stem directly from the logical, not the physical, specifications.

An operational system creates **system management** concerns that are also derived from logical specifications. Personnel, training, and planning considerations depend on the logical structure of a system (Figure 10-9). Understanding the logical specifications may result in life cycle expectations for future maintenance, new interfaces with the new system, and reallocation of functions within the firm, with a corresponding redeployment of staff skills.

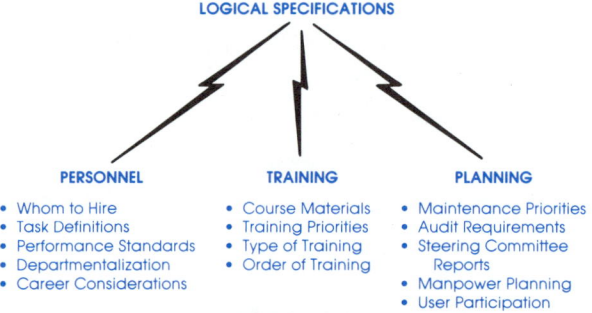

FIGURE 10–9. The Use of Specifications for System Management

Finally, specifications are valuable in planning for and executing **system evaluation**. The *physical specifications* relate closely to throughput, volume, frequency, accuracy, and error, as one would expect them to. The logical specifications spell out how elements of the information resource relate to one another (Figure 10-10), and while the

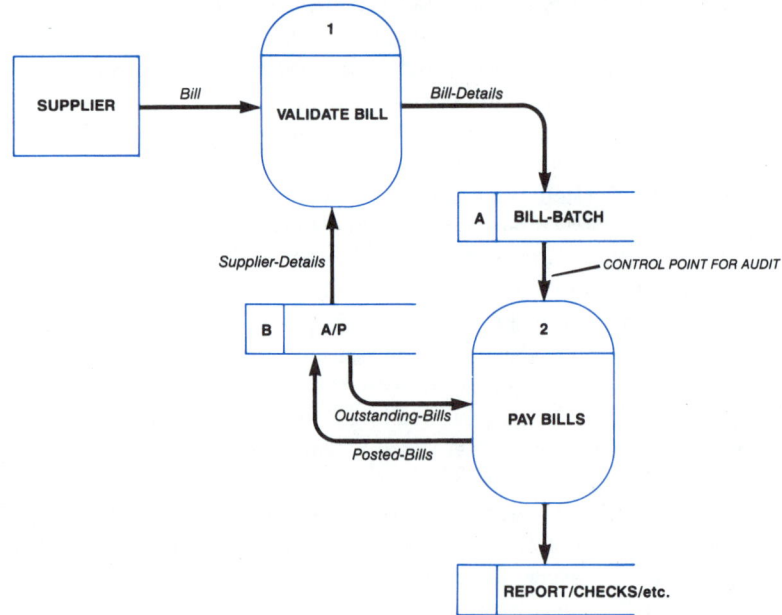

FIGURE 10–10. The Use of Specifications for System Evaluation

logical specifications rarely contain the kinds of numbers that system evaluation seems to require, it does spell out *where* and *when* relationships should function.

For instance, if specifications indicate that bill processing commences only when a sufficiently large batch of bills has been collected, this assists the system auditor in looking for how well this criterion has been met (although, of course, the word "sufficiently" is vague) and when testing of bill processing is reasonable. In addition, details of logical relationships such as input-output or module invocation sequence help the auditor discover where inefficiencies may be coming from; they provide a kind of roadmap for system evaluators.

The following section details how the individuals who work in these areas employ logical system specifications.

10.3 USES OF LOGICAL SYSTEM SPECIFICATIONS

Individuals who use logical system specifications in their work (Figure 10-11) include project managers, system designers, analysts, programmers, and system managers. Because clear lines do not always exist in IRM, some individuals may play several roles: designers are often the same people who performed the analysis; system managers may also be project managers; in small shops one individual may perform all these roles.

INDIVIDUAL	MAJOR USE(S) OF SPECIFICATIONS
Project Manager	A Structure of Implementation Goals
System Designer	Functions to be Achieved by Design Standards Against Which to Measure Design

System Implementors
  Programmers
  Coders
  Trainers
	Implementation-Independent Coding and Testing
	Generation of Test Data
	Functional Goals of Procedures

System Managers	Assignment of (Functional) Responsibility Logical Standard of Performance Audit Procedure

FIGURE 10–11. How Logical Specifications Are Used

As pointed out in Chapter 4, IRMaint takes place in one or more of four modes: **growth, repair, replacement**, and **regeneration**. Growth includes the creation of new elements of an information resource. Repair brings an existing element back up to its original specifications in the face of failure or changed environmental conditions. Replacement creates a new element that performs the same function as an existing flawed element. Finally, regeneration re-creates an element that has become unusable or unavailable. The example of providing access security illustrates all four areas:

- growth: adding modules to request and test passwords
- repair: eliminating the possibility of "back-door" entry through passwords hard-coded in by system implementers

- replacement: putting in a faster password look-up routine
- regeneration: re-creating the password table if it becomes public knowledge or is accidentally or purposely dumped

Specifications provide the working documents through which these modes are activated. Where procedures are structured into **project management** (see chapters 15, 16, and 17 for details), specifications form the basis of what has been termed the *project book*. The specifications are those documents from which everyone works.

The **project manager** sees the logical system specifications as an evolving document during the analysis phase and, thereafter, as a more-or-less cast-in-concrete monument about which all work is structured (see Figure 10-12). Changes can be made to the specifications at any stage, but they require agreement and their cost increases with the passage of time.

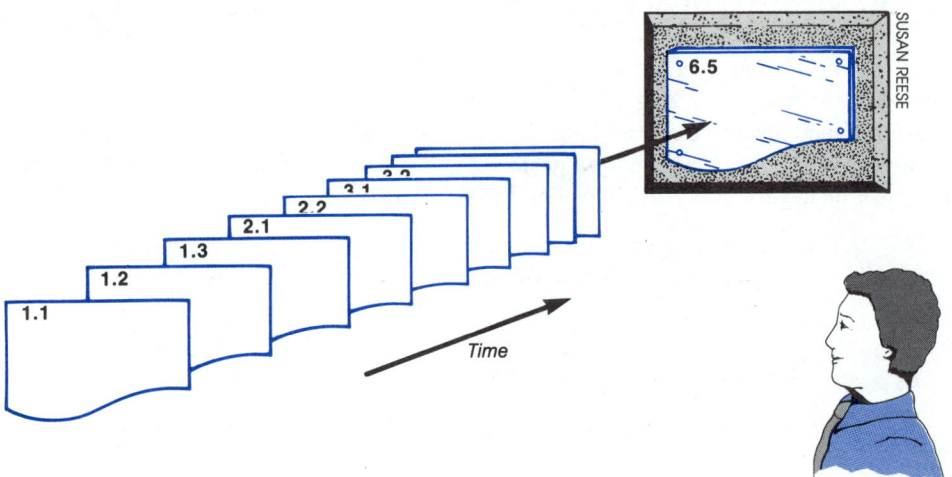

PROJECT MANAGER

FIGURE 10–12. How Project Managers Use Specifications

After the physical design is complete and user involvement is lower, the temptation to make changes based on professional judgment of programmers and operators rises rapidly. Most project management procedures spell out explicitly how changes are to be handled. Figure 10-13 is a facsimile of two memoranda agreeing to a change in logical system specifications. Software production that is not accountable to outsiders is characterized by lower communication, less frequent peer and user review, and greater possibility of well-motivated but poorly coordinated change. Software engineering techniques aim to reduce this possibility. Prototyping retains high user contact, thereby retarding the tendency of implementers to redesign. End-user software development "downloads" the

WESTERN MEDICAL SOFTWARE LIMITED
INTEROFFICE MEMO

FROM: Mark Tallgrass, System Coordinator
TO: Phil Beel, Leader, PHYSINET Project
DATE: 20 July 1984
SUBJECT: Spec Changes

Phil, as we discussed last Wednesday, the relationship between the accounts clerk and the customer is uncontrolled. We need a way to keep track of bills which are paid in the office. I suggested a simple audit trail in an accounts book as each payment is made. Do you think we can make this change? Also, do you think we can purge the trail every three months as we do now with the receipts?

WESTERN MEDICAL SOFTWARE LIMITED
INTEROFFICE MEMO

FROM: Phil Beel, Leader, PHYSINET Project
TO: Mark Tallgrass, System Coordinator
DATE: 22 July 1984
SUBJECT: In-Office Payment Audit Trail

Mark, I share your queasiness with regard to the lack of control over the intra-office money flow coming from customers who pay on the spot. Yes, a simple accounts book will do the job for this system anyway. We should keep the book for a month at a time; this would enable us to use a single sheet of paper rather than a book; maybe loose-leaf would do, but we'll look at that later. The clerk will have to make a separate entry, too, in the ledger for every payment, but we'll have a chronological listing, just in case someone forgets to pay or if we've lost the cash. It's a good idea.

FIGURE 10–13. Memoranda of Agreement

design responsibility to the last possible moment and the most responsible individual: the user. In these cases, logical system specifications may not be written down at all, but may be *implicit* in the prototyping tool (such as a spreadsheet calculator or a screen generator for transactions).

The project manager sees specifications as slots to be filled with physical implementations, goals to be met (in terms of modules to be produced at certain times), and standards to be followed. Analysts, on the other hand, see the specs as the product of their craft and their best judgment of what *should* be done (Figure 10-14). Just as the code turned out by programmers is not their own, logical specifications do not belong to the analysts, But they do reflect their technical judgment. In review meetings, specifications should be treated as technical advice forming a basis for change.

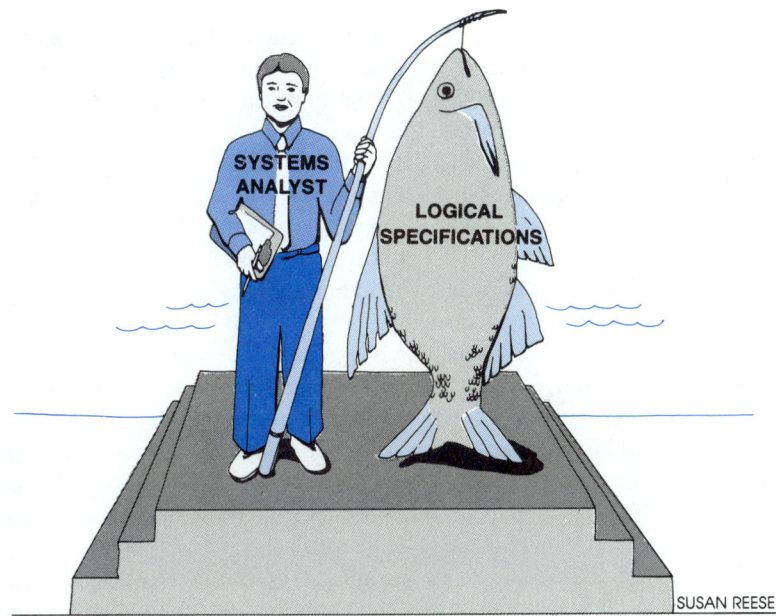

FIGURE 10–14. How Analysts Use Specifications

System designers translate logical specifications into physical specifications. These persons are usually systems analysts who have a great deal of experience with specific hardware or languages pertinent to IRMaint goals. The designers give "form" to the "function" specified by the logical designers. In particular, they determine how programs are to work, what file and report layouts are to look like, how work is to be designed, how office and data processing areas are to be laid out, and which hardware is be be acquired through purchase, rental, or lease (Figure 10-15). They also create an installation schedule.

Logical System Specifications

+

Designer experience
Designer knowledge
Designer discretion

     How programs work
+   File layouts
+   Report layouts
+   Work (procedure) design
+   Office and production layout
+   Hardware acquisition
+   Training programs
+   Installation plan

FIGURE 10–15. How Designers Use Specifications

But physical designers, despite their freedom to interpret, depend heavily on logical specifications. For instance, logical specifications may state that the output from process A is to be the input for process B. The physical specifications detail how that transfer (from A to B) is to be performed. Is the transfer medium cards or tape? Is is in real time or overnight? Is it character-by-character, in blocks of characters, or through a special communications interface? In designing the transfer, however, the designers have to keep the logical specifications in mind: (all) output from process A is to be input to process B. Output from A must conform in format to the expected input format for B. All data necessary to perform the function that process B intends to accomplish must be present in the output from process A.

System implementers include programmers, coders, technicians, clerical and secretarial support staff, program librarians, and trainers. These people use logical specifications as a structure of goals to be met (Figure 10-16 on the following page). Since, logically speaking, functions are placed in a structure to meet *organizational information resource goals*, the structure of the logical design has to mirror the way in which these goals are achieved. Therefore, structure aids implementers in three ways:

1. The logical structure of the means of achieving the goal should *make sense*, making it easy to train users to apply the information resource effectively.
2. The logical structure is determined *before* the physical design. Therefore, the goals of implementation can be separated easily from the limitations of specific programs and hardware.
3. Implementation and testing *may* be rigidly followed from the structure of the logical design. Termed "top-down" implementation and testing, such procedures, although theoretically sound, are difficult. However, the concept is still there: the goals of the system are expressed in the logical structure. As implementation proceeds, these goals should be measurable and met.

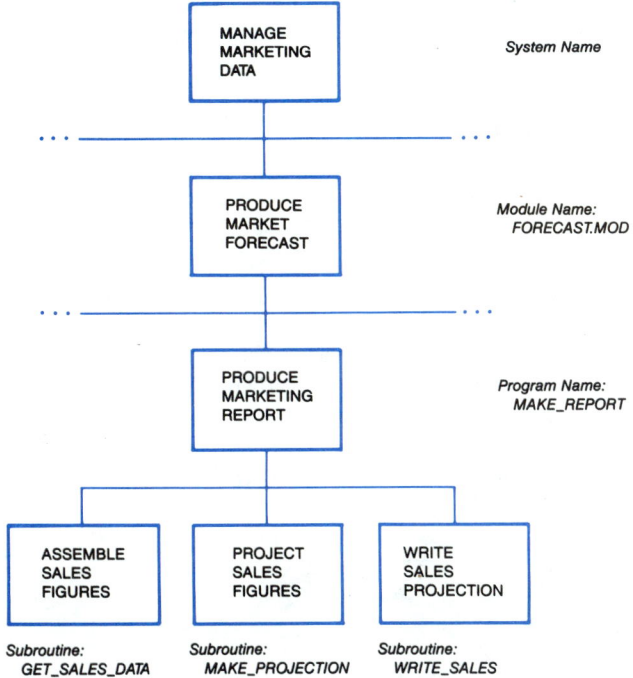

FIGURE 10–16. Specifications as a Structure of Goals

Consider the goal of providing word-processing service as logically dependent on meeting two goals (Figure 10-17): transcribing source documents and distributing processed documents. The former goal is satisfied by obtaining the source documents and transcribing them into processed documents. The latter goal is met by locating the processed documents and distributing them as required. "Processing" includes producing intermediate drafts that may have errors or require revision.

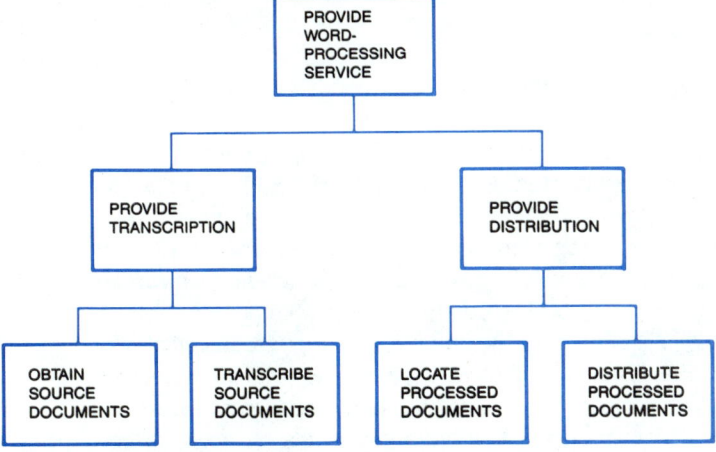

FIGURE 10–17. The Goal Structure of a Word Processing System

This structure is relatively easy to convey to document originators and typists. It is easy to translate the *logical structure of the system goals into understandable procedures* for users and trainers. It is also apparent that if a document cannot be distributed, logical system goals are not being met. (The goal is that *all* documents can be distributed.) Here, then, is an area for improvement. Also, one could read from the logical structure that for a document to be distributed, it must be located. If the user forgets the document name or other identifier, it will be difficult to locate and thus difficult to distribute. The specifications serve the implementers by pointing out these potential problems.

Finally, implementers could build and test programs in a top-down fashion, beginning with one that merely invokes (for example, in COBOL says "PERFORM" or in PL/1 says "CALL") the provide-transcription and provide-distribution modules.

Project managers, trainers, designers, and implementers work to put the system together. Long after systems are built, **system managers** have to keep the system working up to (logical and physical) specification. Logical specifications assist them in three ways (see Figure 10-18).

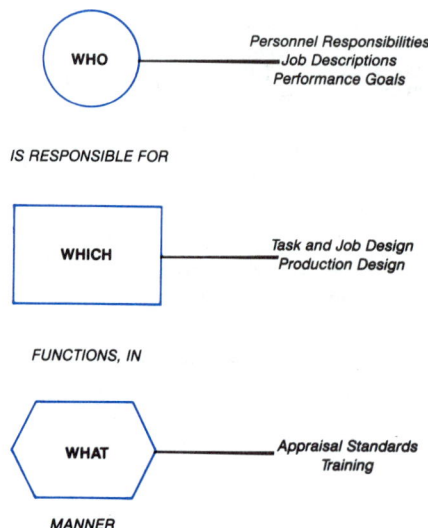

FIGURE 10–18. How Project Manager Use Specifications

First, the system manager can use the specifications to pin down responsibility for errors or deviations. Not only is this important in determining who should fix the problems, but it is important historical information that can prove valuable the next time a similar system is implemented or major changes are made to the current one.

Second, the specifications provide a **logical standard** against which to measure performance. For instance, consider the word-processing system. Suppose it is ultimately used only for final-copy distribution and document originators are unconcerned with distributing intermediate drafts. Once a draft is distributed, the original is deleted from the system. In that case, system performance does not meet the logical specifications, or, alternatively, the logical system design is more complex than it needs to be.

Perhaps there can be some savings by eliminating some of the unused capability from the word-processing system.

The third way in which logical specifications assist system managers is in the continuing **audit** of system performance. Since all systems evolve either through internal growth or decay or from external influence, it is important to maintain a history of what the resource is *actually doing* to be compared to what it is *supposed to do*. When the discrepancy becomes significant, it may well be time to initiate another phase of IRMaintenance and redesign some aspect of the system.

10.4   CREATING CLEAR AND USABLE SPECIFICATIONS

Unless specifications are clear, well-defined statements of what is to be done, they will not be useful. Beyond this basic documentary requirement, specifications should conform to the principles of good technical writing and take into account that they will be used in IRMaintenance and read by people of many different backgrounds.

There are seven criteria by which we can judge logical specifications (Figure 10-19). Clarity is the basic requirement of any document. The next three are requirements of good technical writing, and the final three reflect the IRM orientation of the specifications, which resemble highly fluid building plans.

Clarity is best developed (Figure 10-20) through practice in writing and a keen, open ear to criticism. Clear specifications are precise, with no room for unwarranted, incorrect interpretations. They are not ambiguous. Clear writers eliminate non-standard terms or explicitly define them in a glossary. Clear specifications appeal intellectually to all readers; a Ph.D. is not needed to understand them.

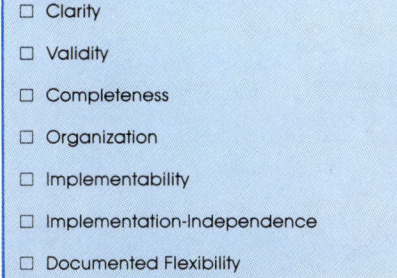

- ☐ Clarity
- ☐ Validity
- ☐ Completeness
- ☐ Organization
- ☐ Implementability
- ☐ Implementation-Independence
- ☐ Documented Flexibility

FIGURE 10–19. The Criteria for Good Specifications

The watchword is *standardization*. Non-standard terms, figures, constructions, diagrams, etc., are well defined in appendixes. Clear writing is stingy with words; however, specifications should not be written in "telegraphese" ("CUSTOMER ACCOUNT TO PROCESS-BILL. STOP. MONTHLY-STATEMENT TO CUSTOMER. STOP."). Dry, boring text is difficult to read, remember, and use. Better writing only means keeping the reader's interest and attention without writing more than is necessary.

For clear specs . . .

SUSAN REESE

The words don't get in the way of the meaning

FIGURE 10–20. Clarity in Specification

Chapter 6 detailed *validity* with respect to models. Valdity is inheritable; specifications are valid only if their (parent) models are. Invalid specifications lead to implemented information resources that are also invalid. Validity is a conformance or correlation between an object under discussion (in this case, logical specifications) and some known, suspected, or acknowledged standard (in our case, the actual running information resource that has yet to be built).

Internal validity relates to the concept of consistency. An internally valid specification is one that contains no conflicts. The specifications in Figure 10-21 are internally invalid because the requirement that passwords be used to allow exclusive access to files is inconsistent with the fact that users can change their passwords without changing those recorded on the files. A simple command (CHANGE PASS = "OUCH") can cause great pain.

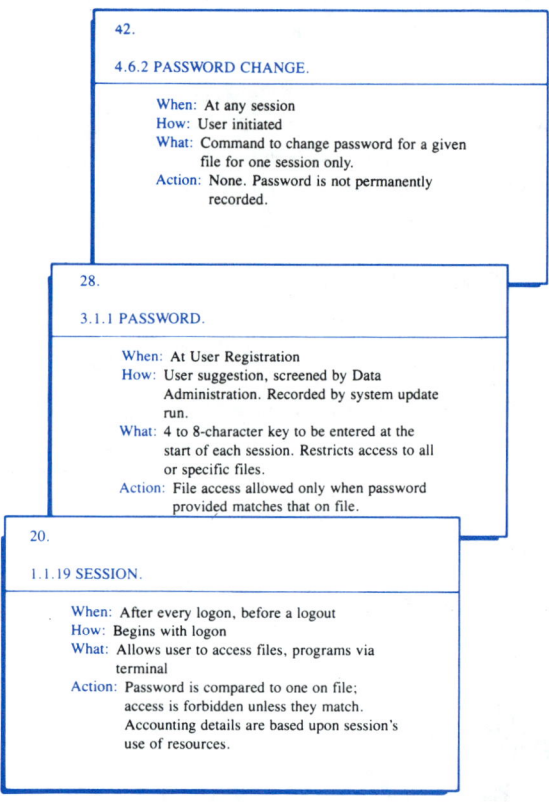

FIGURE 10–21. Internal Validity (Consistency)

External validity (Figure 10-22 opposite) is far more difficult to judge. Specifications must conform to the models from which they are derived. They contain additional information because they are interpretations. This consists of information relating to implementation alternatives. Thus, it is unlikely that mere mechanical comparison of specification against models will uncover all invalidities.

Specifications share the criterion of **completeness** with models, too. All elements of the proposed system should be present in the specifications.

The system boundaries must also be specified completely. This implies that interfaces with other systems and information resources must be specified. This may, in many cases, be quite difficult. Sometimes, especially where manual systems are concerned, inadequate documentation can deter completeness checks and inconsistencies may result. Many a system has failed not because it has not been completely specified but because other, adjoining systems have had no specifications at all.

In **well-organized** specifications, elements fall into place in some useful, understandable way (Figure 10-23). In addition to the usual rules of essay writing, specification writing has some unique organizational aspects. For one thing, logical system specifications are, themselves, structured. In addition, the organization of IRMaint activities requires a sequence of activities and a table of contents for each set of specifications.

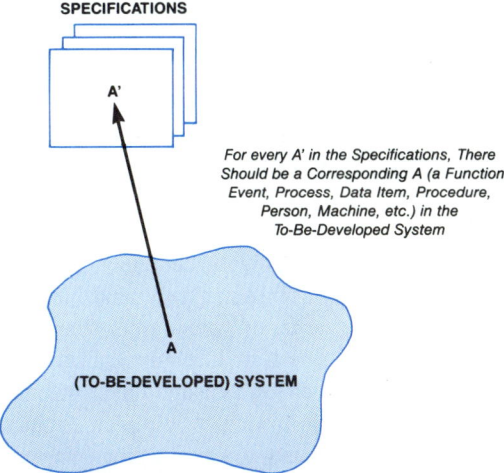

For every A' in the Specifications, There Should be a Corresponding A (a Function, Event, Process, Data Item, Procedure, Person, Machine, etc.) in the To-Be-Developed System

FIGURE 10–22. External Validity

___ Table of contents

___ Executive summary

___ Major sections summarized

___ Paragraphs begun with topic sentences

___ Paragraphs end with summary sentences

___ Signposts placed appropriately

___ References between sections as needed

___ References to pictorial or tabular models in text

___ Glossary (distinct from data dictionary)

___ Index of key terms to pages or sections

___ Highlighting (underscoring, boldfacing) of major concepts

___ Appropriate divisions relating to the system

___ Appendices for tables, models, etc.

___ Structure should relate to system goals structure(s) in the body of the text

FIGURE 10–23. The Organizational Concerns in Specifications

The four criteria discussed so far relate to good, strong technical writing. The last three criteria (Figure 10-24) target implementation of information resources.

Of course, implementation does not determine design, but rather motivates it. In highly structured "classical" top-down design, implementation is totally ignored during logical design. The reality of IRMaint tells a different story, however. In fact, analysts usually end up doing the detailed physical design, too.

Some have formalized these techniques and called them **middle-out** and **outside in** (Hurst and Ness, 1982). The first implies that the *needs* of goals (other than the overriding system goal) determine how implementation takes place (Figure 10-25). "Outside-in" philosophy is a mixture of top-down and bottom-up philosophies. Implementation builds *down* from system goals but must take into account existing implementations available at the bottom (existing programs or hardware). These philosophies excel where analysts possess the required knowledge of the "real" world of technology.

Regardless of the philosophy, **implementation independence** is a major consideration. Specifications should refer minimally to particular elements of the information resource (see Figure 10-26 on the following page). A change in ultimate implementation should not require redesigning the logical specifications.

Suppose a system is designed specifically for an IBM PC/AT. One requirement that

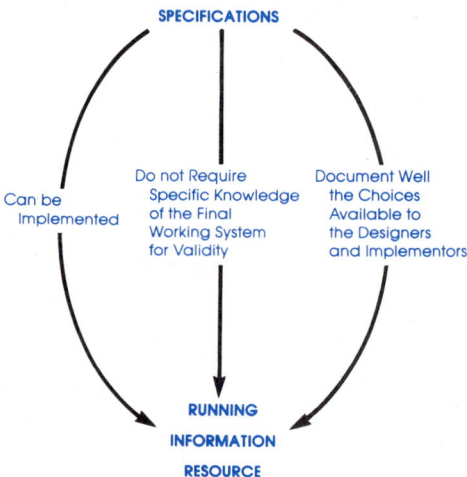

FIGURE 10–24. Implementation Criteria

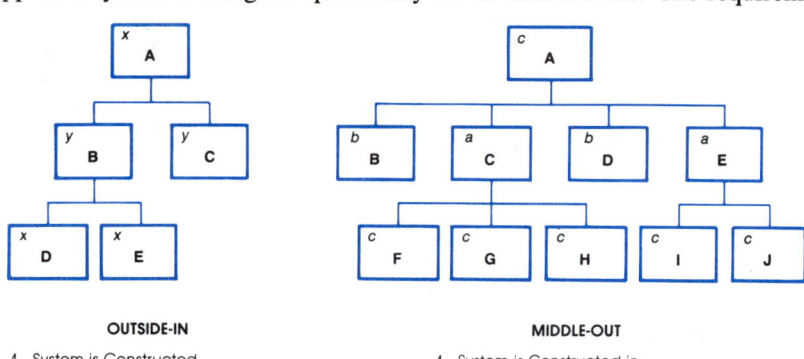

OUTSIDE-IN

1. System is Constructed in x-y Order
2. Advantages: Quick Operation Quick Implementation, Reuse of System Routines
3. Disadvantages: System Routines may be Inappropriate for System Goals (A) Retrofitting may be Difficult
4. Used When Some Work Modules are Available and System is Relatively Well-Known (Many Accounting Systems Can be Constructed From Existing Parts)

MIDDLE-OUT

1. System is Constructed in a-b-c Order
2. Advantages: Subsystems are Working Quickly
3. Disadvantages: Retrofitting of System Goals (A, D) to Already-Implemented Modules (C,E); System Goals may not be Attainable
4. Used When Subsystems are Rather Independent (Office Automation is a Good Example; Word Processing, Electronic Mail, Spreadsheeting are Functionally Independent—often)

FIGURE 10–25. Outside-In and Middle-Out Design

depends upon implementation would be that transaction records be kept on-line at all times. If funds run short or political considerations require purchase of high-density floppy diskette units (with 1.2 Mbytes each of on-line storage as opposed to the 20 Mbytes available on the Winchester drive), everyone will notice the implementation-dependence of the on-line requirement. In reality, *some* implementation considerations always intrude.

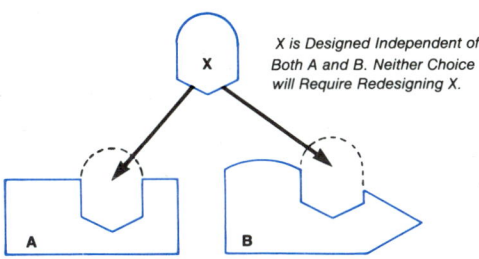

X is Designed Independent of Both A and B. Neither Choice will Require Redesigning X.

FIGURE 10–26. Implementation Independence

In any event, logical system specifications should be **implementable**. By this we mean that specifications refer to possibilities that are relatively well known and generally agreed to be implementable. A clear path from specification to implementation should be apparent to readers. This does not imply that specifications contain implementation plans. It does mean that impossible tasks are obviously going to be left out. For instance, a firm that is absolutely not going to lay out the funds for hard-disk units for decision-support systems simply cannot tolerate specifications that headline on-line access to company transaction information. If the firm compiles sales data on a weekly basis, logical specifications cannot refer to on-line access to individual transactions as they occur. It is easy to say that mental telepathy will be used for input; it is harder to find a manufacturer for the terminals!

Finally, given the need for implementation, **well-documented flexibiity** is necessary. A number of choices should be presented, and the bounds of allowance (what to choose among), the trade-offs (what the costs and benefits are), the effects of making choices (which bridges get burned), and the choice mechanism (how to choose, criteria or checklists) should accompany the text. In the previous examples the choice between on-line and off-line access to current or recent tansactions should be documented like this:

1. CHOICE SET: on-line current versus off-line recent access to data
2. TRADE-OFFS:
   COSTS: (on-line) telecommunication system development, mass storage of several days' transactions; (off-line) lack of currency, inability to take advantage of or change today's events
   BENEFITS: (on-line) Fast response to current events, no lag in making decisions; (off-line) inexpensive, well-known implementation risks and costs

3. EFFECTS OF CHOICE: (on-line) expense, delay in implementation, possibility of failure, need to coordinate sales system; (off-line) none, can always implement on-line processing later
4. HOW TO CHOOSE: (see Figure 10-27 for a checklist)

WEIGHT	FACTOR	ON-LINE	OFF-LINE
____	Lower Cost in the Short Term	__	__
____	Lower Cost in the Long Term	__	__
____	Faster Build Period	__	__
____	Easier to Maintain	__	__
____	Quicker Response to Change in the Data	__	__
____	Lower Staff Requirements	__	__
____	Less Threatening Politically	__	__
____	Better Company Image	__	__
____	Lower Risk of Project Failure	__	__
____	Better Possibility for Expansion in the Future if Business Increases	__	__
____	Greater Change of Shifting to Alternative Mode if Necessary (On to Off; Off to On)	__	__
		__	__

FIGURE 10–27. Weighting of Factors in the Choice Between On-Line vs. Off-Line Transaction Processing

10.5 EVALUATING LOGICAL SPECIFICATIONS

The seven criteria listed in the previous section are necessary, but how can an analyst judge how well given specifications meet the criteria? The following section demonstrates techniques for evaluating specifications.

Clarity and *organization* requirements are best evaluated by reading in design and peer reviews, as these reviews subject the specifications to the scrutiny of people who resemble (and often are) those who will have to read them (See Figure 10-28 opposite).

A **design review** brings in members of the logical design team, one or more members of the physical design team, and possibly management from the IRMaint effort. The purpose of the design review is to examine the text of the specifications to see if it is appropriate. An agenda for design reviews is detailed in Figure 10-36. Simply reading the specifications and discussing them can determine whether or not they are clear and well organized. This form of checking is related to a technique pioneered by Weinberg

(1971) for testing computer programs called "code reading." Code or program reading was developed to examine computer programs to see if they make any sense. Programs that could not be read easily were automatically sent back for reworking. Similarly, specifications that are confusing, poorly structured, or simply difficult to read fail the clarity and organization test.

___ Precision

___ Lack of ambiguity

___ Use of standardized formats (eg. via word processing)

___ Definitions of all terms

___ Parsimony

___ Interest is kept high through use of active verbs, examples

___ Examples bear on the text and clearly relate to it

___ Milestones stand out in the text

___ Writing standards exist and are followed

___ Spelling, punctuation, grammar and good usage are followed

___ Illustrations appear within the text

___ Footnotes are kept to a minimum

___ Consistent format for reference to other sections, materials, etc.

___ Highlighting is used appropriately

FIGURE 10–28. Evaluation of Clarity: A Checklist

Peer reviews examine specifications without the presence of members of other teams or management. In some IRUs, peer reviews provide examination of specifications, models, or code without the psychological threats that management or clients may pose. On the other hand, there is less control over product quality.

Specification standards are important in ensuring clarity and organization. Thus, a guide for specification writers, touching on content, organization, and form for specifications must be written (Van Duyn, 1972). Often, project management packages include forms for specifications and formal procedures for reviewing them. Standards can be implemented easily using word processors, too.

Validity and *completeness* requirements (Figure 10-29 on the following page) are sometimes difficult to meet because external validity is not easy to check. Whereas internal validity (consistency) can be mechanically checked statement by statement, external validity depends on having a valid model to check against. Certain techniques can shed some light on whether the design "couples" and "coheres" the way it is supposed to.

Coupling refers to the reference that modules make to the internal operations of other modules. In Chapter 1, we discussed systems as black boxes within boundaries.

VALIDITY

___ Every module described in the specifications corresponds to a function in the proposed system;

___ Each defined object (for example, a procedure or data definition) corresponds to an essential element in the proposed system;

___ Each item in the specifications is defined in one place in the specifications without ambiguity

___ Each module or item in the specifications serves a unique function

___ Naming conventions exist and are followed

___ Top-down design procedures have been followed

COMPLETENESS

___ Functions in the proposed system are completely defined as modules in the specifications

___ All items (data, processes, materials) of interest in the proposed system have one unique corresponding item in the specifications

___ All situations of interest have been described and compared against the specifications

FIGURE 10–29. Evaluating Validity and Completeness: Checklists

When one system depends on the specific content or activity of elements of another, we say the coupling is *pathological*, or harmful. For instance, if to drive a car, one needed to know the physical position of the pistons within their cylinders at all times, one would not be able to drive. The coupling between the driver's mental system and the engine's combustion system would be **pathological** in this case. Pathological designs can cause trouble (Figure 10-30 is another example).

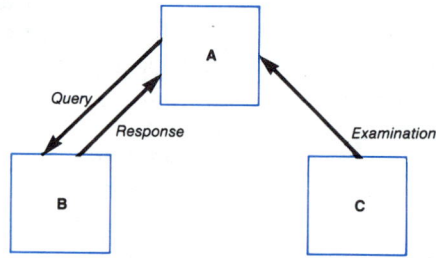

A works with B in query-and-response mode.
A need only know B's input (query) and
output (response) requirements and
formats
A works with C in a too-highly-coupled mode.
A has to know the internal state of C and
<u>how</u> C works in order to work with C.

FIGURE 10–30. An Example of Poor Coupling

Cohesion refers to the extent to which a specific logical function actually performs one and only one task. Figure 10-31 depicts, in structured English, a poorly cohering module that performs perhaps half a scheduling function and half a reporting function. Obviously this module does not "cohere." Since a goal of logical design is actually to create uni-functional modules, such poorly cohering systems need improvement. Further references to coupling and cohesion are detailed in Yourdon and Constantine (1979), DeMarco (1979) and Gane and Sarson (1979).

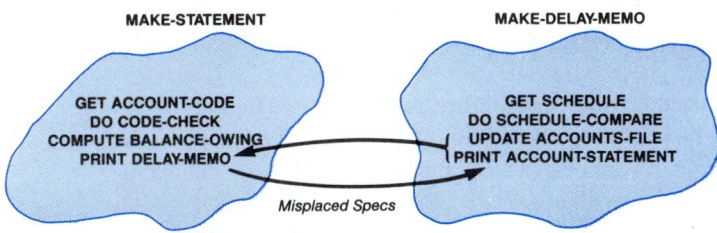

MAKE-STATEMENT

GET ACCOUNT-CODE
DO CODE-CHECK
COMPUTE BALANCE-OWING
PRINT DELAY-MEMO

MAKE-DELAY-MEMO

GET SCHEDULE
DO SCHEDULE-COMPARE
UPDATE ACCOUNTS-FILE
PRINT ACCOUNT-STATEMENT

Misplaced Specs

FIGURE 10–31. An Example of Poor Cohesion

Meeting *completeness* requirements is most easily accomplished using automated and semi-automated documentation aids. In particular, organization standards codified in a specifications manual provide lengthy lists of things to be included in specifications.

Some IRUs have found that adopting a *data dictionary*, even when not used with a database management system, has the advantage of defining important variables, data bases, functions, procedures, and libraries. This list is then available during specification. In this case, the value of the data dictionary is that it motivates the designer to account for a pre-defined list of elements and structures. In particular, system boundaries and interfaces are less likely to be overlooked.

The organization of specifications is also pre-determined when automated and semi-automated documentation systems are in force. Preparing a summary of specifications will tell the analyst how well structured the specifications are. If the analyst cannot find something in the specifications, no one else will either. A summary should indicate the important elements of the design and where they are in the specifications. A statement such as:

> *"A major change in bill processing is the daily update run*
> *outlined in Section 3.4.12"*

is useful during client design reviews. If the analyst cannot find the section in which the update run is described, then the completeness of the specifications is a moot point. What you cannot see, you cannot buy.

Implementation requirements are tested in a variety of ways. Implementation independence is best ensured through structured design techniques. When the design begins to "smell" physical, it is time to cease design work. Design reviews and structured walk-throughs (Yourdon, 1979) of the specifications text subject them to detailed scrutiny. Unlike specification reading, which aims at testing clarity and organization, a structured walk-through concentrates on the logical functions, asking questions such as "Does it make sense to have this module?," "Does this module depend on specific implementations?" and "How can this module work without having this other module work?" The need for documented flexibility is easy to check by reviewing the implications of each alternative and deciding whether or not is has been adequately discussed. If no alternatives exist, find out why. If the choices are not adequately documented in the above sense, it will be difficult for physical designers to make informed choices. Familiarizing analysts with IRU standards is one safeguard against unimplementable systems. Standardizing the configuration of available functional elements is also helpful because it brings about standard file structures, utility packages, application generators, and database managers. Database and prototyping techniques reduce the need for IRMaint analysis and design work.

Other important factors in meeting implementability requirements (Figure 10-32) are software engineering and project management. Software engineering applies the techniques of sound engineering to software design. Reliability, portability, maintainability, testability, and standarization are important aspects. Project management is a skill usually acquired on the job and through study. Recent research has sharpened considerably the tools available to project planners in determining the economic and schedule feasibility of projects. Such estimates can be made from the function specifications produced in logical design. Crossman (1982) has provided empirical support for his estimate that a COBOL programmer required two hours to produce a "function." Of course, defining a "function" may be more difficult.

___ The function being implemented is well known

___ The function being implemented has already been implemented in a similar information resource elsewhere

___ The function refers to data only inside the system

___ The functions can be tested in a top-down fashion

___ The software is specified independent of the hardware

___ Errors in system operation and security are fully reported

___ Incoming data is checked for accuracy and format

___ Outgoing data is checked for validity of destination

___ Standardized modules are available for input and output, access control and system audit

___ If implementation begins before design is complete, only the most inflexible modules are implemented first

FIGURE 10–32. Evaluating Implementability: A Checklist

10.6   WHAT SHOULD BE IN LOGICAL SPECIFICATIONS?

Given that we can measure and therefore improve the *effectiveness* of specifications, what should they contain? In IRUs that have a specifications manual, specification form and content will usually differ in detail from the following recommendations. However, what follows is an amalgamation of concerns at each level of IRMaint mentioned so far in this book.

Figure 10-33 illustrates the six levels of IRMaint and the corresponding concerns toward which specifications of all types are directed. At the top level, IRMaint itself is concerned with providing an information resource to the organization. Here, the concern horizon is the entire firm, cutting across organizational boundaries into areas such as training, physical plant, PR, budgeting, operations, and employee relations.

LEVEL	CONCERNS
I R Maintenance	Service Provision and Planning, Installation, Training and User Relations
System Design	System Specifications, Testing and Evaluation, and Maintenance
Context	System Organization, Functional Constraints Costs, and Access
Products	System Benefits, Outputs, Reports
Procedures	System Operation, User Procedures, System Start-Up, Take-Down, Backup, Restore, Disaster Recovery
Tools	System Data Administration, Operational Decisions, Messages to User, Access and Query Language, Formatting Tools

FIGURE 10–33. Specification Concerns at Six Levels of IRMAINT

At the system design level, only the design group is affected. Here the concern is the creation of a system or system element.

At the third level, the system is detailed in terms of its boundaries or **context**, how it affects the organization, and how it should be operated. At the context level, the primary personnel concern is system management. Moving to the fourth level, concerns now focus on the tangible **products** of the system: reports and other indicators of the system's health. At the products level, the system's customers or clients are known and involved, because it is to them, of course, that the system's effort is directed. The last two levels are concerned with personnel and operation. **Procedures** must be effective during operations. The **tools** through which the system is used must be clear.

Figure 10-34 on the following page, details a table of contents for system specifications. Some of these topics relate to *physical design*, but the bulk of them are actually logical in nature, referring to the goals of the system and the functions that are intended to meet them.

Typically, specifications are in the form of text, supporting tables, and documents peculiar to the function under discussion. For instance, physical layout will be described in words, but more important will be a floor plan and wiring diagram. Documentation standards will be described and supported with sample forms.

1.0 I R MAINTENANCE

1.1 Implementation Criteria
1.2 Installation
1.3 Program Conversion
1.4 File Conversion
1.5 Physical Plant (Lighting, Cooling, Decore)
1.6 Training Needs
1.7 Documentation Standards
1.8 Public Relations, User Relations Impacts
1.9 Maintenance Priorities and Procedures

2.0 SYSTEM DESIGN

2.1 Performance Criteria
2.2 Audit and Evaluation
2.3 Testing Process
2.4 System Interface Performance Standards
2.5 Human Interface (Human Factors)
2.6 Design Variances From Shop Standards
2.7 Maintenance Indicators (What to Watch out for)
2.8 Special Implementation Tools Available or Required

3.0 CONTEXT

3.1 Organizational Impact
3.2 Work (Re)Design
3.3 Organizational Information Flow
3.4 System Boundaries; Boundary Protection
3.5 System Constraints
3.6 Operational Assumptions

4.0 PRODUCTS

4.1 Logical Contents of Reports
4.2 Other Output, Including Intermediate Files
4.3 System Indicators and Status Reports
4.4 Exception and Error Reports
4.5 Report and Other Output Distribution Channels
4.6 Validation and Security on Output

5.0 PROCEDURES

5.1 Startup
    5.1.1 Initial
    5.1.2 For one Session or Daily
    5.1.3 Special Periods (eg. Monthly)
    5.1.4 Emergency
5.2 Normal Operational Procedures
    5.2.1 Peripheral Data Requirements
    5.2.2 User Validation and Rejection
    5.2.3 Login Procedures
    5.2.4 Changes in Normal Procedures
5.3 Exceptional Operational Procedures
    5.3.1 Disaster Recovery
    5.3.2 Normal Backup of System
    5.3.3 Reports to be filed
5.4 Input Procedures, Including Data Validation
5.5 Output Procedures, Including Destination Validation
5.6 Document Distribution (Manual) Procedures
5.7 System Maintenance Procedures
    5.7.1 Normal Maintenance
    5.7.2 Repair of Modules
    5.7.3 Take-Down and Notification for Repair/Maintenance
    5.7.4 Handling Complaints and Suggestions
5.8 Audit
5.9 Data Entry Procedures
5.10 Communication/Telecommunication Procedures
5.11 Security Procedures on Entry, Access, Backup
5.12 System Take-Down and Shut-Down Procedures Apart From 5.6

6.0 TOOLS

6.1 File and Record Layouts and Contents
6.2 Source Document Layouts and Contents
6.3 Source Codes and Coding Procedures
6.4 Special Operational Decisions
6.5 Exceptional Data Conditions and Decision-Making
6.6 Data Dictionary
6.7 Flow Diagrams
6.8 Physical Layout Diagrams
6.9 Record Retention Procedures
6.10 Performance Measurement Techniques
6.11 Auditing Procedures
6.12 Risk Analysis
6.13 System Costing Computations

FIGURE 10–34. Specification Table of Contents

Logical system specifications should also include the following:

1. a **glossary of terms**, preferably in the form of a data dictionary
2. module **structure charts** showing the relationship of logical modules
3. a module-by-module discussion of each function in clear, natural language
4. a **documentation summary** detailing each document's content and function
5. an **executive summary** (Figure 10-35)

Executive Summary

Supercrete, a division of Canfarge, produces a variety of concrete pipe and precast products. This report focuses on the production of concrete for use in the making of pipe products. Bob Chisholm, the pipe division manager, supervises the production of concrete. He is currently experiencing problems in receiving the necessary information to adequately control concrete production. From our investigation, we have determined specific problem areas. These are associated with the presentation of information about labor hours, the consumption of raw materials and the absence of information concerning consumption variances.

To solve one of the problems, we have designed forms which summarize information taken from the more detailed reports Bob is now receiving. New reports were also created to provide the variance information that is desired. Copies of labor distribution reports from the Costing Department are to be distributed and reformatted to provide the required labor hour information. In addition a few minor changes to operations are recommended. Finally, areas of further study are outlined for the purpose of computerization.

FIGURE 10–35. An Executive Summary

10.7 OBTAINING AUTHORITY AND APPROVAL FOR SPECIFICATIONS

The logical specifications represent the last official influence analysts have on the course of IRMaint, unless they participate in physical design. The analysts' responsibilities end with the sign-off meeting, at which the customer accepts the specifications and proceeds into physical design.

This is only the final step, however, in a chain of approvals and acceptances of a less formal nature that the analyst must obtain. This section discusses three important approval meetings:

- the design review
- the design walkthrough
- the sign-off meeting

The purpose of these meetings is to obtain approval for work performed to date and authority to proceed. Checkpoints and milestones stand as points of no return and are a means of checking progress. They allow work to proceed on a firm basis of consensus. In addition to the approval of spending-to-date, these milestones provide an opportunity for the organization to assess the payback on future expenditures. This amounts to approval to spend, or to overspend, if necessary. It is necessary to account for variances and to seek out potential problems at all stages of IRMaint. It is also necessary to assign the burden of responsibilty to those who deserve and require it, making it easier for them to account for their own work and keep costs down.

Design Reviews

The purpose of the **design review** is to obtain approval of the *content* of the specifications to date. Users, analysts, and management attend these meetings. The agenda (Figure 10-36) is essentially a question-and-answer format; participants are

D.R.1 Design submitted several days in advance of meeting

D.R.2 Written questions are submitted before the meeting to the chairperson; questions from D.R.7 arising from previous meetings are included.

D.R.3 The executive summary is distributed and reviewed.

D.R.4 Specific questions submitted under D.R.2 are answered

D.R.5 Additional questions arising from D.R.4 are recorded.

D.R.6 If all questions are answered, approval is granted by IRMaint management to continue.

D.R.7 Where questions are unanswered or answered inadequately, they are deferred to later design reviews

FIGURE 10-36. Design Reviews: An Agenda

expected to be familiar with the content of the specifications to date. A checklist for design review planning is in Figure 10-37.

___ Design distributed at least three days in advance of meeting

___ Written questions received in advance of meeting

___ Executive summary made available at start of meeting

___ All questions from previous reviews addressed

___ High-level audio-visuals prepared and available

___ Presentation prepared and practiced in advance

___ Users, implementors, and managers invited

___ Agenda typed and distributed with checklist

FIGURE 10–37. Design Review Planning Checklist

Attendance might be between five and ten people, and there may be one to three reviews for most projects, depending on their complexity. A design review is scheduled within two weeks prior to a customer design review or sign-off meeting. Meetings may last an hour or more for simple, compact designs in the early stages or up to an entire day near the end of the logical design of complex systems. Design *reviews* are not design *sessions*. Attendees are not expected to contribute to the substance of the design *per se* but to provide, through their questions, critique and advice that the analysts will later incorporate into the design. These suggestions appear in the minutes of the meeting, which result in formal project documentation, distributed with the agenda for the next review.

Design reviews are *not* held for prototyping, DSS development, single-user systems, program maintenance work of a hygienic (fix-up) nature, or for very short projects. Design reviews may prove necessary for complex maintenance work if major problems requiring changes in the logical design of a system occur. An example of this may be the discovery that large error rates are due to poorly designed forms without check digits. Adding check digits and redesigning forms may entail a major change to procedures derived from a new logical design and approval is needed before this "repair" is made.

Design Walk-Throughs

The **design walk-through** is a technical meeting of designers. Its purpose is to understand the content-to-date of the design and weed out errors. The philosophy of a design walk-through is this: every design has errors. The task is to find them. Attendees are the logical design team and invited peers, at least some of whom are not involved in the project. Users are emphatically not invited. The agenda is outlined in Figure 10-38.

D.W.1 Distribute the section of specifications to be reviewed several days in advance of the walkthrough.

D.W.2 Individuals go through design in order of the structure. They may concentrate on action items from previous design walkthroughs mentioned under D.W.4 below.

D.W.3 Goal is to attack the design and find as many flaws as possible.

D.W.4 As each error or opportunity is encountered, it is documented and entered as an action item (see below).

D.W.5 A general consensus is obtained as to whether another design walkthrough is necessary

FIGURE 10–38. Design Walkthrough: An Agenda

The number of such walk-throughs varies with the complexity of the project and the quality of the designs produced. A checklist for design walk-through planning is found in Figure 10-39. As with most meetings, attendance of five to ten and a time limit of a morning or an afternoon are most desirable.

___ Section of design to be walked through distributed before the meeting

___ Invitations to designers and impartial readers

___ Agenda available for discussion

___ Action items on agenda from previous walkthroughs

___ Minutes of previous walkthroughs available

FIGURE 10–39. Design Walkthrough: A Checklist

These are working meetings of a detailed technical nature. Technical support information is valuable, too; a project librarian can handle the accumulation of information for the review. Design walk-throughs are appropriate for all work structures except prototyping. Prototyping *is* sort of a continuous design walk-through involving only the analyst and the user.

Documentation of action items may indicate that major structural changes are necessary. If you have to fix up the order-entry logical design *and* the order-processing logical design *and* the management-reporting function, you may as well redo the entire design. However, it may not be possible to make these decisions during the design walk-throughs, especially if they require management approval; therefore, it may be necessary to hold a full-scale design review if the design walk-through is a failure.

Sign-Off Meeting

The **sign-off meeting** is held to obtain approval from IRMaint management and the customer to release resources and commence physical design. This is a non-technical, managerial meeting, attended by user and IRMaint management and at which only the high-level design is presented. Its simple agenda is presented in Figure 10-40.

S.O.1 An executive summary is distributed several days in advance of the meeting.

S.O.2 An oral presentation of the highlights of the logical system design is given, including history from previous design walkthroughs and reviews (note that approval has been obtained at every prior stage — there should be few surprises in this meeting).

S.O.3 Questions are requested and fielded.

S.O.4 Formal approval is obtained if granted; otherwise, action items are entered into the agenda of another meeting.

S.O.5 In some cases, projects are cancelled at this stage.

FIGURE 10–40. Sign Off Meeting: An Agenda

There is, of course, only one sign-off meeting in a well-run and reasonably lucky project. In some cases, the sign-off meeting results in a stalemate, *almost always because IRMaint management failed to keep user management informed about confusion or change.* In these cases, more design work is usually called for. In extreme instances, the project is canceled.

More commonly, the meeting lasts an hour or two, consensus is reached, minor disagreements are pasted over or improvements promised, and approval is given. Sign-off documentation includes signatures and the specifications that were approved. Figure 10-41 on the following page, is a typical checklist for a sign-off meeting.

Any work structure (project, prototyping, etc.) will require a sign-off meeting, although some may not be very formal. From the analyst's viewpoint, the sign-off meeting is the final curtain on logical design work. From this point, everything is in the hands of the builders.

___ Executive summary distributed at least three
days in advance of meeting

___ Agenda distributed with summary

___ Historical documents available, including all
with signatures

___ User management notified in person of
meeting

___ Sign-off documents prepared and available

FIGURE 10–41. Sign-Off Meeting: A Checklist

10.8 CONCLUSION

The logical system design is used in four ways: for **interpretation** of the models into more tangible forms; for **consolidation** of several trains of system-design thought; for **expression** of the needs of the organization and designers into a systematic form; and for **commmunication** of these to system designers.

Logical system specifications can be valuable to four aspects of Information Resource Maintenance. First, they may aid **system operations** when the sytem is ultimately in production. Second, **system management** is aided in terms of personnel, training, and planning functions. Third, **system implementation** depends critically on the description of the logical functions outlined in the specifications. Finally, **system evaluation** is determined by the functions outlined and the expressed needs of the users and the organization.

Logical system specifications must be **clear** and have a **valid** relationship to the models developed, must *not refer to specific physical attributes* of the system, must represent models as **completely** as possible and be **implementable** in the short term, and must *not be rigidly tied to specific choices*, but should provide documented **flexibility** in implementation and use. A logical specification addresses all levels of the design process: **IRMaint, system design**, system **context**, system **products**, system **procedures**, and the **tools** of system operation and design.

An important aspect of presentation is obtaining the approval and authority to proceed with physical design and implementation. Three **approval meetings** are designed for presentation of these specifications: the **critical design review** of which there may be several; the **design walk-through**, held internally for IRMaint management approval, and the **sign-off meeting**, at which the specifications are finally approved by the customer before physical design begins.

The following four chapters discuss specific tools of the analysis trade. Chapters 11 and 12 explain methods of gathering data. Chapter 13 examines, in detail, over a dozen common diagramming techniques. Chapter 14 looks at the economics of logical and physical system design and serves to support design decision making in those areas.

10.9   A CASE OF THE WILLIES

Will Jenkins is chief systems analyst, and he is sweating. A critical design review is being held for a sales reporting system for the marketing division. The project has been going on for four months and the system is due in six months. At the meeting are Will; his employee, Bill Bowering; Will's boss, the director of MIS; the director of marketing; and two marketing analysts. Bill is about to finish a thirty-minute oral presentation of the design, which is supposed to be nearing completion. The sign-off meeting is scheduled to be held in three weeks.

What has Will worried is that major changes have been made, the nature of which he doesn't know. There was neither indication of this in the logical design summary distributed when the review began nor justification given in the oral presentation. Instead, Bill rambled on at a technical level, discussing data flows, keys, domains, normal forms, list structures, and so forth. Will brought documentation from the three design walk-throughs held in the two months since the last design review and, as far as he knew, none of the changes Bill has alluded to had come up. The specs written so far fill two loose-leaf binders. The director of software development, who has just arrived, is leafing through them, frowning.

The director of MIS raises his hand. "What are the implications of these changes to our operations here? It looks like you're asking for a great deal of capability beyond what we already have, and you know our budget is tight...."

"I don't understand how we'll know if the system is working correctly. With these changes, if I understand them, we'll only get summary reports on a weekly basis," interrupted a marketing analyst.

"I thought our last meeting settled the question of daily control...?" replied Bill tentatively. Will is cringing.

1. What has Will done incorrectly here? What should he have done before the design review?
2. What should Will do at this point, assuming that he is able to do anything?
3. What are the basic problems in this situation concerning the meeting, the attendees, the specifications, and the history of the project?
4. What conflicts are apparent in this situation? Can all of them be avoided? What tools does Will have at his disposal to reduce the level of conflict here?

DISCUSSION QUESTIONS

1. Analyst a produces a set of specifications, s and analyst b produces another set, t. Both a and b work from exactly the same data sources, yet s and t are different. Where could those differences come from, assuming that a and b work for the same IRU?
2. One of the functions of system specifications is consolidation. Assume that two managers in a user area have entirely different ideas about how a system ought to work. Outline a technique for consolidating these two different opinions within the framework of a series of meetings.

3. Analyst *y* has made the following statement: "When I write specifications, they are the product of my own skill and expertise. I do not appreciate having anyone else come along to make recommendations when they haven't been intimately involved in collecting the data and doing the logical design work." Is this attitude helpful in the specification process? What difficulties do you suspect will arise each time analyst *y* works on specifications?

4. The diagram in Figure 10-42 illustrates the flow of information in the process of creating specifications. Analyze this diagram and discuss those elements you feel are the weakest link(s) in this process. What contibutes to this weakness? What techniques discussed in this chapter can aid in strengthening this chain?

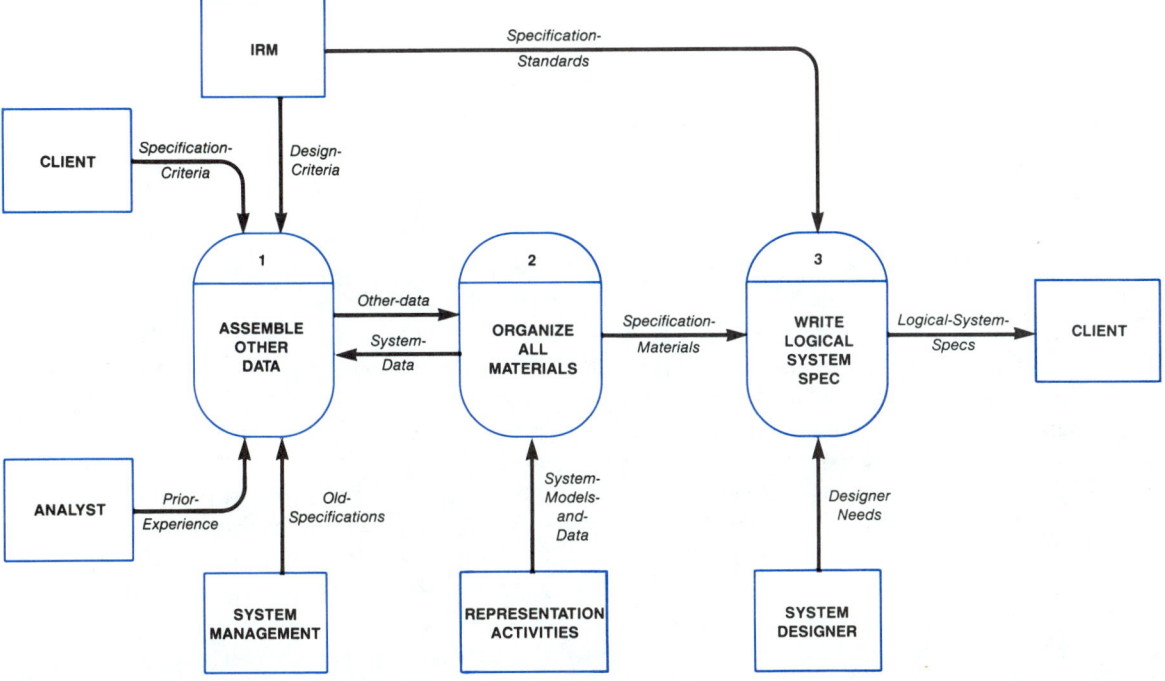

FIGURE 10–42. Creating Logical Specifications

5. This chapter notes that system specifications have value in four areas (implementation, operation, management, and evaluation). How is it possible that a single set of specifications can apply in four areas that have such different purposes? What are the dangers inherent in simply mailing the specifications to individuals in these areas?

6. Discuss ways in which various individuals mentioned in the chapter would evaluate the adequacy of specifications to their assigned tasks. That is, how would a project manager evaluate specifications? (You may want to consult Chapter 15.) A system designer? An analyst? Where do the differences (if any) seem to come from?

7. Look at the following specifications and criticize them based on the seven criteria mentioned in the Chapter. Feel free to assume the worst!
   7.1.1 Operational Procedures
      7.1.1.1 Goals
         7.1.1.1.1 Rapidity
         7.1.1.1.2 Non-defrangibility
         7.1.1.1.3 Low level of repetition due to errors
      7.1.1.2 Actions
         7.1.1.2.1 System start-up
         7.1.1.2.2 System self-awareness of all problems giving warning one hour in advance to operator
         7.1.1.2.3 *Or* system crash with diagnostic follow-up
         7.1.1.2.4 Normal operation
            ...
            ...
         7.1.1.2.5 System take-down (normal end of day)
            ...

8. Logical and physical specifications obviously differ at least in terms of how closely they conform to the boxes and wires that ultimately will be bought, built, and installed. Examine the sample table of contents provided in the chapter. Discuss which sections are more applicable to logical specs as opposed to physical specs and vice versa. Are there additional sections that should be included for each?

9. John, the project manager, under pressure from the user organization, needs to trim the size of the design project quickly by eliminating the part relating to automation of the accounts receivable section. He decides to call an emergency design review. Is this a good idea? What are the advantages and disadvantages of calling an *emergency* design review with little notice? Does John have an alternative move?

10. During a peer-level design walk-through, analyst *x* notes some serious flaws in the logical design and tells *y*, the logical designer, about them. *Y* shrugs off the suggestions by saying they are minor and they do not have enough time to look at all the errors. "Forget it, I'll look at it later; let's go on to something important," says *y*. Is *y* acting wisely here? How should *x* respond?

11. During the sign-off meeting, designer-analyst *d* announces that there have been some major changes since the last design review. "These changes do not appear on your summaries — let me fill you in verbally," she says. *D* then goes into a technical discussion of the design of a module important only to the clerks who will use it. Otherwise, there is no difference in the system function. What, if anything, has *d* done wrong here?

DESIGN EXERCISES
Using dBASE III

1. Communication between programs and users in dBASE III is handled through a number of commands already mentioned. Here they are in detail:

```
PFX = "    "
@ 5,10 SAY "Enter a phone number prefix";
    GET PFX
READ
@ 6,10 SAY "Looking for " + PFX + "-..."
WAIT
```

This sequence shows how to display information at a **specific place** on the screen (5,10 — the fifth character on the tenth row) and to **request input** following it (GET PFX). The program prints out a response to verify the input and then the WAIT statement causes the message "Press any key to continue" to appear; the program will pause for you to press any key.

dBASE III knows how many characters to ask for by the **current length** of the string in PFX. You may have several GETs "pending" before the READ statement. This makes the screen look like a form. The program REQ4SERV paints a Request for Service on the screen first and then asks for input. After all your input is read, a summary is displayed for your approval. DO REQ4SERV and then examine it to see how this is done.

2. Now you can write a **report program**. In exercise 8, you created IDEAS and then filled in the ideas. Write a program named GOODIDEA that will do the following from a menu:

    a. Provide a list of all IDEAs with a PRIORITY greater than some number;

    b. Provide a list of all IDEAs for a specific TOPIC;

    c. List all the FILEs that relate to a specific TOPIC;

    d. List all the TOPICs that relate to a specific FILE. (HINT: A menu normally has a **title**, a list of **choices** identified by a number or letter **key**, and a **prompt** for entry of a key. If an illegal entry occurs (there are only four possible entries), an **error** line may appear. You can make GOODIDEA fancy by CLEARing the screen before providing the answer and re-drawing the menu after WAITing — a DO WHILE loop can be used to repeat this.)

CHAPTER 11

COLLECTING AND ORGANIZING SYSTEM EXPERIENCE DATA: PART I

OBJECTIVES

1. To describe the process of data collection, organization, and storage and to select appropriate techniques for a study
2. To describe and follow the process of selecting collection points relevant to particular studies
3. To describe and create a wide variety of data collection instruments
4. To plan, organize, and conduct an interviewing-collection study critically, so it can be evaluated

COLLECTING AND ORGANIZING SYSTEM EXPERIENCE DATA: PART I

CHAPTER

11.0 THE SYSTEM SUPERSLEUTH

Trevor McGregor is a senior systems analyst for Allied Structures, a manufacturer of mobile and movable construction trailers. Allied has recently made a big venture into the prefabricated housing field. Business has tripled in the past year, along with paperwork and the load on managers.

Trevor has just been asked to improve the sales reporting system. Existing customers are confused about the new line, and new customers are unaware of Allied's other products. New salespeople do not know the procedures, some orders have been lost, and a few customers have noted that they have received improper or substandard products. Meanwhile, marketing people are having a tough time deciding on the next move.

Trevor has given some thought to improving the system. After a preliminary investigation focusing on existing sales reporting problems, Trevor is going to propose an extensive data-gathering effort. However, he has not yet decided the best way to get the information. He knows he has to collect some data on how the sixty people in sales go about their work, how the over nine hundred commercial and retail customers see Allied's marketing efforts, and what specific bottlenecks and sources of error exist. Now, how would a supersleuth go about it?

11.1 PLANNING AND MANAGING THE COLLECTION EFFORT

The data collection plan has six steps (Figure 11-1 on the following page):

1. *Defining the goals* of the investigation
2. *Determining the information needed* to meet the goals
3. *Determining collection points*
4. *Collecting* the data
5. *Organizing* the data
6. *Assessing* the data in terms of the goals (repeating steps 1-5 if necessary)

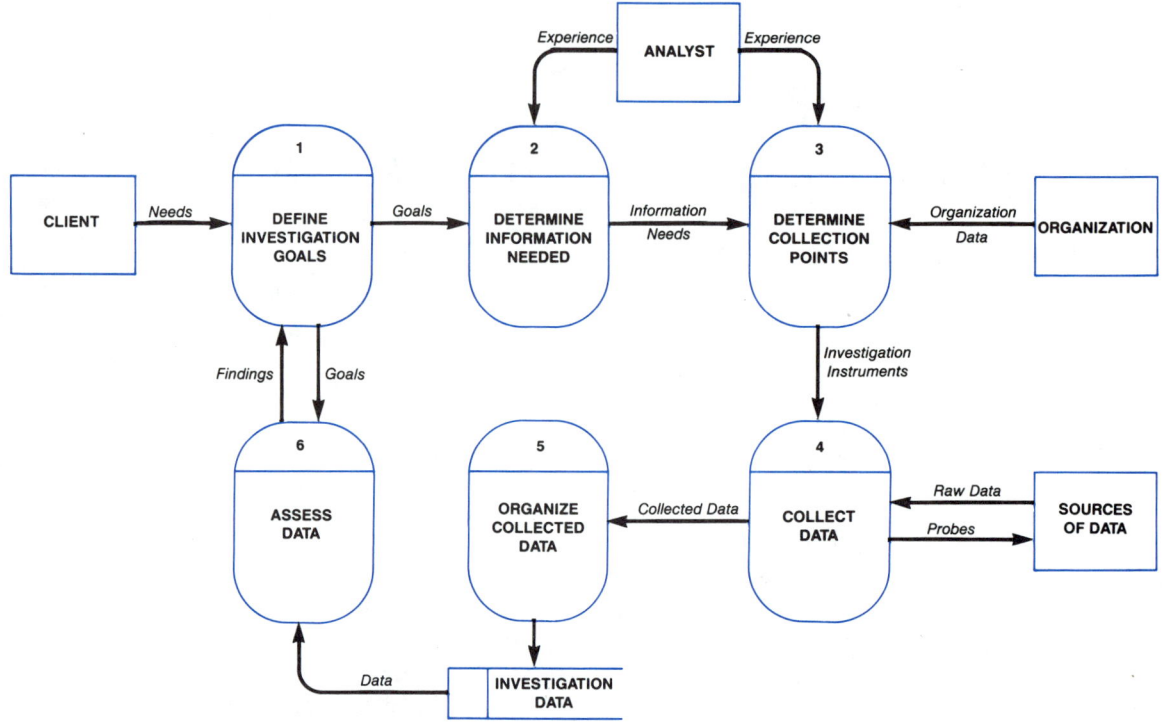

FIGURE 11–1. The Process of Conducting a Data Collection Effort

Using Investigation Goals

Planning for any activity begins with a good statement of the goals in behavioral terms, analyzed in a top-down fashion:

1. Do we know enough now about each term in the goal to feel comfortable about the goal? (If not, find out more.)
2. Can we define a test that tells if this goal has been accomplished? (If not, return to 1.)
3. What are the major tasks involved in reaching this goal?
4. What resources need to be brought to bear in reaching this goal?
5. What skills must be brought to bear upon these tasks to reach this goal?
6. Based on this goal, can we estimate accurately how long it will take to achieve the goal? (If not, return to 1.)

Having a statement of goals is only the first step in planning a data collection. One mediating factor is the cost of collecting the data. Costs vary widely, sensitive to these factors:

- the number of collection points needed
- the relative efficiency of data collection techniques
- the effectiveness (completeness and accuracy) of the data collected by a given technique

The number of points needed depends on the need for accuracy and variety of experience. In general, the more **homogeneous** (similar) the collection points are, the fewer that need to be examined; the more precisely you wish to specify a characteristic, the more points you have to collect. A firm comprised of individuals who perform their work in unique ways requires far more investigation than a firm of individuals all or most of whom work "by the book."

In this and following chapters, each of a dozen data-collection techniques will be described and evaluated in terms of effectiveness, efficiency, and appropriateness for specific kinds of IRMaint activities. The next sections discuss the management of a data-collection effort.

Managing the Data-Collection Effort

Collecting data is a *research* activity, intended to *inform* as well as build. Merely amassing data is not enough. The data have to make sense in the context of answering questions compiled before collection is begun. Without planning, the chances of relevance and coherence are limited.

Data are collected through **instruments**, often appearing as forms for a data collection effort. For interviewing, the instrument is a schedule of questions, or "schedule" for short. For a mailed-out survey, the instument is a questionnaire. Other instruments are diaries, buck slips, and checklists. Whatever the instrument, it must be carefully and completely pretested. A pretest sees if the data can, in fact, be collected accurately and completely. If it cannot, the instrument needs to be redesigned.

The analyst must understand the limitations of collection techniques. Periodic checks should be made on both the data produced and the collection method. This may mean observation of the data collectors, spot checks on the quality and accuracy of the data produced, and staff meetings to bring out problems. A well-designed data collection effort, managed by an experienced leader and coupled with good training and constant quality control, may well serve as additional training for both the IRMaint group and novices, since *all* subsequent design decisions are based on the data.

Data security is also important. The data are about the organization and have value beyond the investigation. That value may be recognized by competitors, unscrupulous employees, or the firm's personnel department. Since the typical investigation lasts a week to several months, secure storage and transcription should be a habit.

The management of an investigation begins with the goals of the investigation and ends with the delivery of well-organized, secure data bearing on the goals. The following sections detail the investigation process.

Determining Collection Points: Sampling

A **collection point** is a single system element: a person, an activity (process), or a data item. Because a system comprises all its elements, the challenge is to describe the system as a whole when we have information on only a selected sample of points.

The specific goals of an investigation need to be kept in mind at all times. Data will have to be described in words. Inferences will have to be made from descriptions as to the causes of problems or the nature of opportunities. Models that adequately represent the experience investigated will have to be constructed based on the collected data. Decisions will have to be made on the course(s) of action available. Finally, new systems and system elements will have to be designed. Ultimately, all of this depends on the quality of the data collected.

There are three classes of collection points. We can collect data about **people**, **activities**, or **information**. People used as collection points are typically information or **knowledge workers**. Activities are usually person-related (such as machine or computer operation) and can be observed directly or indirectly (the latter often through a computer transcript of activities). Collection points may also be combinations, such as a worker's use of information.

Sampling Collection Points

Usually we collect data from a subset of system elements, called a **sample**; making this selection is called *sampling*. Sampling can be either **representative** or **non-representative**, and while representative sampling is greatly preferred, sometimes a non-representative sample must be used and adjustments made or warnings given.

The easiest way to gather representative samples is to sample **randomly**, without apparent systematic bias, giving each (appropriate) element the same chance of being selected for the sample. Random sampling does not, however, *guarantee* a representative sample. It is possible, for instance, to select by chance only male employees when, in fact, the firm is 53 percent female. If a firm has four employees, two men and two women, a random sample of two persons will be half men and half women only 50 percent of the time. The other 50 percent of the samples will be either all male or all female. *Exact* representation is usually impossible. Representation is assisted by taking relatively large samples or high proportions; figures on just how large the samples ought to be are available in most statistics texts.

To sample people randomly, an up-to-date corporate phone directory or payroll list usually can be used. Activities need to be observed over time, and the choice of observation time can be selected by random-number generation. For information sampling, a simple "every *n*th" selection will usually do.

Representation can be increased by a techniques called **stratification**. One would not interview employees of the payroll department without also interviewing at least one of the supervisors and probably the manager. The sample (one manager and one supervisor) may itself involve random selection from the pool available (one manager and two supervisors), but the original decision was not to select from *all* employees but rather from three pools: mangers, supervisors, and everyone else.

People are easily stratified by position, department, or other well-known classification schemes. Activities can be stratified by time or place. Information can be stratified by item type, form, format, and so on. If all management reports cannot be looked at, the analyst will surely want to examine the exception summary that goes to the plant manager.

Working with small groups, such as the single manager above, involves a risk of non-representativeness. The reason for stratified sampling is that, otherwise, important system elements may not be selected. It is important to learn about month-end pressures, even though there is only one month-end each month.

Unfortunately, representative sampling is honored more in the breach than in the act. Three kinds of non-representative sampling occur frequently: *guided*, *ad-hoc*, and *forced* sampling. **Guided sampling** is sometimes called *progressive* or *sequential* sampling; in this procedure, an element is included and, based on the observation of that element, the next element is selected. In this way, a sample is built up that really depends on the original element, which may be a strong limitation.

An **ad-hoc** sample includes whoever or whatever is available or at hand. The analyst may interview people who happen to be around or are not busy, observe what is apparently going on when there is time to visit the payroll department, or select the information that looks interesting. The advantages: quick, dirty, and cheap. The disadvantages: the data include implicit data about the analyst.

In **forced sampling**, the analyst must choose specific elements owing to circumstances. Often only certain people can be interviewed. Many activities may be considered off-limits, or they may be made difficult to observe (people who handle explosives, for example, should not be interviewed on-the-job). If the analyst is told exactly whom to interview, all that can be done is to note that the report reflects only the experience of those interviewed.

By far, the typical situation is a compromise. Limitations are given: this is off-limits, you cannot speak to x, you must observe y, or the data are confidential. Within the subset left, some representative sampling of either random or stratified nature is possible. Sometimes cancellations and fouled-up schedules make it necessary to use referral (that is, guided sampling) for replacements. *Ad-hoc* sampling, however, must always be avoided — it is the surest way to bias your conclusions.

MSA THE SOFTWARE COMPANY

The systems analyst gathers system experience data to become knowledgeable.

General Data Collection and Maintenance Considerations

Four general considerations are involved in data collection: **publicizing** the effort, **creating** the data-collection instruments, **maintaining** and storing the data, and preparing the data for **analysis**.

An analyst venturing out of the IRU should expect some degree of confusion and apprehension as to the purpose of the investigation, unless the purpose has been explained well in non-threatening terms. A strong, incorrect impression in the minds of key individuals can form an insurmountable barrier to the collection of accurate and complete data.

Willoughby (1981) outlines three forms of resistance to change that analysts might encounter: aggression, projection and avoidance. During investigation, **aggression** may take the form of verbal abuse, a lack of cooperation during interviews, or outright faking of data to create a particular image.

Projection usually takes the form of blaming the system for problems and may surface as fear that the system will make work either difficult or impossible. People may fear for their jobs, particularly at the clerical and secretarial levels.

Avoidance means that an employee will do anything not to have to use or think about the system. Collecting experience data from avoiders is difficult since an early form of avoidance is rejection of the study on which the system will be constructed.

Several techniques help to reduce resistance in the study phase. The major technique is to get users involved in the system study at the earliest stages — involvement helps counter the feeling of helplessness which lies at the basis of rejection and hostility.

There are three ways to involve workers and managers in investigation. *Public relations* alerts them to the generalities of the study, without surprises. *Management meetings* build a group identification and share important information that may pay off later in the design work. By *being available* to workers through formal and informal meetings, the analyst dispells rumors, educates the users as to the purposes and techniques of the study, and builds the kind of trust needed later on to get cooperation.

The second major consideration is the design and production of the instruments (forms and schedules) to be used in the study. Forms are generally paper media on which information is recorded or transcribed during the study. A schedule is a list of items or a program of activities. Other materials include letters to respondents or interviewees thanking them for their time and effort, summaries of the data that may be filed or distributed, and master identification sheets, which maintain anonymity in the data.

The creation of instruments will be discussed in later sections related to specific data-collection techniques. It is important to pretest all instruments. Many word processing systems can maintain questionnaires and other lists and produce high-quality, camera-ready copy almost instantly. The days of manual "cutting and pasting" questionnaires are almost over.

These automated systems also aid (but sometimes hinder) the maintenance aspects of a study. Five "terrors" can persecute any data-gathering study. *Anonymity should not be breached.* All efforts should be directed toward anonymity, especially where interview transcripts are maintained. It is true that many interviewees are easily identified by their responses, and in these cases, even greater security efforts should be mounted.

Second, there is the problem of *out-of-date schedules and forms*. Third, *data should be kept secure* so that it cannot be changed or read by those who have no need to access it. The fourth terror is *missing or incorrect data*. Data collection is an expensive activity, and the collected data is a company resource. It should not be subject to the vagaries of wind, weather, or the building maintenance people.

The fifth and final terror is *keeping data unnecessarily*. An investigation may have some residual value to those interested in other pursuits. Data should be shredded or burned after the report is produced.

An important concern is the preparation of data for analysis. Data should be transferred to summary sheets immediately after, or even during, collection. Backup copies of the raw data and summary sheets should be maintained securely in case of accidents. One important and valuable aspect of transfer and backup is that, should difficulties arise during data collection, the summary sheets provide strong indications of what is going wrong. For instance, it may be determined that responses from Department 3 are running 98 percent while Department 2 has scarcely any response to the questionnaire. Before the study is completed, it may be necessary to provide Department 2 with some additional motivation to respond — summary sheets provide this information.

An Overview of Data-Collection Techniques

Fourteen data collection techniques (Lederer, 1981) are discussed in this chapter and the next. They range from the simple but expensive *interview* to the complex but relatively inexpensive *questionnaire*. Figure 11-2 on the following page, illustrates the two major classes of data-collection techniques. A **controlled** technique constrains the activities of people observed and collects information directly from them. Controlled techniques fall into two groups: *in-situ observation* and *laboratory observation*. **In-situ** (on-site) observation (sometimes called "task analysis") includes clipboard-and-pen observation, automated observation, and a technique called *protocol analysis*, which is in-situ observation coupled with an interview of the actor during the activities. In **laboratory observation**, individuals are observed in a controlled environment, working on tasks that may be related to those that will be introduced later.

Uncontrolled techniques fall into three major classes: *individual*, *group*, and *informational*. **Individual** techniques collect data from individuals on their activities, ideas, and use of information. Similarly, **group** techniques elicit responses from groups. **Informational** techniques examine information only, typically performing an analysis on forms and documents.

Data are gathered on individuals through **real-time** or **retrospective reporting**. The former usually involves diaries. Retrospective reporting may be either personal (interviews and critical incident reporting by individuals in the presence of an analyst) or impersonal (pencil-and-paper responses to questionnaires).

Group techniques include personal **meetings** or impersonal **mailouts**. Focus-group interviews and brainstorming are types of meetings. The impersonal technique most widely used is the Delphi method, which is a repeated cycle of questionnaires.

Finally, informational techniques consist mainly of visual or manual **forms tracing** and **documentation analysis** (mostly of forms and reports analysis).

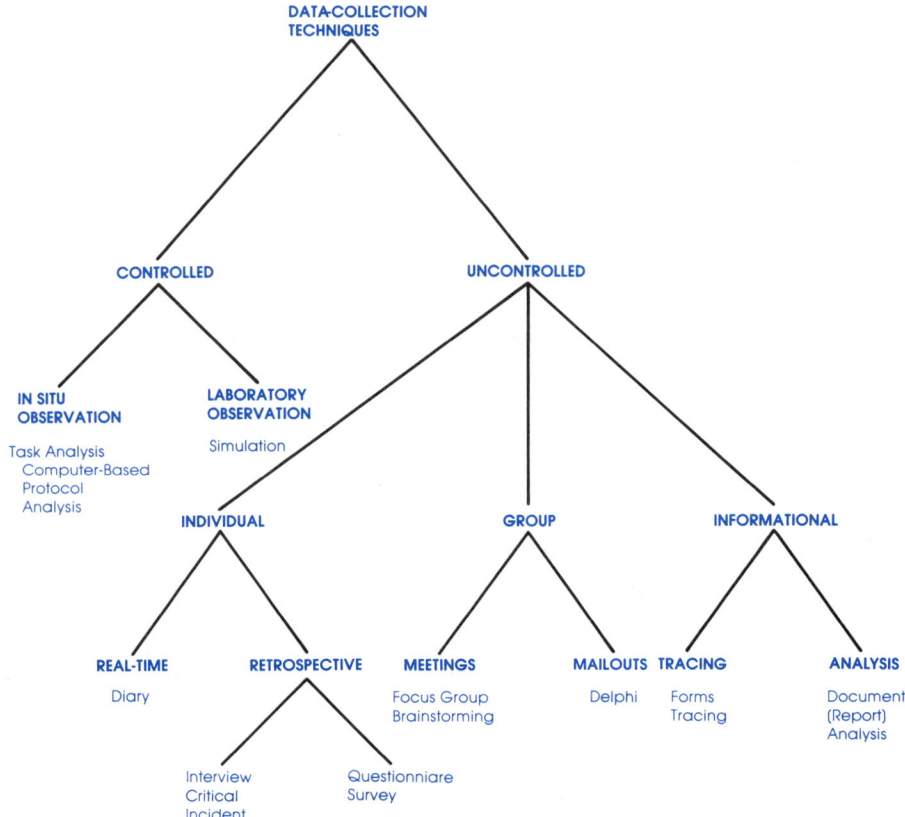

FIGURE 11–2. A Classification of Data Collection Techniques

Chapters 11 and 12 discuss data-collection techniques in detail and provide guidelines for choosing the appropriate technique for your goals.

11.2 INTERVIEWING TECHNIQUES

The interview effort (Molyneaux, 1982, and Stewart, 1985) requires *investigation*, *representation*, and *interpretation*. The investigation is aimed at designing a set of interview questions appropriate for the aims of the data collection. Representation relates to the gathering and recording of information obtained during the interview. Finally, interpretation assesses the validity and reliability of the data collected.

There are six steps in interviewing (Figure 11-3 opposite):

1. Handle the client request
2. Determine the sampling frame from which collection points (individuals) are to be chosen and create the interview list
3. Create the interview questions (schedule)

4. Arrange the interviews
5. Conduct the interviews
6. Analyze the data collected

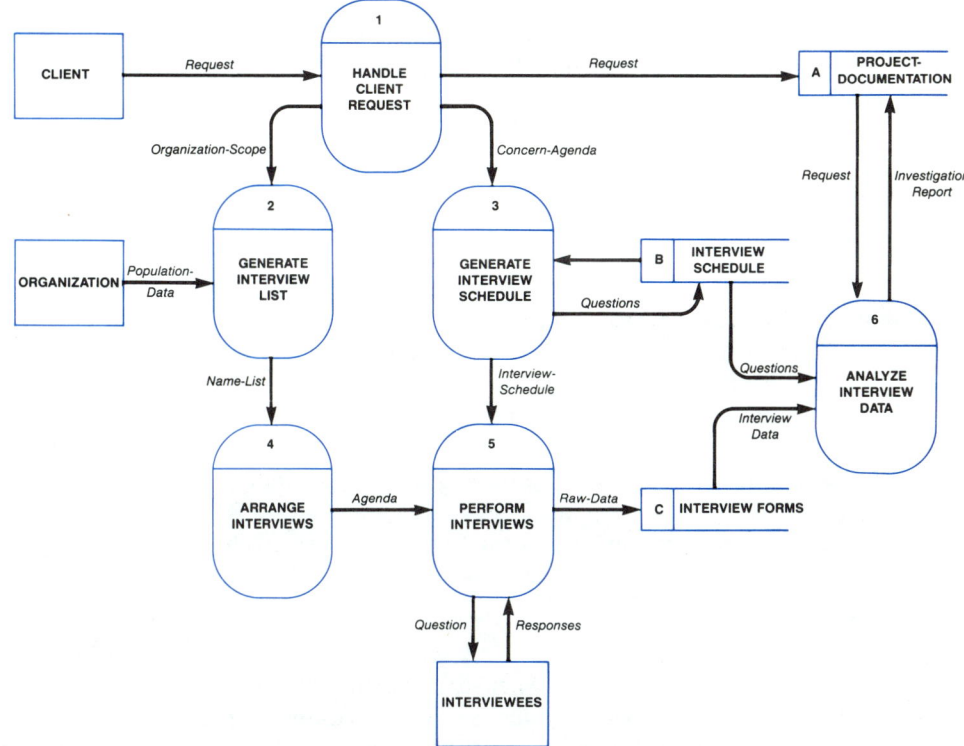

FIGURE 11–3. DFD for Interview Effort

An interview effort can be considered a system with inputs, processes, and outputs. The information that flows into the system is the following:

INTI1: concern agenda
INTI2: organizational population data
INTI3: interview responses

The **concern agenda** stems from the goals of the research. Typically, it is a "wish list" of things that would be valuable to know. If the goal of the investigation is to determine the needs for data processing in the order-entry department, the concern agenda might include the following items:

• What information is currently available?
• Is the information current, accurate, and precise?
• What is the workload in the department?
• What information needs to be updated frequently?

Organizational population data are the **sampling frame** from which collection points (individual employees) are selected. The analyst may use the organizational chart and the telephone directory or payroll list. A series of referrals builds a guided sample, with critical limitations on representativeness.

Interview responses come directly from those interviewed. If care is taken, the responses will be recorded correctly as raw data (see below).

Data flows created during the interview effort are:

> *INTO1*: interview schedule
> *INTO2*: interview agenda
> *INTO3*: raw data
> *INTO4*: report

The **interview schedule** is a list of questions to be asked during the interview and is generated by the analyst directly from the concern agenda. The questions should be pertinent, clear, answerable, and interpretable. The **interview agenda** is a list of individuals who are to be interviewed, the time and place of the interview, and an indication of whether the interview actually took place.

The **raw data** forms are transcriptions of responses to questions on the interview schedule. The raw data and the analysis are used to produce a **report** on the interview campaign. Where details are required, excerpts or comments from interviews (with identifying remarks removed, if necessary) may appear in the report.

Creating the Interview Schedule

There are three important things to be considered when creating the interview schedule: choosing a format for the interview; wording the interview questions for maximum understandability and answerability; and translating the concerns into words.

There are three general formats for interview schedules. The **unstructured** format is used primarily in preliminary work. There is no particular order to the questions asked, and the interviewer skillfully listens to responses to general questions and directs the conversation more or less to the topic at hand. There is a natural tendency to wander. In exploratory work, however, the issues are not clear, the stucture of the problem is probably misunderstood by the analyst, and the respondents are more expert in the difficulties and advantages. Here is an example:

> *SA*: Good afternoon, Mr. Smith. As you know, I'm here today to find out what kinds of problems your department has been experiencing with the sales information system. How has your exposure to the system been?
>
> *S*: Well, I'm not sure what you mean. If you're talking about me personally, I'd say that it's OK. I mean I don't use it. My assistant Mr. Jones does.
>
> *SA*: He uses it for you?
>
> *S*: Yes, because I haven't got the time. I'm too busy tracking down the next month's figures for the data-entry forms, so he sits down for the afternoon and talks to the terminal.
>
> *SA*: How much time does it take for you to track down the figures?

S: Well, I've never actually sat down and figured it out, but it seems to fill whole days sometimes. I mean just last week a whole region's reports were missing and I couldn't raise Allen or Bronson. And sometimes the data is incorrect... sometimes? Almost always, so I've got to track the salesmen. You know how hard it is to find a salesman in the field.

SA: Do you know why the data is incorrect?

S: I just assume that the salesmen haven't the time to check it properly, so we have to, against the invoices they mail in...and *that's* a pain, too. I mean, if we have to do so much work to prepare the data for the system, what good is it?

In this interview, the analyst is discovering valuable *attitude* and *procedure* information that he may not have found out had he asked specific, detailed, and structured questions. Of course, it is still unclear just how the system is being used and what the real user, Mr. Jones, does and feels. The apparent cost of the interview is offset somewhat by the knowledge gained concerning Mr. Smith's relationship to the system: mistrust, fear, and ignorance. We also discover that Mr. Jones is the actual user.

In a **semi-structured** interview, specific questions are asked, usually in a set order, but tangents are followed and probes are inserted where the analyst feels more information is necessary. The advantages are clear: important questions are always asked, but leads to new questions can be followed for a while. Here is the interview with Mr. Jones.

SA: Good morning, Mr. Jones. Mr. Smith, your boss, suggested that I speak with you about problems with the sales information system. I've got a few questions I hope you can help us with.

J: Sure, fire away.

SA: First, Mr. Jones stated that he never uses the system, but that you use it for him. What does that mean?

J: He writes down a series of questions and I sit at the terminal and punch in the appropriate commands. After I'm done, I take the printout and write up a brief report.

SA: How often do you use the system?

J: Oh, Smith's got a question about every other day.

SA: So you use the system about ten times a month?

J: Well, it varies. Sometimes he's very busy and I'll go a week without using it and sometimes it's every day. And I've got my own work to do as a marketing analyst...

SA: So you use it for your own purposes, too?

J: Yes, and I'd say that I easily spend an hour each day using it to retrieve and analyze sales data for myself. So that makes it about two hours a day, including Smith's requests and my work.

Note that the analyst has two specific questions to ask: who actually uses the system (who receives information) and precisely how often the system is used. To the first question, the analyst got a direct answer and continued with the schedule. For the second question, however, a probe was used ("So you use the system about ten times a month?'') to clarify Jones's estimate and allow him time to think about other uses. This was followed by a second probe in the form of a summary of Jones's previous response, ("So you use it for your own purposes, too?") in which the analyst discovers a second user of the system: Jones. Presumably the analyst will now continue with appropriate questions, although the focus may shift to Jones' requests rather than Smith's.

The analyst may return later to ask questions in a **structured** format. In this format, a specific set of questions is asked in a predetermined order and the responses are typically crystal clear. The purpose of the structured interview is to collect the maximum amount of data bearing on specific design points. The interviewer is present merely to clarify questions. Here is Mr. Jones's follow-up interview.

> *SA*: OK, last time we spoke, we discussed your reasons for using the sales information system. Today I'd like to get some specific information on how you use it.
>
> *J*: OK.
>
> *SA*: First, do you use a password?
>
> *J*: Yes.
>
> *SA*: Do you ever change the password?
>
> *J*: Yes, in fact I changed it last week.
>
> *SA*: How often do you change the password?
>
> *J*: Three, four times a year.
>
> *SA*: Can you be more specific? Is it more like three or more like four?
>
> *J*: (Thinking) Well...I'd say every season, so it's like four.
>
> *SA*: How many people know your password?
>
> *J*: Only me (quickly) and Mrs. Thompson, my research assistant.
>
> *SA*: So that's just two people, yourself and Mrs. Thompson.
>
> *J*: Yes.
>
> *SA*: Do you have any trouble remembering your password?
>
> *J*: No, and if I do, I ask Mrs. Thompson.
>
> *SA*: How does she remember it?
>
> *J*: She writes it down in code in her appointments book.

In this interview, the questions come one after another in a strict order, with only short excursions into clarifying questions or summaries. The analyst is attempting to discover how secure Mr. Jones's password is by asking a series of questions that go ever deeper into Mr. Jones's security concerns. The advantages of the structured interview are speed, volume of information collected, and clarity. Note that the analyst is not probing, rather he is continuing to draw out pertinent details about something he is already familiar with, namely, passwords. In the semi-structured interview, the analyst knows a lot about Mr. Smith and his needs, and therefore something about Mr. Jones. However, there is still a lot of room to probe Mr. Jones for his concerns, too. In general, the semi-structured interview combines the best features of the structured and the unstructured interviews.

Structured and semi-structured interviews fall into three general categories:

- The funnel: questions narrow in focus as the interview progresses

- The fan: questions become broader in range and longer and deeper in response during the interview
- The flip-flop: questions narrow to a focus and then broaden toward the end.

The **funnel** is useful in interviews in which respondents need to be taught how to respond (that is, in what terms to answer questions) and may be unsure about the direction of the interview. Questioning begins very generally ("What is your major responsibility?," "What are you trying to accomplish by using the system?," and "What do you do with the information you gather?") and gradually narrow in focus to a single topic in this case by the tenth question:

> ...
> 10. What do you do with the information after you get it printed?
> 11. To whom do you send it?
> 12. How do you send it?
> 13. Is the information received on time?
> 14. If not, why not?

In the **fan** format, questions become progressively broader. This structure is useful when dealing with specialists who have detailed and accurate knowledge about their field but who probably have not thought much about the implications of their work or the greater organizational context. In effect, the detailed questions "loosen them up" by providing them with the confidence to answer later, more general questions. Here is a fan starting to open at question four:

> ...
> 4. Is this form easy to fill out?
> 5. What kinds of errors do you make?
> 6. How do you find out about the errors?
> 7. What do you think causes these errors?
> 8. Do you think something should be done about the causes?
> 9. What should be done?
> 10. Who should do it? Why this person?

The **flip-flop** format is useful when dealing with a person who has some knowledge of details and some valid broader views. Watch this flip-flop:

> 2. What is your job function here?
> 3. Can you explain a bit about how you go about this?
> 4. What parts are the most difficult?
> ...
> 7. Do you understand the instructions on this form?
> 8. If not, why not?
> 9. What could be done to make them clearer to you?
> 10. Would this make the job easier?
> 11. In what ways would the job be improved by these changes?

12. Is this typical of the problems you have in this job?
13. Are these important problems?
14. Do these problems make your work difficult?
15. What do you think could be improved on this job?

It is unlikely that this clerk would be willing to volunteer accurate information to question 15 without it first having been directed to specific aspects of the job. By narrowing the focus of the questions and then later opening the questioning up to broader considerations such as difficulty and improvement, the analyst teaches the worker to focus on aspects of the work (as opposed to management style, social life, compensation, and so forth — these may be important considerations, but not for this particular interview).

The second area of concern is creating clear, concise, and easy-to-answer questions. Ambiguous, threatening, or insulting questions can ruin an interview.

Questions can be phrased in either an open ended or closed manner. An **open-ended** question can be answered in any terms; typically, answers are long.

- "What is your job function here?"
- "What do you think could be improved on this job?"

Answers may tend to ramble, expecially if the respondent is unsure of the terms to use in the answer. The alternative to the use of open ended questions is **closed questions**, which limit the number of possible answers. The respondent will give a short, precise answer.

- "What is your job title?"
- "How many times do you log onto the system each day?"

Another dimension is projective/subjective. A question is usually worded **subjectively**, in reference to the person answering it: "How satisfied are *you* with the turn-around time?" Sometimes individuals may be hesitant about answering for themselves; it would be unprofitable to pursue them to get what may be an unreliable answer. It is often easier to phrase the question in a **projective** fashion: "How satisfied are *people* with the turn-around time here?" The respondent *projects* his or her answer onto the group and avoids being tagged as different or unusual (when in fact the person may be typical). "Do you like your job?" is typically met with a blank stare, whereas the question "Do people around here like their jobs?" may be answered more readily.

Many questions fail on clarity and comprehensibility. Could you answer the following questions?

- "About how often does your program halt because of MUX failure?"
- "Management thinks that most workers make errors because of lack of training. Is this true in your opinion?"
- "How often do you make an error in preparing form *x*?"

The first question is incomprehensible because the interviewee has no idea what a MUX is. The analyst will have to explain what a MUX is, either before asking the question or immediately after the expected response: "What's a MUX?" The second question is ambiguous: is the respondent being asked what management thinks or what the reason for error is? The last question seems understandable and clear, but does everyone recognize his or her own errors? Improved versions of these questions might be:

A MUX is a device that puts a number of signals together into a single stream of messages. You don't normally deal with the MUX directly, but if it makes an error and detects it, the operating system will tell you about the MUX error. About how often would you say you have received this kind of message? Management thinks that most workers make errors because of lack of training. Do you agree with this theory? How often do you detect that you've made an error in form x, including errors caused by mistakes others have made before you get the form?

The guiding principle here is to put yourself in the chair of the individual answering the question. Will that person understand the wording of the question, especially the technical terms *you* are familiar with? Will the words be understood the same way you understand them? (What is "comfortable" for a worker might be "luxurious" to the person paying for the work station.) Can the question be interpreted in a variety of ways? Can the respondent possibly know the answer? Can the respondent answer the question in words he or she feels comfortable with? Here is a list of criteria regarding wording that analysts should keep in mind:

1. Does each word have a precise meaning?
2. Is the meaning the same to the respondent and the analyst?
3. Is the question itself unambiguous?
4. Is there an answer to the question?
5. Can this respondent know an answer to the question?
6. Can the respondent be trusted to provide the right answer?
7. Can the respondent find the proper terms in which to answer the question?
8. Can the analyst unambiguously understand the answer the respondent gives?

Here is a transcript of an interview that violates each of the above criteria. (The number in parentheses refers to the criterion in the preceding list that the question or response violates.)

Analyst [A]: Can you tell me what your security code is? (3)
Respondent [R]: Well, I'm not sure about that....(8)
A: What I mean is, do you have a public K23? (2)
R: I know mine, and so does Tommy — is that public? (7)
A: No, there are public and private K23s. Anyway, who's got public K23s around here?
R: I don't know. (5)
A: Do most people have private ones, then? (6)

R: I guess they must.

A: So the system's pretty secure, right? (1)

R: What does that mean?

A: Nobody gets in or out without permission? (4)

R: What on earth are you talking about?

There are several kinds of questions based on their *purpose* or function. The interview given above contains two kinds of questions that stand out. They are the **probe question** ("Who's got public K23s?") and the **leading question** ("So the system's pretty secure, right?"). These are direct attempts to force information from a reluctant respondent and must be used with care. The leading question is basically dishonest, since it is not a question at all, but a statement that puts words in the respondent's mouth. Under pressure, individuals react differently and sometimes detrimentally to the goals of the interview. They can avoid answering, lie, or divert attention; in many cases the interviewee just does not know and the leading question merely infuriates. In addition, it departs from the purpose of the structured interview schedule and may provide a misleading "clue" to something that really is not important to the study.

The **criterion question** is one that directly relates to the goals of the study. "How many errors do you catch yourself making, on the average, per terminal session?" directly queries error behavior in a study of man/machine interaction.

The criterion question is often unclear without explanation. The **lead-in question** provides the respondent with the opportunity to think about the activity or data the criterion question tests. Lead-in questions such as, "Do you find yourself making any errors during a typical terminal session?" and "Does the system inform you of the errors you make during a terminal session?" begin to focus attention on the errors in a terminal session. The lead-in questions provide a context for the criterion question.

Where the respondent's answers could be unreliable, **check questions** are helpful. A check question repeats the criterion question, using different wording. When the answers differ, the interviewer can quickly focus attention on that difference and attempt to find the true answer. If the analyst knows that the respondent participates in two terminal sessions daily, a check question for the criterion question above might be: "In a week, how many errors would you catch yourself making using the terminal?" The answer should be about ten times (two sessions/day × five days/week) that given in response to the criterion question. If it is not, the analyst should probe. Check questions should be used judiciously, however, since they add to the total interview time.

To provide transitions from topic to topic or to allow interviewers to rest during a long session, **filler questions** are used. Typically, filler questions are about personal data of a non-threatening variety, and they are easy to answer and do not require much thought. Such questions as "How long have you held this position?" and "Do you work with Sally and Joe?" might do. In general, filler questions form about 10 percent of the questions in a one-hour interview, mostly between topics. Criterion questions form about half the interview; lead-in questions, 30 percent; and the other question types, the remaining 10 percent. In a one-hour interview, an analyst might ask as many as thirty questions, of which fifteen are criterion; nine, lead-in; and two each of check, probe, and (if necessary) filler.

Determining the Sampling Frame

Several *sampling frames* are available in an organization:

- the corporate telephone directory
- the payroll list
- the organizational chart - distribution lists for memos
- project management forms

Where an entire department is to act as the sampling frame, the sample typically is drawn in a stratified random fashion. That is, you will want to include *some* managers, *some* supervisors, and *some* workers, rather than a sample that can, by chance, exclude (typically) all managers and supervisors. Suppose we have an organizational chart that describes the billing department of a retail firm. The chart is your sampling frame. If your goal is to determine new procedures for billing, you will have to speak to individuals in each department and at each level of the firm.

In general you will have to observe organizational considerations quite closely — managers like to know what their employees are saying to systems analysts and will want to be interviewed unless the matter is extremely clerical in nature. On the other hand, managers are busy people and scheduling interviews on short notice may not be possible.

Typically, interviewing efforts depart markedly from scientific studies regarding the validity of statistical sampling because both time and the time-resource of individuals are limited.

How to Handle Face-to-Face Interviews

The following subjects provide only an overview of the concerns an analyst should have when conducting interviews.

RAPPORT

Rapport is essential in interviewing and the major component that distinguishes interviewing from the questionnaire survey. Rapport is the sympathetic, harmonious relationship established between interviewer and respondent. It is most easily recognized by its absence: halting answers, long silences, rising tension, confusion, and hostility. Rapport does not necessarily mean agreement, but it does indicate a sharing of purpose. One cannot expect an employee to want to be questioned on the job by a stranger. Often, the analyst is mistaken for that distant cousin, the time-and-motion "efficiency expert." IRMaint aims to improve the tools with which information workers work, not to than replace workers, as is prevalently believed. The analyst typically begins by establishing rapport with the respondent to provide a positive start.

The best technique to establish rapport is to explain both the purpose of the interview and the agenda to be followed. The opening, in particular, should be friendly: "Hi, I'm Dick Thompson from Systems. I think that Mrs. Jacobson's already briefed you on why I'm here, but I think it's important that you know I understand how difficult it is to work on a system that breaks down every day. Anyway, our interview itself isn't going to fix the system, but we hope that by understanding problems workers have been having, we can build a good case for improvement. How does that sound?''

Avoid the third degree. Face the respondent and provide visual feedback through appropriate eye contact. Do not allow distractions and interruptions, if at all possible. Speak in terms that the respondent understands; nothing shows more lack of respect than talking down to someone in technical jargon. Keep the objectives in mind; prepare a written agenda if you feel that will help and share it with the respondent. Be courteous at all times and sensitive to the needs of the respondent to maintain silence and avoid certain questions.

EMPATHY

Empathy is a feeling of identification, a feeling that someone understands you and how you work. Workers often think of systems analysts who conduct cold interviews as spies for management. This may stem from analysts' extensive work with machines, but it is more probably due to lack of training and a tendency to fall back on the "professional" cover of computer specialist. The respondent wants to know there is a person out there who understands. The analyst who wants "just the facts" will have trouble.

Empathy arises naturally from letting the respondent know that you are listening and that you understand. If you monopolize the conversation or appear to be cross-examining, you have little chance to show understanding. Avoid making judgmental statements. ("That's no good" is no good!) Avoid snap judgments about a worker's motives or skills. It is not necessary to diagnose at every stage of an interview; that is why you take notes. Allow the respondent to speak freely and to vent frustrations. Even if you do not record this information, let the respondent know you have heard. Without empathy, people will be closed about their responses, as if they were talking into a telephone.

ACTIVE LISTENING

The term **active listening** seems like a contradiction, but good interviewers listen actively whether or not they are aware they are doing so. Listening is not merely a passive soaking up of information. It is, instead, the other side of speaking; it finishes the circuit from one person to another and back again. An active listener provides **feedback** to the speaker and, at the same time, guides the speaker into appropriate modes of interaction. The passive listener allows the speaker to drift and perhaps become insecure about whether responses are getting through clearly.

The techniques of active listening include *summarizing*, *paraphrasing*, *reflecting*, and *encouraging*.

Summarizing means summing up or providing a synopsis of the respondent's comments so far. It is worthwhile to stop every once in a while and summarize what the respondent has said. This allows the respondent to check your perception of his responses, to be sure the responses were complete, and to clarify where there is a chance for mis-comprehension on your part.

SA: Are you satisfied with the timeliness of your reports?
R: Well, yes and no.

SA: How so?

R: Well, they were adequate a few years ago when we were small and we dealt with a few suppliers, but now, well, I just don't know....

SA: (probe) What is the shortcoming now?

R: The reports still get to me by the eighth of each month, but some of our suppliers want their new orders placed by the first, because they don't want to maintain huge inventories of the special parts we require. And how can I tell them what we need on the first if all I've got to go by is six-week old data....

SA: Is the data that old?

R: Yeah, because a report on the eighth can contain information on a part sold on the thirtieth of the second previous month — that's up to 40 days! In 40 days our seasonal markets can dry up or boom.

SA: (summarizing) Let me see if I can summarize your feelings about timeliness of reports. You say reports that were timely five years ago aren't any longer because you can't work with data that may be almost six weeks old. Your suppliers want you to make decisions almost a week before you have the data. Does that mean that you'd be happy with a report on the first of each month?

R: Sure. By the way, five years ago we didn't have *any* reporting system. The one I use is only two years old...that should tell you how fast we're growing around here.

Paraphrasing is related to summarizing. It means retelling the respondent's side in the interviewer's words. This allows the respondent to find out what his or her responses sound like to others. Then errors or misapprehensions can be corrected. Since you are going to rewrite the responses anyway and re-interpret them, paraphrasing is an excellent way of having the respondent check interpretations.

R: Sure. By the way, five years ago we didn't have *any* reporting system. The one I use is only two years old....that should tell you how fast we're growing around here.

SA: Just how has that growth affected your job other than with respect to your suppliers?

R: Well, I've got seven men reporting to me now instead of the single fellow who worked for me three years ago. And one of these guys just checks the forms as they come in from the warehouse. Meanwhile, if things continue to grow the way they are, I'll have to have an assistant just to handle the paperwork.

SA: (paraphrasing) You're managing a much larger and more diversified staff now than a few years ago?

R: Yes. I'll tell you, sometimes that paperwork piles up two or three days on my desk. Who's got the time to do it? I just give it to my desk boss to check over and hand me the important stuff.

SA: (paraphrasing) Currently you filter your paperwork through one of your staffers to cut down on the time you spend with it?

R: Yeah, but it doesn't work! He tells me everything is *very* important...besides, he makes too many mistakes.

Reflecting is like paraphrasing in that you tell the respondent what you have heard. But in this case, you are dealing with perceptions and feelings rather than facts. It is unlikely that a respondent will want to deal openly with feelings of aggression, anger, alienation, or boredom. Such feelings, however, are commonly associated with lack of information in an organization or poorly designed information-processing jobs. Therefore, it is imporant that you listen for and reflect back on the respondent feelings that you think are being expressed. Again, this is for verification purposes.

R: Yeah, but it doesn't work! He tells me everything is *very* important...besides, he makes too many mistakes.

SA: Let's get back to timeliness for a moment. Where does the information for your reports come from?

R: Well, it comes right from my people. The guy I mentioned is in charge of checking all the forms and summarizing the orders and receipts each day and at the end of the month. But he's got to track down the errors, too; that makes him late sometimes.

SA: Where do the errors come from?

R: He makes a lot of them, and so do the other guys. Let's face it, they're all in this for a while and then they move on to another department. I wouldn't say that the job of warehouse clerk is exceptionally interesting, y'know?

SA: (reflecting) You think it's a boring job?

R: You bet. And the kind of people it attracts, well, you know this company doesn't invest a lot of money training those people...I try...I give 'em a two-day orientation...well my desk boss does, but sometimes they don't learn well and I'm not really satisfied with this system.

SA: (reflecting) You're unhappy with the training you give the clerks?

R: Not unhappy, really, but I sure feel frustrated. I mean, the average warehouse clerk lasts eight months here. So why train 'em...they just leave. But what a mess they make of those forms. I can't get anyone in the office to do anything about it, though. So my guy spends a week, usually, running down the errors on the forms and trying to read the handwriting.

SA: (reflecting) You feel the company has some responsibility here to train your clerks?

R: Yes.

A final active listening technique is to prompt interviewees by countering their responses with **encouragers**. These range from a simple nod of the head, to "uh-huh," "yes," "Can you elaborate?," and "I'd like to hear more." Encouragers tell the respondent that you are listening and that what they have to say is important. They establish a pact between interviewer and interviewee: Yes, I am listening; you can tell me more and I'll continue to listen.

Encouragers are mostly noted by their absence or the presence of **discouragers** like "uh-uh," "Well, I don't know about that," and "Let's talk about something else." While it may sometimes be necessary to discourage if time is pressing, discouragers tell the respondent that you do not care, when you really do. You care because your entire analysis relies on getting complete, accurate data from respondents based on their experiences, wishes, needs, and illusions.

In summary, active listening allows the respondent to feel that there is a circuit established, that lines of communication are open, and that you are listening. Active listening techniques provide feedback and check on the quality of the information you receive. Remember that a respondent may not always say precisely what is on his or her mind. A variety of forces mold a respondent's words. You must be aware of and able to adapt to these forces. Active listening is a primary tool for hearing the real meaning behind the words.

SILENCE

You can, of course, listen actively if things are said, but what if there is **silence**? If instead of obtaining responses, you are met with silence, what can you do?

In most cases, you should do nothing. Silence *is* a response. Do not fear it, as it usually means that the respondent is thinking. Many interviewers assume that silence means ignorance. This is usually not the case. Instead, the respondent may be trying to formulate a response. There are several possible reasons for silence:

1. The answer may be complicated.
2. The respondent may be trying to figure out something (such as converting how often delays occurred last month to your request for "the number of delays per day, on the average").
3. The respondent may be embarrassed to answer for fear of appearing incompetent, stupid, or lazy.
4. The respondent may not remember or be willing to guess under the circumstances, given the way the question was asked.
5. The respondent may not be aware that you asked a question, as opposed to making a statement.

The last explanation is "technical" in the sense that it implies something is wrong with the interview technique. Tone of voice and eye contact are the major mechanisms of what communication experts call "turn-taking" in conversation. It is unlikely that you will cause a turn to be missed in a good interview. Novice interviewers tend to adopt a flat tone of voice, which provides few clues that it is the respondent's turn to speak. An interviewer who stares at a schedule of questions has no opportunity to shift eye contact to the respondent when it is time for a response.

The other explanations listed above are non-technical. A simple "Shall I rephrase that?" or "Shall I repeat that?" will typically oil the mechanism. Give the respondent time to respond, however. Be aware, too, that although *you* are an expert interviewer, the respondent may not be an expert interviewee. It is probable that the respondent, not you, will make a mistake in handing over the conversation by averting the gaze or using a confusing voice tone. Many people end their statements with a rising tone which, for most native speakers of English, indicates a question. This can confuse an interviewer. A questioning response often indicates nervousness or a lack of confidence in the answer, implying missing information.

In any event, avoid at all costs completing sentences for respondents. You can never be sure that *your* completion was the *right* one. Although slow or halting speakers can be irritating, control the urge to supply answers to your own questions. When used correctly, silence can aid, rather than prevent, reaching your goal of obtaining complete and accurate information.

HANDLING HOSTILITY

The aspect of interviewing that almost everyone finds hard to handle is **hostility**. On some occasions, interviewees may have strongly negative feeling about the investigation, the analyst, the IRU, computers, or technology. Relating to those who resent your presence is difficult. The situation is made worse by the fact that many do not recognize their own hostility or express it directly.

There are two kinds of hostility: *open* and *silent*. **Open hostility** is easier to handle, because, in this case, you will recognize the tone. In addition, open hostility toward a new or changed system *is* information. What is causing the hostility? Is it the existing system? Is it the fact that an already well-learned system might be changed beyond recognition? Is it that the respondent was not contacted before the system investigation was begun? Does the respondent fear changes or any change at all? Was experience with a previous situation disastrous? Will the respondent feel less well informed or capable if the system changes? These are questions you want to have answered. In the case of hostility, the answers are filtered through anger, fear, and resentment.

With open hostility, the best technique is active listening with a great deal of reflecting. Because you are not a therapist and your time is limited, you must quickly expose the root of the emotions, reassure the respondent, and get on to the questions. This does not imply a brusque manner, nor does it mean dismissing the respondent's feelings. If you listen actively and let the respondent know that (1) you are listening, but (2) you have your own goals, then you stand a good chance of achieving the goals. Since the typical reaction to automated systems is a feeling of powerlessness and loss of control, you should come to the interview prepared with some facts about where such feelings often come from and what you can say to alleviate them.

Unfortunately, it is not always possible to be aware of angry feelings. Most often, hostility is silent. "Just you try to get me to cooperate" is a common hostile attitude. Rather than complain or argue openly, the silently hostile respondent turns inward, cutting off communication. It is difficult to get the silently hostile person to answer any question.

The reasons for **silent hostility** vary, but there are two major causes. First, the respondent may fear that you are collecting performance data for appraisal. This fear is naturally expressed in a lowered desire to cooperate and answer questions. The other cause is bad experience with automated systems or analysts. The person is therefore not merely suspicious, but resistant to any change that may come from the study.

The silently hostile person's responses may be irrelevant or misleading. To make matters even more difficult, the truly silently hostile person can easily "fake" an interview and answer consistently but incorrectly. What can be done?

Two useful tactics depend on your recognizing the signs of silent hostility and bringing these feelings into the open where they can be dealt with more easily. The first method is **confrontation**. By confronting the silently hostile person, you divert attention from their fear or dislike to the interview process itself. In shifting focus, you can assert control over the interview without pressing on fears about control over jobs. Often, confrontation causes true feelings of anger to surface. Be prepared to deal quickly with this *open* hostility toward you and the interview. Afterward, you may be able to steer the interview back to the topic.

There is a danger, however, that the interview may have to be terminated — come prepared with a quick exit. Confrontation should be used sparingly. Confronted respondents often complain about the interview to their managers, so you should be prepared to take some flak about a hastily terminated interview. And, of course, you can try to avoid silently hostile people by finding out something about them before you interview them, especially if your sampling is "guided."

Below is a sample transcript of a confrontation with a silently hostile respondent. There really is no "typical" interview, but here is how one systems analyst successfully handled the situation:

SA: OK, let's look at your communication system. Would you say that you are satisfied with it?

R: Sure.

SA: Completely?

R: No problems.

SA: How many phone calls do you place each day?

R: Oh, I don't know, maybe thirty?

SA: You say you place over three an hour?

R: Sure, but I don't see what this has to do with anything.

SA: We're trying to get a picture of the phone traffic here. Do all your calls get through?

R: Sure.

SA: All of them?

R: Didn't I say so?

SA: Do you never get busy signals or no answer?

R: Nope, never.

SA: That's unusual; almost everyone I speak with experiences some problems in this regard. Are you absolutely certain that all your calls go through all the time?

R: Listen. I know when and who to call and I don't need any advice on that from you.

SA: I'm not here to give advice...

R: Great!

SA: ...but I wonder why you think I'm giving any. Is there some reason why you don't want to answer questions about your phone system?

R: I think this study's a waste of the company's money, that's what I think. And idiotic questions about how many phone calls I make waste my time!

SA: As you know, in our letter to you — which you returned in agreement — we mentioned improving the phone system as one of our goals, but now you tell me that improvements would be a waste of your time. Why is that?

R: I know what this study is, and I want to tell you that nothing you do can improve on Clara. She gets my phone calls through and does my typing and I can't see what any computerized phone can do better than she can. (R is finally getting to the true problem.)

SA: I think I can set your mind at ease about one thing. Our study will have little to do with secretarial staffing policies. It's quite clearly spelled out that any improvements to the phone system will in fact have to be to the secretaries' benefit. No one will lose any secretaries, but most of us, you and I, will spend a lot less time trying to get through to everyone else. Which brings me back to my original question. Do you ever get busy signals?

R: No, of course not. Clara places all my calls for me. If she can't get through, she walks over there. What do I need a better phone system for?

The analyst now knows that (1) there is some misinformation about the goals of the investigation, (2) at least one secretary is employed in an unusual and inefficient manner, and (3) at least one executive feels that because of the secretary's efforts, there is no problem. Had the analyst continued the interview in the non-productive fashion of the first several questions or terminated when R showed irritation, none of this would have been discovered.

A second, more difficult technique of dealing with silent hostility is to guess the problem and make promises concerning it. For instance, in the previous interview, the analyst could have said this:

SA: I realize that you are quite busy and that you may see this investigation as an inefficient use of your time. However, we see things a bit differently. You are a valuable executive in this firm. Things that displease, confuse, or mislead you are costly to the firm and shouldn't be ignored. If my questions seem obscure or obvious, that may be because you don't see the benefits. If we can't get information about problems with the communication system, then whatever problems there are will continue indefinitely; only the information resources unit can make the sort of improvements that have to be made. You can refuse to work with us, that's your right, but in the end we will all suffer a bit longer. That's a steep price to pay for half an hour of your time, wouldn't you say?

Reaction to this message may range from open hostility to instant dismissal to cooperation, depending on the individual and that individual's position in the firm. If, indeed, you are dealing with a valued executive known to be hostile toward the IRU, it is likely that you will not have the power necessary to conduct the interview there in the first place. On the other hand, attitudes of this sort are rarely chronic at high levels in the firm. Without being obsequious, it is possible to state things well, respectfully, and clearly. After all, you also are an employee of the firm.

Other causes of hostility include fear that jobs will be eliminated; fear that there will be major, painful changes in the job; fear of computers; dislike for the analyst and the interview itself (here, factors such as appearance, dress, and manner are important); and conflict between the respondent and the organization.

Not every interview is a success. But you should always know why a particular interview failed and be alert to future, similar circumstances.

In summary, the interview process depends on establishing rapport, showing empathy, listening, providing feedback, knowing the value of silence, and handling hostility. Not all interviews succeed, but with proper planning, the analyst at least can determine what went wrong and avoid similar failures in the future.

LOCATION

The interview should be conducted in the respondent's working environment, unless that area is noisy, distracting, threatening, or too public.

OPENINGS AND CLOSINGS

Interview openings ought to be friendly, but they should not have a used-car-salesman approach. Make no promises, but make the interviewee comfortable. Introduce yourself, your company, the study, its goals, and the agenda for the interview.

Usually, a simple summary of the interview and a "thank you" suffices for a closing. No concrete promises ("Things will get better." or "The computer will be here next month.") should be made. Let the respondent know that if there are questions later — especially second thoughts about responses given in the interview — there is a way to contact you.

MOTIVATING RESPONSES

Generally, respondents will be motivated to try to answer questions truthfully and completely given the the following:

- honesty on the part of the interviewers
- neat and attractive interviewers
- articulate questioning without the "third degree"
- prompt, unguarded answers to their questions
- active listening
- genuineness in questioning
- comfort
- understanding of the questions and the rationale behind them
- a feeling of being involved

Respondents usually will clam up in situations in which:

- they feel they are being pumped for information
- they feel helpless
- the interviewer appears unkempt, inarticulate, impolite, or uninterested
- questions are unintelligible, full of jargon, or confusing
- there is discomfort, rushing, or lack of privacy
- interviewers arrive late, have to leave early, or appear in a hurry to get to the next interview

SCHEDULING INTERVIEWS

One-hour interviews should be the ordinary maximum (except when coupled with protocol analysis or observation). Planning and rehearsal of the interview schedule can help determine the actual length. Figure 11-4 shows the time course of a typical interview. The opening may take ten minutes with handshakes, introductions, finishing a bit of paperwork, or tardiness. Similarly, the final five minutes may be taken up with the protocol of ending the session. This leaves about 45 minutes for questioning, which may actually tire many people who rarely have to discuss their jobs with others. Given interruptions, background, and debriefing of the interviewer by the respondent, the 45 minutes may actually only be 40.

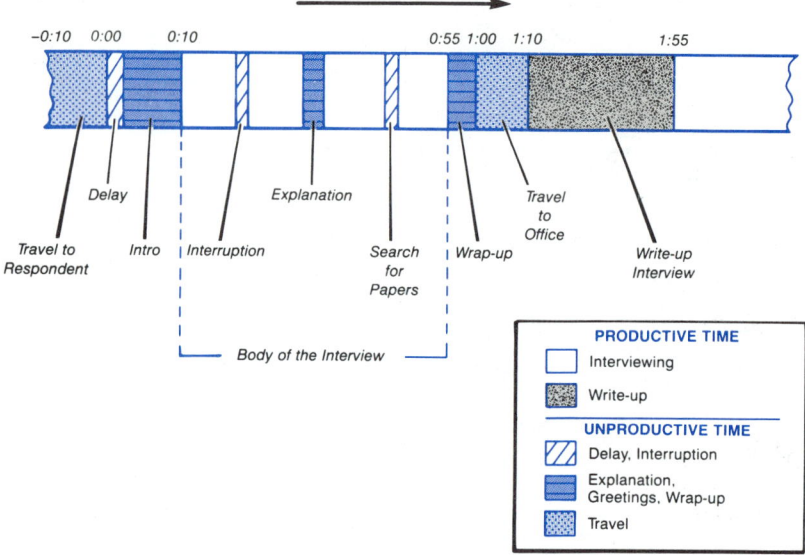

FIGURE 11–4. The Time-Course of a Typical One-Hour Interview

TAKING NOTES

While some interviewers utilize tape recordings effectively, there are several reasons why taking notes is better:

- Many people feel intimidated by a tape recorder.
- Although relistening to responses can increase accuracy, it, in effect, doubles the time spent listening to respondents.
- Serious questions about privacy and security arise when an individual's voice is recorded.
- Most small cassette units are of low quality. The typical hand-held recorder is inadequate in any but the most acoustically perfect environment. It is likely that many responses will actually be *lost* on the tape.
- Finally, using tape really does not stretch listening skills. The interviewer may well become lazy and forget to probe, explain, and assist.

To some people, note taking may also be intrusive and intimidating. In addition, many individuals simply cannot write rapidly and legibly. Despite these drawbacks, taking notes is preferred if only because individuals can *note* characteristics of the interview and the responses when they occur; it is difficult to record these on tape.

Telephone Interviews

While most interviews take place face to face, it is often necessary to interview by telephone to decrease travel costs without any loss of immediacy.

There are drawbacks, however. Most important, it is hard to judge non-verbal reactions (facial expressions, for example) and almost impossible to go over documents (unless copies are available at both ends). Because eye contact is lost, participants have to rely on voice tone to control conversation and turn taking, which may lead to a stilted and cool interaction.

It is more appropriate to have highly structured, closed-answer formats for the telephone interview. Errors in listening and transcription are more likely. Fatigue is more likely at both ends. Thus, conversations should be brief and prescheduled. Relevant information should be mailed out in advance. In fact, questions can even be sent out in advance.

Cost-Effectiveness Issues

Interviewing is expensive. Since there are alternatives, it is worthwhile to ask questions about cost-effectiveness. On the cost side, each one-hour interview costs three man-hours: one hour each for interviewer and respondent and one hour of overhead for the interview. The overhead consists of travel, write-up, and data management time. To this we can add a variable cost for the respondent's preparation time and those who depend on the respondent's availability. As a rule of thumb, three to four hours are spent on each interview; a twenty-interview effort can require two man-weeks of time. Because some interviews may be canceled too late to reschedule, a maximum of two interviews should be completed daily. In many IRUs, direct costs for a small interview effort can run to $2,000, with an additional $1,000 in buried respondent time costs.

To this we should also add labor costs for developing the interview schedule and idle time resulting from an inability to schedule into each time slot.

Effectiveness questions revolve around whether interviews produce good data and valid insights. Four gremlins that dog the heels of interviewers tend to reduce the effectiveness of interviews. They are: social facilitation, evaluation apprehension, experimenter expectancy, and interviewing the interviewer. Most stem from the fact that the process of interviewing is intrusive, shaping the respondent's answers.

Social facilitation means performing better when other people are around. In the case of interviews, this means that actual or remembered tasks may seem easier than they actually are (or were). This can obviously bias the study.

Evaluation apprehension occurs when a respondent feels he is being evaluated and tries hard to create a good impression. In the 45-minute hour available, the interviewer has little chance to discover that this behavior is not typical.

Experimenter expectancy occurs during experiments in which the researcher is "looking for something." Experimental subjects often try to *guess* what that is — and supply it! For the analyst, "analyst expectancy" can be a disaster, since IRMaint relies on accurate and reliable data, not guesses on the part of the respondents as to what the analyst is trying to "prove." The analyst can avoid this by explaining the process of gathering data, thus disposing of any preconceptions the subjects may have. This can result in an ethical dilemma, however, when there really are preconceived plans. Professional detachment is desirable but may not always be possible.

Finally, the problem of **interviewing the interviewer** occurs when interviewers supply answers to or finish sentences for the respondent. The amount of background supplied may be too great, thereby "teaching" *desired* answers to the respondent, or the analyst may simply interpret answers in a "pro-system" or "pro-development" fashion. The analyst can avoid putting words in the respondents mouth by applying the following: active listening, the appropriate use of silence, and the supply of appropriate levels of background. The transcript below shows some of these problems:

> **SA**: Good morning. I'm glad to be able to have a few minutes of your time to ask you some questions that are very important to us up in the information resources unit. As you may know, we are strongly considering upgrading our computer resource by installing a dial-in facility. To give you an idea of why this is an upgrade, just consider what this might do for your salesmen in the field. They could call in every evening or during the day and enter their sales data on-line and get inventory and order-lag information on the spot. Now I'd like to ask you a few questions, if I may.
>
> **SM [Sales Manager]**: Sure, go ahead.
>
> **SA**: Fine. Now, how does work seem to be going in this department?
>
> **SM**: Oh, not too badly. No, really, things are going quite well. We've met our sales goals for November and this month also seems to be shaping up quite well. (evaluation apprehension)
>
> **SA**: Oh, I see. Let me get more specific. Does the information you get on sales seem accurate, up to date, and reliable?
>
> **SM**: Sure...oh, there's an occasional slip-up, but nothing I'm not able to correct with a few phone calls. Sometimes the information's a bit...uh...
>
> **SA**: Inaccurate? (supplying answers)
>
> **SM**: Yes, but nothing really important. I just phone 'em and find out that ten thousand dollars in sales was really eleven thousand or something like that ...really I can fix it up fast (social facilitation).
>
> **SA**: About how often do you detect this sort of error?
>
> **SM**: Oh, very rarely, very, very rarely. But y'know, it's curious. The errors seem the largest when my salesmen are the farthest away. I never thought about it before, but that's true. It's strange, isn't it? (experimenter expectancy)
>
> **SA**: No, not really, maybe these individuals feel more removed from their data or actually have the wrong figures?
>
> **SM**: That they could get from the computer in-line? (guessing the answer)
>
> **SA**: *On*-line. Yes, they could. So you might say that you see a great need for automation and remote dial-in access because of accuracy problems with your salesmen in the field?
>
> **SM**: Oh, certainly.

This exchange need never have taken place. The analyst took advantage of the interviewer effects to show how dial-in access could improve a shop that was probably already functioning well. Needless to say, the truth, whatever it is, lies elsewhere.

On the credit side of effectiveness, an interview is almost indispensible for obtaining personal information from workers where an impersonal format, such as that of a questionnaire, would be unreliable, insulting, or unclear. The interview is more flexible and complete, and can be used in a wider variety of situations than questionnaires can. Someone describing the memory of an activity is not as accurate a measure as directly observing it, but the implications of that activity to the person cannot be observed directly in any event.

Using interview techniques well provides a fine degree of control to the adept interviewer. Because an interview is a conversation and not merely squeezing of information, the respondent can feel more involved and more like a stake-holder and has a clearer picture of what the investigation is about. The interviewer can be questioned. While interview effects lessen some of these advantages, the payoff in control, flexibility, and reliability make the interview more beneficial.

Interviewing is an excellent technique for gathering information in early phases of IRMaint, for gathering data from executives, for completely understanding how a small group functions, and as a follow-up to other techniques (Kendall and Losee, 1986). It is not cost-effective for collecting a large amount of data over a large geographical area or for obtaining complete data quickly.

Interviewing is often combined with other techniques (primarily observation) and has some specific refinements that capitalize its advantages. The critical-incident technique, discussed below, is one such adaptation.

The Critical-Incident Technique

The **critical-incident technique** uses the interview to concentrate on "critical" events rather than day-to-day operations. It is ideally suited for interviews with managers and supervisors who manage by exception. Those in supervisory positions know about events in which a poorly functioning information resource may have led to catastrophic events in terms of (1) cost, (2) procedure, or (3) morale, image, and decorum.

The advantage of this focused technique is that it helps ferret out "bottom line" concerns. The new or improved information resource had better not encourage situations like these:

"We almost went bankrupt when we lost those sales tapes for a weekend."

"I just couldn't find anyone around here who knew what was going on or how to get the system started again."

"It was personally embarassing not having that report read and signed when the client arrived."

The following transcript indicates how the critical incident technique focuses on specific incidents rather than general process and needs:

SA: Your employees handle the payroll requisitions?

PM [Payroll Manager]: Yes, that's correct.

SA: Can you relate any recent event that seemed as though everything in your department was going wrong? I'm not concerned with, say a fire or an office row, but something that involved your use of information...

PM: (Thinking) Hmmm...oh, yes, I can. About three weeks ago, one of my clerks called in sick. Normally that's not a problem, but on this particular day, he'd had a stack of about eighty payroll requisitions on his desk. Or at least he was supposed to! They weren't on his desk or in the locked cabinet. I called him back, but he really was sick and his doctor had sedated him. Meanwhile, I had eighteen managers phoning me asking for the payroll checks due to be printed on Thursday afternoon. I don't mind telling you that there were a lot of angry people around here. Turns out he'd known about his illness and had given the requisitions to another clerk the afternoon before, but nobody told me! And the other clerk was out sick, too.

SA: (probing) So what do you think was the problem?

PM: The problem was that we really have no way to control the work flow through here. The requisitions are signed into the area and the checks are signed out, but when it's in here, we really can't tell who's got what until it's done. I can't get them to follow procedures, especially the new guys. Let me tell you about another incident, which happened last month. That time we lost a check after it was printed. Usually they're stuffed into envelopes right after being printed and since they're printed by department, we just rubber-band them and deliver them after checking that they're all there. Somehow, and I'm still not sure why or how, the rubber band broke and a check got lost on the floor. We found it two days later. Meanwhile we had to make sure it *was* lost and stayed lost while we printed another one. Do you know that we'd never considered how to make a check stay lost? It's possible that an unscrupulous employee might have tried to cash both checks and *that* would have been a problem!

After these extended monologues, the analyst has learned two criteria for an improved payroll system: work-flow control and check control.

The disadvantage of the critical-incident technique is that it encourages elaboration and exaggeration because it avoids the routine and the ordinary. Most organizations do not have only critical incidents. Usually, events proceed normally. But the goal in this technique is not to model the system so much as to determine the basic level of functioning tolerable. Later, other techniques are used to capture the routine.

Summary and a Suggestion

All data collection begins with a plan, the process of which consists of **goals**, determining the **information** necessary to meet the goals, **selecting** collection points, **collecting** the data, **organizing** the data, and **assessing** the data and the data collection effort.

A **collection point** is a single system element that is observed or measured in some way. There are three classes of collection points: *persons*, *activities*, and *information*.

For a given collection study, one, two, or all three might be relevant.

Collection points must, in some sense, be representative of the entire system being investigated. Typically, either points are **randomly sampled** (where large sets of collection points must be accounted for), or analysts are intelligently "guided" to appropriate or important points. The typical compromise is a blend.

Data collection requires five important concerns: the way the study is *introduced* to the subjects, the preparation of the data collection **instruments**, the *maintenance* of data in a secure fashion, the *storage* of data during the investigation, and the *preparation* of the data for subsequent analysis.

The interview technique is an extended conversation with the goal of discovering facts that only the respondent knows. Choosing respondents carefully and systematically and creating an interview schedule and format that are appropriate should ensure that the data collected are as reliable and valid as possible. Because interviewing is an interpersonal skill, becoming good at it requires practice and attention to the process. Skill building should be on an organization's agenda in active ways: courses, seminars, and observation by experts.

Case Study

The City of Bigtown's social service agency is concerned that current manual intake methods of child welfare clients and data gathering are inefficient, prone to inaccuracy, and uncomfortable for clients. There have been reports of critical inaccuracies that have put children at risk. These situations arise when data are incompletely collected or clients move and information cannot be merged correctly (as, for instance, when a client moves to another city). It is felt that a study of automation possibilities is in order.

The goal is to design an automated procedure for data gathering and referral of child welfare clients (children who go into foster homes or group homes paid out of public funds on a per diem basis). The city's concern is that expenses be kept to a minimum (consistent with the safety of the children) by avoiding duplication or unnecessary stay in care.

Data gathering will be by interview, observation of work flow (primarily paperwork), analysis of existing procedures, and critical-incident technique. In the preliminary stages, a number of interviews will be conducted with key personnel in the intake unit of one neighborhood. The individuals to be interviewed in this stage are:

1. the director of social services (at headquarters)
2. the neighborhood unit manager
3. one office administrator
4. two intake social workers
5. two clerks in the office
6. two clients (and perhaps others by referral)

The goal of the investigation is to refine the nature of the problem for the preliminary investigation report. Is it merely a question of efficiency of paper flow? Are the forms completed correctly? Is there a problem with training or supervision? What kinds of errors have occurred? Are the errors due to lack of training, time pressures, lack of care taken in following instructions, or difficulty in filling out the forms? Are the perceptions of management, professionals, support staff, and clients the same? If not, how do they differ? What potential costs are associated with the current system that could be avoided?

The sample is produced through forced sampling (the director and the neighborhood unit manager *must* be interviewed), random sampling of the social workers, guided sampling of the two most articulate clerks, and *ad-hoc* sampling of cooperative clients taken from a list of current clients. One-hour interviews are scheduled. All interviews are semi-structured, with ample opportunity for restructuring based on client, worker, and management perceptions to define the problem as they see it.

The schedule for the office administrator's interview looks like this:

1. What is your perception of the problems and opportunities in the current intake recording system?
2. What aspects of the current system are highly desirable and should be retained?
3. What do you think is the most pressing problem facing the system now?
4. What are your perceptions of the opportunities for improvement?
5. Can you think of any drawbacks to automation of this system other than cost?
6. What criteria would you expect an automated system to meet that the current one doesn't?

The social workers' interview looks like this:

1. What are your major job functions?
2. Let's consider the intake aspects of your job. About how many clients do you see each day for intake?
3. How long does a typical intake procedure take? Are there standards for this procedure? Are there any time pressures?
4. Can you give me an idea of this procedure? (Follow it with the worker.)
5. Are there any aspects of this procedure you find difficult, confusing, tiring, or error-prone? (Probe for each one to get a clear picture.)
6. How do you go about checking the validity of the data you collect?
   6.a. (If no procedure for validating) Have you ever had a problem with incorrect data?
   6.b. If so, how did you go about correcting it?
7. What would help you make the intake procedure more efficient or effective from your viewpoint? (Probe.)
8. What training did you receive for this specific aspect of your work?
9. What would you improve in this job if you could change the procedure?

The social worker interview is flip-flop style, narrowing to the specific problems of errors and efficiency. Questions 2, 3, and 9 provide a check on question 7. Most questions are open-ended, other than the estimates in questions 2 and 3. In addition, certain demographic data are collected: number of years in the agency, years in current position, years in neighborhood unit, professional training and degrees, previously-held positions, and salary grade. The criterion questions are numbers 5 through 8. Question 9 allows the respondent to muse over the job (rather than the procedure) to allow venting of frustrations and expose expectations of job-related information processing.

Responses are transcribed onto response sheets. In the case of the management-level interview, the analyst must be prepared to write quickly, summarize, paraphrase, and reflect, if only to slow down the respondent and check whether the transcription is correct.

The following is an excerpt from the clerical interview:

SA: What if you have trouble reading the hand-written intake form?

CL [Clerk]: I call the social worker.

SA: And if the worker isn't there or is busy?

CL: I put it aside. There's a lot of work to be done here and I can't hold up on it. So I just put it down in the "In process" basket and go on to the next form.

SA: How long does a form stay in that basket?

CL: Oh, I don't know. Maybe a day? Never more than a few days, anyway.

SA: As I hear it, sometimes forms end up in that basket for up to three days and children are sent to group homes in that period of time.

CL: Yes, but that's very rare, maybe once a month. Usually I can read them all right. Besides there's not that much activity in here.

SA: Now, other than requests for specific information on a specific form, what other kinds of requests do you get?

CL: Well, first the workers sometimes misfile their originals and we have to get copies for them. And there's always some official who wants to know how many runaway kids we place in group homes in the city in March or how many girls under the age of 10 we've got in care now, so guess who has to go through all the files?

SA: And are these requests met easily?

CL: (looking surprised) Hah! You try running through these files looking for three teenagers in group homes in March!

SA: So what do you do?

CL: Well, we give it a try...at least *I* do. Sometimes the other clerks, well, they just make a guess or they phone around to the workers. I mean, we make copies for them when they lose theirs, so we expect them to help us out when we have to answer these unusual questions.

Here the analyst follows a general structure but probes when an interesting lead is discovered. Obviously there are some problems maintaining and retrieving information from paper files.

The analysis and report should concentrate on a general flow diagram of paperwork, pointing out where the possibilities for error or delay are the highest. In the preliminary investigation report, definition of the problem as one of error and low efficiency is important, because it is only a detailed and usable definition that enables a reader of the report to look at the flow diagram and assess whether additional work would help put a price tag on the conversion. The conversion goal will be taken later from the wish list of improvements expressed by worker, manager, clerk, and client alike. In this investigation, it is too early to speculate about specific automation techniques. Instead, concerns about paperwork errors and the efficiency of storage and retrieval of information from a logical point of view are most important. Figure 11-5 is a DFD that illuminates areas of concern for a detailed analysis.

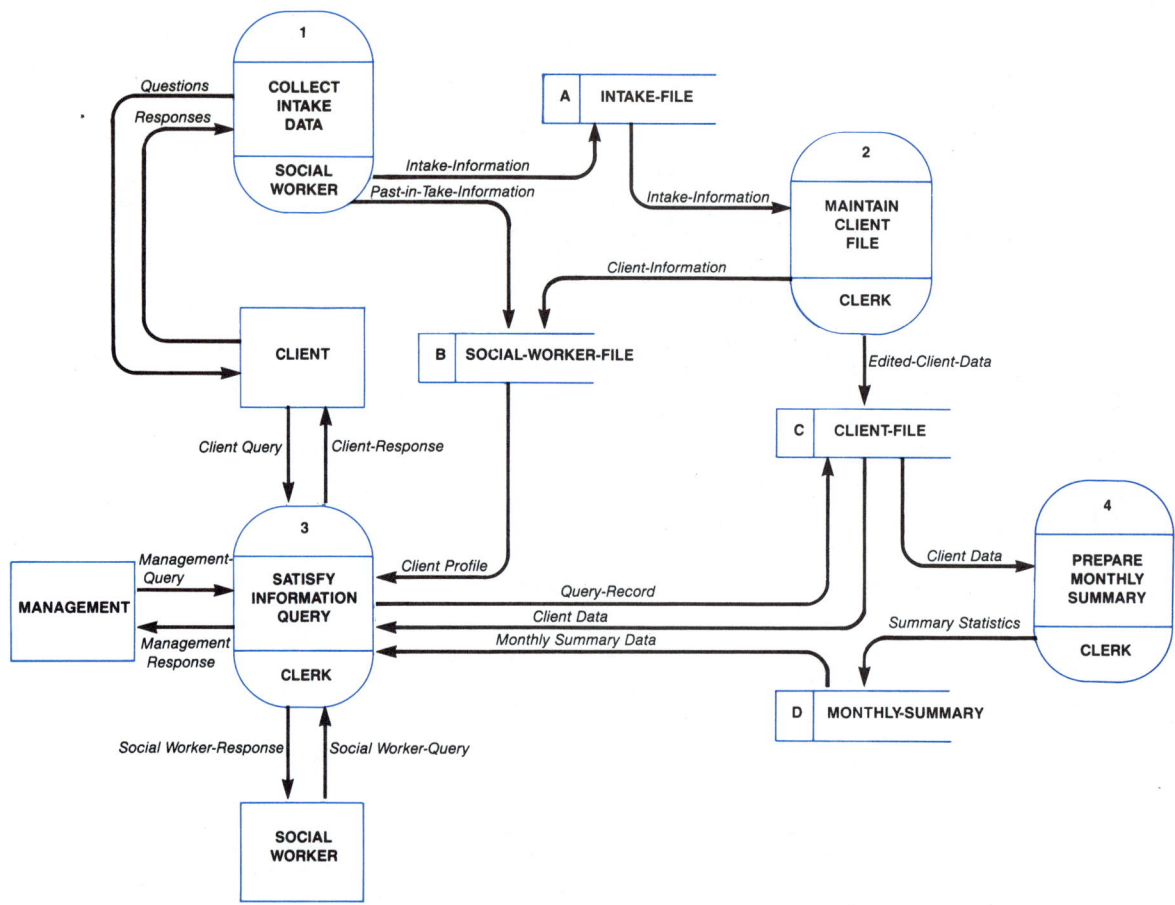

FIGURE 11–5. Child Welfare Intake DFD

Discussion Questions

1. Data collection is the major goal of a data collection *system*. Designing the system is, like other design efforts, both logical and physical. Discuss the steps in collecting data in terms of logical and physical design. Given this, what constitutes the "investigation," "representation," and "interpretation" activities of the analyst?

2. The text points out the need to sample persons, activities, and information in seven combinations. Discuss the persons, activities, and information sampled in each of the following. Recommend a technique to ensure representative sampling in each case:

   2.1. how a bank processes a customer's transaction

   2.2. how customers' purchases are handled in a supermarket

   2.3. how applicants for an executive position are handled in a large management consulting firm

   2.4. what the contract arbitration process is for handling collective agreements at a small, private university.

3. How can a compromise sampling situation be justified when it will certainly result in a non-random sample? That is, under what circumstances would guided, *ad-hoc*, or forced sampling be justified? What is so special about random sampling, anyway?

4. The text contrasts controlled and uncontrolled data-collection techniques as a major distinction among these techniques. Consider also a distinction between *real-time* techniques (such as direct observation, protocol analysis, and forms tracing) and *retrospective* ones (such as interviews, questionnaires, and diaries). What are the relative advantages and disadvantages of each class of technique? In which situations would one class, rather than the other, be most appropriate?

5. Critique the following interview and suggest improvements to the analyst's technique:

   SA: (seated in his own office, to which Mr. Dobbin was called 30 minutes ago) Hi, Dobbin. Let's get down to business about the disappearing shipments. You know who's stealing them?

   Dobbin [D]: (obviously taken aback) Now, see here, young man, no one's ever said that shipments were being stolen!

   SA: No? Well, you explain how the figures always come up short.

   D: There could be a lot of reasons, I suppose, but anyway, is that why you've called me here?

   SA: No, not just that. The whole procedure around this shipping department seems very loose to me and I'd …

   D: Wait a minute. Let's get one thing straight, OK? I'm night manager in the shipping department. Our procedures are taken right from the book. Our people are *very* methodical. We check everything twice, once when a shipment is picked and once before we let it leave the building. No, sir. Not a chance of what you call looseness. How do you come to that conclusion?

   SA: Looseness means letting a number of shipments leave without controls. Now I want to ask you a few questions about those controls. What controls *do* you have down there?

   D: Controls?

SA: See what I mean? You don't have any. OK, Dobbin, what I want to know is what records you keep to ensure that shipments are not lost or stolen before they are put on the trucks.

D: Oh, yes. Well, we keep the warehouse copy of the invoices.

SA: And the picking slips?

D: Yes.

SA: And what do you *do* with them?

D: Why, we check them against each other. And...

SA: If they don't match you call the order desk, right?

D: Yes, and...

SA: Don't you think it's strange that shipments could be lost if you're so careful?

D: No, and anyway, no one said the shipments were *lost* or *stolen*, as you so rudely put it before. We're not incompetent or crooked. If you already know the answers, why are you asking questions, anyway?

6. Explain, in your own words, why active listening is so important in interviewing for data collection. What might happen if you don't listen actively?

7. There is a brief transcript in the chapter in the section "Cost-Effectiveness Issues" (pg. 398) that illustrates the intrusiveness of interviewing. Indicate what you might do to improve the interview.

DESIGN EXERCISES
Using dBASE III

1. So far we have worked with only one file at a time through the USE statement. The real power of dBASE III comes from having several files — up to 9 — active at once. To do this, we SELECT an area by number (between 1 and 9) and then USE a file. That file is then "assigned" to that area and we can switch among the files by SELECTing areas. By default, area 1 is implicitly SELECTed each time we bring dBASE III up. Here is an example which switches between two areas:

SELECT 1	Select Area 1;
USE CLIENTS	CLIENTS is assigned to Area 1.
SELECT 2	Then select Area 2;
USE STYLISTS	STYLISTS is assigned to Area 2;
SET INDEX TO STYLISTS	The index file is STYLISTS.NDX.
SELECT 1	Now in CLIENTS,
LOCATE RECORD 4	Go to the fourth record, but
SNAME = STYLNAME	Save this client's favorite STYLIST into a variable, because STYLNAME will not mean what it does now after we do a ...

DESIGN EXERCISES
Using dBASE III *continued*

SELECT 2	And look at the STYLISTS file to
FIND SNAME	Find the record for this STYLIST.
? TUESDAY	Is the STYLIST available Tuesday?

First we assign files to areas by SELECTing the area and then USEing a file. Then when we SELECT an area we are also USEing the assigned file. We can move from file to file this way, just as a navigator moves across the waves. In fact, this kind of file access is called "navigation." Write a program called WRITEUP that "navigates" by listing all the transactions (CLIENT name, DATE, RCPTNO, item or service and VALUE) for a given stylist's STYLNAME. Here is the plan:

> RECEIPTS contains all the receipts. Each record has a receipt number RCPTNO (which is the index for SESSIONS, each record of which identifies a client by NAME and PHONE, and a stylist by STYLNAME), a TYPE (I for item, A for activity), and a four-character code that is either an ITEMCODE (the index for PRODUCTS) or an ACTIVITY (the index for SERVICES). We can pick up each receipt in RECEIPTS and find out the DATE and STYLNAME from SESSIONS. If it is the stylist we are looking for, we can then look up the item via ITEMCODE in PRODUCTS or the service in SERVICES via ACTIVITY. Each has a price and we can then write out a line indicating the item or service and the price.

2. A file called INTERVU contains an interview with Doug van Vliet. You can DISPLAY STRUCTURE to find out what the fields are. There is no data in the FILE1, FILE2, FILE3 or TOPIC fields. After you read through the interview, go back and EDIT each record of Doug's comments, putting in the FILEs or TOPIC that Doug is making reference to.

Now, write a program to DISPLAY a transcript of the interview line by line with your IDEAS on each line that matches FILE1, FILE2, FILE3, or TOPIC. For example, if one of Doug's comments refers to STYLISTS and four of your comments also refer to STYLISTS, list all four IDEAs and their PRIORITYs below the question and Doug's response. Which of Doug's comments have the highest AVERAGE priority?

CHAPTER 12

COLLECTING AND ORGANIZING SYSTEM EXPERIENCE DATA: PART II

OBJECTIVES

1. To plan, organize, and conduct a number of work observation studies, including task analysis, forms tracing, report and document analysis, and computer-based data collection
2. To plan, organize, and conduct a questionnaire collection study in a critical way so that it can be evaluated
3. To plan, organize, and conduct protocol analyses and simulations
4. To plan, organize, and conduct a diary data collection study
5. To plan, organize, and conduct a variety of group data collection studies

COLLECTING AND ORGANIZING SYSTEM EXPERIENCE DATA: PART II

12

CHAPTER

12.0 INTRODUCTION

Chapter 11 introduced methods of interviewing; this chapter concentrates on the remaining techniques. Of major interest are (1) questionnaire surveys, (2) work observation, (3) diaries, and (4) group interviewing.

12.1 QUESTIONNAIRE TECHNIQUES

Questionnaires can be thought of as interviews without the interviewer. The observer, who may also be an actor, notes and records observations in the absence of the investigator. Many important interviewing considerations (especially those concerning the construction of questions) are the same, but there are important differences that make creating the question schedule and interpreting responses more problematic.

General Process

Figure 12-1 on the following page, illustrates the general process. Based on the concern agenda, questions are created and questionnaire forms printed. Meanwhile, a sampling frame is developed and a sample is prepared in the form of a mailing list. The questionnaire is sent to the potential respondents; responses are returned and collected. This is called a **wave**. Many questionnaire surveys require two or more waves to increase the response rate. Since most questionnaires are anonymous, each wave is sent to the entire list of people in the sample.

The information necessary includes the following incoming data:

QSTI1: concern agenda
QSTI2: organizational population data, phone directory, etc.
QSTI3: completed questionnaires

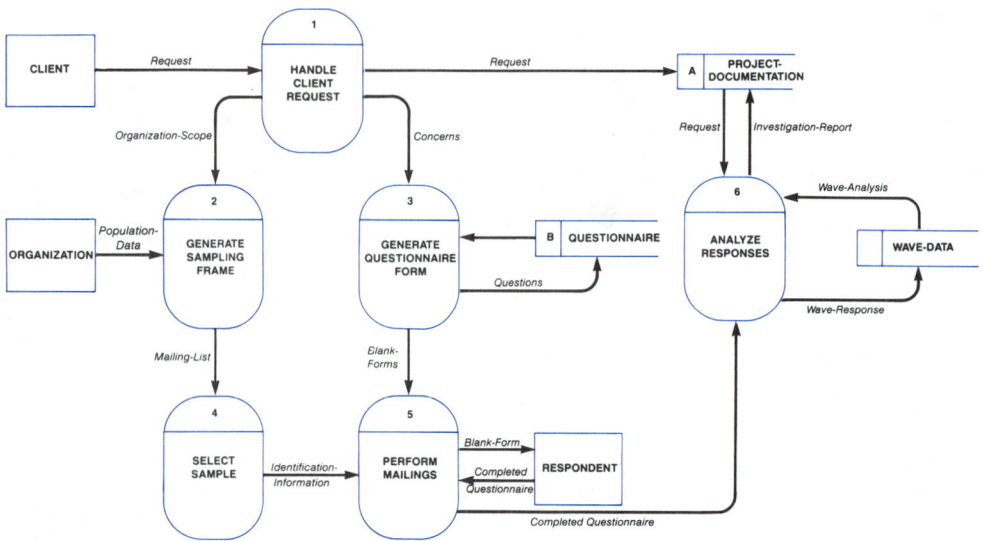

FIGURE 12–1. DFD for Questionnaire Effort

Data stores and flows created by the questionnaire process include the following:

> **QSTO1**: schedule of questions
> **QST02**: blank questionnaire forms
> **QST03**: mailing list
> **QST04**: respondent list (if not anonymous)
> **QST05**: wave analyses
> **QST06**: investigation report

The schedule of questions comes from client concerns; blank forms are produced from this schedule. The mailing list comes from the sampling frame of organizational lists (phone directories, lists of customers or suppliers, etc.). If questionnaires are not anonymous, a list of respondents can be created. Each wave is analyzed; especially important are the response rate and the completion rate (the proportion of forms returned completed). The investigation report is based on these wave analyses.

Creating the Questionnaire Form

The questionnaire form comes directly from the concern agenda. The major concern in creating questions for a questionnaire is getting correct responses. There are several reasons why a response may be incorrect:

- The question may not be readable because of printing problems.
- It may not be understandable. ("How often do you have to flummox and triangulate a T54?")
- It may be ambiguous or imprecise. ("What is your status?")

- There may not be sufficient space to write the response.
- The respondent may not know the response or may not want to give you the right response.

Questionnaires, like interviews, have to be structured. Usually this means setting off different sections of the questionnaire for particular topics. In addition, a number of considerations stem from the fact that the investigator is not present to guide the answers or request more complete ones:

- *Order*. Respondents may skip around in different orders, perhaps biasing their answers. This is an "observer" effect similar to "experimenter expectancy."
- *Omissions*. A respondent may miss a question or ignore a possible answer because of the layout of the questionnaire, or fail to turn a sheet over and answer those questions.
- *Fatigue*. Respondents may tire of answering questions. Make certain the most important questions come first.
- *Filtering* (see below). Complex instructions for skipping questions lead to mistakes. ("If you have not recently purchased a product at the store or have not been dissatisfied with your merchandise and have not returned anything, then skip to question 11.")

The layout of the form should control the sequence of questions. The fan structure, focusing on specifics first, is most common. Filtering should be avoided since it can be confusing. It is possible to word questions so that those to whom it does not apply can answer "n.a."

A great deal of research into **form design** considerations cannot be repeated here. Following, though, are some guidelines. Avoid small sheets of paper — they get lost. A good rule of thumb is to create a questionnaire that can be printed on two sides of a single 8 1/2" × 11" sheet of paper. A paper that can be folded so that it is pre-addressed, and stamped, and ready for mailing ensures against losing the envelope or, worse, forcing respondents to find and address their own envelopes. Allow sufficient room for answers. Pay attention to layouts. Those running across the page are confusing, as this example shows:

Check your POLICY CLASS below:
__A __B __C __D __E __F __G __H __I

In general, it is best to use questions that can be answered in a checkoff format. This prevents problems of interpreting handwriting and language. Most questionnaire efforts concern questions for which the range of answers is well known. Questionnaires have very little discovery value, therefore, and open-ended questions (other than "what is your age") are of questionable worth.

The questionnaire form must also introduce the questionnaire and "debrief" at the end. It must tell people what is expected of them in order to get them in the right frame of mind, but at the same time, it should not bias their answers. A brief, simple paragraph of fewer than a hundred words should serve to introduce the questionnaire. The end of the form should motivate respondents to check over their answers and return the form. The return address should be clearly featured at both the beginning and the end of the questionnaire.

Determining the Sample

The reason for using questionnaires is to decrease the per-unit cost (over interviewing); a random sample will probably suffice without many qualifications. On the other hand, because the populations are usually large, there is *always* a sample involved. In fact, there are *two* samples.

The first sample is the sample drawn to *receive* questionnaires. The other sample is the group that actually *responds*. Both are often biased.

The polled sample is often drawn from a sampling frame that may be out of date or biased in some way. Consider an attempt to investigate customer concerns over the billing system. If the sampling frame is drawn from payment records, those who do not pay will not be sampled — and this may be the group with the most valid, pressing complaints. A more reliable sampling frame is all customers billed in the past three months. Even then you will not sample customers who quit shopping at the firm because of problems with billing.

The biases in the second sample are far more damaging, though. Consider these "biasing" events:

- The respondent never received the blank form.
- The respondent never examined the form.
- The respondent never tried to complete the form.
- The respondent tried, but did not complete the form.
- The respondent either decided not to look at the form or, after answering some questions, decided not to return it.
- The respondent returned the form, but it was never delivered to the analyst.

Problems with delivery are common in large mail-out questionnaires. An in-house survey using company mail facilities generally does not have these problems. A mailout to the general public can count on up to 5 percent of the forms being lost in the mail in either direction.

Forms must be looked at before anyone responds. Presentation of the survey in a distinctive envelope and an introductory letter from an executive may motivate interest. Often there are problems finding the time, place, or energy to complete a questionnaire. In the past, pollsters have motivated response by including remuneration (which would be inappropriate where respondents are employees); others have found that including a small pencil both motivates response and makes it possible to respond immediately.

The sample of respondents is thus limited by a number of factors, not all of which can be controlled. Taking all of these factors into account, professionals consider response rates of less than 40 percent for a public mail-out or less than 75 percent for an in-house survey to be unacceptable.

Why are higher response rates necessary? Where response rates are low, the investigator has to ask why respondents actually *did* respond. Is there something special about this minority? In the case of the in-house survey, the opposite question has to be asked. Because response is easy in-house, what makes those who *did not* respond different? Are they the busiest people? If so, maybe their responses would have been different.

To build response rates, most investigators try several waves of questionnaires. The second wave is sent on distinctively colored paper, and respondents are asked to ignore the questionnaire if they have already answered. The second and subsequent waves are analyzed separately from the first; then all are pooled for statistical analysis. Late responders — those who respond to later waves — are probably different from early responders.

This merely underscores the major problem with questionnaires: the investigator has no idea *who* filled out the questionnaire and *what* that person was thinking, doing, or expecting at the time. During an interview, the interviewer can inquire directly about those concerns and thereby understand why certain responses were made; this is impossible in a questionnaire.

Commonly, a small random sample of potential respondents may be telephoned and asked about the questionnaire: Was it received? Was it completed and returned? Are there any concerns? This way, low or unacceptable response rates can be explained or anticipated.

Handling Questionnaire Data

Because questionnaire data spends a lot of time out of the direct control of the analyst, there are some specific packaging, handling, and distribution concerns.

- Contents of the package: Cover letter, questionnaire form, return envelope with address label, inducement.
- Distribution: (1) Company mail or secretarial fan-out; a predistribution memo may improve visibility of the survey; (2) post office; predistribution letters may be appropriate; survey may be sent with regular billing or reports.
- Handling responses: Determine end-of-wave-response date before distribution; file each response as it arrives; decide on need for new wave.
- Problems of nonresponse: Sample respondent pool and interview about questionnaire before end of wave date; determine systematic biases.
- Motivating response: Use legitimate authority; provide inducements if appropriate; make response handling automatic by supplying a pencil, addressed response envelope; put form on distinctive stationery.
- Multi-wave efforts: Analyze each wave for peculiarities in response patterns.
- Analyzing data: Prepare coding instructions *before* sending out questionnaires.

Advantages and Disadvantages

The major advantage of a survey questionnaire is **cost**. On a per-response basis, questionnaires run at only 5 to 10 percent of the cost of an interview. Considering the postage, printing, and data-handling costs of a two-wave survey resulting in a 50-percent return rate from waves of 1000, the cost of each respondent would range from $2 to $3. Factoring in employee time at $30 per hour estimating 20 minutes to complete the survey brings the cost to about $12 per questionnaire. Compare this with a cost per interview of over $100 plus employee time (an executive's hour may be worth $100).

A questionnaire effort can be completed in a matter of weeks with little time clocked against IRU salaries other than clerical costs. On the other hand, the disadvantages relate to the inability to ask and interpret lengthy questions, the need to know the nature of the responses in advance of the investigation, the lack of control over who answers and why, and the problem of handling nonresponse to specific questions.

As a general rule, questionnaires should be used only with large groups and only after all issues have been raised through interviews. Large volumes of data can be collected in order to make precise qualitative statements such as "customers feel..." or "employees want...."

Case: A Questionnaire Study of Distributed Processing Needs

A large, national publisher wants to determine the information needs of its 300 direct sales personnel in 14 regions as part of an attempt to modernize its information resource via distributed processing. An analyst has been seconded from the IRU to the marketing department to design a questionnaire to discover the needs, capabilities, and expectations of this group.

The analyst designs a checklist-oriented questionnaire form for distribution to the sales staff (Figure 12-2). The questions focus on information required; the value of this information; frequency, currency, and accessibility levels; and the current reporting methods.

Because each region has unique problems in accessing and reporting sales data, a random sample stratified by region is selected, sampling one-third of the sales people. This results in oversampling of the "government and military" region and undersam-

---

3. How many sales contracts do you report monthly?
   ☐ Fewer than 5
   ☐ 5 to 9
   ☐ 10 to 19
   ☐ 20 or more

4. How do you obtain booklist information?
   ☐ By mail from Headquarters
   ☐ From my sales manager
   ☐ Training Department suppliers
   ☐ Other (Explain)
   _____
   _____

5. How useful would you rate the booklist information you get
   A. For informing yourself of the contents of the booklist
   ☐ Useless
   ☐ of little value
   ☐ of some value
   ☐ Very valuable

8. How often do you obtain updates on sales volumes and booklist activity
   ☐ Weekly or more often
   ☐ Twice monthly
   ☐ Monthly
   ☐ Quarterly
   ☐ Less often than quarterly

9. How current do you feel the booklist information you get is?
   ☐ Very much out of date
   ☐ Somewhat out of date
   ☐ Mostly current
   ☐ Up-to-the-moment

10. How do you NOW report your current sales contracts? (check as appropriate)
   ☐ Mailed form
   ☐ Phoned in to Manager
   ☐ Phoned in to Headquarters
   ☐ Personal contact
   ☐ Other (Explain) _____

FIGURE 12–2. Checklist-Oriented Questionnaire

pling of the headquarters region. Forms are distributed through the company mail, preceded by a distinctive memo from the VP of Sales. Respondents are told that they have two weeks to reply to the questionnaire.

At the end of two weeks, 80 of the 100 have responded. A telephone sample of 20 of the 100 is drawn, and it is discovered that 3 of the 20 are on vacation. This means that it is likely that only 10 of the 100 have ignored the questionnaire, an acceptable rate.

As the forms arrive, they are coded. A statistical analysis results in a prioritized list of information needs, access modes, and sources. In addition, the analyst can make an estimate of the volume of information needed for each salesperson on a daily, weekly, and monthly basis and thereby assess the need and value of distributed, as opposed to centralized, information dissemination.

12.2 WORK OBSERVATION DATA GATHERING

The techniques discussed in this section all relate to observation of the work individuals do, with and without their verbal comments. They range from the completely impersonal, mechanical computer-based work analysis to the highly personal and very interactive protocol analysis technique. Two of the techniques (forms tracing and report and document analysis) examine the information that individuals use in their work.

Types of Work Observation

Task analysis is a form of on-site (*in-situ*) observation directed toward an individual or a group of people performing a task as part of their job. It is suited only for well-defined tasks for which there is an accepted definition of how a task *should* be performed. Task analysis has not been applied successfully to managerial and professional work.

Analyses are based on job and task descriptions. The goal of task analysis is to describe each task or activity in terms such as time to perform; frequency of performance; steps actually employed; problems and bottlenecks encountered; number of errors per unit time; error type; use of materials and information; interaction with others; and information input, output, and storage necessary for the task. This shows task analysis to be a descendant of time-and-motion studies; however, the emphasis is not on measuring employees but on determining the *effectiveness* of existing procedures as performed by those employees.

Report and document analysis focuses on the content of reports and forms. Typically, an information-processing task involves acceptance of incoming information from forms or reports and production of new or amalgamated forms and reports. Report analysis documents the content of reports in terms of the data items and types of characters found therein, amount of information, the probability of errors in particular fields, the number of reports produced, and their destinations.

This analysis also looks at forms, concentrating on the source of information for each field, its usefulness, and the probability of an error for each field and the document as a whole. Forms analysis is an essential first step in logical design. Report analysis is likewise a preliminary step in the interviewing of the managers who read the reports. Interviews can shed light on the value and validity of reports and forms after the analyst has become familiar with the contents.

Forms tracing is related to report and document analysis. Here, however, the emphasis is on the station-to-station flow of forms and then on to reports. The goal is to describe information flow in terms of volume, time between processes, time within process, number of hands and eyes involved in the processing, location of specific bottlenecks, need for additional information sources, and the external sink for the data (or disposal of the media).

Protocol analysis combines features of interviewing and task analysis. It combines observation of a worker with an interview during task performance. This technique is useful for determining the performer's state of mind and for finding out what tasks are difficult or easy from the performer's viewpoint.

Unlike an interview, however, performers typically speak aloud while performing and describe what is being done. As **participant-observers**, they observe their own behavior and report from their unique perspectives rather than that of the isolated observer who is not familiar with the demands of the task. The major advantage of protocol analysis is discovering the kinds of conscious decisions users of information resources make, decisions that do not show up immediately in behavior. Protocol analysis can uncover: critical decision points, choices made during the procedure, sources of error or difficulty, and attitudes toward the task.

Here is an example of protocol analysis in action:

SA: Let's see how this task is performed.

AC [accounts clerk]: OK. Well, first I take the sales receipt....

SA: Where did you get it from?

AC: Oh, I get a stack of these receipts each morning and I just take them one at a time from the top. Anyway, I take the sales receipt and I first check the arithmetic. You want me to do that now?

SA: Yes, just carry out the procedure as you normally would, but think out loud so that I know what you're doing.

AC: Well, I go down the column of total amounts. Hmmm...yes, this one's right. Seventy-six dollars and forty cents. Then I make sure every field on the receipt is filled in properly. Let's see...department, yes, that's right...it's a cash purchase...I know it's a cash purchase if I don't see a credit copy attached...the clerk's initials are J.B., I think...it's sometimes hard to make these out, y'know...date is...oops, here's a mistake, this clerk's put Tuesday's date on here, I hope J.B. hasn't made the same dumb mistake on all of these...now I have to correct it in red pencil and note the correction here on this form...slip number 97450...date-...corrected to...let's just peek ahead, aha, he must have just forgotten to use the date of Wednesday...ok...so I make an entry here for the gross...97450...gross of seventy-six forty and go on to the next one.

SA: What do you do with that sales slip?

AC: It goes on a stack I build up here of checked slips.

In this example the accounts clerk is merely going through the sales slips from yesterday's sales, verifying totals and recording them on a form to make sure the cash totals for the day balance receipts from each department. The protocol analysis uncovers a technique for judging the volume of work: peeking ahead to see if the other dates are incorrect. It also shows that the clerk does not check individual items (gross = unit price × units) and is satisfied if his calculation matches that of the sales clerk. It is unlikely that either an interview or task analysis would have discovered these anomalies.

Simulation differs from the other techniques because its focus is the planned or designed activities of the modified or newly designed information resource-to-be. None of the activities simulated actually exist. Simulation is the only way one can gather observational data on tasks that have never been performed. It is also valuable in exploring the ways that small changes in work can change work performance; it may be valuable in designing new tasks.

Simulation involves a written exercise in which "fake" data are used under controlled circumstances by persons similar to those who will be performing the actual task when it is part of the new information system. Also, mechanical or computerized situations may be used; it is easy to design these using a simulation language or a screen-driven prototype generator.

The goal of simulation is to parametrize and evaluate procedures that might be used in the future in terms of time to complete, sources of errors, clarity of instructions for the procedure, need for additional information, the sequence of activities to be performed, and subjective evaluation of the task by the performer.

General Procedures

The goals of the observation generally dictate both the form and the content of the data-gathering activities. The appropriate work observational technique is selected; if it is simulation, the simulated task is designed and implemented (more on this later). The sampling frame is determined and, from this, participants are selected. Observational tools (typically instructions to observers and forms to be filled out) are created. The observation is performed and data are gathered and maintained.

Information flowing into the observational process consists of the following:

OBSI1: concern agenda
OBSI2: organizational population data
OBSI3: list of reports, forms, and documents
OBSI4: raw data from observation (which can take many forms, depending
        on the technique chosen)

While a random sample is highly desirable, typically one will wish to sample the population selectively, to observe both individuals who perform well (but not necessarily according to the established procedures) and those who are unable to perform well (for reasons that may not be apparent until they are observed). Other factors, including age, years in position, training, and physical location may be used to create the sample, depending on the task. Since most observation will be of clerks, performance data may be fairly easy to obtain, if only of a narrative nature from supervisors. It is important, in any event, not to observe *only* the best or the worst, but to have a range of individuals to observe.

Information that is produced by or necessary for completion of the observation includes:

OBSO1: observation scheme or program
OBSO2: observation scheduling form
OBSO3: recorded data
OBSO4: report

The **observation scheme** or program is the technique for observing. It may be a list of points to be observed and recorded (time to complete each activity, number of errors observed in each unit of time, name of each activity, etc.). It may also be a list of data items discovered, errors observed and noted, and the source of each data item.

Complementing the observation scheme is a list of times, with appropriate observations given for each time concerning activity engaged in or information used. The scheme may be an actual program (when using computer-based observation) that records each input, output, function used, time between each function, length of input lines, and so forth.

The typical scheme is a form which must be filled out with raw data. The observation-scheduling form is used to schedule observations, much in the way that a similar form was used to schedule interviews. Since individuals have to be observed, it is important to note which individuals were subjects. For protocol analysis and simulation, an appointment usually will have to be made to ensure that selected individuals know when and where to participate. Where direct observation of an on-going process is to be performed by the analyst, times and locations of observation should be determined in advance to avoid *ad-hoc* sampling, which can spoil the representativeness of the study sample.

Choosing Techniques

The following discussion relates to specific techniques, their advantages, and their drawbacks relative to one another.

TASK ANALYSIS

Task analysis is best suited for jobs that are well known and well documented. One must carefully watch out for observer effects. The selection of individuals to be observed is also critical. By choosing only good performers, one can never really discover how average performers work or what kinds of errors those who are ill trained or ill suited to the job make.

In task analysis, one must be careful about obtrusiveness. The sight of a lone observer, clipboard in hand, is intimidating to all but the most confident worker. There is large individual performance variation, even in highly mechanical tasks, and the analyst's presence may provide the opportunity for highly aberrant activities.

Observers must be carefully trained; otherwise, individual biases and confusion about what is being observed may taint the data. A poorly trained observer will miss much of the activity with too narrow a focus. It is important to understand the "official" description of the activity being observed. Protocol analysis can aid in discovering the "unofficial" (but often more effective) activities necessary to get a task accomplished. For this reason, protocol analysis often precedes task analysis.

In addition, data can be contaminated by the observer, the conditions under which specific observations are made (time of day, for example), and specific individuals observed. Given a more or less random sample of workers and times, the major factors under control of the investigation director are the training of the observer and the design of the observational scheme.

Figure 12-3 illustrates a typical scheme for observing clerical work. The form directly reflects the understanding of the task by the analysis team: the parts of the task are predefined and sequenced; the types of errors that can be made are predefined and in a check list; and the information items used are predefined, too. If for some reason this understanding of the task is incorrect, all observations will be invalid.

BILL POSTING (one form per event)

1. ☐ Receive Bill
   ☐ Opened Envelope
      Verify Correct Department
   ☐ Unopened Envelope
      Opened Envelope
   ☐ Stamp date and time on Bill stub

2. ☐ Verify Bill

3. ☐ Validate Contents
   ☐ Read stub
   ☐ Check money order or check
   ☐ Read any text (present)
   ☐ Verify check or money order amount
   ☐ Verify check is signed

4. ☐ Locate Customer File

5. ☐ Prepare posting entry onto data sheets

6. ☐ Post Bill
   ☐ Stamp "PAID IN FULL" or "RECEIVED"
   ☐ File in "Paid" Envelope

FIGURE 12–3. An Observation Scheme for Bill Posting

FORMS TRACING

The general method of forms tracing is to model or map the flow of information through stages of transformation. Since a form travels well on its own and may not travel so well when carried by an analyst manually stepping through the procedure, a non-obtrusive means such as **buck slips** are used to trace forms. Buck slips resemble forms themselves and are used to record the time of arrival and time of departure of a form or document, with the name of the process or the intial of the person performing the process recorded next to the times. When the document has been completely processed, the buck slip contains a history of that processing.

Forms tracing does not, however, tell anything about the processing itself. For instance, information that the form arrived at station A at 8:15 and left at 9:45 does not tell anything about what went on for an hour and a half. And while each form carries its own history, there is no real sense of how forms contribute to one another or whom they inform along the way. A feeling for this limitation can be obtained by looking at the "true" history of a form, outlined in Figure 12-4.

Arrive	Depart	Station	
8:15	9:45	X	X examines the form and notes that customer ID looks wrong. She compares it with ID on a form she filed last week and confirms her suspicion; she makes the change. A call to the warehouse confirms that one of the items ordered is on back-order and only part of the order can be filled. She completes the packing form, attaches it, completes the buck slip, and passes the package along.
10:00	10:30	Y	Y returns from coffee at 10:00. One copy of the packing slip is sent to the warehouse. She makes a note in her order book about the back-order and types a form to the customer explaining the delay and sends the package to the phone desk.
10:30	2:30	Z	Z looks up the customer phone number in the customer files and tries to get the purchasing agent who returns his call at 2:15. Since a partial order is fine, Z approves the partial order and completes a packing slip, attaching it to the package with a paperclip. He fills out the buck slip and passes the package along.

FIGURE 12-4. A Typical Form History

None of the information on the buck slip relates the other information necessary (such as customer ID, customer telephone number, inventory data) to complete or correct the form or decide on appropriate action. Interpreting a buck slip is very much like trying to find out why people drive the way they do by looking at traffic patterns; only part of the story is in the pattern.

REPORT AND DOCUMENT ANALYSIS

Unlike forms tracing and task analysis, report and document analysis is a static description of the content of a document. In any type of IRMaint work, report and document analysis are essential in determining what information is useful, what is useless, and what is related to what (in terms of appearing together on documents) to construct DFDs. There are two major drawbacks, however. As with forms tracing, the static picture relates only to the existing system; it cannot speak directly to the design of a new system.

The second objection relates to the *number* of forms and reports. A typical business firm will have dozens, perhaps hundreds of forms. Even a manual information resource can produce several reports. Many of the reports (and not a few of the forms) likely will be nonstandard or even "bootleg," in the sense of being designed by the staff in an *ad-hoc* fashion and not registered with some sort of central forms-control bureau.

Forms tracing is an important partner to report and document analysis because logically related items should have logically related sources and destinations. That is, information that appears together on a form to complete a concept (say customer name, address, and telephone number) should have sources that do not vary independently. Otherwise errors could occur frequently. An example is habitually obtaining a customer's telephone number from an out-of-date directory, rather than directly from the customer. Employee data should come from the employee. Sales reports should be sent to sales people, summary reports sent to upper management, and financial analyses to financial analysts.

Since there is no process being observed, there is no chance of observer effects. On the other hand, there may be hazardous **selection effects** to contend with: one may choose the wrong documents to examine or select an ill-informed source to clarify meanings of specific item names.

A reasonable use of document and forms analysis would be with extended interviews, forms tracing, and, where appropriate, protocol analysis and simulation. In many information resources, the trend toward database management systems has introduced the concept of the data dictionary, which makes document and forms analysis easier. Where data dictionaries are not employed, document and report analysis is not only necessary for IRMaint, but should be performed regularly in conjunction with a forms-control program that catalogs and standardizes reports, forms, and other documents.

COMPUTER-BASED WORK ANALYSIS

Computer-based work analysis is a formalized version of task analysis in which the human observer is replaced by the computer. It shares many of the same goals and techniques of task analysis. Since the computer is able to exercise far less discretion (but will make far fewer observation errors because of this), the observations are typically confined to programmed items relating to speed, correctable errors, throughput, and use of specific program and hardware features.

Observation is confined to well-known parameters of the work under observation. Either the software controlling the tasks is **patched** or a **front-end** is created to monitor activity using the software. In some instances, feature utilization and performance are already measured on a continual basis.

The major advantage of computer observation is reliability: the computer rarely makes errors, never tires, and can be easily programmed to analyze its own data on the spot. This, however, is also the major disadvantage. The computer can observe only that which it is programmed to observe. Frustration, tension, confusion, and determination cannot be measured; instead they must be inferred from more objective measures such as input errors, variability in input rate, search errors, and other usage patterns, based on some behavioral theory.

The other drawback to computer observation is the cost of programming the observation. While a well-developed policy of evaluation of software effectiveness will often include continual production of effectiveness and efficiency, the cost of this software, like the cost of all software, is difficult to predict and control. On the other hand, once such software is in place, it is greatly effective in producing data in the limited areas in which the data is valuable.

Few individuals like being observed, and even fewer appreciate the computer doing the observing. The notion of "Big Brother" is seldom comfortable for anyone. Before a program of computer observation is instituted, a great deal of public relations ground-work must be done. In addition, the analyst must be honest: if the data are to be used for performance judgments, this *must* be announced.

General Issues in Work Observation

There are three general areas of concern in a model of observation (see Figure 12-5) that separate the observer from the actor. These are (1) observer effects, (2) sampling effects, and (3) boundary effects.

OBSERVER EFFECTS

Social Facilitation
Inadvertent Reinforcement
Group Norms
Self-Fulfilling Prophecy
Importation of Values
Evaluation Apprehension
Experimenter Expectancy
Physical Intrusion
Hawthorne Effect
Effects on the Observer (Halo, Recency, Primacy, von Restdorff)

SAMPLING EFFECTS

Unrepresentative Sampling
Ill-Chosen Observational Mesh

BOUNDARY EFFECTS

Irrelevant Observational Scheme
Ignoring the Relevance of the System Boundary
Failing to Observe Coupled Systems

FIGURE 12–5. Issues in Observation

Observer effects come about when the observer somehow intrudes into the system under observation. The resulting data are a function of not only the action and the actor, but also of the observer. This can severely damage the validity of the data produced, since the observer should not normally be present during the operation of the system.

There are ten important observer problems:

1. Social facilitation.
2. The observer may unknowingly inform the worker about how the task is to be performed through subtle, unconscious nods of the head, glances, encouragers, or gestures.
3. Group norms may become more rigid during observation. People who "show off" or "slack off" might be punished by ostracism or verbal criticism from the group.
4. Observed workers who feel threatened and fearful may be less attentive to their work, thus causing errors, in a sort of **self-fulfilling prophecy**.
5. The observer brings a unique and potentially foreign set of values and biases to the scene, which the worker may pick up, learn, and imitate unconsciously.

These include a business-like manner, a systematic approach, a humorless attitude, or an overly observant approach to work. None of these may be appropriate to the specific work setting.

6. Evaluation apprehension.
7. Experimenter expectancy.
8. The existence of an observer may physically affect how the work is done if the observer takes up space or a desk normally used for work.
9. Merely observing work may increase productivity. The possible reasons for this include work performance improvement based on feelings of importance from gaining attention or a refocusing of group attention away from work problems toward the observer. This is often called the **Hawthorne Effect**.
10. Finally there are effects *on* the observer. A **halo effect** may cause a single observed action (a stupid mistake or a heroic save) to overwhelm all others; a **recency effect** may make more recent observations (those toward the end of the session) more memorable; a **primacy effect** may make those toward the start of the session more memorable; and a **von Restdorff effect** may make more memorable events seem more important (and hence more critical to the process).

Sampling effects come about because either too few or unrepresentative points are selected or the observational mesh misses regularly occurring important events. For example, observation of bill posting makes little sense when performed only at end-of-month if most bills come in at mid-month. Fridays and Mondays are always special days on the job; ignoring that consideration will probably invalidate conclusions.

Boundary effects occur because the observer is looking at the wrong things. This can occur in three ways. First, the observational scheme must be relevant to the analysis goals. If the problem is speed, then speed must be observed and repetitive actions must be recorded. If transit time reduction (time to process a document) is the critical design goal, then how stacks of pending work are accessed should be included.

Second, the observer should look at and just beyond the system boundary to avoid missing interfaces between the observed system and others. It may not be sufficient to note that documents leave department X at such and such a rate if they are merely piling up in department Y.

Finally, if tuning the system is the goal, the problem often lies at the system boundary, and the form, format, and manner of handing over data need to be examined. Observation of the handover may be more necessary than observation of the process.

Examples of Work Observation

TASK ANALYSIS OF FORMS COMPLETION AND TRANSCRIPTION

The goal of the observation is to collect data on forms completion and retyping in the benefits and claims (B&C) department of Bigtown U. Staff travel claims, hospitalization, and reimbursement for entertainment expenses are typical claims that this department handles. Forms are sent in and need to be evaluated for completeness, reasonableness, and conformance to university regulations.

In-situ observation of the B&C clerks takes place. The observer attends to: the sources of information and how they are checked, the procedures for checking forms and completing evaluations, completion and turnaround time, errors noted, the return of forms to the employees for correction, the retyping procedure turnaround time, and the return of forms to the clerks when information is illegible or missing.

Both times and people are sampled. On each of five randomly selected days (including one at the beginning, one at the end, and three in the middle of a nonsummer month), two clerks are chosen at random, one in the morning and one in the afternoon. The ten observation sessions last 1 1/2 hours each (form processing typically takes less than 20 minutes, as noted from preliminary observation).

Observations are guided by an observation sheet, which directs the observer's attention to the work. At the end of each session, the clerk is debriefed by the observer in an interview that focuses on error detection. Secretarial typing is also observed during the day at half-hour intervals, and the work-pending and work-completed baskets are examined at the end of each session before the work-completed basket is removed.

The analysis and report concentrate on throughput, error rate, types of errors noted and corrected, decisions necessary to complete the forms, sources and destinations of the forms, transit time at the secretarial desk, and iterations of work because of missing or unreasonable information.

FORMS TRACING OF CUSTOMER INVOICES

The goals of this investigation are (1) to discover how long it takes to process customer invoices in an order-taking department, (2) to locate bottlenecks in processing of the invoices, and (3) to find out why some orders are lost. Some customers have complained that orders have never been received, while others note that it often takes a long time to receive their orders. Investigation of the warehousing and shipping departments, however, has revealed no bottlenecks or lost data.

The technique to be used is forms tracing. Buck slips are attached to each form as it is initiated. Each form is numbered and individual forms are accounted for at the end of each day. A buck slip (see Figure 12-6) is stapled to each form. Upon receiving an order form, each processing clerk fills in the date and time and a process ID (O for initial order taking, I for inventory checking, C for credit checking, and W for warehouse instructions). At the end of each process, the clerk notes the date and time and initials

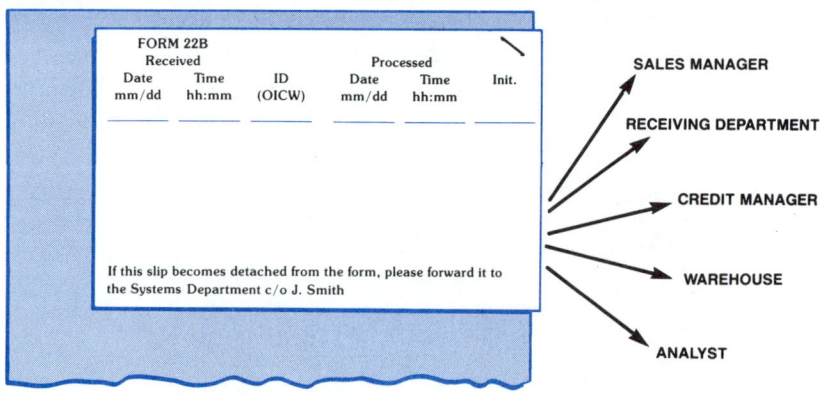

FIGURE 12–6. Buck Slip

the buck slip. At this point, an action form is filled out (see Figure 12-7), indicating the unique order form number, date, and time.

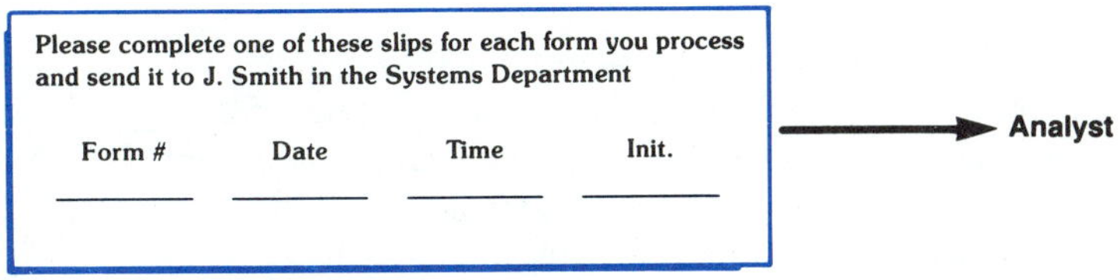

> **Please complete one of these slips for each form you process and send it to J. Smith in the Systems Department**
>
> Form #   Date   Time   Init.

→ **Analyst**

FIGURE 12–7. An Action Slip

When the form leaves the order-taking department, the receiving department removes the buck slip and returns it to the analyst. Forms can be sent: (1) to the sales manager if an order cannot be filled, (2) to the credit manager if a customer cannot purchase on credit, and (3) to the warehouse for shipping. The buck slip has instructions concerning return to the analyst in the event that someone outside these three areas receives a form by mistake.

At the end of a month, all the buck slips and action forms are collected and each order form accounted for. The analyst then creates an order form transit model using a data flow diagram augmented with transit times for each process. Based on these models, the analyst will report on throughput, bottlenecks, and probability of loss at each process.

COMPUTER-BASED WORK MONITORING

A data-entry department has been receiving complaints about errors on source documents and slow turnaround time. While the manager of the department feels that productivity is high, there seems to be no ready explanation for the complaints and no data to refute or confirm them. An analyst has been contacted to redesign the data-entry system to monitor the clerks' activities.

After a number of discussions with the manager, Doris Porter, and an after-hours group discussion with nine of the eleven data-entry clerks, the analyst has received consent from the workers to alter the program. The clerks have been assured that data will *not* identify clerks, that results will not be used in any way to appraise the performance of individuals, that all data will be expressed as averages, that a rationale for errors will be provided (for example, a rise in errors during a busy period would be noted), and that the analyst will also note environmental factors such as lighting, work spaces, distractions, and seating in interpreting the data.

The goal is to discover needed facilities for key-data-entry verification and work control to (1) reduce errors, (2) reduce repetition of work already keyed in, (3) balance work assignments, (4) implement a work-load schedule allowing for more frequent breaks, and (5) more accurately and fairly monitor output and fatigue factors. Implicit in the design of the monitoring software is a need to improve the quality of working life for the key-data-entry clerks while fairly monitoring productivity.

Referring to existing system specifications, the analyst modifies the logical design to monitor keystrokes, verify input data, highlight errors in reasonableness and format, assist in editing erroneous data (rather than cancel and reinput), time sessions and remind clerks of break periods, note trends in possibly fatigue-related errors (such as transpositions), and maintain statistics on errors noted and completion time for particular forms, clerks, and source departments.

To provide data on the error and throughput rates, the analyst builds a simple patch to the existing system to count keystrokes entered; document openings (starting a form), closings (finishing a form), and cancellations (discarding a form and starting over); and trends across time. The computer records all of this data in a file for later analysis.

The analysis produces a report on: keystroke volume per unit of time by time of day and shift; throughput and transit time by document type, source department, and clerk; cancellations by document type, source department, clerk, and time of day. Because Mondays and Fridays are often quite busy and Wednesdays are often days off for accumulated overtime, the day of the week is also important in the analysis.

SIMULATION OF A PROPOSED MARKETING TRIAL DATABASE INQUIRY PACKAGE

Charlotte Reid, VP of Marketing, is interested in a database of marketing trial facts that would be available from a terminal by inquiry. The firm, a major producer of cosmetics, has recently had an important failure in a major market, and Reid is nervous about how the trials were done. An analyst has been brought in to design a computer-based system to enable market research analysts to learn from their mistakes.

The goals of the project are: (1) to create a database of market trial data, (2) to create a number of easy query-and-report procedures for marketing analysts, and (3) to design data-capture procedures to keep the database current. The analyst begins the investigation by noting the availability of a fourth generation language (4GL) on the firm's mainframe (currently used only by factory general managers to monitor production). This would provide the basis for the system. Marketing analysts are forms oriented, however, and the 4GL is oriented toward English-language dialog.

The analyst has decided to use pencil-and-paper simulation of the inquiry procedures. Over a two-week period, she has captured live data from recent market trials and has examined the kinds of reports that the marketing analysts use.

The simulation is set up as follows (see Figure 12-8 on the opposite page):

1. A marketing analyst (subject) is asked to write on slips of paper requests for information needed to create a specific report.
2. The subject then passes the slips through a slit in a cardboard barrier to the systems analyst.
3. The systems analyst provides the data, while retaining the slip of paper, on a hand-written "screen" form of 25 lines by 80 characters; this is passed back to the subject. Each slip of paper is time stamped when received; each returned screen is time-stamped before being passed back.
4. When the subject completes the required report, it, too, is passed through the slit to the analyst, who time stamps it and stores it away for later examination.

Five marketing analysts and three reports are selected for this simulation. Each subject writes each of the three reports. Each session requires the morning, with a fifteen-minute break between reports.

FIGURE 12–8. Simulating a Marketing Database Inquiry Situation

After receiving the three reports, the analyst collects all slips and sits down with the subject to discuss the simulation. A short interview concentrates on accuracy and relevance and the wording used by the marketing analyst in the "queries." The format of the "screens" is also important. The five subjects fill out attitude scales; their written comments also are solicited.

The systems analyst produces a report on the vocabulary used, the time necessary to complete reports, feelings about the "screens," and an assessment of each of the fifteen reports by a marketing manager.

PROTOCOL ANALYSIS OF INVENTORY CONTROL

A synthetic-fiber-processing plant is automating, and control over semi-finished inventory is necessary. Because of the high temperatures and difficult working conditions the employees face, maintaining accurate control over semi-finished inventory has proven difficult in the past.

The analyst has decided to use protocol analysis. Despite the potentially dangerous shop floor conditions, the production manager agrees that it would be best to know what workers are doing. Any new automated procedure should not be difficult, dangerous, or distracting. The analyst observes an experienced worker at four different times: (1) start of shift, (2) middle of shift, (3) end of shift, and (4) batch transfer. The analyst also has chosen to make these observations a set of four fifteen-minute sessions on each of three days: (1) start of week (normally very busy), (2) mid-week, and (3) end of week (normally a high-pressure time). Thus, a total of twelve quarter-hour sessions take place.

The analysis requires the worker to think out loud about what he is doing while he completes forms controlling semi-finished inventory. The analyst writes down these thoughts, occasionally prompting the worker to articulate reasons for specific judgments or decisions. The analyst concentrates on the procedure by which the worker notes the weight of each batch, its apparent quality, and waste considerations. For example, a batch is weighed before and after processing but waste is determined independently during a high-temperature run. The condition of the batch changes during processing and may not be finally noted until the batch is completed, although sometimes workers make premature judgments to discard all or part of a batch prior to completion.

The transcription is reread after each session and the facts are distilled. Considerations of accuracy, reliability, availability of data, reliance upon intuition, and the need to remember things are most important in this distillation. The analyst will also attempt to re-create the process by which the worker noted the data necessary to complete the inventory control tickets and forms and then fills them out.

The result of the analysis is a description of how an experienced worker goes about filling an inventory control responsibility. The analyst notes potential sources for error, unreliability, and loss of data and takes special pains to point out how an automated system might have to "bend" to accommodate working conditions.

12.3   DIARY TECHNIQUES

Often used as a compromise between the expensive, in-depth interview and the inexpensive, poorly controlled questionnaire, the **diary** utilizes the concept of the participant-observer, who also records observations.

The Participant-Observer

The diary depends on individuals observing their own behavior and recording aspects of it at various times. Because one individual serves three roles, there are many opportunities for intrusion and confusion.

The **participant-observer** is an individual who observes his own behavior (and, in the case of the diary, records the observations). Normally, we are not very observant of our own behavior. We observe ourselves only at special times — when we are on stage, so to speak. At these times, we may also notice that our activities become less fluid or capable. We may stutter or stumble. In addition, we become very much aware of how *others* see us, although we may have a distorted view of their vision.

The participant-observer must deal with the lack of concentration on the task at hand as well as the expanded sense of self-awareness and the difficulties that accompany it.

On the other hand, participant observation overcomes a major problem regarding the point of view of most trained observers — they are "outside" the system. Outside observers must find and use a "valid" observation scheme, but may not be capable of applying it correctly.

The diary technique relies on the use of actors as observers. To capitalize on the benefits of observers who "understand" their own behavior, without falling prey to the problems of participant-observation, analysts provide training sessions for those filling out diaries. The instructions for completing diaries are critical, as are pretesting and training. It is important to create a diary that does not make actors overly aware of their own behavior.

General Process

A diary effort (Figure 12-9) does not differ much from questionnaire usage except in the training of the participant-observers. The first step is to determine the observation goals. Usually, these relate to time, frequency, and volume of activities or data.

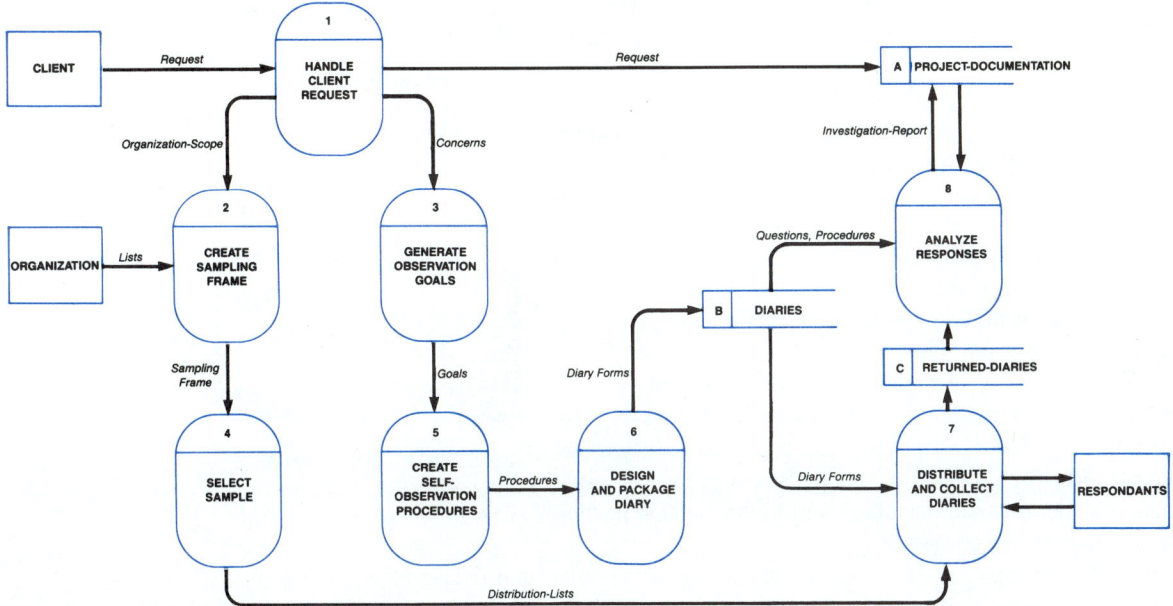

FIGURE 12–9. DFD for Diary Effort

Next, feasible self-observation procedures have to be created along three dimensions. First, will the recording take place for each *event*, or can it summarize a *period of time*? The first is called **demand** recording; the second, **scheduled** recording. Next, will the diary pertain to an *individual's* activities or will it pertain to a particular document or *item*? In the first case, we say the diary is **fixed-person**, while in the other case, the diary is **fixed-object**.

Third, will we require the observer to record **narrative** data, or can the diary be a **checklist**? A narrative recording can range from an essay to a fill-in-the-blanks recording. Checklists indicate choices.

The choices depend on a number of factors: (1) the suspected frequency and duration of events to be recorded, (2) the complexity of the investigation goals, (3) the skills of the participant-observer, and (4) how well we understand the system that is to be observed. For infrequently occurring events involving an articulate participant-observer, we would normally use demand recording and may rely on narrative. When events occur rapidly or in bunches, a scheduled recording involving checklists might be more appropriate. Most diary efforts are fixed-person.

The diary is usually in the form of booklet. There are two general formats. Each page may be devoted to a single event (Figure 12-10), much as a telephone message pad records each message on a single sheet. The alternative is a spreadsheet format (Figure 12-11), where each event is described on a single line according to a number of categories.

Microcomputer systems make it easy to install desktop organizers (such as Sidekick™), which can be brought in with a single keystroke. Records can then be picked up automatically by software and analyzed without being kost.

The diary "package" consists of the diary itself (paper form or computer program), instructions, the return envelope, and information or assistance telephone numbers.

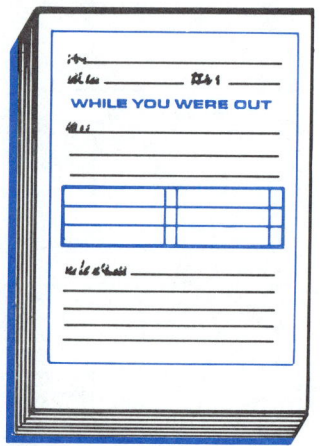

FIGURE 12–10. Tear-Off, Single-Event-Oriented Diary Format

Instructions should clarify how to fill out the diary, but no diary effort should be undertaken without training sessions, unless the form contains only a few simple questions. Participants will always have questions about gray areas, problems with observation, and concerns about intrusion into their work.

Most diary efforts are not anonymous; thus, it is appropriate for participants to be able to contact the analyst during data collection. Because diaries often are employed over a period of weeks, there are always special considerations and problems in judgment that participants may have trouble with.

After the package is designed, diaries are distributed and data collection is begun. Sampling of time periods is important, especially if demand reporting is used. Busy and quiet periods should be represented in a stratified fashion. Many diary efforts take place over a one- to four-week period. The landmark study of managerial office automation opportunities performed by Booz, Allen & Hamilton Inc. (1980) asked managers to report on activities every quarter-hour for a three-week period. Three hundred managers provided 300 observations of their own activities, resulting in 90,000 data values. A three-week period can capture important month-end, month-beginning, and mid-month ranges of time.

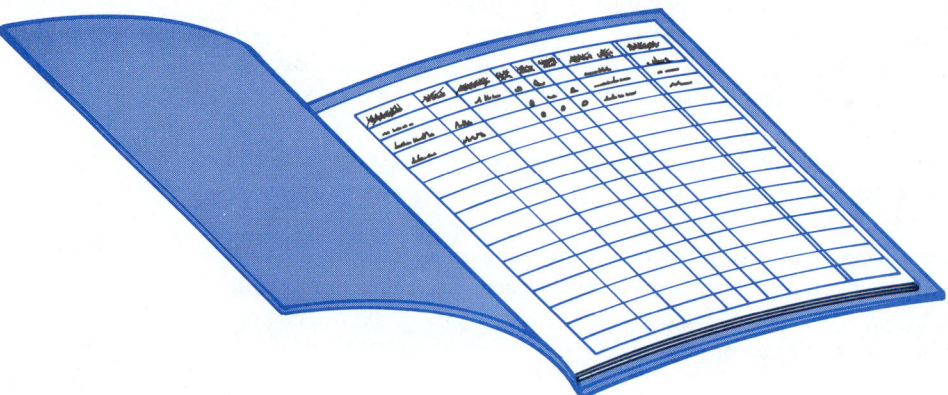

FIGURE 12–11. Spreadsheet, Multiple-Event-Oriented Diary

At the end of the observation period, diaries are collected, analyzed, and distilled. A report on the diary effort is produced.

The input data required in a diary effort include:

DRYI1: concern agenda
DRYI2: organizational lists
DRYI3: completed diaries

The output produced consists of:

DRYO1: self-observation scheme
DRYO2: diary forms
DRYO3: instructions
DRYO4: distribution lists

The self-observation scheme is the prototype diary, equivalent to the schedules of questions created for interviews and questionnaires. This scheme indicates what will be observed, how often observations will occur, and what characteristics of the event will be recorded.

Advantages and Disadvantages

Diaries should be used when a large amount of data needs is needed about a single phenomenon or type of event. In particular, diaries are well suited for discovering the frequency and pattern of these events. When well planned, diaries can be less obtrusive, more valid, and less expensive than direct observation.

On the other hand, when the phenomenon or event is not well understood, the possibilities for confusion, contamination, and lack of reliability are high. If an observer cannot tell whether an event of interest has occurred or cannot describe it according to a consistent description scheme, the validity of observation is greatly reduced. If individuals have to make unique decisions about observing, noting, and recording, reliability may be very low. Because the participant is also an observer, there is a good chance for "measuring the diary" if it is poorly designed or if participants are poorly trained or unadequately motivated.

In truth, the diary is a compromise between the questionnaire and the interview, and it shares most of their disadvantages. For well-understood events, the cost of a well-designed diary is also midway between that of an interview and that of a questionnaire. Diaries, depending on their length, can often be designed, printed, and executed for $5 to $20 per participant. Considering that several hundred observations may be obtained, the per-event cost makes diaries a good compromise.

Case: Using a Diary to Build Models of Document Flow Through a Typing Pool

In preparation for the introduction of a major office-automation system in an insurance company regional office, diaries will be used to count and categorize document flow through the existing typing pool.

The goals of the data collection effort are to determine the type, volume, through-put, and transit time of documents flowing through the sixteen-person typing pool work-ing in the office. Because documents come from remote geographical areas, it is more reasonable to chart them through the typing pool than to tag them at sources and note them at sinks.

The observation technique is to create and distribute a diary that notes "transcription events," the arrival and departure of documents from typists' desks. Demand scheduling is used. The line-per-event and check-off formats are used. (Figure 12-12 shows the diary.) The nature of the document, arrival time, work done, departure time, variance from established procedures and times require more than a check.

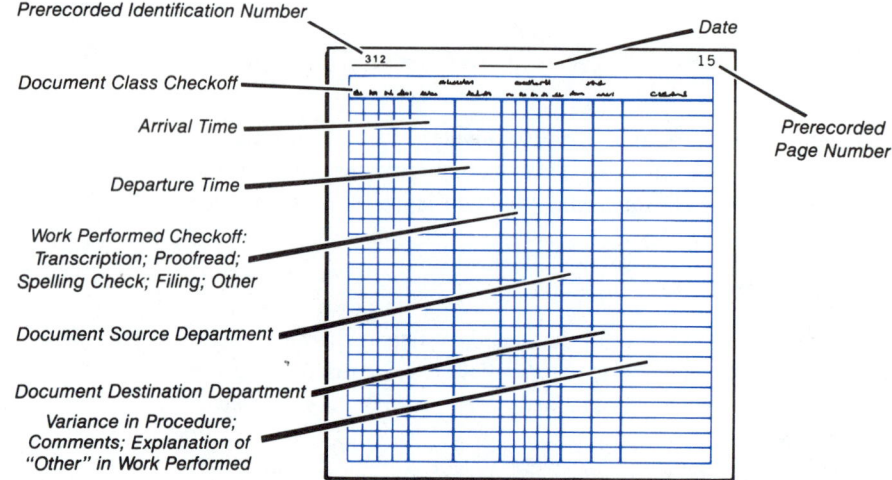

FIGURE 12–12. Diary for Document Flow Investigation

The sample is the entire sixteen-person typing pool; the time period is three weeks, beginning with the third week of March and extending through the first week of April. Because it is estimated that over nine hundred documents pass through the pool each week, each participant will supply information on 150 events per week, or 450 events over the three-week period (each document generates two events).

The diary is organized with thirty lines per page, simulating a log book. The head clerk currently keeps a log book, noting each document as it arrives, assigning it a number and a typist. The log book will also be used as a source of information, but it is often incomplete or inaccurate. In addition, several kinds of documents sent to the typing pool (including claims reports) consist of several "sub-documents," each of which must be recorded by the typists. It is estimated that each event will take only a minute to record, which will "cost" 150 minutes per week and will mean a reduction of about 6 percent in productivity for the three weeks. Direct costs for the accumulated 150 x 3 x 12 minutes of "lost" time come to $1,440 over the three-week period plus the cost of developing and pretesting the diary (an additional $960). Exclusive of data analysis, data collection costs come to $2,400, or precisely $10 per diary per day.

The analysis is aimed at determining gross document volume per week, volume per typist, variance across typists, volume by type, volume by source and sink, transit time by type and week, source by destination, volume and type by time of day, and volume by type by day of week. These figures will be used in scheduling forms creation during the physical design phase of the project and in determining the kinds of records that the system will have to keep to monitor production and distribute costs back to document sources.

12.4 GROUP-OBSERVATION TECHNIQUES

Thus far, we have looked only at individual data collection efforts. We can collect data from groups of people, too. These techniques include the focus-group interview, brainstorming, and the Delphi technique. They are quite different in process and employment.

Types of Group Techniques

The major reason for employing group, as opposed to individual, techniques is to capitalize on the **interaction** of a group. In a group, people often act differently from the way they do as individuals. Because group techniques are normally retrospective — that is, they concentrate on past events — the interaction present is normally unrelated to the "acts" people report on. Group interaction, however, does bring a number of benefits:

1. Individuals can listen to others' ideas and enlarge upon them, thinking about approaches they might not have normally considered.
2. The group may approximate the social situation on the job, thereby providing a context for what may otherwise become "academic" conversation.
3. Groups provide a kind of check on the validity of statements made by individuals.
4. In some contexts, groups may "take off" on flights of fancy that free them from too tight an observance of reality.

These advantages have corresponding potential disadvantages: a new approach might not be appropriate or better; the social situation may actually inhibit conversation or may introduce conflict that poisons the atmosphere; negative statements may eliminate good ideas; and flights of fancy may dominate discussion that is better kept at the level of reality.

FOCUS-GROUP INTERVIEW

The **focus-group interview** has its origins in market research for consumer goods. Intended to elicit responses or reactions to ideas, the technique has grown to be employed as a kind of "expert" panel. The group is exposed to ideas, and the ideas are discussed, evaluated, and enlarged upon. Because it relies on the question-and-answer format, it retains much of the manner of an interview. Many variations on the group interview exist, some looking less like an interview than a therapy session.

Generally, six to twelve people participate under the guidance of a group leader or **animator**. Participants are chosen for particular skills, jobs, or group experience. They should be articulate and forthcoming. Often a group is an entire department or work unit.

The group interview may include questionnaires to gather individual data and should be followed with a post-session questionnaire on pertinent topics. The goal is to allow group dynamics to generate as much feeling and experience as possible early in IRMaintenance. Because the group is assumed to be knowledgeable, it is important that its members be allowed to discuss concepts as openly as possible. Some liken the group interview to a training session with questions; individuals should leave feeling they know more than they did when they arrived. The new knowledge generated is the prize the analyst is looking for, since individual interviews can usually uncover existing knowledge. For example, in individual interviews, an analyst may determine the "official" procedures and goals; from the group, the analyst may find out the folklore and the *real* way to get things done.

A group interview should last no longer than three hours, with one or two breaks. The animator should also have an observer, since it is possible to "over-animate" and overcontrol the meeting.

Because the group interview is also a meeting, care should be taken to avoid the drawbacks of meetings. Participants should be comfortable; appropriate refreshments should be available; there should be no interruptions; no individual should be present with his or her direct supervisor; the agenda should be carefully laid out and publicized; and individuals should be treated as expert resources, not as complainers, petitioners, or culprits.

Because the group interview is an *interview*, the analyst should prepare questions, agendas, and the participants just as he would for an interview. In particular, the goals of the investigation must be made clear and security of comments promised. The use of tape recorders is not advised because of the privacy problem and because the quality of most tape recorders is too poor to pick up group comments clearly.

The disadvantage of the group interview is scheduling. Costs are lower for group interviews than they are for individual interviews because analyst time is not duplicated. But preparation time is increased, and it is very difficult to get six to twelve busy people together. The narrow focus on specific problems or issues necessary to propel the group forward can be countered by a well-made slide or transparency presentation and by giving the group a chance to ventilate on general issues. A good animator knows when to let the group loose and when to rein it in. *Most analysts are not trained as group animators.* It may be advantageous to bring in an organizational consultant who has experience in running groups. In this case, the analyst can prepare the animator and participate as an observer.

BRAINSTORMING

Usually limited to an hour or two, the purpose of **brainstorming** is to generate as many ideas as possible on a single topic. Usually the topic is a problem, such as finding out how to decrease the delay on month-end reporting.

As with the focus group, individuals are chosen for their communication skills or because of their exposure to the problem under discussion. The animator does not train the group in the topic, as would be done in a focus group, but an individual is appointed as the **client** who has the "problem." The client describes the problem and the group is charged with suggesting solutions.

As solutions are suggested, they are recorded. Negative comments are discouraged. In one variation of brainstorming, the client takes the list and prioritizes it, perhaps choosing the top two or three as topics for later sessions. These later sessions will have the goal of generating ways of arriving at the proposed solution. For example, if a proposed solution to the slowness of month-end reporting is to gather month-end data continuously during the month, that names a new problem: how to gather that data.

Another model of brainstorming merely creates the list of solutions, which is distributed to participants. Analysts can use brainstorming early in the logical design phase to uncover hidden needs that surface in the list of solutions.

The primary benefit is the large number of solutions generated in a short time; the drawback is the lack of a real-world focus, providing many unfeasible solutions. If the emphasis is less on the quality of solutions and more on what makes those solutions desirable, the analyst can benefit from this technique.

THE DELPHI TECHNIQUE

The **Delphi** technique is named after the Oracle at Delphi, who, when asked questions by the ancient Greeks, would reply in puzzling terms. Unlike the ancient Oracle, the modern Delphi technique is a group-oriented data gathering exercise. Its goal is similar, however: to obtain expert opinion.

Delphi is a cyclical questionnaire technique in which participants' responses are pooled and returned to the participants for further thought. The cycle repeats until the investigator is convinced that no further knowledge can be gathered.

Delphi may involve a either a few or a few hundred participants chosen for their expertise or involvement with the issue being investigated. Questionnaires are distributed and responses collected. Analysis of the responses results in another questionnaire. This one is identical to the first, except that means and standard deviations for the entire set of respondents, as well as this particular respondent's answers, are also recorded on the questionnaire. Figure 12-13 illustrates the difference between first-round and second-round questions.

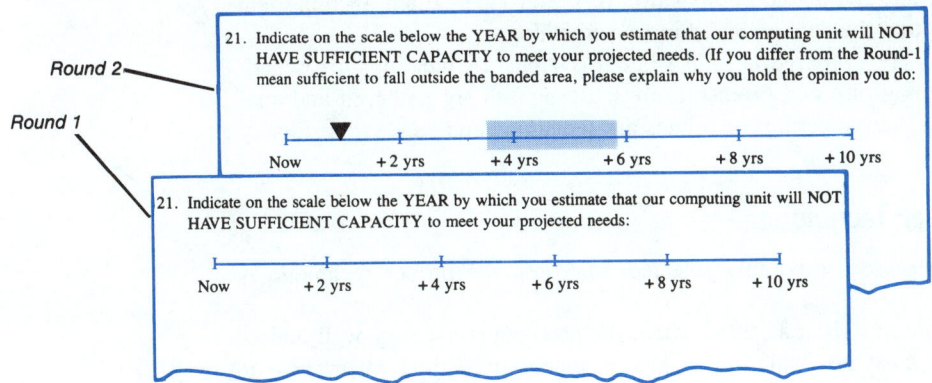

FIGURE 12–13. Delphi: First vs. Second Round Materials

Respondents are asked to reconsider their responses in light of the group's average value. This round of answers is again analyzed and, if necessary, a third round is sent out, with second-round means, standard deviations, and responses. The procedure is repeated until a consensus is reached or until round-to-round differences fall below a preset margin.

The primary benefits of Delphi are distributed expertise without the expense of interviewing and a consensus without the need for expensive group interviews. A large, geographically dispersed group can be contacted in this manner within a relatively short period of time at very low cost. Delphi can also be computerized (it is part of many computer conferencing systems; see Hiltz and Turoff's 1978 book, *The Network Nation*, to get an idea of how this works) to speed up data collection and analysis.

The major drawback of Delphi beyond those inherent in questionnaire techniques is the interpretation of the generated consensus. What does it mean? Can the average of 100 individuals' opinions mean anything, especially if it is arrived at through this indirect process?

Yes, because respondents are selected for their special skills. They are experts in the topic or system under scrutiny. Most Delphi efforts ask respondents who fail to alter their responses to indicate why, since they probably have special knowledge or exposure to the system that makes them more confident of their judgment.

General Techniques

As usual, the goals of the investigation have to be spelled out in detail. Next, the appropriate method is selected. Most group techniques are valuable during the earliest stages of investigation, although group interviews may have some use during system audit to determine installation, training, or maintenance problems.

Next, an appropriate sample is chosen. For each of these techniques, only knowledgeable individuals are selected. Where a group is to be gathered, two additional considerations are important. First, individuals have to be **accessible**. Second, they have to be **articulate**. They must contribute to the group's efforts. These two requirements often produce an *ad-hoc* sample of people who happen to be available. In most cases, group techniques neither require, nor pretend to use, random samples. Representativeness is always an issue, and the actual composition of the group is often an important consideration in arguing that the data gathered mean something.

Next, questionnaires are prepared and distributed (Delphi) or meetings are scheduled and conducted (focus group and brainstorming). Responses are gathered and analyzed. If necessary, subsequent meetings or rounds are scheduled and conducted. Finally a report is produced.

Choosing the Proper Technique

Because groups are seldom randomly selected, choosing the proper technique is twice as important.

Focus-group techniques are best selected when: the problem is not very well understood by individuals or must be "sold" to a group, where individual interviews or questionnaires cannot be conducted, where the group may know things that individuals do not know, and where a group can actually be put together and worked with.

Brainstorming works best: where a lot of solutions need to be generated for the analyst's later scrutiny, where trained facilitators can be located, where a series of meetings can be put together, where articulate and experienced group members can be found, and only in situations where threats to individuals can be minimized.

Delphi depends upon finding "experts." It is appropriate: for predicting future needs, for putting limits to situations that are already well defined, where the requirements for questionnaire-type studies can be met, and where sufficient time and calculating facilities are available to process several rounds of questionnaires.

Remember that group techniques are valuable only if the analyst feels that the group knows things that individuals do not know, things that can be drawn out in a group.

Conducting Meetings: Face-to-Face

The skill of **conducting meetings** is one analysts should cultivate. A group interview is merely one kind of meeting in the analyst's repertoire. The group interview differs from others in its intense focus on data collection rather than negotiation or fact sharing.

Five considerations are important. First, like any meeting, a group interview has to be scheduled, and two-to-three hour chunks of time are hard to find during production shifts. After-hours meetings may be most common. Alternates should be appointed to make sure enough people show up.

Second, threats should be lowered. Avoid conflict, disallow criticism of individuals, discourage labeling ideas as the "property" of specific persons. Managers should not be invited along with their employees.

Third, participation must be motivated without cross-examination. The analyst should not talk too much, but should encourage responses. The active listening techniques employed in individual interviews should be used.

Fourth, relevance is important. Analysts should choose participants who have some knowledge and let participants train each other.

Finally, conduct a good meeting. Study the topic in advance and describe it well to participants. Introduce them to each other. Keep comfort levels high. Draw conclusions at the end and thank attendees. Don't expect miracles.

Case Studies of Group Techniques

A DELPHI STUDY FOR PLANNING EDP EXPANSION

The goal of this study is to determine when EDP capacity will be insufficient for demand. A sample of user management is selected based on charge-back costs. Some heavy users and some light users are invited to participate through a preliminary memo. Two rounds of Delphi are sent out, asking for plans for EDP usage, feelings about capacity relative to needs, and projected needs for specific EDP services, including microcomputers, office automation, and supercomputing.

The analysis of the data shows that there is some degree of consensus on the date by which capacity will be insufficient, but a great range on the types of facilities required. Two groups (accounting and personnel) see little need for new facilities, but engineering and facilities management see capacity exceeded in several areas within the decade. Both accounting and personnel indicated that they plan no expansion of activities or change in type of activities over the next ten years.

FOCUS-GROUP INTERVIEW ON OFFICE LANDSCAPING

The land titles department at City Hall is going to be automated over the next three years to improve the paperwork flow. The analyst's goal is to understand how the group views its office landscaping needs.

A focus-group interview is to be conducted with this group. In addition, several aldermanic assistants will be present, since this group has already been automated. An animator is obtained from a local management consulting firm to conduct the meeting.

The analyst prepares several slides of offices, office furniture, and decorations to illustrate the concept of landscaping. He also has some diagrams that show why cable runs, work stations, and document flow influence, and are influenced by, office landscaping. The agenda will concentrate on reactions to the idea of automation, current office landscaping and problems, reaction to a number of possible approaches to landscaping, and preferences measured by questionnaire scales.

The analyst is prepared to collect a list of concerns, a list of the advantages of existing landscaping, reactions to proposed approaches, and new ideas the group generates. These will be listed on flip-chart. Questionnaire data will be collected after the session.

A total of twelve people will attend: the analyst, the animator, seven members of the land titles unit, and the three aldermanic assistants. The meeting will take place from 4 to 7 p.m. on a Thursday, with light refreshments and coffee available.

The analyst's report will contain the group priorities, perceived tradeoffs, preferences from the questionnaire data, problems to watch for in the transition to automation, a possible implementation/landscaping schedule, and some comments on the value of this technique for later work.

BRAINSTORMING ON CUTTING DOWN ERROR RATES IN DATA ENTRY

Southern Gas & Power has a trouble number to phone for gas-line and electrical problems. Consumers have always complained that it takes too long for repair personnel to arrive at their homes. Company officials have long felt that response time is reasonable, but a combination of occasional errors in recording the address and the nature of the problem and the high anxiety levels generated by potentially lethal gas or threatening darkness leads customers to feel that response time is long.

Because errors in data entry by phone clerks do occur, the company is willing to look at ways of cutting errors to zero. The analyst has drawn together a brainstorming group to ponder the problem. The goal is to conduct a series of weekly meetings over the next month aimed at improving error rates and making sure that no request goes unheeded.

Nine people, in addition to the analyst, are attending this, the second of the meetings. They are three phone clerks, a dispatcher, two repair persons, the assistant director of customer relations, a representative from the employee's union, and an installation engineer. All except the union rep were chosen because of their involvement and their supervisors' recommendations about their abilities to contribute to a group of this sort. The union rep is required to be present during any meeting involving working conditions.

The first session determined three possible solutions to the problem: (1) an improved form for taking phone messages, (2) a computerized system tied into dispatching, and (3) additional training for the phone clerks. Today's agenda is to examine the first and third solutions and to find out how to get them.

The analyst is using this one-hour session to elicit ideas from a group that is working remarkably well together. He lists the ideas on flip-chart paper and facilitates clarification of the ideas put forward. By keeping criticism to a minimum, the list grows rapidly. Many of the ideas are obviously infeasible (one suggestion was to analyze all the calls made over the past year for errors), but each contains the nucleus of an idea (in this case, spend some time trying to find the conditions under which the errors took place).

The analyst's report will be made available at the end of the third meeting; a fourth meeting will be called to discuss the report. The analysis will focus on a prioritized and scheduled list of solution implementations. Each meeting's output is typed and distributed to respondents before the next meeting. Based on the final consensus, the analyst will recommend a detailed investigation of one or more of the proposed solutions.

12.5 CONCLUSION: SYSTEM SUPERSLEUTH OR SPY?

Systems analysis has its origins in industrial engineering, the notorious field of "efficiency experts." Thoreau once said, "If you see a person coming toward you with the obvious intention of doing you good, run for your life." Many shrink from the withering gaze of the analyst-"spy" with just such an attitude.

The analyst needs data to draw valid conclusions. Part of the reason that the level of cooperation is often low and the level of fear high is that in the past, system development did not involve users at all, other than for a little bit of data gathering. Like the wheels of government, the reels of tape of the information resources group seemed to grind on in blissful ignorance of the users' needs, expectations, or attitudes.

That is changing. End-user software, applications generators, the information center, and a fresh cadre of business-trained analysts bring a new attitude to the user-analyst interaction. The techniques for data gathering discussed in this chapter stress such an interaction and that data gathering is a shared responsibility of mutual benefit. The analyst is a supersleuth, and no spy.

Investigation is the first of the triad of analyst activities. This chapter brings the reader to the tools level of investigation. Now, attention turns to the use of the investigation data in constructing models. Chapter 13 details a variety of models built based on investigation data. Chapter 13 depends on the discussion of procedures in Chapter 9 and draws the introduction to the analyst's major design tools to a close. Chapter 14 approaches logical design from the point of critical decision factors, shifting emphasis from design activities to controlling the IRMaint effort.

DISCUSSION QUESTIONS

1. A civic government procurement department is going to change its vendor form to make it more appropriate for computer data entry. Before this happens, a questionnaire is to be sent out to find out what potential suppliers think about the proposed changes. Outline a plan for conducting this questionnaire survey. What considerations will be important in sampling, distribution, and analysis?

2. The major problem with all observational techniques is intrusion. But there is another problem that comes about when the technique is to be used to assist in designing a new or improved system. What is this problem and how can it be overcome?

3. Protocol analysis combines features of interviewing with those of direct observation. It also combines many of the drawbacks. Which of the following situations would *not* be considered appropriate for using protocol analysis:
   a. finding out how fast someone can sort application forms into classes
   b. discovering which aspects of customer complaint registration are most likely to introduce informational errors
   c. determining which stations a form goes through after being filled out by a customer
   d. understanding what is on a customer's mind while he waits for service on a malfunctioning component

4. Consider the following costs: $150/interview; $100/morning observation of a worker; $50/participant in a group interview; $25/diary; $2.50/respondent to a questionnaire. The costs are in a 60/40/20/10/1 ratio. Is the data you might gather in the more expensive ways really worth 60 or 40 times as much as the data you could collect in the less expensive manners? In what ways?

5. A professor is interested in learning about how well students utilize a course syllabus at the beginning of each term. Design data collection efforts to prepare the professor to redesign the syllabus in the following ways:
   a. a short interview (prepare the interview schedule)
   b. a short questionnaire (prepare the form)
   c. a diary (prepare the diary — what could you hope to observe?)
   d. protocol analysis (prepare the task)
   e. a group interview (prepare the interview format)

DESIGN EXERCISES
Using dBASE III

1. You may have noticed that there are three types of navigation. The first type, called **direct** is simple. You supply the record number and dBASE III supplies the record:

DESIGN EXERCISES
Using dBASE III *continued*

LOCATE RECORD 3	Go to the third record.
GO TOP	Go to the first record in the file.
GO BOTTOM	Go to the last record in the file.
SKIP	Go to the next record.

The second type of navigation is called **keyed**. You supply a key to an indexed file and dBASE III gets the record:

USE CLIENTS	Examine the CLIENTS file.
SET INDEX TO CLIENTS	CLIENTS.NDX is the index file.
FIND YOURNAME +	Get the record matching these keys.
YOURPHONE	

Note that the length of YOURNAME + YOURPHONE has to match the length of NAME concatenated with PHONE. (dBASEIII does not know where to put blanks.) FIND "MARY 555-1212" will try to find a person named MARY 555-1212 and a phone number of blanks!

The third type of navigation, which you used in exercises for Chapter 11, is called **linked**. Two files are linked when a field in one file is a key into the second. For example, the CLIENTS file contains a field with a stylist's name in it (STYLNAME). dBASE III has a fairly simple way of tying files together when they are linked like that:

SELECT 1	In Area 1...
USE CLIENTS	Examine CLIENTS
SELECT 2	And in Area 2...
USE STYLISTS	Examine STYLISTS, with...
SET INDEX TO STYLISTS	Index file STYLISTS.NDX; then
SELECT 1	Go back to Area 1
SET RELATION TO STYLNAME INTO STYLISTS	

Whenever you access a record in CLIENTS, you can also refer to the fields in the related STYLISTS record, like this:

? NAME + "PREFERS " + STYLNAME + B->LASTNAME

B-> tells dBASE III that the field called "LASTNAME" is found in the file in Area 2 (A means Area 1, B means Area 2 and so forth).

Using this technique, print out a table of CLIENTS and the days of the week that their favorite stylist is available in the following format:

CLIENT	MON TUE WED THU FRI SAT
Name + Phone #	M, A, E, D, or N

DESIGN EXERCISES
Using dBASE III *continued*

2.   Using linked files makes it possible to store data without redundancy or waste. HAIR, for example, stores information on each session as a series of purchases, each identified by the receipt number (RCPTNO) for the session (in SESSIONS). Thus, you can go through RECEIPTS, extract all the records for a particular session (through RCPTNO), link into SERVICES and PRODUCTS to find the costs, and print out a statement. The pattern would be the same as 12.1 above, except that you want to look at every receipt for a specific RCPTNO:

```
SELECT 1
USE RECEIPTS
SELECT 2
USE SERVICES
SET INDEX TO SERVICES
SELECT 3
USE PRODUCTS
SET INDEX TO PRODUCTS
SELECT 4
USE SESSIONS
SET INDEX TO SESSIONS
SELECT 1
SET RELATION TO ITEMACT INTO SERVICES
SET RELATION TO ITEMACT INTO PRODUCTS
SET RELATION TO RCPTNO INTO SESSIONS
RNO = " "
ACCEPT "WHICH RECEIPT NUMBER? "TO RNO
LOCATE FOR RCPTNO = RNO
DISPLAY D- >NAME, D- >DATE, TYPE,
 B- >ACTIVITY, B- >CHARGE,
 C- >ITEMNAME, C- >PRICE
```

> This prints out the name and date of the session, the activity and its charge
> and the item and its price; but there are problems!

When TYPE is "I," the service part (Area 2, prefix "B- >") is meaningless; if TYPE is "S," the item part (Area 3, prefix C- >") is meaningless; if ITEMACT is blank or incorrect, links to the other areas are meaningless; and if the RCPTNO is illegal, the whole thing is meaningless!

Write a program called STATMENT that will query for a receipt number, do the appropriate checking, sum the charges, and print out an itemized statement.

CHAPTER 13

PROCESS, DATA, AND MATERIALS FLOW DIAGRAMMING

OBJECTIVES

1. To choose the most appropriate charting technique(s) for a given modeling situation
2. To chart system processes through system and program flowcharts, decision tables, decision trees, and Nassi-Schneiderman charts
3. To chart logical system information-processing relationships through the use of data flow diagrams and documentation flow diagrams
4. To chart functional system relationships through the use of structure charts, Warnier-Orr charts, and HIPO charts (including pseudocode process logic)
5. To construct a materials flowchart, a work distribution chart, and an office layout
6. To produce data definitions through the use of data dictionary specifications
7. To assess the value and the correctness of particular examples of system models

PROCESS, DATA AND MATERIALS FLOW DIAGRAMMING

13

CHAPTER

13.0 INTRODUCTION

"A Picture Is Worth 1K Words"

Caleb was in a tight spot. Somewhere off to his right was a pack of remarkably well-preserved Allosauruses who didn't understand that they had become extinct 60,000,000 years ago. Ahead of him thrust up the wall of the Gorge of Chan, six hundred feet of mixed sand and obsidian, atop which lay the Barren Plains, almost one hundred miles of flat, featureless sand. With few footholds and no vines or roots to grasp, it would be a difficult task for anyone to climb that wall. Despite his six years as a commando squadron leader, Caleb doubted he'd make it up ten feet before he became dino dinner.

So it was downriver for Caleb. Now let's see, on his map, didn't that cave over behind the spray lead to the Chamber of Death? Or was it past that bend? Gosh, wondered Caleb, where *was* his map? Ye gods of Goshen! He'd left it in his wallet, still in his dress pants on a hanger at the Ritz! Well, there was no choice, he thought. Without a chart, he was lost. Over there was a boulder. Jump! Hey, why is that alligator smiling?

Few systems analysts find themselves in difficulties like this. But the value of a map or diagram cannot be doubted. Diagramming is the meat and potatoes of the systems analysis craft.

In this chapter we will review *process charts*, which were alluded to in Chapter 9. *Logical relationship diagrams* show the logical dependency of system elements and *functional relationship diagrams* show functional dependencies. *Materials flow diagrams* are important where physical media must move through a network of processing stations, or, alternately, people must move through among these stations. Finally, *data models*, crucial to structured systems analysis techniques, round out the list of diagrams. At the end of the chapter, we will show the relationship among these kinds of charts. In a practical vein, some guidelines for production of these models are provided.

13.1   PROCESS CHARTS

In Chapter 9, we discussed procedure writing and we defined *procedures* as "programs for people" to carry out processes. This section discusses some of the most common process charts.

Charting Processes: Assumptions and Goals

Most charting techniques are based on the assumptions in Chapter 9: structured programming as the model, little or no concurrency, no indeterminacy, no truly random processes, no turbulence, and repeatability of activities. Each of the techniques reviewed in this section is equivalent to a pseudocode representation; the reason for producing a chart is to aid in communication.

Recently, process charting has been promoted as a tool consistent with logical, structured design. This has more closely integrated several tools that are essentially logical in nature (such as decision tables) with the logical design phase than it has with the physical design phase. Program flowcharting is still primarily confined to the latter stages of documentation. Newer tools, such as Nassi-Schneiderman (NS) charts have largely replaced program flowcharting because of their suitability to structured, top-down approaches.

This section looks at system and program flowcharting, decision tables, decision trees, and Nassi-Schneiderman charts. The first two are useful in documenting existing systems' actual operation. The next two are useful in modeling complex decisions. Finally, the Nassi-Schneiderman chart combines some of the best features of program flowcharts and decision tables into a graphic form that, although somewhat clumsy to draw, complements the structured approach to logical design.

System Flowchart

The **system flowchart** is a venerable diagram, having its origins in operations and methods (O&M). This diagram attempts to show, in a single diagram, both *control* and *data* flow, as well as some important *materials* flows (primarily those media that carry data).

The distinction between control flow and data flow is subtle, since in the DFD, data flow determines logical relationships. The concept of control, however, is strongly related to the idea of a von Neuman processor.

As you may recall, in Chapter 9 we discussed the concept of proceduralization. A procedure goes in order through a prescribed series of activities until it runs out of things to do. Some activities may temporarily alter the nature of the **flow of control** with choices of actions based on conditions. These correspond to the IF-THEN-ELSE (alternation) structure.

Control structure differs from logical structure in two important ways. First, control structures are strictly linear. If activity A precedes activity B in the control structure, then *B may not begin until A has completed*. That is, the beginning conditions for B are completely determined by the termination conditions of A. This is unlike logical structure, for which we would say that B may not *complete* until A has completed (see Figure 13-1), (opposite).

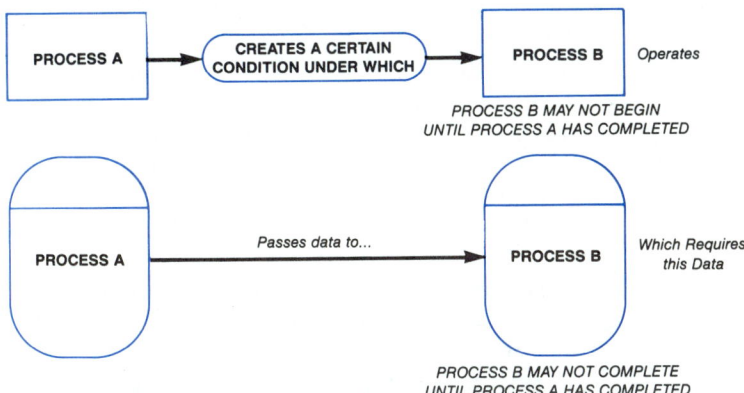

FIGURE 13–1. Comparison of a Control Structure and a Logical Structure

The second important difference is that one could say at any moment in time, control is *in* a specific activity. In other words, at each moment in time, a single process *has* control and may, in effect, do anything. Other processes may not interact, limit, modify, or control the process that *has* control. Analogous to a popular children's game, at each moment in time one activity is *it*.

The implications are many. Should the process fail, it fails at a specific point. Using structured techniques, it is theoretically possible to trace and absolutely determine which processes created the environment that motivated the error. In other words, *control* can be used to locate problems.

On the other hand, data do *not* exist in only one process at a time. Problems with operations can be traced to specific points using control structures, but problems with data seldom can. Because control flow is at least partly determined by data values, control can "flow" to activities that perform strange operations on "non-expectant" data. If structured conventions are not followed to the letter, control flow diagrams can prove useless. Figure 13-2 illustrates just such a situation. Where did the error come from? How can it be corrected?

Finally, control flow diagrams are strictly synchronous because of their linear nature. Therefore, it is natural to be concerned with time. However, during logical design, questions of timing are less important because the physical design usually determines speed.

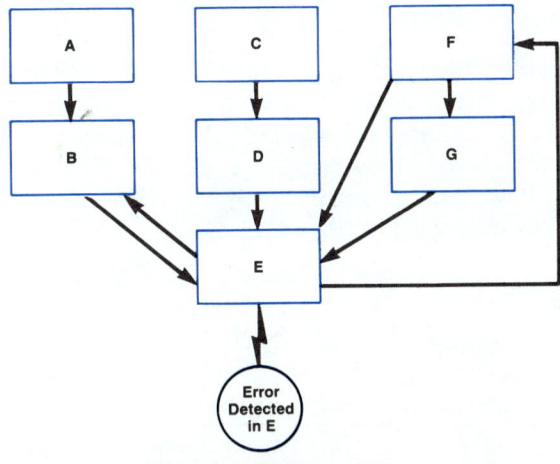

Possible paths: ABE, ABEBE, ABEBEBE, CDE, FGE, FE, FGEFE, FGEFGE FGEBEBEFEFEBEFEG....

FIGURE 13–2. The Control Flow Enigma: Where Did the Error Come From?

System flowcharts put data and control flows together. Figure 13-3 illustrates the system flowchart. There are standard symbols and standard drawing techniques. To an extent, the system flowchart resembles a data flow diagram. But the data structures represented in a system flowchart are large groups of data, usually files. Logical dependencies may be guessed at, since the processes shown are often entire systems rather than simple procedures. It is difficult to clarify statements about logical dependencies from a system flowchart.

On the other hand, physical dependency is easily determined. In Figure 13-3, the *response tape* is manufactured by a program called RESPAN based on the *data tape*. We could therefore infer that the response tape depends somehow on the data tape and that both the response and data tapes have to be present when RESPAN is run.

The system flowchart also tells us something about media. Certain kinds of data are stored on tape. Other data appear in printed reports. "Operator messages" appear on the computer console. This model is specific to a particular kind of technology, namely a computer with tapes, an operator's console, and a printer.

Why have system flowcharts? They serve a number of useful purposes, even for logical system design:

1. By documenting existing procedures, system flowcharts are useful in discussions with operators of these procedures.
2. System flowcharts are useful for noting physical bottlenecks and for putting numbers on throughput, turnaround, and traffic.
3. Where physical requirements are given *before* logical design (which is an unfortunate truth in real life), the system flowchart documents many of them.
4. Much of IRMaint is, in fact, maintenance: clean up, fix up, and so on. The system flowchart may be a given that cannot change.

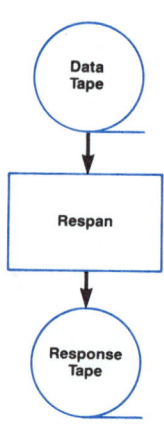

FIGURE 13.3. Control Flows Illustrate SOME Data Dependencies

Program Flowchart

The **program flowchart** merely represents flow of control among activities. Where structured programming conventions are used, the program flowchart is the equivalent of the pseudocode discussed in Chapter 9. The structures illustrated in that chapter were, in fact, drawn as program flowcharts.

Program flowcharts are strictly linear, dynamic models of processes. Here are the most commonly used symbols (Figure 13-4):

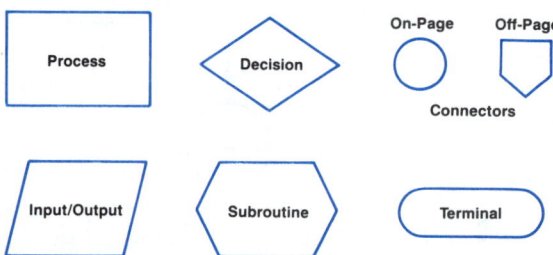

FIGURE 13–4. Program Flowchart Symbols

- A **process** or activity is represented by an oblong box.
- A **decision** that affects the flow of control is represented by a diamond. In standard form, decisions are drawn as two-valued: either a condition is true or it is not. If it is, one pathway is followed; otherwise, the other is followed.
- **Flow of control** is represented by the directed arrows.
- Specialized **input-output** processes are represented by the parallelogram; data structures participating in these processes are mentioned within the parallelogram, preceded by the words *Read* (or *Get*) and *Write* (or *Put* or *Print*).
- **Terminators** (entry to the process or end of process) are represented by ovals
- **Connectors** to other diagrams are represented by small circles
- **Subroutines** (see the note below) are represented by hexagonal boxes.

Figure 13-5 illustrates the process of updating a record in an inventory file based on a retail transaction and providing a report on the results. The concept of a subroutine is closely related to that of a subsystem. In programming terminology, a subroutine is a process that has **parameters**. A parameter is a peculiar kind of value that is determined before the subroutine is used and that exists only during subroutine execution. A subroutine, like a subsystem, interacts with the environment. However, it is *under the control* of the procedure that invokes it (in the manner of a control box as discussed in Chapter 1). In the inventory situation, four subroutines (ADD-INVENTORY, REDUCE-INVENTORY, QUERY-INVENTORY, and REPORT-INVENTORY) are used.

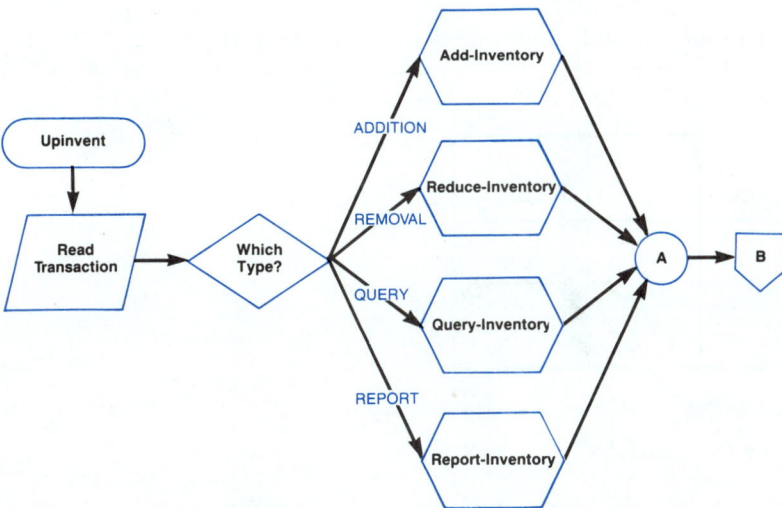

FIGURE 13–5. Program Flowchart for Updating an Inventory File

Program flowcharts may be drawn for any process. In fact, they are useful tools for teaching procedures, documenting the "right" way to perform something. For logical system design, however, the pseudocode equivalent is more compact and often less confusing, since program flowcharts may spill over onto many pages.

Decision Tables

A **decision table** models a single complex decision. Because it models the IF-THEN-ELSE structure, the decision table carries all the assumptions of structured programming. In particular, decision tables such as that illustrated in Figure 13-6 imply the following:

1. All possible conditions of the environment that *can* exist are already known.
2. All possible activities that are supposed to be performed *can*, in fact, be performed if needed.
3. The relationship between prior conditions and consequent activities is fixed.

	LOAN > 25,000				LOAN < = 25,000			
	Not Depositor		Depositor		Not Depositor		Depositor	
	< 65 yr	≥ 65 yr	< 65 yr	≥ 65 yr	< 65 yr	≥ 65 yr	< 65 yr	≥ 65 yr
LOAN RATE = 10.3%							X	
LOAN RATE = 10.625%					X			X
LOAN RATE = 10.75%			X			X		
LOAN RATE = 10.875%	X			X				
LOAN RATE = 11%		X						

FIGURE 13–6. An Example of a Decision Table for a Loan Situation

The decision table format is not standard. Figure 13-7 illustrates the parts of the decision table used in this text.

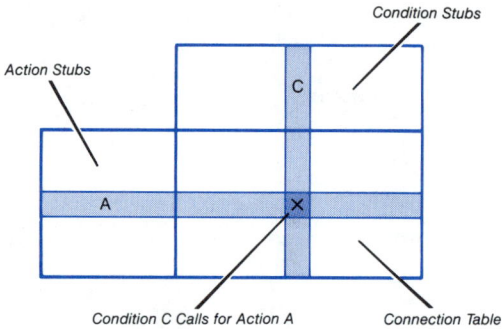

FIGURE 13–7. The Anatomy of a Decision Table

Condition stubs appear across the top of the table, describing conditions in the environment of the decision. Because the world is often observed in terms of supposed independent factors, condition stubs are often *generated* by factorial combinations. For example, if there are two possible values of *sex*, three possible values of *age*, and four possible values of *position*, then there are 2 × 3 × 4 = 24 possible conditions. Many of these combinations result in the same decision. This means that there are fewer than twenty-four *relevant* different conditions. However the condition stubs are generated, they describe the conditions of the environment at the time of decision in *mutually exclusive and exhaustive* fashion. No two conditions can occur at the same time and the list exhausts all the possibilities.

The **action stubs** appear down the left side of the diagram. These describe, in exhaustive fashion, all the possible, relevant sets of subsequent actions that we could choose to engage in. They may, however, overlap in the following sense: two actions can contain some activities in common, but no pair can be exactly the same.

The **connection table** indicates which conditions will be met with which actions. We show this with an X in the appropriate row and column. Each column can have only one X; a row can have several Xs. The reason for this is that, in the assumptions of structured programming, each condition of the world must be followed by a subsequent activity — there can be no indeterminate situations. Then, too, a condition cannot imply either of two subsequent actions — this introduces concurrency, which is not allowed. Therefore, each column must have one X. If two activities are to be engaged in *in sequence* for a given condition, we simply label the sequence and put that label in the left column as an action stub. Figure 13-7 has such a stub (Process A) which will be described elsewhere.

Decision tables have limited usefulness, since they describe only single decisions. However, they can be nested. For example, an action stub can actually be another decision table.

Decision tables can be modeled as program flowcharts, of course. However, they do serve additional purposes. They *list* antecedent conditions and consequent actions more neatly than a flowchart might. They can be used to *generate* all possible conditions and raise questions about completeness that program flowcharts cannot. And they may be made computer-readable to *generate code*, which is a far more difficult task using program flowcharts.

Decision Trees

A **decision tree** is another expression of the content of a decision table and the two are fundamentally equivalent. Whereas a decision table emphasizes the logic of connection (IF a certain condition THEN do something), the decision table focuses on the structure of the conditions.

Figure 13-8 illustrates a typical decision tree:

- The **root** is the point at which we begin reading the tree. Most decision trees are read from left to right (probably we ought to call them *decision vines*).

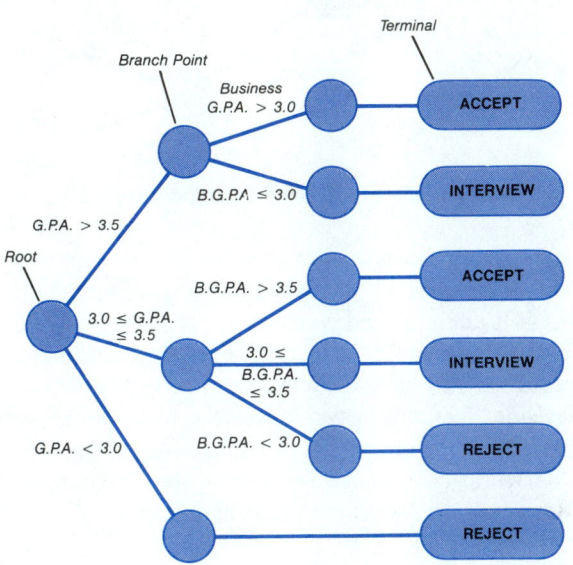

FIGURE 13.8. A Decision Tree for Applicants to an Exclusive Business Major

- **Branch points** represent decisions about a condition. Unlike a decision table, however, we break complex decisions into simpler parts and make independent decisions on each part.
- **Terminals** are points at which no further decisions are to be made; terminals include descriptions of actions to be taken.

The points on the decision tree are joined by lines that indicate paths to be followed. Resembling control flow diagrams, only one path can be "active" at each moment in time. There is one important warning, however: control does *not* "flow" in a decision tree. While it may appear that we are making a series of sequential decisions over time, in fact, we are really first gathering data about the environment and *then* making *all* decisions simultaneously. Because we cannot read the diagram all at once, we have to read component decisions one by one.

Consider a simple buy-sell decision table (Figure 13-9). Suppose we had $100,000,000 to spend on stock. It would appear that our first decision would be on the price. We decide to buy an inexpensive stock. It would appear as though some time passes before the next decision is made (whether we want an industrial or a utility stock. This passage of time would be disastrous, however.

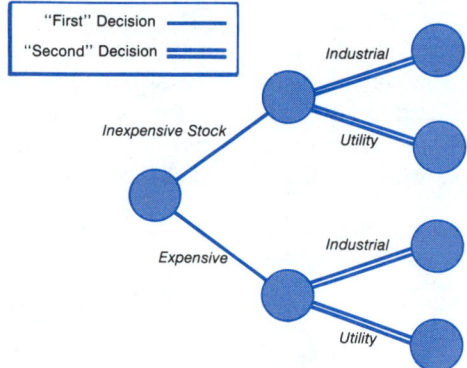

FIGURE 13–9. A "Disastrous" Time-Linked Decision Tree

Imagine the impact of the news that an important investor was looking around for an "inexpensive" stock to invest $100M! That might drive up the price of some marginally inexpensive stocks. It might indeed *change* the environment while we were beginning to make the next decision. It would, in fact, invalidate our original decision since the stock we ultimately purchase may have risen in price to the "moderate" category.

Reading decision trees this way violates the principle of structured programming which states that only one event occurs at a time. We presume that activities (processes) take time and the world can change because of them. After all, we process things to change them. But decisions are supposed to be instantaneous. Breaking a decision into components contradicts this.

Decision trees are really discussion devices to help us understand the structure of the environment. They focus attention on our assumptions about the environment. They also assist in other ways.

A decision tree focuses upon the way we observe and describe situations in the environment. We can also consider the probabilities that these situations exist as we describe them, placing a value on the likelihood of each condition's possibilities. Consider the decision table in Figure 13-10. Suppose we assign the probabilities of 0.3, 0.6, and 0.1 to the three branches. Examine the top branch ("expensive"). We know that half of these are utilities, a quarter are industrials, and the remainder are transportations. What is the probability that a given stock is an expensive transportation? It is $0.3 \times 0.25 = 0.075$. In other words, our decision making will take us to action A ("Wait a week.") about one time out of 13. Action B ("Buy \$50,000,000.") will result about 0.3×0.5 or 15 percent of the time. What is the probability of performing action C?

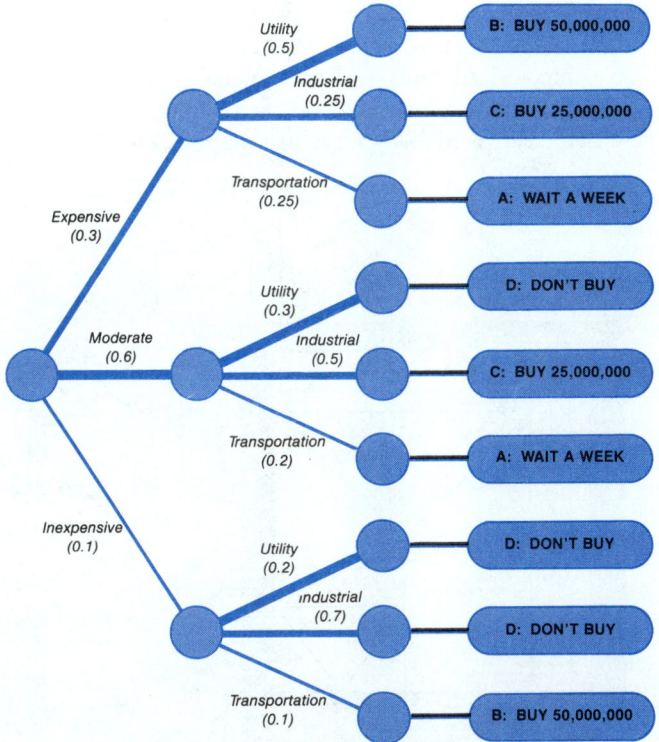

FIGURE 13–10. A Decision Tree with Probabilities Added

Some actions follow from combinations of several conditions. We can compute these by *adding* the probabilities because the conditions leading to the subsequent action are mutually exclusive (independent). If there were some dependence, we would have to reduce the computed probability by *subtracting* the probability of both happening simultaneously. Using this decision tree, what is the probability of doing D?

Nassi-Schneiderman Charts

The **Nassi-Schneiderman** (NS) chart is a compact graphic expression of structured programs. Each structure has a unique geometric symbol, resulting in a box that captures the procedure.

The advantage of the NS chart is its graphic appearance. On one hand, it resembles a flowchart. On the other hand, the NS chart allows you to follow the structure of a sequential set of conditions as you would in a decision tree. Remember, however, that a decision tree is a **static model** of a single, perhaps complex, decision; whereas, the NS chart is a **dynamic model** of a procedure.

Decision tables, decision trees, and NS charts suffer from the same drawbacks: the need to plan them in advance, a certain fussiness in drawing them, and a two-dimensional layout that spreads rapidly over large sheets of paper. These disadvantages can be overcome with planning, some computerized tools (microcomputer graphics packages are ably suited to drawing them), and an understanding of their limitations. The advantages are those that our hero Caleb discovered — they are like maps and may be worth thousands of words.

All process charts are, to an extent, equivalent (except decision trees and tables, which represent only the IF-THEN-ELSE structure). They depend on the assumptions that every condition is known in advance and that only decisions that can be made during the time the procedure is used are important. In other words, careful analysis is necessary to use process charts reliably. That is why their use is recommended for physical, rather than logical, design.

Figure 13-11 shows the equivalent pseudocode, program flowchart, and NS chart and the decision table and decision tree that represent the first major complex decision. Note how easy it is to convert each of these representations into a computer program.

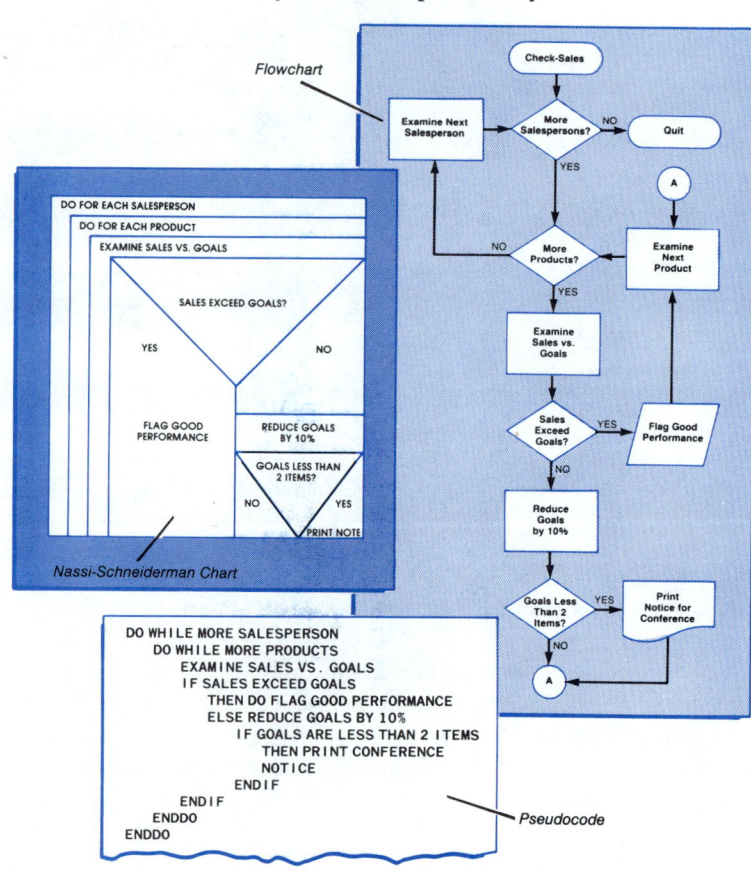

FIGURE 13–11. Three Charts of the Same Process

13.2   LOGICAL RELATIONSHIP DIAGRAMS

This section concentrates on charts that show logical, rather than physical, relationships. For their logic, information relationships generally are defined by **data dependency** among processes. Unlike process charts, logical relationship diagrams do not represent flow of control and generally do not address concerns of timing and sequencing. In this section, we will see how PERT charts show both logical and temporal relationships among processes, making them extremely useful in project management.

Logical Analysis Revisited

Chapter 4 discussed top-down logical analysis as it relates to the goals of systems and their subsystems. Chapters 5 through 7 detailed the products of logical analysis. In this section, we will review some of the concepts of output-driven, top-down logical analysis in preparation for discussing dependency charts.

Logical analysis begins with the desired system goal ("What are we trying to produce or accomplish?") and works *backward*, asking what is logically necessary to accomplish that goal. These steps are repeated until simple, easy-to-describe goals are detailed. As illustrated in Figure 4-3, this set of goals, in the structure derived, completes the logical analysis. Because these goals are easy to represent graphically, the *structure chart* is the major functional tool of logical analysis regarding subsystem goals.

The goals of an information resource are generally expressed in terms of data. For instance, a major goal might be to assist an executive in making monthly sales goals by providing rolling averages on product performance over three months. Minor goals are to accumulate sales data, compute rolling averages, and display averages in tables by product. In expressing these goals, it is evident that the data logically determines relationships among processes. In other words, the task of computing rolling averages depends on accumulated sales data.

The data dependency relationship is key to logical analysis because it enables us to work backward from desired products. The analyst does not have to work from raw materials as givens and design the whole process *in the hope of producing something desired*.

On the other hand, the Law of Utilization of Information compels the analyst to reverse the logical relationships discovered during investigation of the physical system to find out which processes or data relationships are unnecessary. The analyst can easily test the relationships to verify this.

Analysts commonly use three kinds of dependency charts: data flow diagrams (DFDs), document flowcharts (DFCs), and scheduling charts, such as PERT diagrams and CPM charts, that are not data related.

Data Flow Diagrams

The **DFD** shows logical relationships among processes that are dictated by data transformations performed by the processes. There are two forms of DFDs (Yourdon (1979) and Gane and Sarson (1979, used in this text)). Computerized packages such as Excelerator™ let users diagram either way. Both kinds of DFDs express data relationships among processes, but they can say much more.

DFDs express subsystem structure, since each process is a subsystem. This can be seen clearly from the example in Figure 13-12. Process 4.2 ("Produce year-end report") clearly has a goal distinct from process 4 ("Produce Reports"), although it is obviously a crucial subgoal of process 4. Were we to look inside process 4.2 (as we are doing within process 4), we would see additional elements (4.2.1, 4.2.2, 4.2.3) and data relationships.

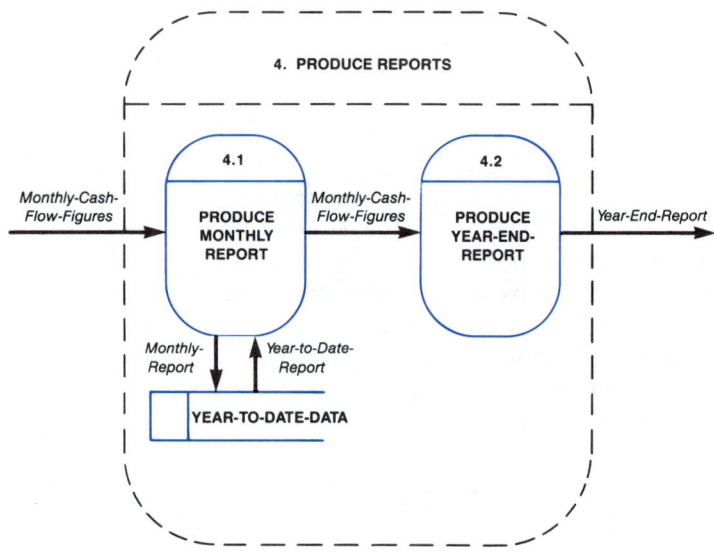

FIGURE 13–12. Subsystem Structure Apparent from a DFD

Data relationships are logical dependencies of a particular nature. Saying that "Produce year-end report" transforms "Monthly-Cash-Flow-Figures" into "Year-end-report" ignores the important dependency that process 4.2 has with 4.1 ("Produce Monthly Report"). We can accumulate year-end data throughout the year (and, in fact, easily produce a year-to-date report as well) as long as "Produce Monthly Report" produces valid monthly cash-flow figures. The year-end report therefore exists, although incompletely, throughout the year as a year-to-date report, updated each month from process 4.1.

As the discussion of morphs in Chapter 1 detailed, the DFD is valuable for isolating boundary elements, functional subunits (repair cells, for example), and logical loopholes. Consider the DFD in Figure 13-13 on the opposite page. The diagram asserts that the year-end report depends on monthly cash-flow figures. But it also depends on something else. How does Process 4.2 know when the end of the year has occurred?

Document Flowcharts

Similar to DFDs, **document flowcharts** (DFCs) illustrate logical dependencies, but they relate only to documents. The DFC is ideally suited to documentation of an existing manual records-management system.

Figure 13-14 gives an example of a document flowchart. Because a document may be represented by a data store symbol in a DFD, the DFC can be equivalently expressed as a DFD. Some aspects of the DFC, however, are unique, especially the use of specific symbols for merging, sorting, collating, and binding. Remember that models also serve a documentation purpose. Understanding the existing system is the first step toward understanding how (or whether) to improve it.

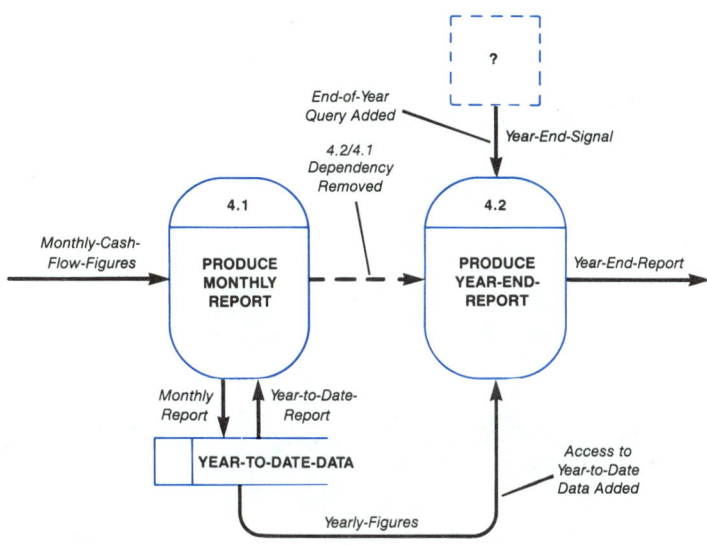

FIGURE 13–13. Results of Analyzing Figure 13–12.

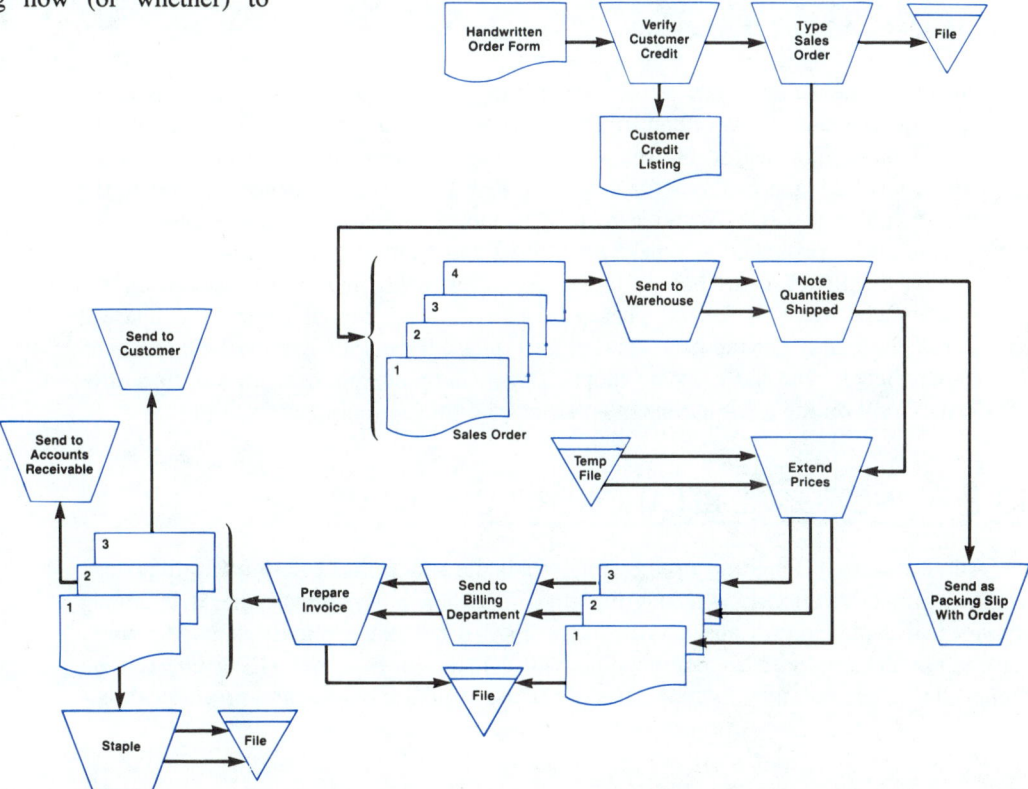

FIGURE 13–14. A Document Flowcharts Diagram (Powers, Adam, Mills, 1984; pg.88)

Other Dependency Charts

Logical diagrams show logical, rather than physical, dependencies. A number of other charts are also logical in nature, although they do not concern data. Many of the charts used in project management fall into this class.

Foremost among them are **PERT** (Figure 13-15) and **CPM** charts. These diagrams show how phases of a project depend on one another, without reference to their products. In other words, we read the arrows in the diagram as relating two activities or phases logically: phase B cannot begin until phase A has ended.

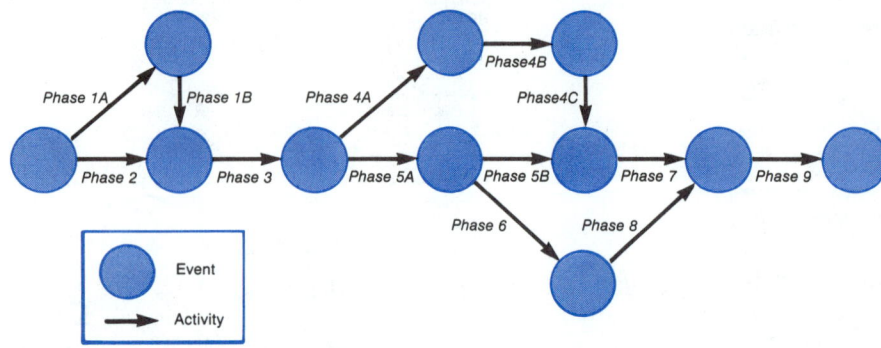

FIGURE 13–15. A PERT Diagram

How does the PERT chart differ from a program or system flowchart? First, the arrows uniquely mean "flow of control"; data or materials are not modeled. Second, the models exhibit a high degree of concurrency: several activities can be ongoing at the same time. Critical-path methods for analyzing PERT charts were developed due to the nature of complex projects. Without them, it would be difficult to see how phases of a project could be coordinated to "make everything come out on time, together."

Finally, the PERT chart has no loops. A certain phase may be necessary for the proper commencement of several phases (phases 4 and 5 cannot begin until phase 3 ends), or a phase may depend on several phases (phase 9 cannot begin until phases 7 and 8 have completed). But there are no loops. These charts are not drawn according to — nor do they rely on the assumptions of — structured programming conventions.

13.3 FUNCTIONAL RELATIONSHIP DIAGRAMS

Another class of diagrams used by analysts is the **functional relationship diagram**, which expresses relationships among functions (between subsystems) rather than among activities or data. Control and data flow are implied but not explicitly detailed. Often, these diagrams are used in conjunction with DFDs; in fact, the HIPO chart is an "umbrella" chart that may include parts of DFDs, structure charts, and process charts.

Subsystem Structure: The Texture of a System

In Chapter 4, we discussed goal analysis and looked at a system's imputed goals, those that the analyst postulates are the actual goals of the system. This set of goals is analyzed for subgoals. As each goal is analyzed in turn, the question to be answered is: "What must happen or be accomplished for us to say that this goal has been met?"

Goal analysis produces a structure chart of goals. Next, each goal is identified with a function to be performed. By making this identification, we say that goals are neatly separable, not only intellectually (in our analysis), but also in fact (in the doing). Ultimately, we will want to implement (build) a system with a set of subsystems, each of which will accomplish the desired goal by performing the designated function.

This three-step process (goal → function → subsystem) was called structural-functional design in Chapter 4. The tools we have for representing these three steps are functional relationship diagrams. As models, they represent the "texture" of a system in the sense that they show both logical and functional relationships among subsystems. They show "bumps" and "crevasses" in these relationships and give us a feel for how the system works. Because they contain so much information in a small form, functional relationship diagrams are the ideal first models of systems and the best ones to use in early stages of both logical and physical design.

In another sense, the "texture" of a system resembles the texture of an environment. In Figure 13-16, subsystem A1 influences system A. If A1 is fairly unpredictable for some reason (suppose that process A1 is error-prone), then subsystem A will bear the brunt of that unpredictability. Turbulence is inherited upward through the structure chart. Remember, the purpose of goal analysis is to determine what is necessary to achieve a goal. A goal can also *not* be met; given a constant environment, one of the reasons for this would be *internal* turbulence in the form of errors.

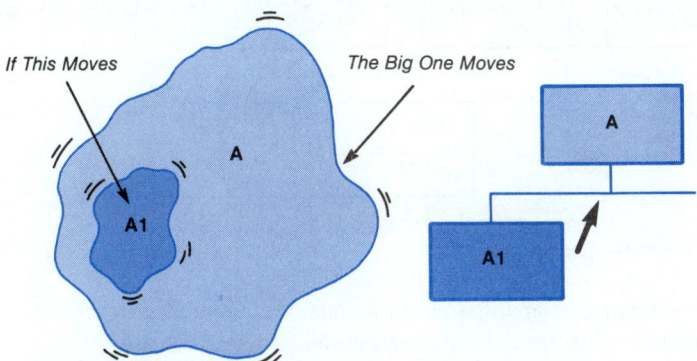

FIGURE 13–16. Inheriting Turbulence Upwards

The functional relationship chart is an important tool in debugging designs and, in later stages, for debugging modules of computer programs or procedures. Turbulence at one level is either intrinsic or inherited from lower levels.

Turbulence may also be passed *down* the functional relationship. The translation of goal → function → subsystem is that the operations of the subsystems performing functions to meet goals have to be coordinated at the higher levels. The shift supervisor coordinates the activities of the workers between midnight and 8. The operations manager coordinates the activities of the shift supervisors. The plant manager coordinates the activities of the operations manager, the warehouse manager, and the administrative manager. By "coordination," we mean both sequencing and directing. In the goal analysis, we were only a bit concerned with how goals are put together. In the physical design of activities, there is great concern about what precedes what and what information or materials need to be passed from operation to operation. If coordination fails, lower-level subsystems may be unable to perform their activities correctly.

Consider the operation of an accounts-receivable department. As payment is received (Figure 13-17), check and bill are matched, entries are posted, and a summary report of the day's activities is produced. Weekly lists of outstanding accounts are created, monthly rebilling is tabulated, and ninety-day overdue accounts are sent to collections. Normally the work is routine. When anything goes wrong (for example, if a check is illegible or erroneous), the supervisor is consulted. If he cannot figure out what to do, he asks the chief accountant and she tries to help. Suppose, though, that the supervisor informs a clerk that illegible checks have to be processed and he will have to "guess the amount — probably the amount due anyway." Because this will surely introduce error into the posting process, turbulence will be passed down from the supervisor and recorded into account statements, daily and weekly logs, and even ninety-day-overdue reports.

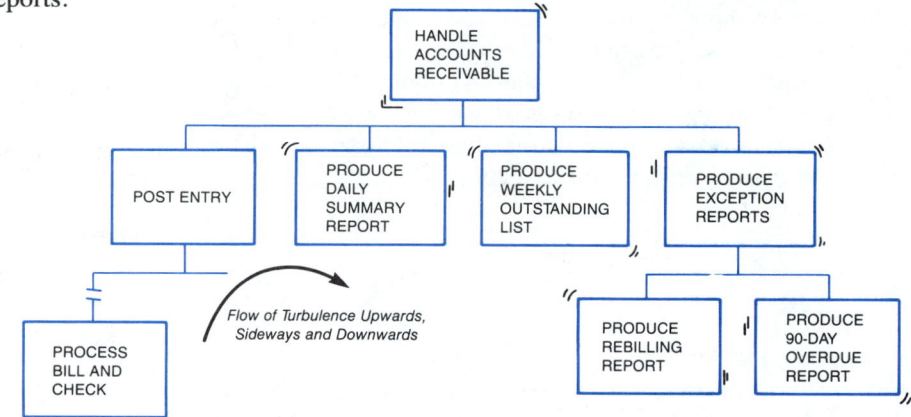

FIGURE 13–17. What Turbulence Achieves

The subsystem structure derived from goal analysis is not unique. As mentioned in Chapter 1, imputing goals to a system is a function of the observer and may not actually reflect the true goals of the system. *There are many ways to accomplish each goal.* This principle, called **equifinality**, is the bugbear of analysts. Although an analyst could prove that a particular design is adequate, there is no proof of uniqueness. A better design may appear tomorrow. "Better" might mean easier to implement, more reliable or efficient in operation, or conceptually easier to discuss. In any event, functional relationship diagrams assist in the discussion.

Structure Charts

Structure charts were introduced in Chapter 4. They are derived from structural-functional analysis and may represent goals, functions, or operating subsystems. There are two ways to represent structure charts (Figure 13-18): notation similar to organization charts indicates a **functional design**; radiating lines indicate **operational structure**. This is not a universal distinction, but it makes it easier to distinguish function from operation.

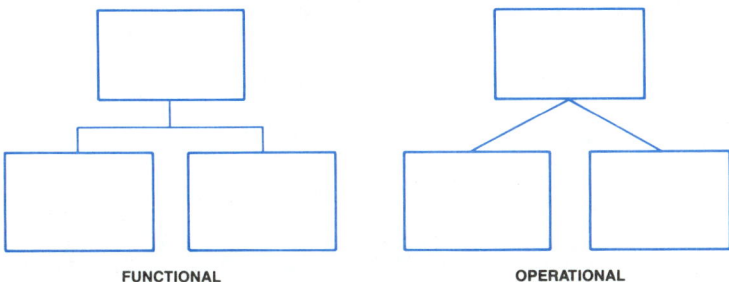

FIGURE 13–18. Two Ways of Denoting Structure Charts

Figure 13-19 indicates the structure of the functions of word processing. Here we do not represent the order of the functions or the way in which functions coordinate other functions. Figure 13-20 shows the operational subsystem structure of the operating word-processing system. Here we note several new symbols, derived from structured programming conventions:

1. The radiating lines are read from left to right in **sequence**.
2. The small diamond indicates **alternation**. Only one of the subsystems at the ends of lines leading down from the diamond is activated at a time.
3. The curved line indicates **iteration** of a sequence.

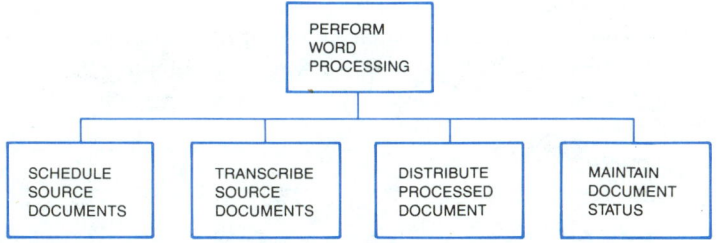

FIGURE 13–19. Functional Representation for Word Processing

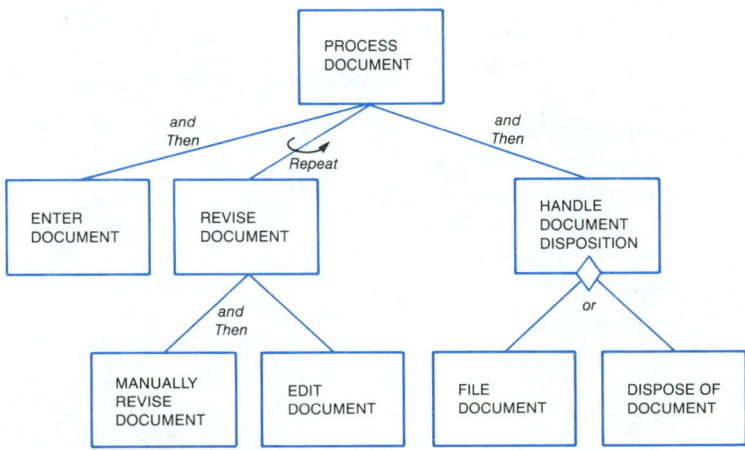

FIGURE 13–20. Structure Chart for Operating Word Processing System

We then read the diagram as follows:

> Word processing is performed by entering a document and then repeating a cycle of manual revision and document editing after which the document is either filed or discarded.

This chart does not indicate how many revision-editing cycles are performed, how we know when to stop the cycle, or how we choose between filing and discarding. Furthermore, we have no idea of the speed of performance of these functions. This is part of the logic of each subsystem's operation.

The structure chart can be an important design tool. First, we represent goals. Next, we equate goals with functions. Finally, we implement the functions as subsystems. More detailed "texture" of operations is supplied by the process charts that describe each subsystem.

HIPO Diagrams

We can put together hierarchy (structure chart), process (process charts), and input-output relationships between subsystems and between subsystems and the environment in a **HIPO** (hierarchy-input-process-output) chart (Figure 13-21). While the HIPO (sometimes called IPO) chart contains no information not found in separate diagrams, it is a handy, easy-to-visualize, compact way of combining the information. This chart generally consists of sections devoted to the following:

- hierarchy: which modules control ("invoke") this one; which modules this one invokes
- input: which data elements are needed as input from other subsystems or the environment
- process: the process "logic" of this module
- output: which data elements are exchanged as output to other subsystems or the environment
- notes: commentary on the goals, functions, or processes

ID Revise Document (2.0)	
CALLED BY Process Document	**CALLS** Manually Revise Document Edit Document
INPUTS Document-Name Keyboarded-Document	**OUTPUT** Edited-Document
PROCESS Get Keyboarded-Document, Document-Name DO Manually-Revise-Document IF revisions-exist THEN DO Edit Document ENDIF	
NOTE: Revisions-exist is set by Manually-Revise Document	

FIGURE 13–21. HIPO Chart

In this single chart, we can indicate how a function or subsystem works, what it is working for or toward, what functional resources it needs (invokes), what data items are used in transactions with other systems, and design notes from the analyst.

Warnier-Orr Diagrams

The **Warnier-Orr diagram** (Figure 13-22) illustrates in tree structure the functional relationships in a design. This chart depicts structure and function and can be extended, as can the HIPO diagram, to physical design simply by carrying the analysis further to physical operations. The symbols used in the Warnier-Orr diagram are again derived from structured programming and include the following:

- **sequence**: vertical order
- **alternation**: the " + " within a circle indicating a choice among the vertical list
- **iteration**: the number in parentheses (or a variable, if the number of iterations is to be determined during operation)
- **negation**: the horizontal bar over a condition indicates the negative of that condition

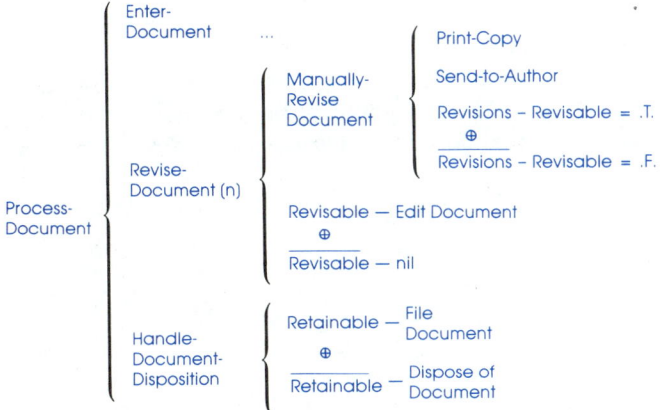

FIGURE 13–22. Warnier-Orr Diagram of the Word processing System

Summary

Functional relationship diagrams indicate the ways in which functions go together to meet system goals. They may be combined with process charts and even elements from data flow diagrams into a kind of all-purpose diagram. These diagrams are useful from the earliest stages of investigation through implementation.

13.4 MATERIALS FLOW DIAGRAMS

Thus far, we have concentrated on logical design and the processes that effect changes in data. This section shifts the emphasis from logical, conceptual modeling tools to physical materials, which have to be allocated, altered, transported, and stored. An information system is more than just information. It also includes the means to transport the information and the media on which information is recorded. The relationship between information and the recording media is complex. It is necessary to understand this relationship to design a physical information resource. We will now look at the relationship between information and the information resource.

Media and Materials

All information is **encoded**. That is, information *represents* facts about the environment or the system in a code. The code usually consists of arbitrary symbols (a humorist once noted that computer science must be a waste of time, since half of its content is zeros — the other half, of course, is ones). The symbols themselves are physical things. Even the letters you see now are deposits of ink with some weight and dimension. They take up space and have some mass; they affect or are part of the **media** on which they are recorded.

Recording medium is only part of the story, however. Not only is information encoded into recordable items on media, but these media themselves have to be purchased, stored, transported, cataloged, protected, accounted for, and ultimately, destroyed. The relationship between the symbol and the medium may be different from that between the medium and, say, its storage vault. For example, punched cards have information permanently encoded on them in the form of punches. But the cards themselves may have a temporary life span, moved from location to location and then unceremoniously dumped into a trash bin.

Information and media can have several kinds of relationships (see Figure 13-23):

INFORMATION-MEDIA	MEDIA-ENVIRONMENT
RECORDING PERMANENCE	ACCESS
Permanent	On-Line
Temporary	Off-Line
RESIDENCE	EXPANDABILITY
Resident	Expandable
Transient	Fixed-Size
EDITABLE	SHELF LIFE REPRODUCIBILITY
Editable	Manual
(Read/Write)	Optical
Uneditable	Copy-Protected
(Read-Only)	Machine-Readable
EDITING LEVEL	TRANSPORTABILITY
Character	
Field	
Record	
File	
Point (Graphics)	
Line (Graphics, Display)	
Screen (Graphics, Display)	

FIGURE 13–23. Information-Media-Environment Relationships

1. Information can be recorded **permanently** or **temporarily**.
2. Temporary information can be either **resident** or **transient**. Resident information changes from time to time, but the format and intention of the information remains the same. For example, tapes can be formatted so that records appear in a fixed structure: ID number (8 characters) followed by name (30 characters)

followed by job class (1 character), and so on. While the actual values of the name may change, the fact that characters 9-38 represent the name does not. Transient information is not guaranteed to be in a fixed location on a particular medium. A form is an example of information with a resident characteristic; a filing cabinet houses transient information.

3. Information can be **editable** or **uneditable**. Editable information can be changed without damaging the medium or invalidating the record of information. Uneditable information cannot be changed.

4. Where editing is possible, there are a variety of *levels* at which it can take place. Disks and diskettes are normally editable at the character level, although this usually means reading a record into RAM, changing it there, and rewriting the whole record back to disk. Tape is far less easily edited at that level. Usually a character edit means rewriting the entire file to another tape. RAM is editable at the character level (and often specific bits can be changed using machine-language instructions). All-points-addressable displays such as those found in most graphics systems and microcomputers can edit a single point on the screen. Intelligent terminals can alter single letters, while "dumb" terminals provide no facility to edit even a line on the screen; instead the entire screen must be rewritten.

Permanence, residence, editability, and level of editability are the four major relationships between information and its media. The media themselves have these characteristics:

1. **On-Line** or **off-line**. On-line media are available at any moment, while off-line media have to be manually accessed and mounted or switched into the system. Tapes, diskettes, and disk packs are normally off-line; the machinery that channels and records information may be on-line but the media themselves stored elsewhere. Fixed disks on microcomputers are normally on-line.

2. **Expandable** or **fixed size**. Expandable media can be enlarged to accommodate more information. Fixed-size media can hold a fixed number of characters or records. Reports can be any size when printed, while a fixed disk in a microcomputer normally is not expandable. Many operating systems can simulate large "volumes" that extend over a number of media units. Two common examples are multi-reel tape files on mainframes and multi-diskette backups on a microcomputer. In each case, the operating system shields the user from having to program for the contingency of running out of space.

3. **Shelf life**. Most media have limited shelf lives. Cards and tapes must be stored in relatively cool, mildly humid environments. Magnetic recording media must be kept free from dust and magnetic flux. RAM, of course, loses its information when the power is turned off; ROM and bubble memory do not. The paper in reports may last ten to twenty years on a shelf in an air-conditioned building but less than a week in an industrial setting or on a sales counter.

4. **Reproducibility**. It is easy to reproduce a tape or diskette, although it may take time and tie up a number of units. Paper, being non-machine readable, may have to be manually reproduced on a duplicator. It may be cheaper or faster to produce many originals at once rather than to resort to a manual process. Reproduc-

tion of data may be inhibited by using certain kinds of materials to enhance security. Diskettes holding expensive software may be reproduced, but a number of schemes exist to make it hard to execute the copied programs. Common techniques include entry of passwords matched against highly encrypted keys, burning serial numbers into the medium, or more exotic hardware/software schemes.

5. **Transportability**. Some media can be carried by hand from place to place. Other media require heavy or precision-made cases, which make moving them difficult or dangerous. Disk packs are quite bulky. RAM obviously is not movable, but ROM can be (this is the basis of many plug-in cartridges used on video games).

Accessibility, expandability, shelf life, reproducibility, and transportability are five of the more important data-independent characteristics of media. Materials flow diagrams show how the information-bearing media are moved from processing station to processing station. The data-dependent and -independent characteristics we have mentioned dictate how rapidly, efficiently, reliably, and carefully these movements can be performed.

In Figure 13-24, a **materials flow diagram** depicts the flow of materials through a fabric-production plant. At every stage of production, various labels and forms are detached, read, altered, and reattached to batches as they move through the line. Information, media, and materials accompany each other through the processing. While, in a logical sense, the fabric materials are irrelevant to the information processing, they and their manufacturing processes are critical to the information resource design. A person cannot fill out a form with his hands full. Another person cannot verify the contents of a batch if she cannot find the batch waybill.

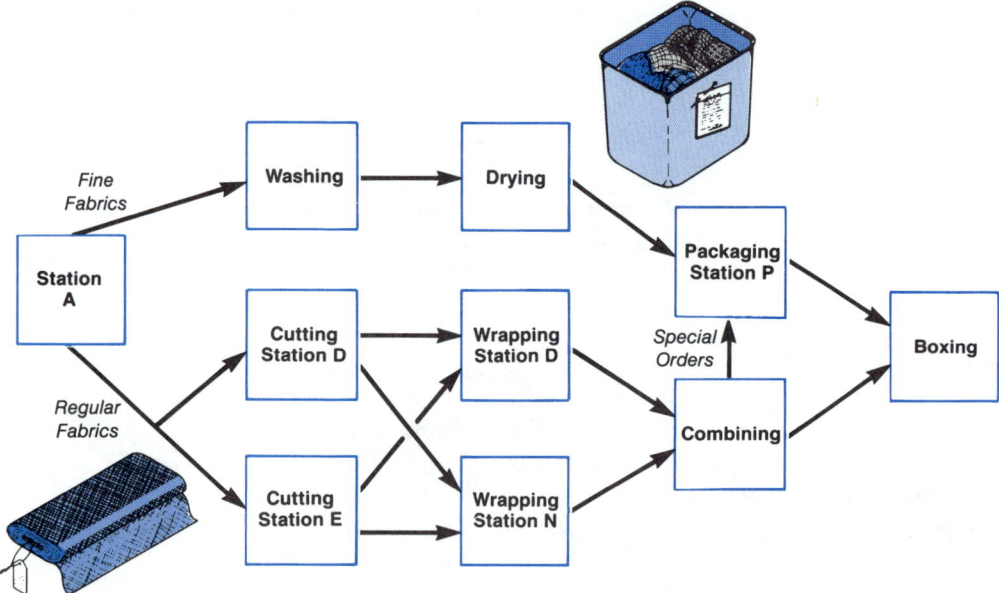

FIGURE 13–24. Materials Flow Implies Media Flow Implies Information Flow

Document flow diagrams show how information flows among recorded media, and forms distribution charts show how forms are distributed across departments. These diagrams involve physical media only in an incidental way. They can usually be created directly from data flow diagrams with little consideration for the media, primarily because we have a tacit agreement about what a "document" looks like and the heft of a form. But if the "document" turns out to be a seven-volume set of company financial statements, some consideration may have to be given to packaging.

Work Distribution Charts

A business may run on information, but financial and human resources are the energy that make the engine go. Analysts are often called upon to model distribution of work (projects and time) across individuals when managing an IR project. They may have to design procedures that illustrate in graphical form how work is to be performed. **Work distribution charts** show this.

The **GANTT** chart in Figure 13-25 has two axes. Across the horizontal axis, a time line is marked out in appropriate units (days, weeks, months, quarters). The vertical axis lists individuals or projects. Information is placed in the resulting grid indicating that individuals or projects are "active" across specific periods. "Design marketing system" is active in April, May, and June. "Develop marketing system" takes place in July and August.

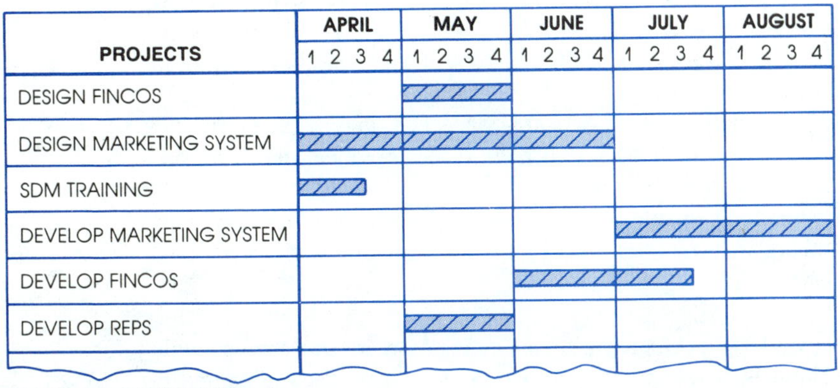

PROJECTS	APRIL				MAY				JUNE				JULY				AUGUST			
	1	2	3	4	1	2	3	4	1	2	3	4	1	2	3	4	1	2	3	4
DESIGN FINCOS					/////															
DESIGN MARKETING SYSTEM	/////	///	///	///	///	///	///	///	///	///	///	///								
SDM TRAINING	/////																			
DEVELOP MARKETING SYSTEM													/////	///	///	///	///	///	///	
DEVELOP FINCOS									/////	///	///	///								
DEVELOP REPS					/////															

FIGURE 13–25. GANTT Chart

Another way to draw the chart is seen in Figure 13-26 on the following page. Here is a detailed look at the "Design marketing system" project (April through June) indicating the thirteen weeks involved. Listed are the four people involved and the periods of their involvement. Shaded boxes mean analysis, dark boxes mean design, and open boxes mean supervision. Shirley supervises all thirteen weeks. Thomas and Evelyn perform analysis work through April, at which point Veronique joins the staff and works with Evelyn on the physical design through June.

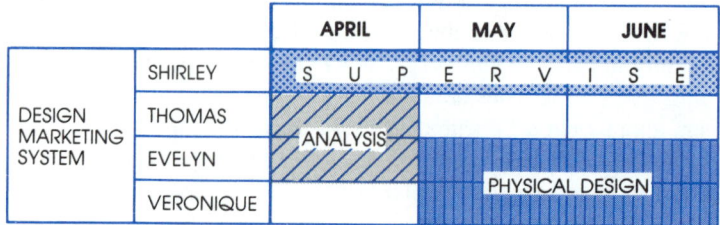

		APRIL	MAY	JUNE
DESIGN MARKETING SYSTEM	SHIRLEY	S U P E R V I S E		
	THOMAS	ANALYSIS		
	EVELYN		PHYSICAL DESIGN	
	VERONIQUE			

FIGURE 13–26. Another View of the GANTT Chart

Office Layouts

Often, there is a considerable office landscaping effort attached to information resource maintenance. Before 1970, most computer systems were designed for utility rather than beauty. Since then, there has been a great deal of interest in laying out computer-based systems (and some manual systems) that maximize efficiency while pleasing the eye and ear. Here are some conditions:

1. Traffic patterns: who has to walk where, in how much time, carrying what?
2. Cable runs: which cables have to run from where to where?
3. Noise intrusion: how intrusive will noise be to workers who require concentration?
4. Lighting levels: is appropriate light available to perform the work? What sources are available for natural or artificial lighting?
5. Terminal sharing: if terminals are to be shared, who will share and how will they know when their turn is?
6. Privacy: some work, especially if customers are concerned, may require privacy (meaning walls). If so, doors or baffles will have to be provided and these, in turn, will influence traffic patterns.
7. Foliage, decoration, windows, displays: some interior decorating is conducive to productive work; others may distract. A window is great to see through, but glare creates headaches.

An **office layout** is essentially a blueprint of an office. On it, the analyst will attempt to place desks, terminals, cable runs, lighting, and interior decoration most appropriate to the work to be performed. Often, an analyst will work with an interior decorator during physical design. For example, it may be too expensive to place cable runs in such a way that the terminals depicted in Figure 13-27 can be shielded from glare. In that case, special glare control measures will have to be taken concerning the windows (perhaps curtains or glare-reducing coatings).

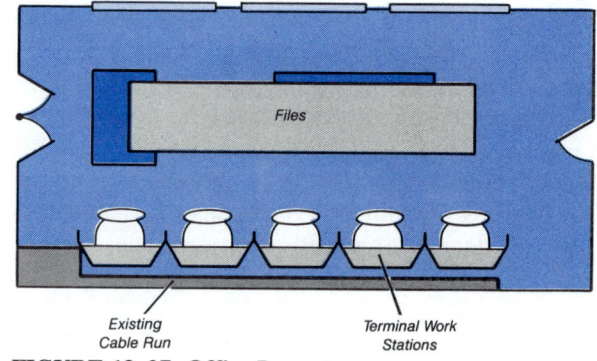

FIGURE 13–27. Office Layout

Forms Distribution Charts

A final materials flow diagram shows how particular forms move through an organization. Very closely related to the DFD, the **forms distribution chart** depicted in Figure 13-28 shows that the request for transfer form travels from forms management to the worker, to the worker's supervisor, back to the worker, to the division manager, and then to Personnel, where copies (with appropriate information on decisions) are sent back to the worker, the supervisor, and the division manager, and back to the requesting supervisor. The numbers in the circles represent the copy number (#1 is the original or "top" copy). One copy is retained in the personnel files.

The forms distribution chart is often the starting point for a systems study, since late or error-ridden forms are often the impetus for the study. While procedures may say one thing, practice may be quite different. In Figure 13-28, the worker is often not given the form. Instead, the manager and worker jointly complete the form, sometimes saving time. Of course, if the manager is busy and the worker has an odd schedule, the meeting might be delayed. Without privacy, the worker may indicate the wrong things on the forms. Here is an example of the informal process replacing the formal one to the potential detriment of everyone.

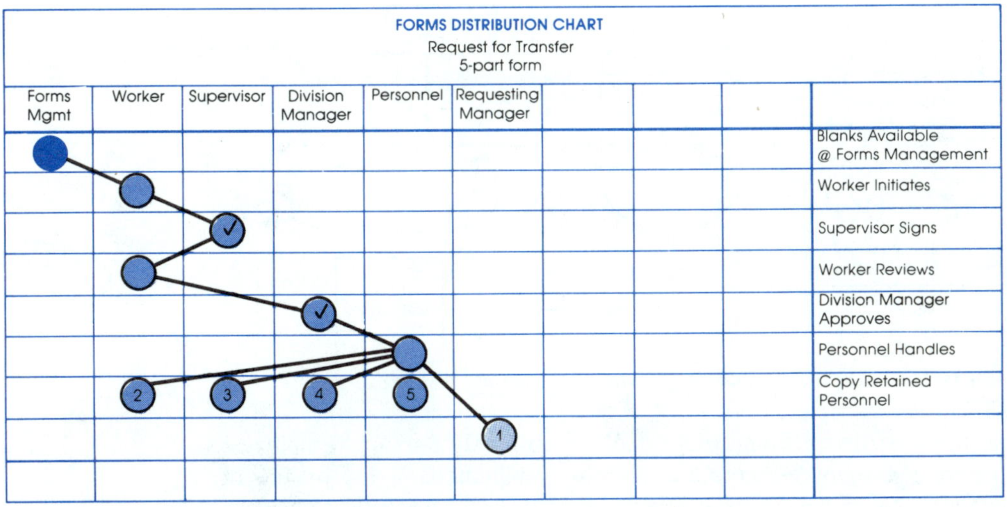

FIGURE 13–28. Forms Distribution Chart

13.5 DATA MODELS

Logical system design expresses logical relationships in terms of data exchange. That is, the system under discussion owes its internal dynamics to the relationships inherent in the exchange of data. DFDs illustrate these relationships directly by showing how data flows interrelate processes and data stores. This section is concerned with models that show the relationships among data items themselves.

The central concept is that of a **data dictionary**, which is like a dictionary because data are defined in simpler terms. It is also like a glossary because terms are defined only in relation to the system rather than in terms common to the field of information resource management. Finally, because the data dictionary is limited to data that is in a specific system, each system has its own data dictionary.

This section will look at the structure and function of the data dictionary. But first, here are some thoughts on why static models of data are necessary in the first place.

The Need to Provide Static Models of Data

Look at the DFD in Figure 13-29. There are seven data flows, two data stores, two processes, and two external entities. The DFD is a dynamic model showing relationships brought about by the flow of data. But there are also fixed, or **static**, relationships among the data items.

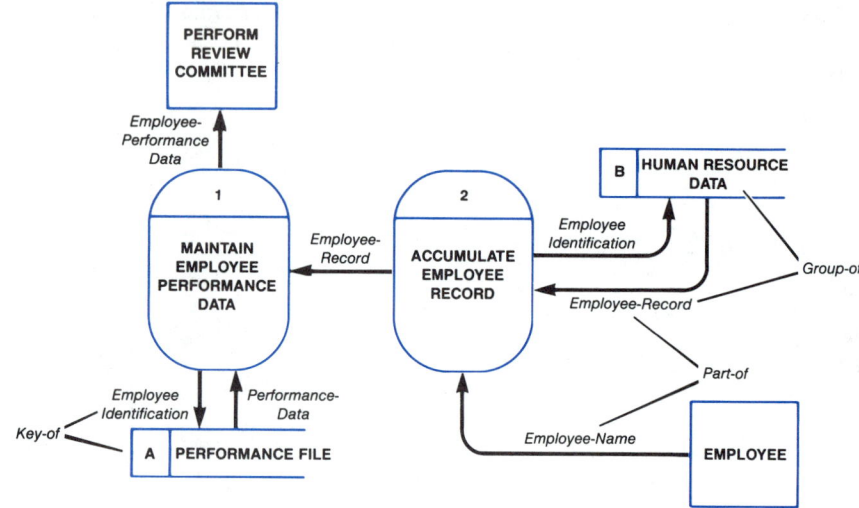

FIGURE 13–29. Three Static Data Relationships (Group of, Part of, Key of)

For instance, the data flow "Employee-Name" is obviously *part of* "Employee-record." Furthermore, "Employee-record" is simply one item on a list or file of "Human resource data." No matter what the data is, no matter what particular values the data take, these relationships hold across time. We call this a static relationship, one that doesn't change.

Understanding the DFD in Figure 13-29 requires more than merely noting which data are necessary for which processes. Fuller understanding really depends on knowing the static relationships. Why?

First, one data item being composed of a set of other data items tells us that we produce the first by successive entries of the second. Human resource data is, in a very real sense, nothing more than the accumulation of employee records.

Second, because one data item is logically related to others, we know that if the second is incorrect, then logically the first is, too. If the employee name is wrong, so, too, must be the employee record.

Third, because one data item may "stand for," "signal," or "key" another, we know that if we have the wrong key, we cannot hope to obtain the right item. If employee identification is wrong, then we have no way to retrieve the correct employee record.

These relationships (group of, part of, key of) are static data relationships expressed by data models. The data models do not indicate how data are to be processed or transformed. They do, however, tell how data are to be treated *with respect to each other*. Is an item merely a *group of* other items? Is an item a *part of* another? Is one item a *key of* another?

Data Dictionaries

A **data dictionary** is a list of data items with their definitions. A data definition is a description of the data in the following terms:

1. Structure: does this data item have some static relationship to other data item(s)?
2. Form, format: does this data item have some specific format? (For instance, is "Billing-Date" expressed as month-day-year or day-month-year? Is "Rate-of-Return" expressed as a number with two, three, or four decimal places of precision?)
3. Type: is this a number? A character string? A bit?
4. Name: does this data item have other names or synonyms (sometimes called "aliases")? Is "Billing-Date" the same as "Invoice-Date"?
5. Values, ranges: does this item have an restricted set of values? Must "Customer Name" begin with a capital letter? Is "Balance-Due" restricted to positive values less than $1,0000,000.00? What are the allowable values for "State-Code"?
6. Usage: is "Monthly-Billing-Summary" produced monthly? How often is "Sales-Data" entered? Approximately how many sets of "Sales-Data" are entered in a batch?
7. Access: is there a restriction on access to this item? Are the data encrypted? Who "owns" the data? Who "produced" it? Where is it ultimately headed?
8. Location: is this item found in any specific location in the firm or on a particular medium? (These are physical requirements, but if they are known during logical design, there is little sense asking the designer to find all this out again.)
9. Volatility, life cycle: how permanent is the item? Is it "temporary," "working data," "permanent," or "updated?" How long can the data be expected to last between updates? Is the item around at the termination of the process?
10. Address: which processes, data stores, and external entities use this data item?

It is not necessary (or always possible) to describe each data item completely. Most of the physical information will be added later, during physical design. However, name(s), address, form, type, and access are logical characteristics, important during logical design.

What kinds of "data" are put into data dictionaries? Certainly data flows, data stores, processes, and external entities will be defined. Data stores are complex aggregations of data. Usually data stores are created to slow data down into a "batch." The batch may be created for efficiency (it is easier to do an operation repetitively if set-up has to be done only once rather than once for each item), aggregation (it is easier to find all of J. Smith's personnel records if they are in the same place at the same time), or security. Chapter 1 showed a number of morphs that create, peruse, and utilize a batch. Because data stores are composed of a number of data items, they are usually data "structures."

It may seem unusual to consider a process to be a "data item," but it is handy to define processes within the data dictionary, too, in the following terms:

1. Name, aliases, ID: the name and aliases of the process and a coded ID number of name. For example, the process "Produce annual report" might also be called "Make shareholders report" and be assigned the unique ID number "8.3."
2. Inputs and Outputs: data flows that enter and leave the process
3. A description of the process's goals in English
4. Process logic: the logic of the process expressed in flowchart form (or equivalent pseudocode or decision tree)
5. Supra-system: Those processes of which this process is a subsystem. "Produce annual report," process 8.3, is a subsystem of "Inform shareholders," process 8.

The process entry in the data dictionary is a HIPO chart. External entities are normally defined in a narrative style. This is logical since external entities are outside the study boundary. Information we have about them is not systematically collected and analyzed. No formal notation is needed to define external entities such as "Management." The simple phrase "The departmental manager" will probably suffice.

Data Element Definitions

A data dictionary may be computer readable, hard copy, or both. The advantage of a computer-readable data dictionary is that definitions can be turned into software during implementation. Most database managers routinely construct simple data dictionaries for each file created. These can be generated automatically from logical designs expressed as **data element definitions**. Systems such as PSL/PSA work from these kinds of definitions to assist automatic software generation.

Figures 13-31 through 13-33 illustrate paper forms for creating data dictionary entries. These are easily produced on word-processing systems for on-line entry, checking, and filing. Computerized systems can check for inconsistencies, undefined data elements, and incorrect usage of the conventions of definition. Figure 13-30 on the opposite page, illustrates a simple microcomputer form for entering a data element definition using Excelerator, which has complex dictionary facilities linked to DFDs and structure charts.

These definitions are then accumulated into a data dictionary and referred to during logical and physical design. Since they may be changed during design, it is helpful to cross-reference the dictionary so that a change in one entry can be checked against others that refer to it. For example, the entry "Employee-record" contains information on current position and each promotion. Obviously, the last promotion will describe the current position; there is little logical reason to save the redundant information on current position. Eliminating "Current-position" means finding all entries that refer to this as part of their definitions and changing the entry to refer to the last promotion position.

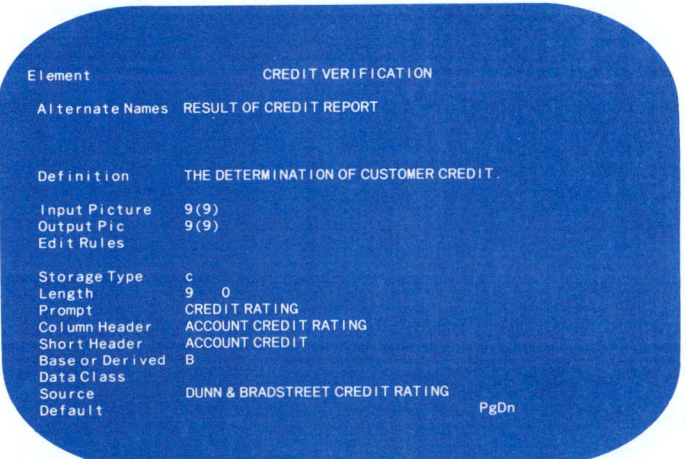

FIGURE 13–30. A Screen for Entering Data Dictionary Information (Excelerator)

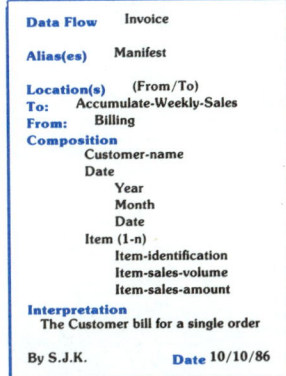

FIGURE 13.31. Data Dictionary Entry for a Data Flow

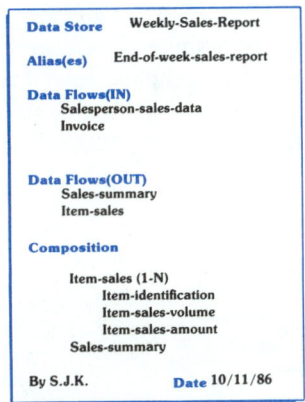

FIGURE 13.32. Data Dictionary Entry for a Data Store

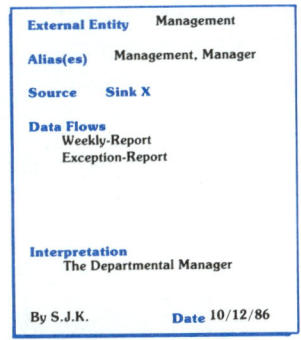

FIGURE 13.33. Data Dictionary Entry for an External Entity

With regard to naming data entries, names should reflect content. "Employee-date-hired" is more descriptive and easier to understand than "E-d-h"; there is nothing to be gained by using abbreviations or acronyms. Names must also be unique. "Employee-record" must refer to precisely the same data everywhere it is used in a DFD; there can be only one "Employee-record" entry in the Data Dictionary. Avoid terminology such as "Employee-record-1," "Employee-record-2," and so forth. The suffixes are not informative. Instead, use "Basic-employee-record," "Edited-employee-record," and "Purged-employee-record." While the terms "basic," "edited," and "purged" are somewhat vague, they at least indicate different stages of processing or refinement.

There can be several names for the same data element or structure. These are called **aliases,** and they arise because different work groups refer to the same data with different names. Perhaps they use the data differently or in different stages. The terms "manifest" and "waybill" are used interchangeably in business, and customers may refer to their "statement" as a "bill." By using aliases, system designers can converse easily with all clients and users.

Sometimes data entries are logically, but not physically, related to one another in important ways. Consider a typical end-of-week sales report. The entries for "salesperson-sales-data" and "aggregated-sales-data" are related, even though none of their constituent parts have the same names. "Aggregated-sales-data" is derived by adding "salesperson-sales-data." It is important to note this in the definition of both, since decisions made about changing formats, ranges, or access to one will definitely affect formats, ranges, and access for the other. If salesperson-sales-data is accurate to no decimal places (that is, dollar values only), it makes no sense to compute aggregated-sales-data with dollars and cents.

Data Structure Definitions

Entries in the data dictionary can be either simple data items (single values) or complex structures of many values. Figure 13-34 illustrates the composition of data entries. The smallest unit is the data **element,** which has, at any moment in time, a single value. Data **structures** are composed of a number of data elements. These structures may be simple (a time-sheet entry may consist of nothing more than the employee ID and the number of hours worked this week) or quite complex (a typical employee-history file). Data stores and flows are typically structures, but data stores are usually more complex (especially since the stores accumulate sets or series of data flows in batches).

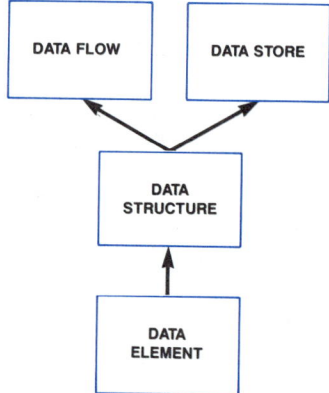

FIGURE 13–34. Composition of Data Entries

Data entries for structures must therefore express this more complex idea. To do this, several formalisms or systems of notation have been developed to assist designers, based on the theory of structured programming. The reasoning goes like this:

If data exist, they must have been created by programs. These programs must have produced them consistent with the logic of structured programming. Furthermore, the use of the data can only be consistent with that logic. Therefore, the same three structures must exist: sequence, alternation, and iteration.

Thus, most notation for data structures begins to look a lot like programming. This is convenient for entering data definitions that later may be read by a computer to create software automatically. A recommended notation is this:

1. **Sequence**: substructures are in sequence when one appears above the other
2. **Alternation**: each instance of the structure consists of one of the substructures listed in sequence, set off by braces ({...}).
3. **Iteration**: there may be zero to *n* repetitions of a substructure if the substructure is followed by a number or a data element in parentheses (that is, ...(*n*)).
4. **Option**: an abbreviation for zero to one repetitions (that is, present or not) is to place the substructure name within brackets ([...]).

The following definition illustrates the examples of a full employee time sheet as well as the daily entry:

Employee-Time-Sheet ::   Employee-Name
                         Week-Ending
                         Employee-ID
                         Salary-Class
                         Daily-Entry (1-7)
                         [Supervisor-Approval
                         Manager-Approval]
                         {Holiday-Days}

Daily-Entry ::   Regular-Hours
                         Overtime-Hours
                         [Vacation-Hours]
                         ([Case-charge
                         Overhead charge]
                         Charge-hours)(1-5)
                         {Explanatory-Note}

Table 13-1. Sample Data Structure Definitions

The employee time sheet contains identifying information, one entry for each day of the week, approval by either the employee's supervisor or manager, and an optional tally of the number of days of statutory holidays that week. The daily entries consist of the number of regular hours worked, an entry of overtime hours, an optional entry of vacation-hours, and either a case number or an overhead charge number against which to assess the accumulated labor costs. There may be up to five of each. If there are more than five, then an explanatory note will be attached.

To complete this part of the data dictionary, entries will have to be created for each term that appears in the definitions above. These entries are all elements, which may then be described in terms of values, ranges, coding, and format.

The data dictionary serves as a central reference document to define the labels that appear in data flow diagrams and structure charts. Entries include all such labels: data flows, data stores, processes, and external entities. Different data items require unique data entries. Aliases are permitted for discussion purposes. Data structures are defined in terms of simpler substructures, which are eventually defined in terms of their data characteristics. Processes are usually defined using a HIPO-like entry. External entities are described in narrative fashion. The use of an automated system to create and store data dictionary entries can facilitate later steps in system design.

Entity-Relationship Diagrams

Data dictionaries are constructed after discovery of data elements. The data "system" consists of these elements plus relationships established between them. For example, consider the relationships among employees, their departments, and their job descriptions. We can say that

1. Each department employs several employees.
2. Each employee follows one job description.

In addition, we might note that each employee is employed in one job class, each department is headed by one manager, and each manager supervises a number of employees.

These relationships determine relationships among the data elements (department-data, employee-data, job-data) which can be represented graphically in an **Entity-relationship diagram** (ERD). These diagrams become the basis for determining data structures and processing relationships among data that become important during physical design.

Figure 13-35 on the opposite page, illustrates the ERD for the fragment discussed above. Entities appear as boxes, and relationships, as circles on lines. The lines that appear to cut through entities indicate the source, actor, or agent. We indicate that a supervisor supervises an employee by exiting the line labeled "supervises" through the center of the supervisor entity and by terminating the "supervises" relationship on the employee entity box.

13.6 PUTTING IT ALL TOGETHER

The variety of charts and diagrams discussed in this chapter may seem bewildering, but most of them fit neatly together into a model documentation package well suited to logical design. This section discusses a logical design package featuring DFDs, structure charts, data dictionaries, and HIPO charts. We will look at detailed descriptions of processes that fit "into" the HIPO chart. This section also examines techniques for assessing the value and correctness of system models and some pointers for the drafting of these models.

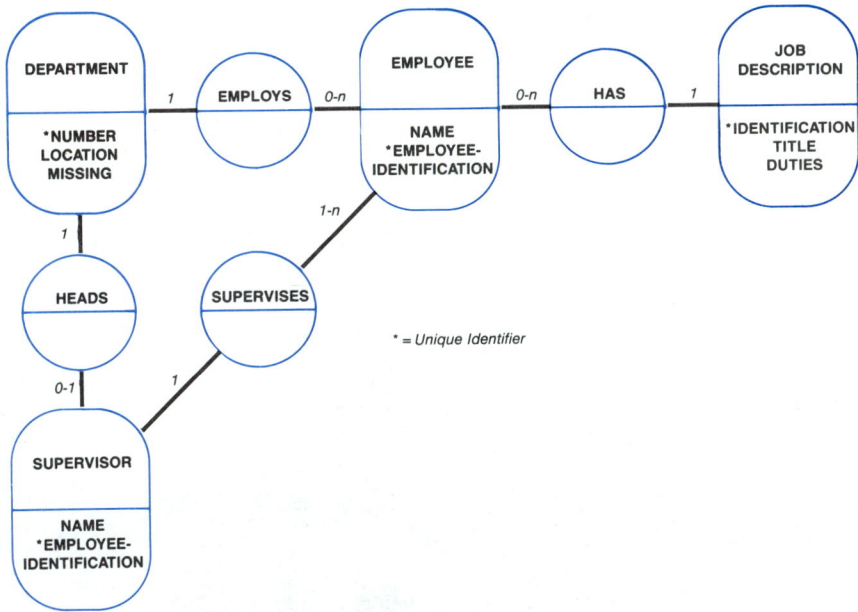

FIGURE 13–35. Entity-Relationship Diagram

The Logical Model Package

The **logical model package** is one of the major deliverables of the logical design phase, consisting of structure charts, a data dictionary, a HIPO diagram set, and a DFD set. Each focuses attention on a different aspect of logical design:

1. Structure chart: logical structure of the subsystems
2. DFD: process-entity, data-relationship structure of each subsystem
3. HIPO: details on the relationship of each process to the DFD in which it resides
4. Data dictionary: definitions of all terms mentioned on the DFDs

The ERD in Figure 13-36 on the following page, illustrates how these fit together. The structure chart shows the subsystem structure at each level and is derived from a top-down goal analysis (utilizing structural-functional design as outlined in Chapter 4). The functions themselves are described in the HIPO chart, which indicates inputs and outputs for each process and has a spot for detailed process logic. The data dictionary defines every term appearing in each subsystem description, while the DFD expresses the elements and relationships of the subsystems. Processes are the elements or entities and data flows are the relationships — remember, logical design is based on *data* rather than *physical* relationships.

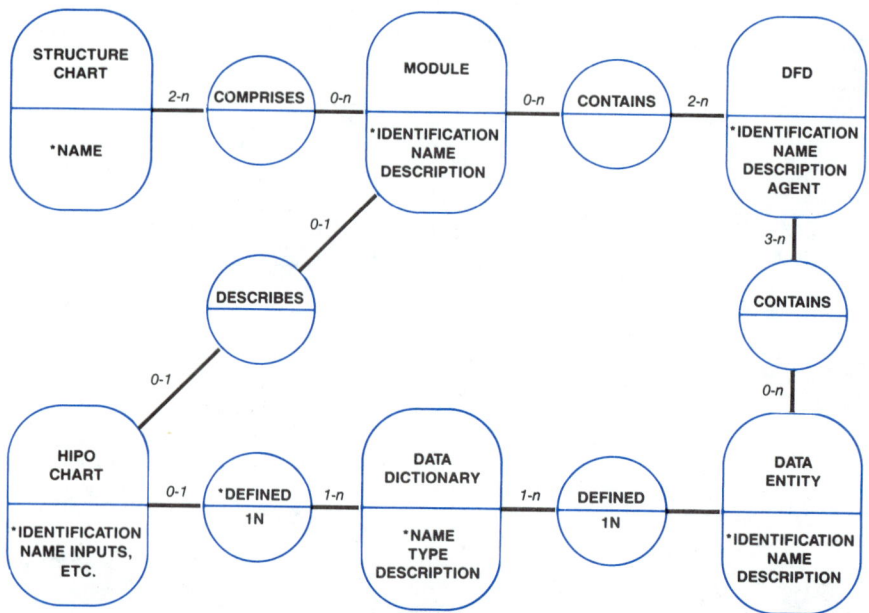

FIGURE 13–36. The Logical Design Package (ERD)

The order suggested has the advantage of a top-down orientation. Subsystems referred to in structure charts are "defined" in the DFDs, processes referred to in DFDs are described in HIPO charts, and all data elements mentioned in HIPO charts are defined in the data dictionary.

Process Logic Package

The DFD processes detailed in HIPO charts are described further in the process logic package. Process logic is most commonly expressed in pseudocode, although other formats are appropriate for specific kinds of processes. For instance, if a process is a single complex decision, then decision tables or trees may be appropriate. Nassi-Schneiderman charts and flowcharts depict complex processes graphically. Warnier-Orr diagrams may depict both logical structure of the system and process; many consider it an effective all-in-one logical design tool. This accolade is well deserved since, unlike pseudocode, physical design can proceed directly from this diagram. However, Warnier-Orr diagrams become unwieldy for large systems. Multi-page diagrams can become as cumbersome as multi-page listings of pseudocode.

The **process logic package** consists of:

1. process logic for each logical subsystem
2. a system flowchart (if available) of the existing system
3. flowcharts for physical systems that must be retained

Often, the process logic is embedded within the HIPO diagram, or it may appear in the data dictionary as a sort of definition of the process. Current usage favors placing process logic in the HIPO as pseudocode unless it is quite complex, in which case separate diagrams can be appended.

Remember, however, that "process" logic does not mean programming. Except at the lowest (work module) level of the structure chart, logic means coordination of subsystems. Processing of information (inputting, outputting, computation, transformation) is the task of work modules; the physical design of the process can be left for later. For example, it would suffice to write

Compute SALES-COMMISSION as RATE * SALES-VOLUME

in the process logic rather than the more complex and physical

IF SALES-VOLUME > 5000
      Then SALES-COMMISSION = SALES-VOLUME * 0.10
      Else SALES-COMMISSION = SALES-VOLUME * 0.075

since the factors 5000, 0.10, and 0.075 and the distinction between the two classes may be a factor in the physical design. That is, physical designers may decide to make these *parameters* for subroutines or data to be read in so that they can be altered without rebuilding the module.

The process of **parameterization** is really one of physical design, since it concerns *how* data are to be accessed as opposed to *what* data relate which processes. The rule of thumb in specifying process logic is always to defer "physical" specifications to later design. Hence, when the question of the actual commission limits and rates comes up, it is clearly a question of how these values are to be determined during processing, obviously a physical design consideration.

Assessing the Value and Correctness of System Models

Several times in this text (Chapters 1, 6, and 9), we have looked at the use of models to test design hypotheses. In addition, we have stressed that each model has a set of rules dictating how the model is actually to be drawn.

There are a few systematic procedures for testing the correctness of system models, especially for consistency. Each diagram should be thoroughly checked for syntax first. Next, each package should be checked in top-down fashion. Is every module in the structure chart diagrammed in its own DFD? Is every DFD detailed in its own HIPO chart? Is every input to each process and each output from each process mentioned in the HIPO chart? Are all data flows, stores, processes and external entities mentioned in the data dictionary?

Next, the three laws of information flow should be checked. Working backward from each external entity, can each data flow exiting a process be determined uniquely by some combination of data entering that process? If not, something is missing — perhaps the process is "intelligent"; that is, perhaps the process is providing information (in which case an additional external entity needs to be created; see Figure 13-37 on the following page).

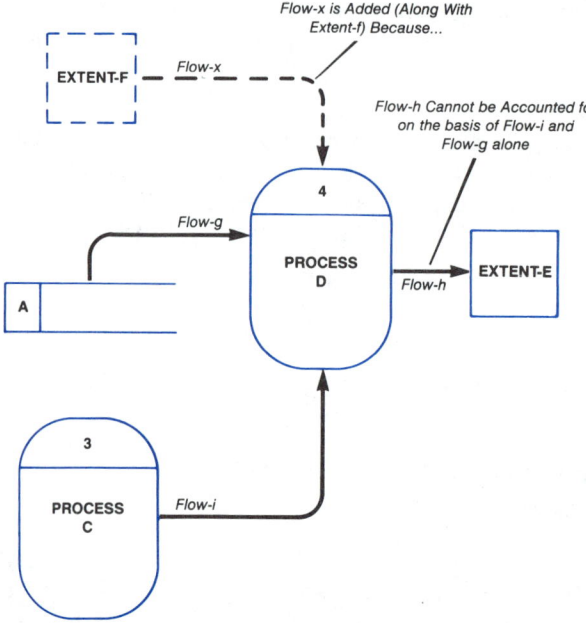

FIGURE 13–37. Applying Logic to Assess Logical Models

Next, working forward from each external source, is each input actually used in determining each process's outputs? If not, information collected or generated is being discarded and data flows can be eliminated. In fact, entire sequences of processes can be removed (see Figure 13-38).

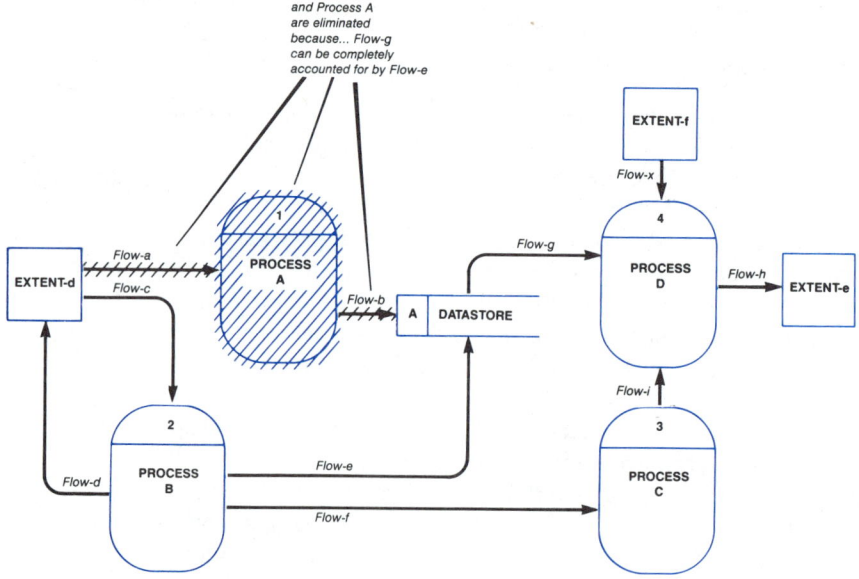

FIGURE 13–38. Continuing to Assess a Logical Model

Our goal in this analysis is to state without fear that each output B to a sink is a function only of inputs from sources A1, A2, ... ,A*n*. This function, of course, is the value added provided by the processes when actually implemented in the structure indicated.

Process logic can be checked in the same top-down fashion, although detailed logic may be difficult to follow (in fact, this provides the same problems as desk-checking programming).

Drawing Models

As Chapters 6 and 9 pointed out, conceptualizing and drawing models are two different things. Most models rapidly grow large and unwieldy and tend to generate paper. There are some packages available for generation of DFDs and data dictionary entries, especially on microcomputers. Other diagrams, such as Nassi-Schneiderman charts, require some planning and a pencil instead of a pen. HIPO charts can be put into forms easily, as can data dictionaries. Structure charts also become unwieldy after three or four levels. A rule of thumb for management is well applied to structure charts: keep the span of control to between three and nine. In fact, the basic information-processing model of input-process-output usually dictates three modules for many subsystems, except those that process a variety of transactions.

13.7 CONCLUSION

Chapter 6 introduced the concept of system modeling as central to the labors of the systems analyst. Chapter 9 followed this by detailing a variety of charts, showing their relationships and uses in specific systems analysis tasks. This chapter looked at the specific charts that systems analysts employ and their uses.

There are five classes of charts in common use. **Process charts** concentrate on the information processes themselves, modeling the transformations that data undergo on their way to becoming information. Process charts include **system** and **program flow-charts**, **decision tables**, **decision trees**, and **Nassi-Schneiderman charts**. Each of these shows the mathematical and logical transformations individual data items undergo and the logic underlying those transformations.

Logical relationships are demonstrated through the use of diagrams that show logical dependencies among processes. **Data flow diagrams** and **document flow diagrams** focus on the data involved in systems rather than control relationships, which are the focus of process diagrams. Since data relationships form the "logic" of the structure of an information system, logical diagrams are the cornerstone of information resource systems analysis.

Functional relationship diagrams show how subsystems depend on one another. **Structure charts** and **HIPO charts** (the latter involves pseudocode to detail process logic) illustrate how subsystems share data and pass control among themselves. The HIPO chart is the logical/functional counterpart to the system flowchart.

Materials flow diagrams include **work distribution charts** and **office layout charts** and show production relationships of a physical nature. While information is an abstract quality, materials often accompany, precede, or follow information; tracking materials as well as office layout is an important aspect of the analyst's craft.

The last class of charts includes a variety of **data models**, **data dictionary forms**, and **data structures**.

13.8 CASE: SCRIMMAGE SYSTEMS' SPORTS ANALYSIS PACKAGE

Sandy Goforth operates Scrimmage Systems, a consulting firm specializing in sports. Sandy employs three people: Mattie, his secretary; Arnold "Tiger" Tremanian, sports consultant; and Richard Hatfield, a researcher. Together the three seek clients such as college football teams, professional hockey clubs, and community sports groups. Scrimmage Systems advises on both the business and activity ends of sports. Recently, Sandy helped a AAA baseball team modernize its bookkeeping and provided the analysis work on an instruction package for community soccer coaches for the Bigtown Civic Soccer Association.

Sandy's latest venture is a play analysis system to support coaching for the local professional football team. The particular assignment is to design a system for storing and retrieving all information on each play run by the Bigtown Bisons during the season. Using this system, the Bisons' coaches can get a better understanding of what works under what circumstances. The more complete the information, the better the chances that the Bisons will be able to plan for each game the following season. In addition, the Bisons want the play analysis system blended with some sort of scoring system for players' performance — this needs to be put together with a system for "building teams" at each moment of the game so that the best eleven players can be on the field at any moment.

Questions

1. Which charts or diagrams would be appropriate for a first look at the system from a coach's viewpoint? What would be best to use to explain to coaches how the system might work for them?
2. Draw a structure chart that would show the functional relationship among the various modules mentioned (there are three).
3. There is currently no document flow because none of the required data are collected systematically. Sandy has discovered that coaching assistants are available to examine videotapes of the games and that it is possible to record, at field level, the play patterns called by the Bisons. Following the games, players' contribution to the plays can be analyzed and a report created from this analysis. Finally, player characteristics can be made available from preseason assessments and post-game advice from coaches, trainers, physicians, and the manager-owner. Draw (1) a DFD to show how this information can flow together; (2) a HIPO chart of each process, and (3) a Warnier-Orr Diagram of the system.

4. Data flows are important. Watch a football game and develop your own terminology to use in creating a series of data models. In particular, create (1) an entity-relationship chart and (2) data dictionary entries for each data flow and store in the system.

DISCUSSION QUESTIONS

1. This chapter distinguished control flow and data flow. Control flows from activity to activity, while data flows between processes. How are control and data flows *related*? How is the concept of flow of control related to the concept of logical structure? Would it be fair to say that a study of control is part of physical design while a study of data is part of logical design?

2. Suppose we were able to introduce the following notation into a decision table:
   A process indicated by a * prefix (*TABLE2) is another decision table whose results are to be computed.
   Using this notation, create a set of decision tables to model the following complex decision:
   If this is an order for a part we have in stock, then attempt to fill the order. Fill out a picking slip and send it to the warehouse. If the item is actually on the shelf, then package it, fill out the packing slip, and send the two together with the picking slip to the mailroom. If the item is missing from its shelf, fill out a back order request, staple it to the picking slip, and send the two together to the order desk. If the item showed up as not in stock, and if it's an "A" prefix item, fill out a back order request and send it to the order desk. If it's a "B" prefix item, fill out an out-of-stock form and send that to the order desk. If it's a "C" prefix, then phone the special order desk. If they've got some in stock there, fill out a picking slip and send that to the warehouse. If they don't know whether there are any in stock, send the order form back to the order desk with a note that there's no chance to fill the order. If there are special order "C" items in stock, find out when they can be shipped. If they will arrive within a week, put the order form back in the ORDERS binder with the expected delivery date recorded. If it will take more than a week, put the order form in the WAITING SPECIAL ORDER binder with a bring-forward date recorded for one week.

3. Model the previous ordering process as (1) a decision tree, (2) a Nassi-Schneiderman chart, and (3) as a Warnier-Orr diagram. Then create a program flowchart and a system flowchart. In what ways can you mechanically translate among these without much thought? How does the system flowchart differ? What is (are) the major mechanical difficulty(ies) in drawing these? Suppose you had to make a major change to the logic of these decisions. Which diagram(s) would be most difficult to change and why?

4. Create a DFD from the description of the process (system) described in question 2. What is "logical" about this diagram as opposed to the system flowchart? What sorts of physical attributes of the processes involved are "transparent" or irrelevant in this diagram?

5. The text says "Turbulence is inherited upward through the structure chart." But we later say that there is the possibility for "internal turbulence in the form of errors." Discuss (1) how turbulence is inherited or passed downward through the structure chart, what the symptoms might be, and what the implications are for logical design and (2) how the structure chart can be used to isolate and "debug" that turbulence.

6. By adding symbols from structured programming (diamonds indicating alternation, curved lines indicating repetition), we introduce aspects of physical design into our logical design and thereby limit the implementation independence of the structure chart. Discuss the value of this augmented structure chart in going from logical to physical design.

7. The following conventions have been introduced to further relate logical and physical design:
   - Arrows pointing down a structure chart indicate information passed to control a module at a lower level.
   - Arrows pointing up in a structure chart indicate information passed in response to requests from control modules.
   - Arrows with open tails ($O \rightarrow$) indicate data elements which are to be or have been processed (for instance, CUSTOMER-RECORD may be passed down to be read and then passed back when it has been processed.
   - Arrows with filled tails ($\bullet \rightarrow$) indicate control information (sometimes called "flags") that is created to control processing without actually entering into the process. For instance, a flag such as READ-STATUS might be sent upward to indicate whether a record was read correctly or not.

   Using these conventions, draw a structure chart to show how a student might "process" a series of tests and papers during the semester given that the student will decide to drop the course if she fails any test or gets a grade of less than "C" on any paper.

8. Draw a Warnier-Orr diagram for the process described in question 7.

9. Review the reasoning behind the notation used to express data structures. Do you see an alternative? What kind of "programming" would that alternative create? Can we, in fact, have programs that operate that way?

10. A mental health record consists of a patient name, address, and telephone information as well as records of visits and a discharge status. The name is considered to be a honorific (Miss, Ms., Mr., Dr.), a last name, a first name, and a middle initial. There are two telephone numbers possible (home and work) and these include area code. An optional entry for next of kin includes name and phone number as well as relationship. Each visit is described by a date, the results of up to five tests on a scale of 0 to 99, a subjective evaluation by the therapist as free text, and the therapist's initials. The discharge status is 0, 1, or 2. If it is a 2, an explanation should be provided. Create data dictionary entries completely defining the data in the mental health record. What facts do you need to know to complete this exercise?

11. Create an entity-relationship diagram for the above description. As a hint, think of all the relationships that are possible between therapists and patients and their friends and relatives.

12. Elaborate on the advice to leave parameters in your definitions for physical designers to supply later on. Why is this good advice even if the parameters are all known during logical design?

DESIGN EXERCISES
Using dBASE III

13.1 dBASE III contains a number of **computational aids** to assist in preparing reports. They are called **functions**. You have used two functions already: RECNO(), which is the record number of the current record and EOF(), which is .T. (true) if you are at the end of the file and .F. (false) otherwise. Below are a number of useful functions. Use the HELP command to examine the complete list.

BOF()	.T. if you are at the first record
DELETED()	.T. if the record is marked for deletion
INT(n)	The integer part of n
LEN(x)	The number of characters in x
LOWER(x)	Convert x to lower case
UPPER(x)	Convert x to upper case
ROUND(n)	Round the numerical expression
SPACE(n)	n spaces (blanks)
SUBSTR(x,a,b)	The b characters of x beginning at the ath character (see below)
STR(n,a,b)	Convert n to an a-character string with b decimal places
VAL(x)	The numerical value of the string x
TRIM(x)	Remove the right-most blanks from x

As an example, assume that AMOUNT = 45.67, BILLNO = " 154 " and PERSON = "Arthur Harris, Sr." The following would result:

Example	Result
LEN(BILLNO)	6
LOWER(PERSON)	arthur harris, sr.
UPPER(PERSON)	ARTHUR HARRIS, SR.
ROUND(AMOUNT)	46
INT(AMOUNT)	45
BILLNO + SPACE(3) + "//"	154 //
SUBSTR(PERSON,1,5)	Arthu
SUBSTR(PERSON,10,7)	rris, S
SUBSTR(PERSON,10)	rris, Sr.
SUBSTR(PERSON,10,20)	rris, Sr.
STR(AMOUNT,6)	45.67
STR(AMOUNT,6,1)	45.6
VAL(BILLNO)	154
VAL(PERSON)	0
"/" + TRIM(BILLNO) + "/"	/ 154/

You may have noticed the symbol & (ampersand) appearing in several programs. dBASE III lets you write commands in which variables stand for things other than data. For example, enter these commands:

USE STYLISTS	Examine the stylists' file
ACCEPT "DAY?" TO DAY	Ask yourself to provide a day
DAY = UPPER(DAY)	Make it upper case
? &DAY	What appears?

What you see is the code for the first stylist for the day of the week that you entered. Try this:

ACCEPT "FILE?" TO FINAME	Ask yourself for a file name...enter a valid file name in the HAIR system; then type
USE &FINAME	And type...
DISPLAY STRUCTURE	What happens?

In dBASE III, whenever you prefix a variable name with the & sign, you are asking to use the value of that **variable** as a *name*. In dBASE III, this is called a *macro*. Macros allow you to create menus that refer to fields without having to use the DO CASE statement to display the fields or to refer to files without DO CASE. What does this produce?

```
C = "B"
B = "A"
ACCEPT "FILE? " TO A
USE &&&C
```

Using macros and functions, write a program called SALESRPT to prepare a report on the value of services or sales broken down by stylist or customer. Use only rounded dollar figures and accumulate the totals. [*Hint*: ask for the value field ("ITEMCODE" or "ACTIVITY") and the breakdown field ("SYTLNAME" or "NAME"; remember that NAME is not enough of a key) and use these data items as the name of fields via a macro].

13.2 Look through the files and programs in HAIR and prepare the following charts:

1. **Structure chart**
2. **Entity-relationship diagram**
3. **Data flow diagrams**
4. **HIPO Charts**

Then, using the information from your role-plays and the interview, prepare amendments to these, including new HIPO diagrams for improved or new modules.

CHAPTER 14

COSTING AND DECISION MAKING IN SYSTEMS ANALYSIS

OBJECTIVES

1. To outline the costs, benefits, markets, and market players of information as a commodity
2. To perform a cost-benefit analysis of a proposed information resource implementation plan by applying techniques of economic analysis, including net present value, and social accounting
3. To perform a value-added analysis of a proposed information resource implementation plan, showing where value is added to data at each process
4. To compute payoffs for a variety of alternatives in information resource implementation through the use of payoff tables and to make projections of payoffs for periods into the future
5. To describe the decision-making points in an information resource maintenance effort, indicating the value of decisions at those point

COSTING AND DECISION MAKING IN SYSTEMS ANALYSIS

14

CHAPTER

14.0 INTRODUCTION

Is It Really Worth Doing?

"I don't know, Leo, it just seems too expensive."

"Tess, you can't reject something just because it has a high price tag."

"Leo, this is the most expensive project we've taken on since I became director of information resources. You're asking me to spend hundreds of thousands on new equipment, software, and services, and I just can't get a feel for what we're going to get out of it."

"I don't know where you're getting your estimates from, Tess. The hardware will be $250,000.00, sure. But the software we don't know about, and what services are there?"

"Well, for one, there have to be maintenance contracts. And for another, I know they charge for their software. It's not bundled into the price of the hardware. And one more thought before you interrupt me, we'll need a great deal of consulting on the database package — as you're no doubt aware, only three days of training is included with the database package."

"OK, Tess. You know, we don't have to go with A-A-Ace Computers. There *are* other vendors."

"Right, Leo, but the problem I've got is this: even if we knew precisely what it would cost, we really don't know what we're getting. I want to know, is it *really* worth doing?"

When the Systems Analyst Has to Make a Decision

Leo and Tess are having a useless argument. No one knows the actual costs or the true benefits of the project they are considering. Without a good estimate, their arguments will fail to convince and they will never reach a reasonable, informed decision.

During the logical design stage of information resource maintenance, analysts concentrate on logical relationships among functions, not on the dollar costs and benefits of the physical implementation. Nonetheless, there are good reasons for analysts to be aware of — and able to justify — the costs and benefits of the development efforts.

First of all, IRM itself is a learning system (Figure 14-1) that requires information. Information from workers, clients, vendors, suppliers, and IRM managers contributes to reaching decisions at crucial points:

- **Work planning**: overall cost and benefit estimates are needed to justify commencing the effort.
- **Initiation**: to begin, initial costs must be known,
- **Preliminary investigation**: estimates of the costs of the detailed investigation are needed.
- **Logical design specification**: the costs to completion and an estimate of cost savings from the project are needed to make the decision to do physical design work.
- **Termination**: especially if the project is prematurely canceled.

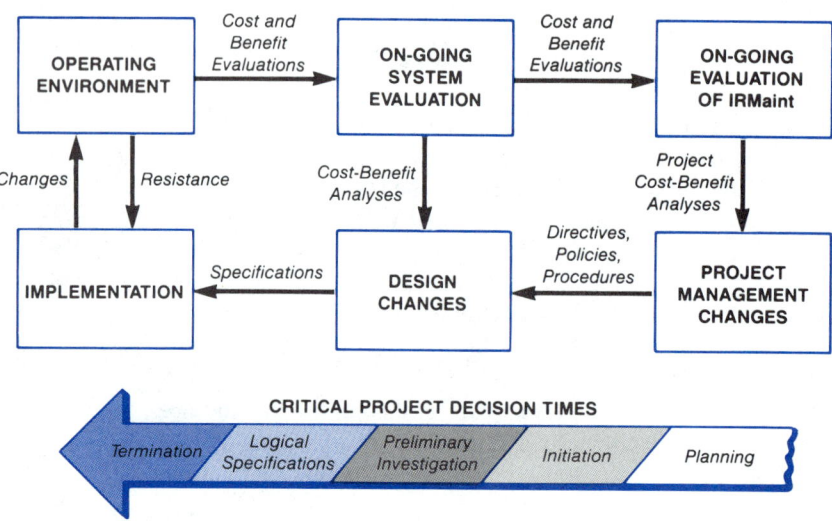

FIGURE 14–1. IRManagement as a Learning System over IRMaint Projects

Economic Analysis Principles

This chapter discusses cost-benefit based decisions in the context of systems analysis. It also stresses decision making based on dollar-related considerations.

This may seem to imply that an analyst's decision making is somehow restricted to situations in which money counts. In fact, most of an analyst's decision making is based on logical considerations (what goes with what, what depends upon what) and some considerations of human factors (what people can do) and ethics (what ought to be done). Cost-benefit-based decision making is a subset of analytic responsibilities.

It would be easy to understate the importance *others* place on dollars. In the past, decision making at critical project points usually was at least partly based, and often with little reliable data, on the dollar costs and benefits of continuing. It was the impossibility of making valid decisions so early in design that provided much of the rationale for the structured approach to systems design. Today, physical design is delayed as long as possible. Once the complete repertoire of functions has been discovered, an accurate estimate of costs and benefits be determined.

In Chapter 14, we will examine cost-benefit decision making and categorizing costs and benefits. One of the thorniest issues here is putting a value on intangibles, since many of these costs and benefits do not show up in balance sheets. Then we will focus on information as a commodity with a market value for the firm. The third section of the chapter puts these ideas together and examines four types of economic analysis: payback, net present value, cash flow, and social accounting.

This chapter sees the IRU as a **profit center** that adds value to the firm's products. Typically, the firm views the IRU as a way of trimming labor-related costs. Recent thought has shifted to the IRU as a "competitive weapon" in the marketplace.

Last, we will look at the tools an analyst has for decision making. Techniques include payoff tables, straight-line and cyclic projections, dynamic and logical modeling, consultation, and Delphi.

14.1   COSTS AND BENEFITS

This section develops a method of cost and benefit classification as groundwork for preparing detailed cost-benefit analyses, a major aspect of which is placing a value on intangibles. Most benefits from IRMaint efforts lie in this realm and many costs are difficult to price.

Out-of-Pocket Costs

It is easy to notice **out-of-pocket costs**, which fall into the following classes: labor, machine rental or amortized costs, consumable supplies, software, services, and the indirect costs of staff time for interviews, meetings, management overhead, and secretarial labor (Figure 14-2).

LABOR
Salaries of Staff
Benefits for Staff
Contracted Labor

CONSUMABLE RESOURCES
Paper, Ink, Forms
Tapes, Disks, Diskettes

SERVICES
Telecommunication
Consultants
Maintenance Contracts

MACHINE RENTAL
Computer Time Rented
Apportioned Charge-Back
Leasing Costs

SOFTWARE
Purchase
Shared or Unshared
   Development Costs
Application/System/Development
   Tools
Manuals, Books

INDIRECT COSTS
Staff Interview and Meeting
   Time
Management Overhead
Secretarial Overhead
Training of Programmers and
   Analysts
Training of Users
Publicity

FIGURE 14–2. Out-of-Pocket Costs for IRMaint Efforts

Firms commonly develop **charge-back policies** for pricing each aspect of these costs. For instance, the costs of purchasing and operating a computer system may be charged back to users on a per-microsecond basis. Many sophisticated schemes charge for the use of every system resource: each line printed, each disk access, each tape read, and so on. The development of a fair and reliable charge-back system is complex.

Finally, with each project, there may be a need to retrain technical staff. If a new machine uses a different COBOL, programmers have to be trained. A new system development methodology may entail sending analysts to a course halfway across the continent.

Intangible Costs

Consider a system built to speed up delivery of products from a warehouse (Figure 14-3). This system has specific operational costs in terms of hardware and software. But

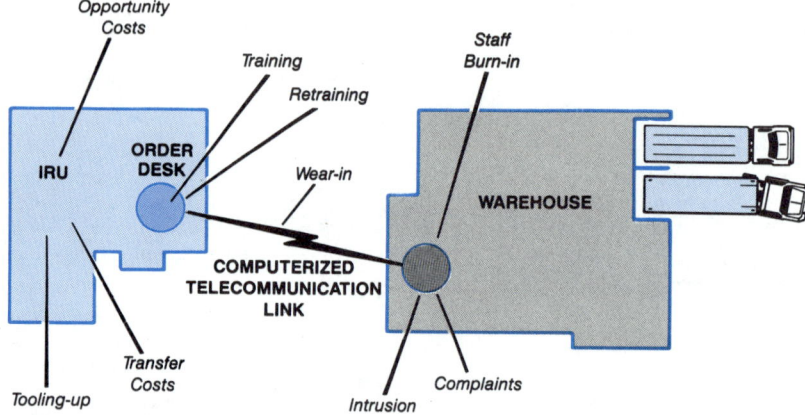

FIGURE 14–3. Some "Hidden" Costs for IRMaint

what about the **intangible costs**?

Most estimates ignore the cost of operation by the warehouse staff, since that is part of their job anyway. Instead, labor costs will include only system maintenance by programmers and analysts. In fact, there are several other costs to be considered:

1. **Training**: the cost of trainers' time *plus* the time away from the job spent by the warehouse staff learning how to use the system
2. **Staff burn-in**: the period of time of less-than-optimal use after learning reduces productivity and increases costs
3. **Complaints**: during early operation, many complaints have to be handled which appear as costs both for users and IRU staff
4. **Intrusion**: many well-running systems become show pieces, the object of tours by school children and head-office personnel
5. **Retraining**: new hires have to learn the system without benefit of the original designers and users. An informal "culture" creates less-than-optimal folklore, which affects productivity.

6. **Wear-in**: software does not wear out, it wears in. Eventually everyone tries to "get around" the system because new demands from a turbulent environment require new techniques from old tools, thus, productivity declines.

7. **Opportunity costs**: because the warehouse system was built, interest on money that might have been invested was lost.

8. **Transfer costs**: because the warehouse system was built, rewiring the warehouse has been delayed, office routines in the order-taking area have been disrupted, and the code produced on the customer filing system is inefficient because the best coders were working on the warehouse system.

9. **Tooling-up**: the firm incurs costs in acculturating the technical staff to the new computer and its operations, and some public relations efforts may have to be initiated.

These intangible costs should be accounted for in estimates of total project cost brought forth at the time of detailed investigation reporting.

Cost-Savings

Traditionally, the information resource was justified because it cut costs. Most of these costs were in staff reduction or moving work to cheaper categories of labor. With the aid of a computer, a staff of two can handle work that twelve could barely do before. Labor cost savings are easy to compute: ten salaries. In other cases, complex managerial work can be routinized and handed to junior professionals or clerks.

Another example is time savings within a job category. Suppose a spreadsheet program helps a manager reallocate staff in a more effective manner and does so more quickly than pencil-and-paper methods. There are two cost savings: the manager's time is saved to do other tasks and the more effective reallocation saves staff time.

Union contracts, personal and political considerations, and good manners, however, may prevent letting staff go or reassigning them to other jobs. The collective agreement may define job classifications so narrowly that computerized work may be seen as upgrading a position — and requiring a raise. Or work levels may be preset by the contract, disallowing layoffs. In many cases even if layoffs were allowed, it would be hard to lay off one-third of a position.

Personal considerations make it difficult to reassign work. A physician's office may purchase a microcomputer to assist with billing, but freeing part of a worker's day may not be a saving at all. A well-publicized study of office automation possibilities (Booz, Allen and Hamilton, 1980) pointed out that a good system of electronic mail could save executives two hours per day. But what are they expected to do during that time? Can a firm afford to lay off a quarter of its executives or employ a three-quarter-time vice president of marketing?

Finally, employees are not wind-up dolls. They have feelings, aspirations, and expectations of their place of employment. Tinkering with job descriptions and assignments may be interpreted as impersonal, bad manners, or aloofness. The IRU does not need to be labeled as a group that justifies its existence by having others laid off.

Nonetheless, until recently, cost reduction or elimination has been the major goal of IRMaint. Two trends have affected this goal: the value of information as a commodity and a view of the IRU as a value-adding production component. Where it may be difficult to justify a project based on cost reduction, it might be easier to find the value that the project will add to the firm's products.

While cost reduction or elimination may not be a feasible goal, cost **transfer** might be. In cost transfer, the costs of an operation are moved to a more controllable category. For example, in moving a task from an executive function to a clerical function, it is easier to supervise and account for the time of clerks than the time of executives. While one may not be able to demonstrate a true cost savings, one could at least argue that costs are more controllable in the clerical area. Moving costs from labor to hardware categories may not actually save money now, but at least the direct costs are more measurable, hence potentially easier to control. We will return to the cost-transfer aspect later.

Putting a Value on Intangible Costs and Benefits

By definition, "intangible" means unmeasurable. Yet many intangibles *are* measurable; analysts just have not had the motivation to measure them in the past.

Most easily valued are staff training and burn-in costs. Examination of past projects can provide an estimate of the learning curve and the costs extrapolated through charge-back for each labor category. If one estimates that it takes four hours of classroom training, eight hours of practice at less than half productivity, and an additional twelve hours of reduced productivity, the cost of training four warehousemen who are paid $20 per hour would be calculated as

4 hrs. × $20/hr.	$ 80	Training
8 hrs. × 0.75 efficiency loss × $20/hr.	$120	Practice
12 hrs. × 0.25 efficiency loss × $20/hr.	$ 60	Burn-in
	$260	Per worker

Thus, training and burn-in would cost $1,040.

Examination of complaints for similar systems may lead to an estimate of productivity losses over time, with the major losses appearing — but not limited to — the time immediately before major IRMaint activity. Some work on software reliability has led to curves such as that in Figure 14-4 (opposite), but the science is still in its infancy and accurate figures are hard to obtain.

Because benefits are *not* achieved on a project not begun, these can be considered opportunity costs, just as purchasing a car with cash negates the benefits one could accrue spending the money on something else.

Transfer costs need to be looked at more in the future. As IRMaint work becomes more complex and affects people at professional ranks, the project costs that "rub off" on other users, departments, customers, and workers have to be factored in. The IRU should be a good corporate member and not spread *its* costs to areas that cannot account or pay for them.

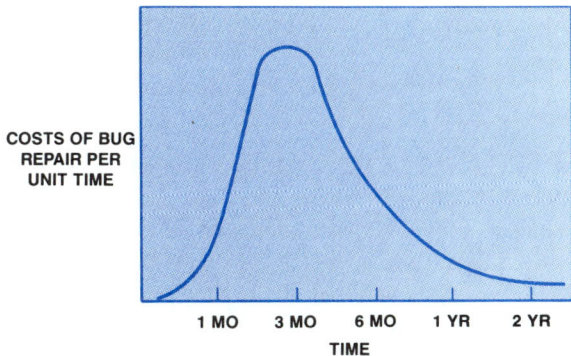

FIGURE 14–4. Typical Software Reliability Cost Curve

Tooling-up costs may be available historically, although a firm may prefer to adopt some rule-of-thumb figure. Many use the idea that any major system change will affect productivity for three to six months; job changes require this long to retool mental and physical processes at a professional level.

It is possible to underestimate intangible costs, often equal to out-of-pocket expenses, which may be shocking. On the other hand, carrying out the exercise may enlighten management on the *actual* cost of a project rather than the rose-tinted estimates that concentrate only on out-of-pocket costs.

14.2 INFORMATION AS A COMMODITY

Attention has been slowly turning from the IRU as a service organization within the firm to the IRU as a vendor of services and products. Porter and Millar's (1985) "How Information Gives You Competitive Advantage" discusses three ways in which the IRU can do more than serve the organization:

1. New technology makes it easier to perform new, valuable services; they cite Federal Express's Zapmail™ and Merrill Lynch's Cash Management Account™.
2. New technology motivates demands for new products. Foremost among these are data supply services, open access to packet-switched communication networks, videotext home shopping, and public electronic billboards.
3. Finally, new businesses can be grown within the shelter of existing ones. Excess computing capacity can be sold to others; detailed expertise can be consulted out. Existing customer data files can be sold, rented, or shared. Porter and Millar point out the examples of Sears, which sells credit-authorization and transaction processing services; A. O. Smith, which runs automated teller machines; Eastman Kodak, which sells out excess voice and data communication capability; and American Can, whose 9-million-customer, direct-mail retailing database was a sweetener in a recent merger.

Information is clearly a commodity that the firm ought to invest in and market for a profit. This section looks at the market for information and information services, pointing out the broker role for the IRU.

The Market for Information

The typical medium-size retail firm might employ 100 workers, bill about 15,000 customers monthly, and push through about $10,000,000 in sales annually. The firm may sell over 400 products and deal with dozens of suppliers. The amount of information on customers, products, and suppliers is enormous. While only a small portion of this information is compatible with automation, it is systematically filed, concise, and accurate. In other words, the information is packaged for sale.

Not only is ordinary business data useful for mailing lists, advertising, and market research, but vendors such as Dow Jones can sell up-to-the-second stock market information to those who feel that speed is important. Academics routinely purchase bibliographic library-search services (and the information involved). Others, such as Dun and Bradstreet, are getting into the software business to support their already lucrative business-reporting business.

Inside the firm, there are hints that the message of information as a commodity is being listened to, if not acted on. The sale of information as a commodity requires several things:

- data administration: control over the existing data, structure and cataloging, and knowledge of what is there
- policies on information sale: Privacy and security policies, limits on sales, ownership of information, and distribution of profits
- market research: knowledge of the marketplace
- legal work: either contracts for purchase or license for buyers to ensure non-redistribution, limited liability, and payment
- software: conversion (if manually distributed), telecommunication, updating, and report generation
- sales organization: salespeople, trainers, distributors, account managers, and product managers

The Market for Information Services: Petroserve

Petroserve is a petroleum products service station (a gas station) repair service organization. Petroserve employs twelve people and has been in business for over twenty-five years. They sell repair services on pumps, hoists, lubrication equipment, and tanks on an annual contract plus service-time fee basis. Most service calls begin with a phone call and end with a skilled technician's trip to the station to repair a piece of the equipment.

Bob Mowry is the finance manager at Petroserve. Initially, Bob purchased a small microcomputer to keep company sales records, process financial information, and do some spread-sheet-based planning. Later he became interested in one of the varieties of advanced decision-support systems that examines sales calls for patterns in enabling companies to cut down on unneeded trips. Working with Clay Cahoon, the owner of Petroserve, Bob became convinced that his fund of information was valuable for even more.

"Why can't we take our database and sell it back to the chain operators? They should be interested in repair patterns for their stations' pumps and hoists to take advantage of their mass purchasing power. If they see that a particular pump is giving a lot of problems in certain kinds of stations, they can purchase another kind of pump and avoid the repair costs and the downtime."

Bob worked for several months with a team of consultants to build his database of service calls in the Bigtown vicinity. Because there was little competition for Petroserve, Bob's database was unique, not only in Bigtown but also in the region. After building a large database, Bob produced a number of sample reports and channeled them to the local chain operators and a few of the head-office people he had met over the years.

They were impressed not only with the quality of the sample reports, but with the potential for cost savings on repairs that Bob's reports could provide. After a few discussions with these industry sources, Bob commissioned the consulting firm to design a data subscription service to be distributed on diskette. Bob's service is marketed mostly by word of mouth, but Lennie Horowitz, his sales manager, includes the data service as part of his initial visits and standard sales literature.

Because Bob charges his twelve customers $400 monthly for data services, the investment in consulting services has proven profitable. Bob is considering expanding the service to include some spreadsheet capabilities based on past performance data. Customers would then have more than just passive reporting; they could actually calculate cost advantages in terms of repairs on a variety of equipment. Bob gets current pricing on all pertinent gear and has been playing with the spreadsheets.

Because his business is small and the data are already part of management decision making, Bob did not have to build a sales organization. But he did look into legalities, software, and policies for data sales. Because the community of service station chain operators is also small, he made his work easier by using personal persuasion in making sales and reaching agreements. Bob's experience could be duplicated in a number of other service businesses in which similar data are important.

The Broker Role for the IRU

Figure 14-5, on the following page, illustrates the role of the IRU as **broker** for information products. Because the IRU processes but does not always own the marketable data, the IRU must broker the data. This means locating units willing to sell and customers who want to buy.

This means that the IRU has to have policies about information sales, especially because it, personally, does not use the data. For example, mailing lists may belong to the billing department. The IRU has no real "right" to sell this information. Other organizations might profit from having this list. Can the IRU approach these organizations? Probably not. Instead, the IRU can approach the billing department, let them know the value of the resource they have accumulated, and introduce them to the concept of sale of information. It will be others in the firm, such as sales managers, corporate executives, and planners, who would know the marketplace for the information.

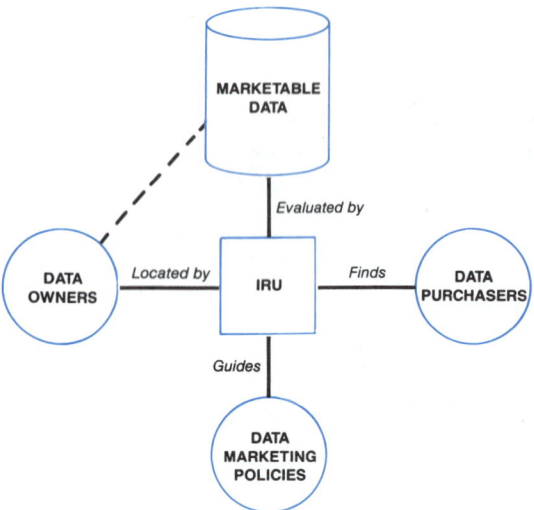

FIGURE 14–5. Data Brokering Role of the IRU

14.3 Cost-Benefit Analysis

The Bases for IRMaint Decisions

The decision to proceed with IRMaint activities is based on several considerations. Figure 14-6 illustrates the components of this decision.

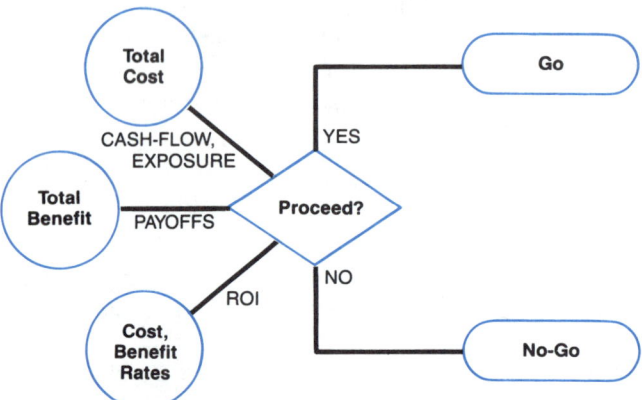

FIGURE 14–6. Components of a Project-Related Economic Decision

Of primary concern is this inequality:

$$COST < BENEFIT.$$

In other words, the cost of an activity should be less than the benefit. The expenditure should *effectively* produce the *benefit*.

Costs and benefits follow different time curves (Figure 14-7) and are compared at

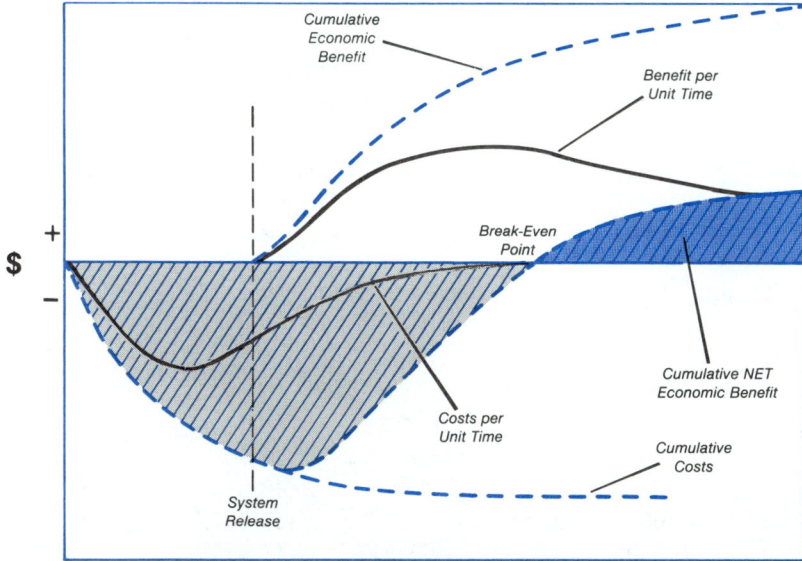

FIGURE 14–7. Costs and Benefits Have Different Time Characteristics

specific times, generally at yearly intervals. The criterion becomes:

There is a *t* such that

$$\sum_{i=0}^{t} COST_i < \sum_{i=0}^{t} BENEFIT_i$$

The criterion states that the sum of benefits will eventually exceed the sum of the costs, which are evaluated using **net present value** (NPV) to compute accrued benefits against costs over time. As a rule of thumb, most major IRMaint efforts should be profitable within five years; often, a more stringent requirement is placed on projects.

Another approach looks at the efficiency of the expenditure. Stated mathematically,

(BENEFIT-COST)/COST > ROI

where **ROI** (**return on investment**) is a predetermined ratio. We wish to know if we can achieve profits more *efficiently* after the project (and its cost) than we could otherwise. If we could make more money by leaving the money in the bank, then it may not be worthwhile (that is, efficient) to spend the money on the project.

ROI computations are confounded, as we will discuss later, by turbulence in the industry, the unpredictability of interest rates (making R difficult to predict in many industries), and the necessity to factor in intangible costs and benefits that are two to five year in the future. For this reason, ROI computations are rarely carried out beyond 5 years.

Net Present Value

The first questions to ask are: What is the value of the effort over time? Does this value exceed the costs? In other words, will we spend the money *effectively* to achieve the benefits?

Consider an effort with projected first-year costs of $70,000, second-year costs of $30,000, and costs of $10,000 annually thereafter. Production will begin in the second year, with projected benefits of $20,000 the second year and $40,000 each year thereafter. There are no benefits in the first year.

Figure 14-8, on the opposite page, shows graphically the facts apparent from Table 14-1, below.

Year	Cost	Benefit	Net	Rate	PV	NPV
1	70,000	0	(70,000)	0.900	(63,000)	(63,000)
2	30,000	20,000	(10,000)	0.810	(8,100)	(71,100)
3	10,000	40,000	30,000	0.729	21,870	(49,330)
4	10,000	40,000	30,000	0.6561	19,683	(29,647)
5	10,000	40,000	30,000	0.5904	17,712	(11,935)
6	10,000	40,000	30,000	0.5313	15,940	4,005
7	10,000	40,000	30,000	0.4782	14,346	18,351

Table 14-1. Accumulated net present value of the project

The "rate" illustrates the time value of money, which is assumed here to be ten percent. That is, a dollar worth 100 cents this year is assumed to have a value of only 90 cents, in today's values, next year. This **discount rate** must be estimated and may not be constant over the lifetime of the project. Spreadsheet calculators make it easy to project net present value over a variety of scenarios. The one above indicates that the project shows a cumulative profit only in the sixth year of its life. It is ineffective for the first five years. However, management may be able to see the losses as necessary over the first several years, justifying this rather long payback period.

Payback Analysis

From Table 14-1, we see that the **break-even point**, using a discounted net present value approach, is in the sixth year. **Payback analysis** looks at the sensitivity of the break-even point to specific expenditures. For instance, suppose that through the expenditure of training funds in the modest amount of $3,000 annually, we could achieve an additional benefit of $10,000. Furthermore, suppose we had to delay $20,000 of the first-year expenditure to the second year when prices might fall, reducing our benefit by $10,000 in the first year only. The result is shown in Table 14-2.

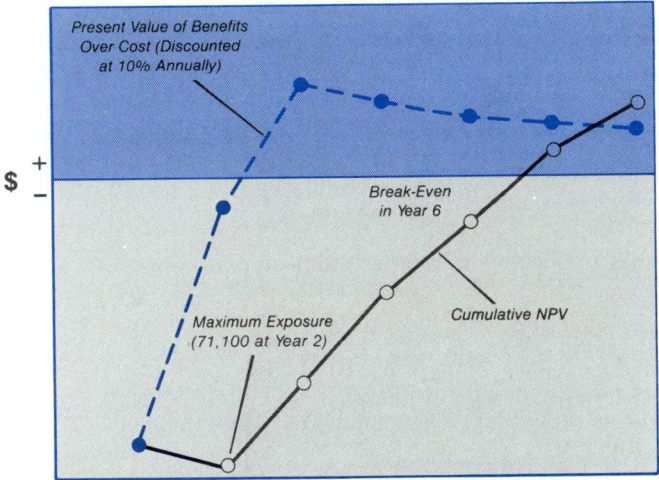

FIGURE 14–8. Net Present Value Employing a 10% Discount Rate

Here, changes in our assumptions move the break-even point up about a year, but the shift in expenditure in the first year has little beneficial effect, increasing maximum exposure to $77,400.

Payback analysis asks questions involving risk, such as the following:

- Given a specified discount rate, in what year will the project pay off its costs?
- What does the discount rate have to be for the project to pay off in a certain number of years?
- What is the maximum exposure (negative NPV) this project will have and in what year will it occur?
- If we incorrectly guess the discount rate and miss by some proportion, what is the worst case for payback period?

The fourth question is not merely a mathematical exercise. Costs usually precede benefits, and discount rates are always more difficult to predict in *later* time periods. This makes our computations fuzzy on the benefit side, while, relatively speaking, costs are more confidently discounted. Figure 14-7 shows typical cost and benefit curves over time.

Year	Cost	Benefit	Net	Rate	PV	NPV
1	50,000	0	(50,000)	0.900	(45,000)	(45,000)
2	50,000	10,000	(40,000)	0.810	(32,400)	(77,400)
3	13,000	50,000	37,000	0.729	26,973	(50,427)
4	13,000	50,000	37,000	0.6561	24,276	(25,151)
5	13,000	50,000	37,000	0.5904	21,847	(3,204)
6	13,000	50,000	37,000	0.5313	19,650	16,446

Table 14-2. Accumulated net present value of the project with changed

Cash-Flow Analyses: Return On Investment (ROI)

Payback analysis concerns the effectiveness of investment; **cash-flow analysis** looks at efficiency. The analyst tries to answer whether or not the funds to be expended on the project increase benefits with maximum effectiveness. Generally the baseline is the prevailing value of money, the bank savings rate. If we could earn money faster in a bank than by funding a project, we probably would not take the risk.

Other standards also should be considered:

- the rate of return on past, similar projects - the rate of return on other projects competing for the same funds
- the overall ROI for the firm over the past quarter, year or years

In each case, we want to know not only how much the project is worth in terms of benefit and profit, but also how "good" we would feel about putting our efforts into this profit venture relative to past and possible efforts.

In simple terms, ROI is the ratio of return to investment. Return is the profit (uncorrected for the cost of money), and investment is the actual expenditure (also uncorrected). The reason for the lack of correction is that our baseline for comparison is often very close to the cost of money. Because ROI is frequently calculated over a number of years, we usually express ROI as **annual ROI**. This involves determining the *annual* rate that would compute to this overall rate over the period considered.

Recalling the previous example (and ignoring the discount rate), the annualized ROI over the first five years of the project would be computed as follows:

$$
\begin{aligned}
&\text{(1) Costs(in \$K)} &&= 70 + 30 + 10 + 10 + 10 = 130 \\
&\text{(2) Benefits} &&= 20 + 40 + 40 + 40 \quad\quad = 140 \\
&\text{(3) ROI (over five years)} &&= \text{(Benefit-Cost)/Cost} \\
& &&= (140{-}130)/130 = 10/130 \\
& &&= 0.077 \\
&\text{(4) Annualized ROI} &&= 1.077^{1/5}{-}1 = 1.015{-}1 = 0.015
\end{aligned}
$$

In other words, investing the funds at 1.5 percent would prove to be as efficient a use of the funds over the five year period.

Of course, extending the period could change the ROI. Consider the ROI over six years:

$$
\text{ROI} = \text{(Benefit-Cost)/(Cost)} = (180{-}140)/140 = 40/140
$$
$$
= .286 \text{ or about 4 percent annual return over six years.}
$$

Because most software-based projects may be assumed to have a true useful life of three to six years, management's decision regarding the efficiency of this investment would be difficult.

Consider the tactic of postponing some hardware purchases and introducing the modest training program. What happens to the ROI over five and six years?

In general, larger projects have higher up-front costs and greater ultimate benefits. Given the turbulence in both money markets and the computer industry, the marginal ROIs that large projects promise are often seen as small in contrast to the relatively larger short-term ROIs inherent in microcomputer software development, especially where technical labor can be converted into semi-professional end-user labor. Whether or not the products of semi-professional labor have use beyond a few months may prove problematic as we acquire more experience with information centers and end-user software.

Social Accounting

In recent years, attention has shifted away from an exclusive focus on hard dollars to the softer areas of quality of working life (QWL), company image, professionalism, accountability, and the social organization of work. Since landmark studies of QWL in the early 1970s, industrial researchers have been pointing out that the value of an investment may be modified — up or down — by the effect that investment has on employees, customers, and suppliers and the effectiveness of management.

Consider a physician who is contemplating automating her practice's billing and bookkeeping. Currently, billing is handled by the receptionist and bookkeeping is performed one evening a week by a part-time bookkeeper. End-of-year accounting and tax-form preparation is contracted to an accounting firm.

The costs of computerizing would amount to $21,000 over a three-year period. Dollar benefits would amount to about $4,500, mostly in accounting costs. Some bills might be brought in faster with reminders, and it is possible that scheduling efficiencies may allow for denser scheduling and a bit more income. Why is the physician automating?

The answer is that it makes good accounting sense. First, the doctor currently has little control over office cash flow because it is difficult to get a good view of the data. The computer can produce financial statements on command. Second, relations between the doctor and the receptionist are not always smooth because the manual system is totally under the receptionist's control, resulting in double-booking, underscheduling, and a great deal of confusion on busy days.

Third, the doctor has only a vague idea of whom she sees for what. She *thinks* she treats mostly sniffles and bruises, but she would really like to know more. Fourth, the doctor's receptionist has been dropping strong hints about relocating to Bigtown's suburbs. Hiring a new receptionist would be time consuming, and training one would be a superhuman task.

Finally, all the major clinics and many physicians in the area have computerized to some extent. Patients have mentioned that their friends are receiving computerized bills and prescription advice and that scheduling seems more flexible. While there are no monetary benefits for these aspects of computerization, obviously the doctor does not appear as "state-of-the-art" as other physicians.

The costs of *not* computerizing may exceed the costs of the computer. Over the long term, **social accounting** recognizes avoiding these "costs" as benefits. Each factor is assigned an arbitrary point value, later translated into dollar figures, as Table 14-3 illustrates:

Item	Cost	Benefit
Image of being "state of the art"		100
Increased control over office processes		400
Increased awareness of practice components		200
Training of new receptionists		400
Lowered receptionist efficiency during learning period	200	
Better social relations with receptionist		300
Time and confusion in learning about computers before purchasing one	500	
Rationalizing and documentation of office processes		400
Totals	700	1,800

Table 14-3. Social accounting of the costs and benefits of automating a physician's office

Now the question is whether 1100 points in social benefit is worth the loss of $16,500. Each point seems to be worth about $15. Is it worth $4,500 to have better relations with the receptionist? Does it really cost $6,000 to train a receptionist? These are questions the physician must answer.

14.4 THE VALUE-ADDED APPROACH

A New Approach

Previous sections have focused on the actual or imputed dollar value of a project. This section takes a different approach. It attempts to answer the question: "What value is added to the firm's products?"

This question differs from "Is the project worth doing?" because it concentrates on the integration of the IRMaint activity with those of the firm. For example, automation of the billing function may displace labor costs to the information resource. These displaced costs can be used in a calculation to justify (or unjustify) the automation based on total dollar value or efficiency. However, there are other effects on the firm's product.

Suppose the firm is a retail department store. Not only will the automated billing system enhance the efficiency of billing, it would also enhance the value of its product — the service of providing merchandise.

Figure 14-9 lists the following value components of the retail firm's product to the customer:

1. the tangible value of the merchandise, *plus*
2. the value of the firm as a source of knowledge of what is available for purchase, *plus*

3. the value of the status attending being a customer of the firm, *plus*
4. the value of the act of shopping itself, *less*
5. the cost of the merchandise

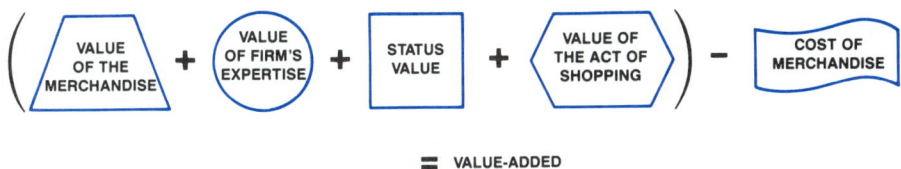

FIGURE 14–9. The Calculus of Value Added in a Shopping Situation

A customer might buy overpriced merchandise if the store has all the latest things and provides entertainment or relaxation as part of the shopping experience.

How does the billing system fit in? It might decrease the cost of the product by reducing overhead. But more likely, a smoothly running billing system will increase the net value of the act of shopping as well as increase the status of the customers. After all, if everyone in Bigtown knows that the store is famous for billing snafus and if, indeed, the snafus are a problem, then the value of shopping at the store will decrease.

The **value-added concept** works in this fashion:

1. Determine the components of the value of the firm's products.
2. Determine how the information resource affects those values.
3. Assign, through research, a value to these impacts.
4. Compute the value added by changes to the information resource.

The Value of the Information Resource

In Figure 14-10 on the following page, the system accepts inputs from the environment, transforms them, and produces outputs. In the process, value is added to the organized inputs — value that the firm hopes is positive and large enough to induce elements of the environment to exchange for resources the system needs. In this view, the system exists to add exchangeable value.

This noted, it becomes important to know what added values that can be modified by changes to the information resource. Classical MIS theory concentrates on the values of timeliness, accuracy, completeness, consistency, and reliability. It is assumed that when these are increased or maximized, the information resource is valuable. However, we need to look beyond these general values to specific impacts and beyond the management perspective to the firm as a whole.

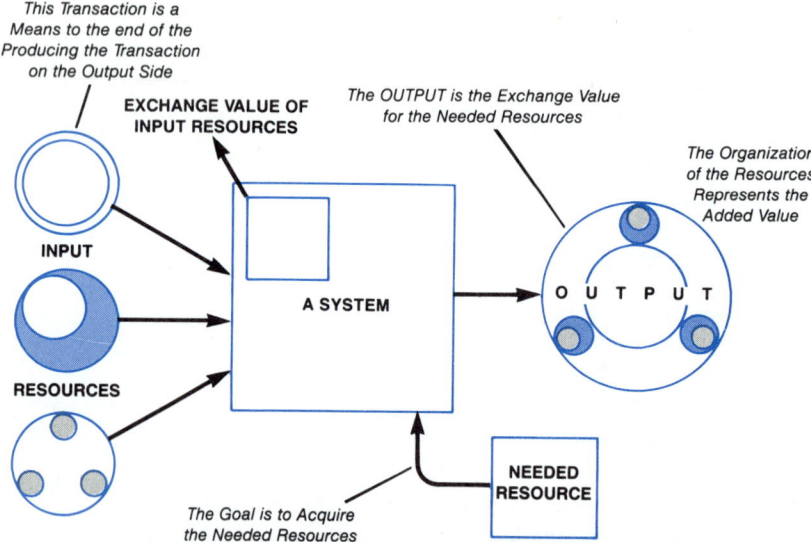

FIGURE 14–10. An Open System Exchanges Resources to Produce Added Value

What exchangeable values does the information resource have and affect? Figure 14-11 lists several, detailed below:

1. **Reliability**: customers normally expect the product to have certain characteristics.
2. **Quality**: the product should do what it is expected to do or serve the advertised function well.
3. **Access**: customers should know about the product and be able to acquire it without undue effort.
4. **Effective exchange**: once purchased (leased, rented, etc.), the product should be effectively controlled by the customer.

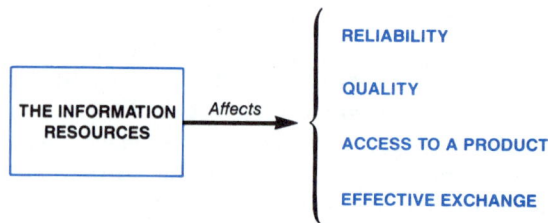

FIGURE 14–11. Exchangeable Values of the Information Resource

In other words, a product's value should not be diminished by incomplete knowledge of the product, ineffectiveness, or inability to find, acquire, or maintain the product. The value of the information resource is therefore determined by its contribution to reliability, effectiveness, accessibility, and maintainability.

Specific Values and Their Worth

We can begin to determine the value added by an information resource by examining higher-level DFDs for proposed systems. While the value-added concept works at all levels, because information systems are usually dedicated to the *control* rather than *locomotor-manipulator* function(s), we will concentrate on the values that an information resource can add to control aspects. In other words, we will consider the firm to be the management of an organization and the "products" to be management decision making.

The question is then posed in the following fashion:

"How does the proposed system or system change add to the value of management control in the form of decision making and execution?"

Figure 14-12 illustrates a model of management decision making, based upon Herbert Simon's ideas (*The New Science of Management Decisions*, 1960, pp. 1-4) and augmented by recent work in decision-support systems. Decision making is seen as a seven-step process:

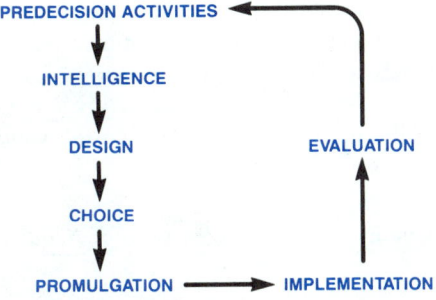

FIGURE 14–12. A Model of Decision-Making

1. **predecision activities**: learning the rules of decision making, deciding on a decision-making procedure, deciding to make a decision, and understanding the context for and role of decision-making in an organization
2. **intelligence**: recognizing a problem in need of a decision
3. **design**: developing models through which decisions will be made, arraying choices, and gathering data
4. **choice**: evaluating alternatives and selecting satisfactory or optimal choices
5. **promulgation**: communicating the result of decision making, achieving consensus, framing the decision in appropriate terms, and informing the appropriate people
6. **implementation**: carrying out the selected course of action
7. **evaluation**: evaluating the effectiveness of the selected alternative and the decision-making process

If decisions are the product of management, we can show how an information resource can increase the value of these products by making them more reliable, effective, accessible, and acquirable.

First, reliability increases through systematizing the intelligence, design, and choice phases and by assisting in predecision activities (in particularly, having information about decision making available through manuals or "help" facilities).

Second, decision making becomes more effective because the information resource can assist in increasing the validity of the decisions made. It makes data gathering easier; helps sift the data for relevance, accuracy, and timeliness; and helps categorize the data as to reliability, depth, and source. It becomes easier to consider more alternatives and to explore them more deeply with "what-if" questions.

Decision making becomes more accessible as the information resource provides assistance in the promulgation phase (for results) and the predecision phases (to motivate decision making). Those who need decisions can make their needs known more rapidly, completely, and understandably. Then later, they can become aware of the decision and its implications more rapidly, completely and understandably.

Finally, the information resource can assist others in "acquiring" the decision by showing how it should be carried out and the implications of variances from the decision. It is fine to know that we should market product X, but what about marketing X.1 and X.2? If we note no improvement in sales figures for the first two months, have we done something wrong?

Table 14-4 illustrates the contribution of the information resource to the value of management decision making in each of six stages:

Stage	Reliability	Quality	Access	Acquisition
Predecision	X		X	
Intelligence	X	X		
Design	X	X		
Choice	X	X		
Promulgation			X	
Implementation				X
Evaluation	X	X	X	X

Table 14-4. Values added by an information resource at seven stages in management decision making

The computation of added valued begins with the data flow diagram and locating points at which value is added in terms of reliability, efficiency, access, and acquisition.

Process 1 in Figure 14-13 (opposite), "Acquire raw sales data," is a morph whose activities consist of basically inputting sales data from the salespeople, checking it for accuracy and timeliness, and batching it. Activities are controlled from two sources. First, the data flow "Sales categories" come from a data store controlled by the sales VP. Second, the data flow "Acquisition list" comes directly from the district sales manager. In other words, the data acquired depend on (1) predetermined, slow-to-change sales categories that come from policy makers and (2) the individual salespeople whose data are requested, based on dispositional, *ad-hoc* requests from the sales manager.

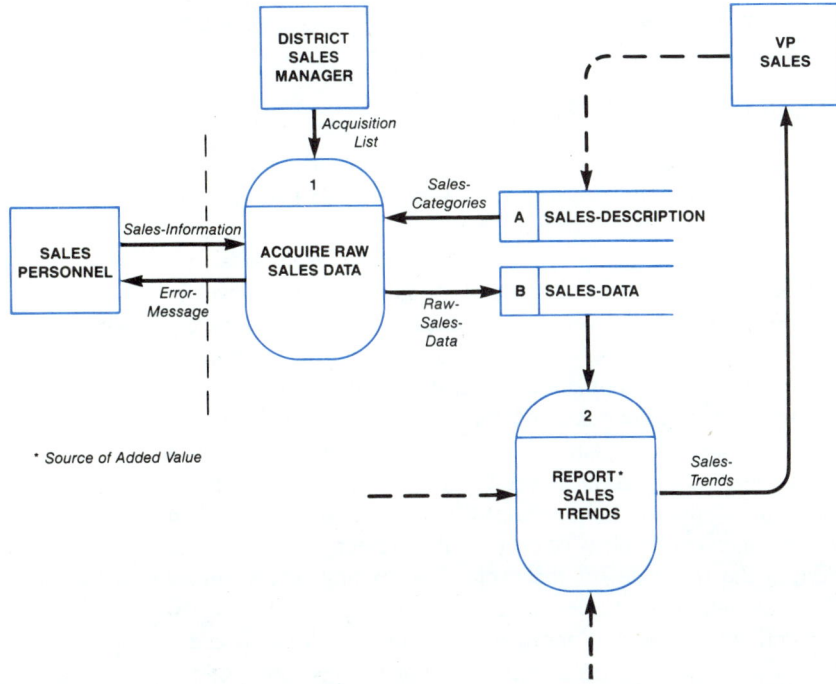

FIGURE 14–13. Finding Added Values

Because reliability and efficiency depend on the source of the data, we can consider that those values are added by the ability of the manager to choose sources. Thus, one added value of the system depends on the link from the sales manager to process 1.

How much value is added? The question may be answered best by looking at how much value is *lost* if the wrong sources are examined. A bad decision may cost $5,000 if a product is advertised in a poor market and an additional $2,500 if the proper product is not advertised. Thus, we can consider that in each cycle, up to $7,500 can be lost by a bad decision. Taking the average (although we may have more exact figures and a better algorithm for making this computation under some circumstances), we can say that $3,750 is added by this acquisition. If there are later checks on the validity of the data that reduce the possibility of a bad decision by 50 percent, then the value added by this check (process 2: "Report sales trends") would be half of $3,750, or $1,875.

The total value added depends on an external estimate of the worth of a properly-functioning information resource. Evaluation of each stage of processing depends, in turn, on a step-by-step factoring of possibilities. The value-added approach essentially reads the DFD and assigns a portion of the total value added to particular processes based on probabilities. These probabilities, in turn, depend on intuition about the behavior of data and people, usually gathered from experience.

Because of their intuitive nature, value-added computations are repeated often during logical design. In the earliest stages, only a very high level DFD is used and gross figures are assigned to each process. During design, both the processes and the figures are refined. The value-added approach can therefore be used at all stages of logical design and is not confined to an initial economic feasibility projection.

Linking Value-Added and Cost-Benefit Approaches

The value-added and cost-benefit approaches are merely two ways to compute the worth of a project. Theoretically, they reflect the same reality and provide two views of the same facts. In practice, however, they play different roles.

Cost-benefit analysis is intended for use early in the IRMaint work. Traditional systems analysis stressed economic feasibility studies at the earliest stages. System development work has become more complex. As relatively routine data processing applications have given way to more difficult management decision-making applications, it has become increasingly difficult to make reasonable development-effort cost estimates.

Intangible benefits, too, seem more important, but they are still hard to measure. Therefore, cost-benefit computations are fuzzier and riskier than ever before. The irony is that automated project-planning tools make it easier to see the folly of precise estimation!

The value-added approach also depends on these imprecise figures to an extent, but it does not attempt to answer the question in terms of costs of the project *per se*. Instead, it attempts to focus attention on the rationale for the project. Assuming that technical problems are solvable (a big assumption), value-added analysis inspects the logical rather than the physical design. This makes it more appropriate and less "pseudo-precise." After all, if errors in the raw data are a major part of the problems of the existing management information system, then correction of these errors should be of some value. During the physical design stages, we can put precise figures on the cost of correcting the errors. Meanwhile, we can scrutinize the process and try to assign figures to those parts of it that are causing the errors. Then, when we know the benefit, we have some idea of the extent to which we can spend dollars to fix it.

14.5   DECISION MAKING IN SYSTEMS ANALYSIS

This brings us directly to how analysts make important decisions about proceeding with work. Several of the most common techniques will be discussed.

Static and Dynamic Decision-making

There are two classes of decision making. **Static decision making** assumes that all the facts are available and that the best decision can be "computed" from these facts. **Dynamic decision making** arrives at decisions iteratively, acquiring facts or new opinions as it progresses.

Static decision making includes payoff tables, straight-line and cyclic projection, and so-called "dynamic" modeling. Dynamic decision making includes logical modeling, consultation, and the Delphi technique.

When we make decisions using a static technique, we array the facts and proceed to boil down the data to a series of **index values**, numbers that indicate values without specific units. The higher the index value, the better the choice. Index values typically reflect expenditure, effort, time, and other resources.

Dynamic techniques also produce index values that prompt us to examine the assumptions that go into our estimates. We recompute index values until we are confident of their validity. Dynamic techniques rely on the principle that stability and validity are related, that if the index values do not change much, we have arrived at the end of our decision-making. This assumption is risky, but because dynamic decision making is open to additional data, earlier errors in assumptions can be corrected.

Payoff Tables

One of the easiest and most intuitive techniques in static decision making is the **payoff table**. The payoff table is used to calculate an "intuitive" net present value for an effort by breaking down the benefits into a series of mutually exclusive factors along which we rate a set of alternatives. In attempting to bring in both tangible and intangible costs and benefits, we assign index values to these factors for each alternative. Then we weigh the factors in terms of their overall contribution to the net benefit and total for each alternative. This total value represents the net present value of the project.

We can perform this calculation for a number of periods in the future, changing both the ratings for alternatives on factors and the weights of the factors. This gives us a projection of the value of alternatives over time.

Finally, we can introduce a certain fuzziness to our computations to indicate the future range of values that can result, making it more difficult to rely on questionable assumptions further into the future.

Consider the possibility of having word processing and electronic mail on executive's desks in a large manufacturing firm. Suppose the executive group is composed of 61 persons in three cities, augmented by 84 executive assistants. This group of 145 individuals would receive the direct benefit of the project, although there are obvious benefits to the firm as a whole.

The analyst charged with examining the possibilities would first attempt to discover the important factors in benefits, perhaps using the value-added approach. In addition, research in office automation has pointed out areas in which benefits are often found.

Next, a **projection** of the weights of these factors over the next ten years is attempted, probably through interviews with a sample of the executives (Figure 14-14).

Factor	Year 1	Year 2	Year 3	Year 5	Year 10
Faster Decision-Making	40	40	30	30	20
Better Integration of Efforts	20	30	40	40	30
More Concise Information and Reports	20	20	20	20	20
Automation of Routine Decision-Making	0	0	0	10	30
Cost of Executive Group*	20	10	10	0	0
Totals	100	100	100	100	100

FIGURE 14–14. Factors and Weights in a Payoff Table

Then, the analyst attempts to put figures on the value of each of several alternatives. Three are considered: (a) the status quo, (b) an integrated office automation system, and (c) introduction of word processing for the executive assistants alone. The values gathered are on a scale of 0 to 9, with 0 indicating no net benefit and 9 indicating quite a lot. The weights are multiplied by the values and summed for each alternative. The totals appear in the box at the bottom of Figure 14-15 and are graphed in Figure 14-16.

Factor	← — — — — — Benefits — — — — — — →														
	Year 1				Year 2				Year 3				Year 5				Year 10		
		A	B	C		A	B	C		A	B	C		A	B	C		A	B	C
Ease of Decision-Making	40	5	4	4	40	4	5	4	30	3	6	5	30	2	6	6	20	1	7	7
Integration of Efforts	20	3	3	3	30	5	4	4	40	6	4	6	40	5	5	5	30	4	6	4
Conciseness of Information and Reports	20	7	2	7	20	7	3	8	20	7	4	9	20	7	4	9	20	7	4	9
Automation of Routine Decision-Making	0	n. a.			0	n. a.			0	n. a.			10	0	8	6	30	0	9	7
Cost of Executive Group*	20	8	3	3	10	8	2	2	10	8	3	3	0	n. a.			0	n. a.		
Totals	100				100				100				100				100			
Totals for Alternatives	A	560				530				550				400				280		
	B	320				400				450				540				670		
	C	420				460				600				620				650		

FIGURE 14–15. Computation of Payoffs for Three Alternatives

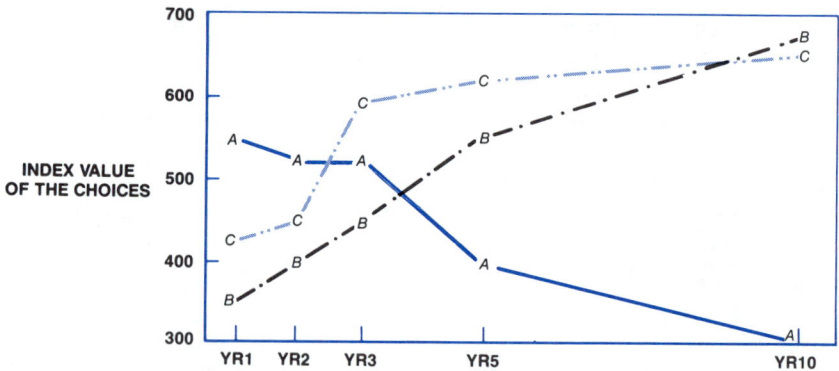

FIGURE 14–16. Graphical Display of Payoff Trends over Time for Three Alternatives (A, B, C)

It is apparent from this table that neither of the two computerized alternatives exceeds the benefit of the status quo until the third year; the integrated automated office does not seem better until the fourth year. In other words, it seems as though computerizing is still three or four years off. Given the lead time to develop such a system, efforts could be delayed until year two or three without any loss. In fact, accumulated benefits

of the status quo are not surpassed by alternative C (automation of the executive assistants) until the end of the fourth year or by alternative B (automation of the executive
group) until the seventh year. The best choice seems to be to wait two years, automate
the executive assistants, and then consider phasing in the executive group, based on this
data.

Although our data seem to indicate that automating the executive assistants in year
three is the best choice, what if our data for alternatives B and C are off by a factor of 25
percent in various places? Figure 14-17 shows the results in terms of our estimates and
the resulting "fuzziness."

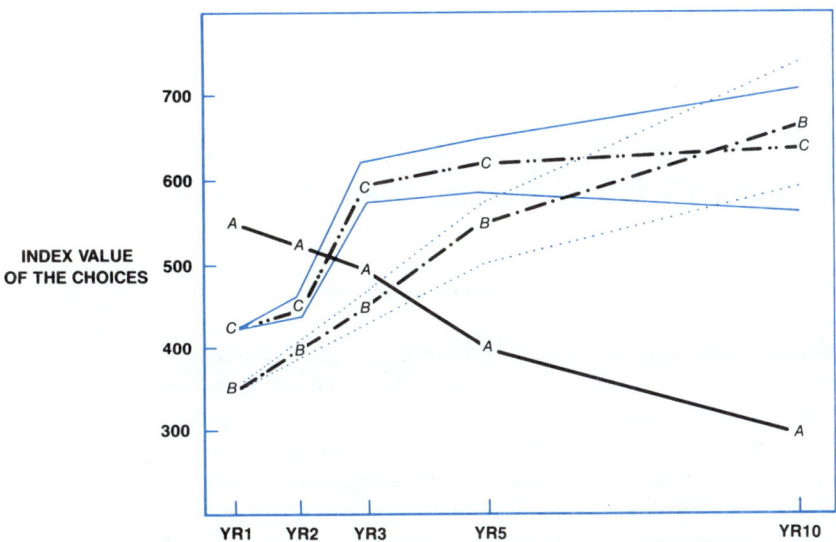

**FIGURE 14–17. Payoffs for Three Alternatives (A, B, C) over Ten Years with Estimates
of Impreciseness**

Forecasting Techniques

The payoff tables have an element of **forecasting** in them, since they attempt to look into the
future. Three other techniques are used for forecasting: straight-line projection, cyclic projection,
and dynamic modeling. These are static techniques
because they do not attempt to refine data after
forecasting (they are not iterative).

Straight-line projection looks at recent trends
and extrapolates them using a straight line. Figure
14-18 illustrates cost and benefit estimates for a
customer information system for marketing managers. Costs are estimated at $25,000 for the first
year and $5,000 for each subsequent year. Benefits
are delayed until the second year, at which time

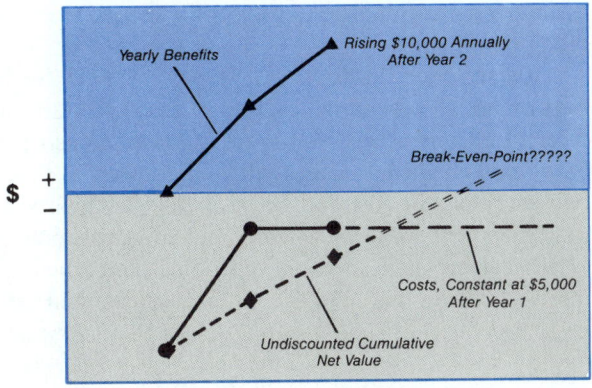

FIGURE 14–18. Straight-Line Projection

they jump to $15,000 and rise $10,000 annually. Ignoring the time cost of money, when will accumulated benefits exceed accumulated costs?

Cyclic projections are useful where there are known cycles of events. These include monthly, seasonal, and yearly business events, daily and weekly employee behaviors (When are the telephones the busiest during the day? When is the office least crowded? When do the greatest pressures for decisions occur?), and longer cycles such as the three-year business expansion/contraction cycle. Examining the data collected for requests for citizen assistance at the information booth in City Hall, what might be predicted for the first Friday in March as indicated in Figure 14-19?

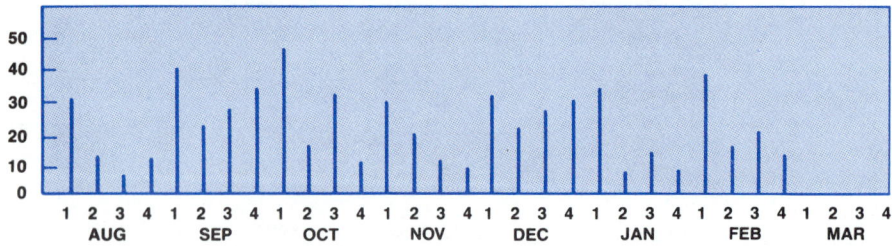

FIGURE 14–19. Cyclical Phenomena Require Cyclical Projection Techniques

Finally, **dynamic modeling** approaches the decision-making question by building mathematical models of the situation, posing what-if questions. Assumptions built into the model are the relationships between variables. Data input are parameters that evaluate those variables. We are familiar with these models in business because they are essential to most planning processes, but the development of spreadsheet calculators has made decision making both easier (in the choice phase) and harder (in the design phase). Choices are easier because we can more quickly produce far more precise estimates of outcomes. But design is more difficult because we have the capability to create many different models with different assumptions. To further complicate the situation, these assumptions are often complex and difficult to debug. Recently, attention has turned to the value of models developed by nonprofessionals. Concerns include poor or hidden assumptions, incomplete testing of the model, poor documentation, and poor input data reliability.

Regardless of these considerations, dynamic modeling is useful, whether computerized or not. Examine the budget in Figure 14-20 on the opposite page. Here we can forecast expenses based on specific assumptions of labor and hardware costs. We can build explicit models that predict, for instance, that employing two programmers and two analysts will cost $62,500 over the first two years, increasing the number of programmers will speed up delivery, but at significantly increased costs. By asking a series of what-if questions, we can determine the earliest delivery time, with the lowest cost increase. Given this "best" situation, we can then ask, "Is the result worth the cost?"

	A	**B**	**C**	**D**	**E**	**F**	
1	JOB CLASS	CT	UNIT SALARY	%	PROD	COST	
2	PROGRAMMER	2	$24,000	33		$32,000	$\leftarrow B2 \cdot C2 \cdot 2 \cdot D2$
3	ANALYST	2	$30,000	20		24,000	$\leftarrow B3 \cdot C3 \cdot 2 \cdot D3$
4	SECRETARY	1	$19,500	33		6,500	$\leftarrow B4 \cdot C4 \cdot 2 \cdot D4$
5							
6							
7							
8							
9							
10							
	TOTALS	5	DEVT SPEED		0.5	62,500	$\leftarrow$ @SUM(F2..F4)

@SUM(B2..B4)

$$\frac{4 \cdot B2 \cdot D2 + 5 \cdot B3 \cdot D3 + B4 \cdot D4}{10}$$

FIGURE 14–20. A Budget as a Dynamic Model (Spreadsheet Format)

Dynamic Decision-Making

Three techniques of dynamic decision making are logical modeling, consultation, and Delphi. These techniques are dynamic because they focus attention on the assumptions of the decision-making process, refining both assumptions and data until a palatable decision-making situation is reached.

Logical modeling is based on research in artificial intelligence. As shown in Figure 14-21, a network of assertions is created, and a set of data is used to evaluate those assertions. Finally, a set of rules tells how the data and the assertions go together. These are the model base, the database, and the rule base.

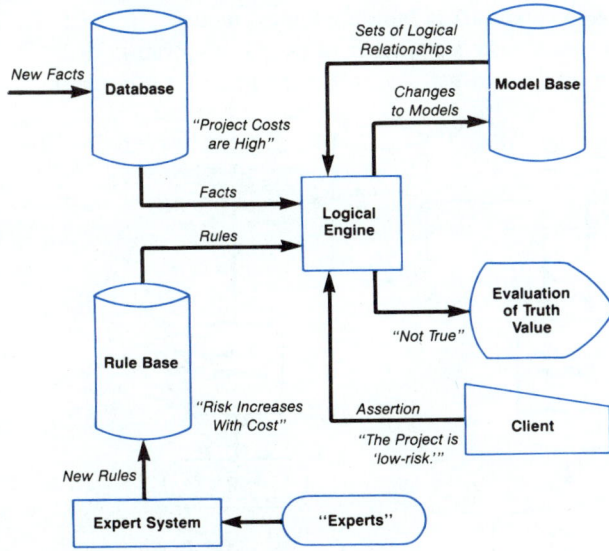

FIGURE 14–21. Building and Applying Logical Modeling

Suppose we want to examine the risk of continuing a project to build a complaints database for City Hall. Some of the assertions of interest to us are these:

1. The project is low risk.
2. The project will be completed on time.
3. Benefits will exceed costs within three years.

The data consist of cost and benefit values already observed or predicted. The rule base defines the terms and their relationships:

1. Cost = Project-cost + operational-cost + ...
2. Benefit = Citizen-benefit + alderman-benefit + ...
3. Completion-date follows starting-date.
4. Completion-date depends upon delays.
5. Hiring staff is a delay.
6. Training operators is a delay.
7. Risk increases with cost.
8. Risk increases with number of modules.
9. ....

The logical model is then computed. The result is an evaluation of the truth of various assertions. Suppose the statement, "The project is low risk," evaluates false. Does this mean the project should halt?

Not necessarily. Most logical models also indicate the path taken in determining a statement to be false. Therefore, we can examine why "The project is low risk," is false (why the project is high risk). The logical model can help us uncover those factors that increase risk, and we may be able to move to affect these factors. Packages such as Guru™ and languages such as PROLOG can assist in building logical models.

Consultation is commonly used to increase the number of people involved in decision making. A common consultation model is illustrated in Figure 14-22. A consultant

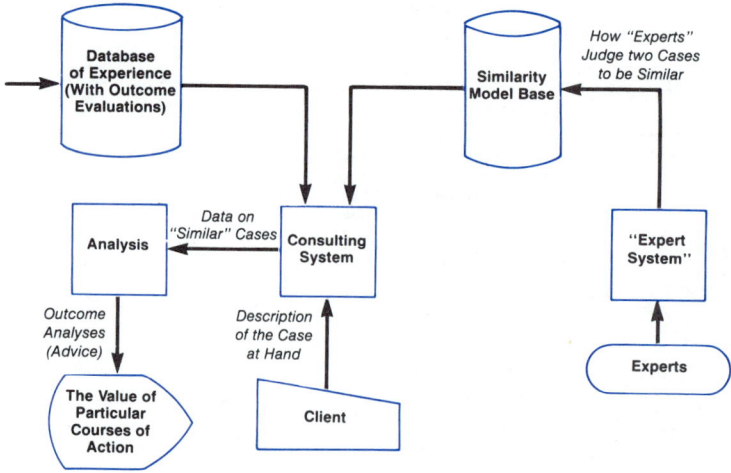

FIGURE 14–22. Building and Applying a Consulting System

is an "expert system" memory for events. The consultant also knows how to judge the similarity of two events. Based on similarity, the consultant picks out events from the past that are similar to the one needing advice; the set is then analyzed for appropriate tactics.

Consider the City Hall project. A consultant may remember several similar projects and advise project personnel on likely courses of action at this point. Of course the consultant who judges similarity based on the age of project personnel may prove to be less valuable than one who takes factors such as total expenditure, time frame, number of project personnel available, their skills, and so forth into account. In other words, a sophisticated consultant makes similarity judgments on a large number of relevant factors; a naive consultant relies on a small number of irrelevant factors.

Consultation is a dynamic process because consultants focus attention on the problem definition in an iterative fashion, querying the client about assumptions and facts. Perhaps the most attractive aspect of human consultants is the feeling clients get of conversations being complete and relevant. Of course, many clients cleave to poor consultants because they are falsely persuaded to feel confident. Automation of consultation aims at pooling advice from several consultants, thereby introducing a panel aspect.

The panel is the primary instrument of the **Delphi** technique (reviewed in Chapter 12). The technique is an iterated questionnaire. Experts are asked to provide estimates, make predictions, or give advice. The results are pooled and then returned to the experts for comments or revision. The results of this "wave" are again pooled and returned. The process ends when confidence in the stability of the results is high (when the analyst can predict the result of the next wave with great accuracy).

Delphi is most useful for long-range planning, but it can be applied to decision making, too. The lengthy process of mailing, compiling, data reduction, report generation, and redistribution makes it inappropriate for quick or secure decision making. However, automated Delphi systems (Hiltz and Turoff, 1981) can provide quick turnaround. Coupled with a secure communication system, Delphi could be used in many sensitive decision-making areas.

Function-Point Analysis

A technique of program and design measurement known as **function point analysis** is valuable when determining system complexity. Function point analysis has been employed to document the increased effectiveness of fourth-generation languages over more conventional program development tools. By assigning index values or "points" to specific input, output, and processing "functions," this analytic technique can be used to compare programs based on the number of function points. In this manner, researchers have determined that certain fourth-generation languages are twenty times as productive (in terms of function points coded per hour) as COBOL and PL/1. This ratio increases with the size of the coding effort.

We can also employ function-point analysis to determine the costs of program development. As a first step, we can calculate the number of function points in representative projects. Next, we can derive a rule of thumb for the number of hours required to code each function point. Finally, we can estimate the number of function points our logical design requires. Multiplying this number by the estimated cost per function point gives us the total cost for coding.

The technique assigns points to functions as follows:

1. Isolate each function. These are external inputs, external outputs, external queries, and data stores.
2. Assign a number to each function based upon its complexity: 3 for simple ones, 4 for average ones, and 6 for complex ones.
3. Sum up the values for all functions.
4. Adjust the sum. For each of the following 14 characteristics, assign a number between 0 and 5 indicating the degree of influence with 0 meaning no influence and 5 meaning a strong influence. Then sum these influence numbers:

> Use of data communications
> Distributed data or functions
> Specific performance objectives
> Heavily-used configuration
> High transaction rate
> On-line data entry
> End-user efficiency concerns
> On-line updating
> Complexity of processing
> This application to be used in other applications
> Conversion and installation concerns
> Importance of ease of operation
> Use of multiple sites for operation
> Application intended to facilitate change

The maximum score for this adjustment is 70.

5. Compute the function point value by this formula:

$$FP = \text{Function sum [step 3]} * (0.65 + (0.01 * \text{adjustment sum [step 4]}))$$

Many adjustment factors depend on physical design and installation considerations — these will have to be estimated. It is possible to adjust function point scores up or down by 35%. Thus, this estimation can have an important influence on the final function point score.

Consider a logical design involving 4 complex outputs, 6 average queries, 2 simple and 2 complex inputs, and 8 data stores. The unadjusted function score is computed as

$$4 * 6 + 6 * 4 + 2 * 3 + 2 * 6 + 8 * 4 = 98$$

Suppose there are important specific performance objectives, a busy computer to be used, a moderately high rate of transactions, a strong need for end-user efficiency, some on-line updating, simple processing, and a lot of emphasis on facilitating change. This gives an adjustment computed as follows (assuming unmentioned factors are unimportant):

$$5 + 5 + 3 + 5 + 5 + 3 + 0 + 5 = 31$$

The number of function points is then computed as

$$98 * (0.65 + 0.01 * 31) = 98 * 0.96 = 94.1$$

If we will be employing a powerful fourth-generation language to create this system, we might find that, as a general rule, it takes about one hour for each function point. Therefore, our estimate for programming expenditure will be about 94 hours. Of course, these estimates depend upon our abilities to create valid rules of thumb, understand environmental and operational considerations, and predict complexity and importance of values. Under these constraints, function point analysis is a valuable way of comparing development situations and deriving predictions of costs.

14.6 CONCLUSION

Information resource maintenance requires information and decisions at specific times during its operation. It is a learning system that modifies policies as it goes along, based on information from workers, clients, vendors, suppliers, and managers. There are several critical points in the management of an IRM effort at which decisions must be made on available information. These points include: **project planning**, **project initiation**, **project termination**, and **audit points** during operation of the information resource. Data needed are typically cost- and benefit-related information. Specifically, these data relate to out-of-pocket costs; cost-saving and revenue benefits; and intangible benefits such as image, comfort, status, and organizational relations.

Before a cost-benefit analysis can begin, the value of information as a commodity must be taken into account. Information can be sold, bartered, and repackaged. Many firms consider information a by-product of work; others consider it a marketable commodity. Market players include software vendors and suppliers, hardware manufacturers, competitors, trade associations, academics, government regulators, and others for whom information makes a difference. The IRU can play the role of broker, initiator of products, and consumer of information.

Cost-benefit analysis techniques include computation of **net present value** and **payback periods** — these concern only tangible costs and benefits. Two other techniques are uncommon but increasingly important as IRM examines increasingly complex systems. **Social accounting** — attempts to assign numbers to social values and costs without actual dollar figure assessments, taking into account factors such as image of the organization, employee relations, customer relations, and market positioning. Organizational communication network analysis looks at the values of changed relationships on the job.

Value-added analysis examines DFDs and attempts to discover which specific value is added to information at every processing step. Values include faster, more accurate, or more confident decisions, higher-quality data with fewer errors, faster access to more data, and so forth. The value-added concept may then be combined with a cost-benefit analysis to demonstrate that certain aspects of a proposed IRM effort are not worth doing or will have unexpected benefits. Value-added analysis also focuses on transactions between (sub)systems, something economic cost-benefit analysis tends to ignore.

Making decisions requires choosing between alternatives. A number of decision-making techniques fall into two classes: static and projective. **Static** techniques compare existing characteristics of alternatives. Constructing a **payoff table** is the primary technique in static analysis. There are several projective or **forecasting** techniques that systems analysts can employ: **straight-line projection**, **cyclic projection**, **dynamic modeling**, **logical modeling** (borrowed from artificial intelligence), and **consultation**. The **Delphi** technique has been employed successfully in making forecasts for a variety of "futurist" projects.

This chapter ends our concentration on analysis tools and techniques. The following chapters look closely at how to organize IRM efforts, generally in the context of projects and alternatives to projects.

14.7   CASE: DECIDING ON THE COSTS OF CROSSING THE LINE OF SCRIMMAGE

Scrimmage Systems has presented its preliminary analysis of the sports package required by the Bigtown Bisons. Sandy has proposed an extensive data collection system based on videotape, sideline recording of plays, a number of preseason and pregame assessment strategies, and a series of post-game, post-season, and monthly meetings for analysis by the coaches, manager, and owner.

Sandy figures that the better the Bisons play, the better their game receipts will be. Tickets sell for $14.50 each (average) and average attendance has been 22,000. Broadcast rights are not to be figured in since the Bisons are locked into a five-year contract — anyway, local games are blacked out. Sandy figures that the use of his system will increase gate attendance by 5 percent for each of the seven home games and three exhibition games. Currently, 100 percent of the $14.50 is used to cover expenses.

Sandy figures that the cost of development will be: $24,000 for consulting fees, an additional $16,500 for computer hardware, $26,000 for software to be purchased and developed over a two-year period (half in each year), and $12,000 in annual labor and management costs. Sandy includes certain continuing consulting costs into his fee.

Questions

1. Calculate the net present value of the sports system over a five-year period given discount rates of 5, 10, and 12 percent.
2. Calculate the payback period of the system.
3. Calculate the return on investment over the five-year period.

4. Create a list of out-of-pocket costs for the project and its operation over five years. What are additional intangible costs? Are there any cost savings that this system may bring? Are these available to bring into the computations for questions 1-3?

5. What might be learned from a social accounting exercise with regard to the Sports System? How might this affect the management of the project and its products?

6. Consider the value added to the existing "information resource" (which is quite informal). What values are added to this resource and what are they worth?

7. Use your friends and colleagues as the source of information for constructing a payoff table for this kind of system versus using the coaches' intuition and experience alone. Perform these calculations over a ten-year period using today, +5 years, and +10 years.

8. Write up an informal audit plan for the Bisons to use in evaluating the effectiveness of the sports system. Given the cyclical nature of the system, what periods would you propose to perform measurements in?

DISCUSSION QUESTIONS

1. The text divides costs into out-of-pocket costs and intangible costs. For an information resource, the out-of-pocket costs can be neatly built into a charge-back policy, but the intangible costs are borne, too. How can intangible costs be built into a charge-back policy? What would be the justification for this? Can an information resource make a "profit" through charging for intangible costs?

2. Cost-savings are traditionally the justification for building or improving an information resource. The text indicates a number of sources for cost savings and difficulties in making economic arguments in many cases. Is there a pattern here? In what sense might it be better to say that cost-saving arguments avoid the issue of *effective*, rather than merely *efficient*, use of information resources? In the future, if cost savings are not apparent, what cost-benefit arguments may prove more important?

3. Locate and point out opportunities for information as a commodity in your environment. How would you go about determining the value of information in this environment? If an information-supplying organization *failed* to prosper, what would be the likely causes?

4. The rationale for cost-benefit analysis is to show that a project is (or is not) worth doing. What assumptions are these analyses based on? What do we have to assume about money, the environment, technology, people, and work? In what ways does social accounting differ from economic analyses (NPV, payback, ROI) in terms of these assumptions?

5. A number of questions were asked about the payback analysis example on page 503. What are the answers? What use can we make of those answers in decision making? Which answers are important for which decisions? Suppose the discount rate were predicted to fall from 10 percent the first year by 1 percent in each subsequent year. What is the payback period? What is the maximum exposure?

6. A builder in Bigtown has long used his "good judgment" to estimate home-improvement projects. Usually he gets the estimate correct, but recent wide fluctuation in building materials costs has caught him a few times and reduced profit. Discuss the specific values that might be added through systematic cost-determination procedures and data collection and analysis of suppliers to builders in the home-improvements market.

7. Contrast static and dynamic decision-making techniques in terms of their assumptions about the environment. What value are the index values that both techniques produce?

8. A payoff table is really an algorithm or process with a specific program that is easy to put into spreadsheet format. The table of benefits is easy to create and modify using a spreadsheet processor such as Lotus 1-2-3™ or Multiplan™. Locate such a program and process and graph the data provided in the chapter. Make modifications in the weights and note the effect they might have on your decisions.

9. Contact a management or business consultant and find out how they provide consultation help to others. You might begin this effort by first asking how *you* provide advice to others.

10. The major deliverable of an audit plan is a regular evaluation of information resource effectiveness and efficiency. However, choosing the *time* of the audit is critical. Discuss the effect of factors such as seasonal and general business cycles, human resource considerations, technological development, and company policies on the timing and terms of reference of the IR audit.

DESIGN EXERCISES
Using dBASE III

1. In dBASE III, there are two ways to produce reports. You can write programs that examine files, accumulate data, and then produce reports, either on the screen or via a printer. An alternative to this is the **dBASE III report facility**. For details on how to produce reports using the report facility, consult *Learning to Use dBASE III: An Introduction* by Shelly and Cashman (Boyd & Fraser, 1986). Space permits only a brief introduction to what the REPORT command does.

A **report form** is entered through the CREATE REPORT command. Create a report called "MARGINS" that provides a breakdown of sales margins by product class (the first two characters of ITEMCODE) in PRODUCTS.

Enter the following in response to prompts as given:

USE PRODUCTS	We will report on PRODUCTS
CREATE REPORT MARGINS	The report is called MARGINS
PRODUCT MARGINS BY CLASS	Goes into page heading

DESIGN EXERCISES
Using dBASE III *continued*

Press enter until the next screen comes up. It is identified by "Group/Subtotal on:" near the middle of the left side of the screen. Enter the following:

SUBSTR(ITEMCODE,1,2)   Product class is identified by the first two characters

Press ENTER until the next screen, identified by the words "Field Contents" near the middle of the left side of the screen, comes up. Enter the following:

DESCRIPT ~ ITEM ~   (*Remember*: ~ means enter.)

The field DESCRIPT will be displayed with the column header "ITEM." Press enter until the screen reappears with an empty form. Enter the following:

SALESPRICE – PRICE   The margin comes next, with 2 decimal places and a
~ 2 ~ Y ~ MARGIN ~   total, the column headed by "MARGIN"

Press ENTER until another empty form appears. Note the line across the middle of the screen. This indicates that there will be a 15-character field followed by a number between 0.00 and 99.99. Press ENTER instead of any field contents to end the creation of the report form.
    Now, type the following:

SET INDEX TO PRODUCTS   Files must be **indexed** or **sorted** to have subtotals by
      groups.
REPORT FORM MARGINS   Then ask for the report.

You could print the report out by typing the following:

REPORT FORM MARGINS TO PRINT

Suppose you wish to produce a report on margins by supplier. You will have to sort the file first on this field:

SORT ON SUPPLIER TO PRODSUP
USE PRODSUP

Now modify the report by entering the following:

MODIFY REPORT MARGINS

DESIGN EXERCISES
Using dBASE III *continued*

Follow the prompts to modify the report. You will have to change the "Group/Subtotal on:" area. Now, do the report.

2. dBASE III has a few other commands and features that help you write reports. The first is a **selective copy** command. Let us take copies of each record on order in PRODUCTS:

COPY TO ORDERS FOR ONORDER = "Y"	Copy each record for which the ONORDER field is a Y.
DISPLAY ALL ITEMCODE, ONHAND, DESCRIPT	Tell us.

Do you notice something missing in this file that could be of assistance? (*Hint:* How effective is it merely to record that something is on order?)

3. Of course, the dBASE III Report facility cannot produce every report. In particular, **Logical Design Specifications** cannot be produced that way. Look at INTERVU and IDEAS, your program GOODIDEA, and the data in REQ4SERV. Write out a report indicating what changes should be made to HAIR to help Doug run his business better. What influences come from dBASE III and its way of processing information? How does knowledge that the system runs on a microcomputer affect your judgment of the logical functions and their structures? As you write your report, you can use the files and programs you created to check your ideas.

CHAPTER 15

PROJECT MANAGEMENT: PLANNING, CONTROL, AND REPORTING

OBJECTIVES

1. To outline the steps involved in a computer-based information system implementation project
2. To outline the resources needed to engage in a CBIS project
3. To carry through examples of planning and control techniques including CPM, labor distribution, and work assignment in a case example
4. To list several techniques for monitoring and reporting project progress, expenses, and documentation
5. To discuss the important issues in selection, deployment, and management of programming personnel on projects
6. To plan and present an oral project report
7. To plan and present a written project report

PROJECT MANAGEMENT: PLANNING, CONTROL, AND REPORTING

15

CHAPTER

15.0 AND I WANT IT *YESTERDAY*!

Ed Waller is a senior systems analyst for a large media firm in Bigtown. Ed's responsibilities include the direction of three analysts and one junior analyst, and occasional project management. Currently, Ed is working on two major projects. The first is a simple manpower planning program for the personnel department. The other is a more complex inventory system for tape clips that television reporters have been clamoring for. This morning Ed acquired a third project.

Ed was summoned to Virna Bright's office just after sunup on this cold December day. Virna, corporate information systems director, told Ed that she had been having discussions with Twig O'Neil, the sales VP, about putting up an on-line sales system for their media salespeople. As luck would have it, Charlie Kowalski, Twig's assistant VP, just happened to be in Virna's office with a proposal.

Briefly, the proposal ("Ultimatum," thought Ed) was that all sales leads, time-slot allocations, advertising rates, categories, and some sample media scripts would be available to the salespeople, who would use cellular radio and portable microcomputers to communicate directly with the computer, a mid-size, aging, minicomputer of late 1970s vintage.

"Just think about the *image*, Ed," Charlie crooned in a salesman-like tone. "We're the first, the best, and now we're the latest! High tech, even in sales. Who could fail to sign if we can commit *and* enter script ideas right from their office?"

Ed wasn't so sure, but because Virna was obviously in favor of it and because there seemed to be no immediate technical problems, he agreed to look at the project and get back to Virna in two days. Charlie danced his way out of Virna's office. Virna looked at Ed. Ed looked at Virna. Then Virna asked, "Well, Ed, where do we go from here?"

This chapter is the first of three on *project management*, which is where Ed "goes from here." Project management involves coordinating staff efforts to produce a viable, useful product at the end of a limited time. Because most projects function in turbulent environments with limited resources, and because projects are themselves growing, living organizations, many of the characteristics of *systems* apply directly to projects. We will look at the project as a system with two different emphases. First, systems analysts are often pressed into service as *project leaders*, individuals who, like Ed, take charge of projects. Second, project leaders are involved in all phases of a project, not just systems analysis; these three final chapters allow us to put logical systems design back into its full context.

We will begin with project planning. As Figure 15-1 illustrates, project planning is one of four subgoals of project management. Project planning precedes actual project work. The second, project monitoring is continuous, as it is essentially a set of reporting activities. Project managers must be aware of current information regarding the project's progress and make it available to others. Project control, used here in the larger sense of "directing and controlling", includes a number of techniques for getting the project accomplished on time and within budget. Finally, project documentation is an integral part of the previous phases. Reporting on project work — already discussed in Chapters 7, 9, 10, and 13 with respect to logical system design — completes the job of delivering the product. Several kinds of project documentation are discussed.

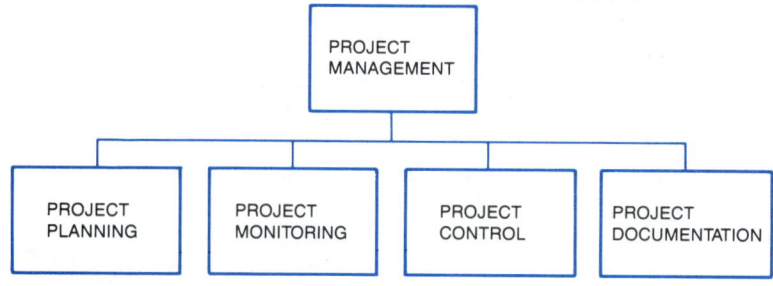

FIGURE 15–1. Phases of Project Management

In addition, Chapter 15 discusses the difficult area of human resource management in a project setting, including selection of personnel, motivation, training, and career development. A special section on managing programmers also is included.

15.1 PROJECT PLANNING

The project "is a temporary organization of individuals working on a single product or system for a specific customer" (Licker, 1985, pg. 48). Projects are *temporary*. They are meant to begin at one point in time and end at another. The pressure of schedules is real. Resource expenditure is limited, as are budgets. Projects are usually run on the model of engineering construction work, implying that an estimated timetable and budget are required in advance. Historically, it has been difficult to meet these promises. Not only is the hardware complex, but the *people* in the project can be difficult to manage.

And that is the second major component. A project is an organization of people more than it is an organization of tools or computers. Project management cannot be separated from human resource management. Good *general* management skills are necessary, which implies communication, coordination, planning, motivation, and judgment.

Projects work to produce specific products. These products are rarely simple single-module software units; more often they embody complex interactions of people, procedures, software, and hardware. A project is often expected to produce not only a running system but trained users, manpower plans, and financial plans as well. Nonetheless, it is product orientation that identifies individual projects: the personnel file project, the market analysis system project, the executive DSS project.

Finally, and often ignored, is the political aspect of projects. A project exists for a customer who ultimately has to claim and use the product. Traditionally, the IRU has not paid much attention to customer relations or communication (Guinan and Bostrom, 1986). Interest in alternatives to project organization has grown because this interface has been badly managed in the past.

Due to the time, resource, product, and political limitations of projects, planning is an important function. Project planning includes project selection; financial, human, and technical resource need forecasting; and work and manpower assignment.

Aspects of Project Selection

Kronke (1980, pp. 100-101) speaks about three aspects of **feasibility** that are relevant to project selection: can the project be done (1) on time, (2) within budgeted resources, and (3) with current technology? There is one more: *should* it be done?

Time and resource needs are difficult to predict; obviously, projects that are too long or too expensive are often rejected out-of-hand before commencement. Making detailed economic projections is infeasible before the data-collection phase, but some projects should never even begin because of their enormous expense:

> *CASE 1.* A large multinational oil firm wanted to put all seismic information on its minicomputer. Because the analyst knew that the mini's capacity was only a tenth of that necessary, he vetoed the work at the start because it was overly expensive.
>
> *CASE 2.* A transportation firm wanted to build a database system to handle on-line queries for parcel and shipment location to be accessed from any point in its interstate network. The analyst in charge nixed the project at the start because there were no employees on staff who had had any experience working with telecommunications. Merely tooling up the staff would have required six months and a doubling of staff.

Finally, a project may be rejected because of undesirability with respect to developers, users, or clients.

> *CASE 3.* A forest products manufacturer wanted a system to keep track of forestry lands. The IRU, however, felt that it could not spare anyone to learn graphics software development techniques for at least a year, considering that this project would be unique and would not produce skills that could be reused by staffers.
>
> *CASE 4.* A scheme by a retail organization to computerize its executive offices and put "a terminal on every desk" was rejected when it was discovered that executive secretaries could too easily access confidential records. While they were trustworthy, no one could guarantee that their passwords would not be discovered.

CASE 5. A state bureaucracy wanted to automate the handling of citizen complaints for easier response (granted, by form letter, but a response nonetheless). The MIS director said "No" when he learned that one legislator proposed using the files as a way of learning which political pressure groups were pushing the strongest.

A final consideration is the overall business plan of the IRU and the goals of the organization in general. Many IRUs have a plan for desirable projects, especially where the IRU is autonomous and treated as a profit center. Organizational goals may include state-of-the-art computing, but short-term limitations may restrict exploratory work and pioneering. The older and more established the IRU, the greater the chance that the information resource requires significant human resources for its own maintenance, thereby precluding new development. As Martin (1982, pp. 83-84) pointed out, maintenance work — system maintenance foremost in this — will form an increasingly larger proportion of information resource work.

In most organizations, IRU budgets are of two types. In service-oriented units, user requests are gathered annually, prioritized, priced out, and compared to resources. Projections of resource needs are based on these requests and company history. Smaller or autonomous IRUs can express their own needs and budgets according to an overall business plan in an annual budgeting exercise.

Financial Resource Planning

A number of costs have to be estimated at the start of a project. Direct costs are those that tend to appear in budgets, but all costs ultimately have to be accounted for. A *loaded labor rate* often includes allowances for benefits, shared office costs, and other indirect costs. To cover the rest, project managers often utilize a rule of thumb to multiply fully productive labor times to get a more realistic estimate.

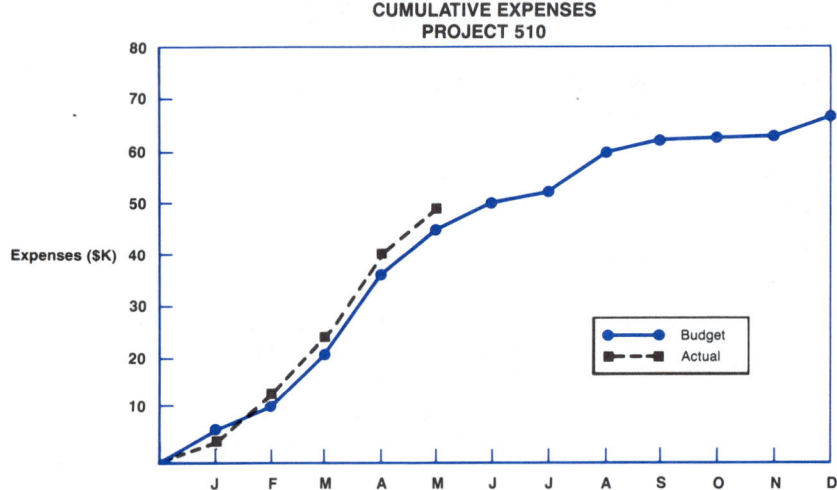

FIGURE 15-2. Expense Tracking

Figure 15-2 illustrates a common vehicle for expressing the budget of a project. The x-axis represents time; the y-axis, either expenses by period (usually months) or cumulative expenses through a period. This diagram is also used to make periodic variance reports between actual expenditures and budgeted amounts.

Human Resource Planning

The right people have to be in the right place at the right time. The primary tool for this kind of planning is the **Gantt Chart** (see Figure 12-26), which shows who is where doing what and when. The chart in Figure 15-3 shows that M. Slivovitz will work on analysis from 1 July through 15 August; R. Shane, on analysis from 15 July through 15 August; N. Markoff and S. Linker, on programming from 15 August through 15 November; and T. Fountain on project management from 15 June through 30 November.

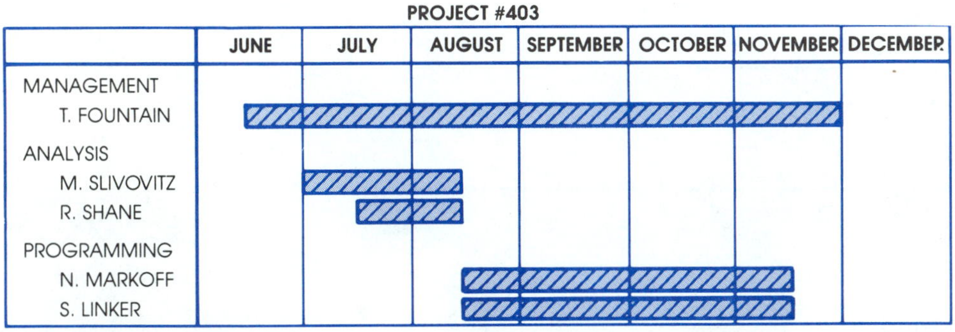

FIGURE 15–3. A Gantt Chart

The Gantt chart is also useful for determining costs, since most organizations have general figures for labor costs broken down by worker type. Consider that a senior analyst may charge $250 daily (Slivovitz will work 33 days); a junior analyst, $200 (Shane will work 22 days); a programmer, $175 (Markoff and Linker will work a total of 132 days); and a project manager, $300 (110 days, one-third time). Total daily labor costs come to $46,750 over 5.5 calendar months.

Technical Resource Planning

Technical resources include the special hardware and software that have to be procured for a project and existing resources that have to be shared with other project or production work. One thorn in the side of development work in small organizations is the necessity to overutilize an underpowered micro- or minicomputer for both production and development work.

Technical resource scheduling is often illustrated, along with completion dates for phases of IRM and software development, in a **milestone chart**, illustrated in Figure 15-4. Here, a variety of milestones in hardware procurement and installation, systems analysis and design, program development, customer relations, and project management are charted simultaneously across time (x-axis). A milestone chart is the basis for use of critical path methods.

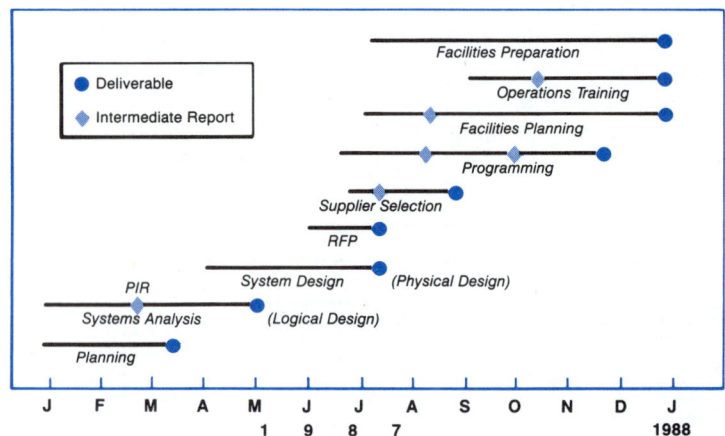

FIGURE 15–4. Milestone Chart

Critical Path Methods

Critical path methods (CPM and PERT — Program Evaluation and Review Technique) are concerned with project **activity scheduling**. They are based on three assumptions. First, all activities to be engaged in are known about in advance. Second, before we begin planning, we know, within a range, how long it will take to perform each activity. Finally, we know the *structure* of these activities (which are necessary for which).

These methods aim to discover the **critical path** of activities whose timing is crucial to the overall completion date of a project. Activities *off* the critical path may slip a bit without affecting the overall termination date of the project; those *on* the critical path cannot slip without delaying delivery of the product.

To begin the critical path method, we create an activity list based on the milestone chart. Figure 15-5, on the opposite page, gives an example, showing seven activities involved in collecting interview data. The circles indicate the activities; the lines, the dependency relationships among them. Thus, we cannot schedule interviews until we have created an interview list. We cannot perform the interviews until we've scheduled the interviews and pretested the interview schedule. In some cases, we may wish to perform a few interviews while we schedule.

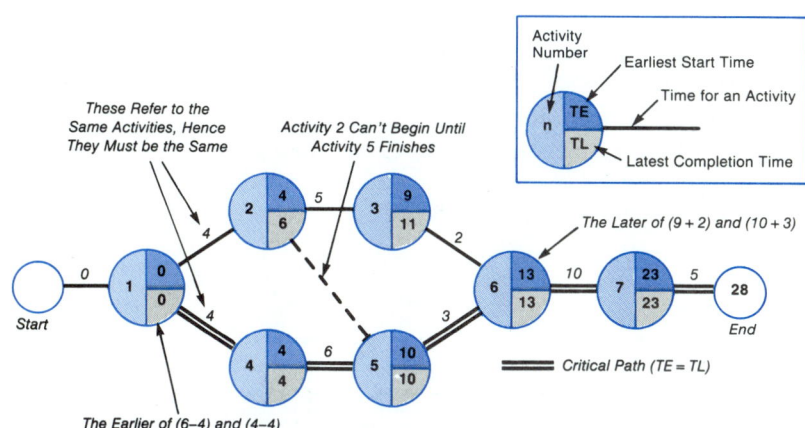

FIGURE 15–5. Using CPM

Along each line, we indicate the calendar span each activity will require. We then assign day zero to the *start* task and accumulate **earliest completion times** (TE) and **latest completion times** (TL) for each task. Since no task on the diagram can begin before all antecedent tasks are completed, the TE at any point is the latest TE for all tasks with arrows leading into it plus the time to complete that activity. We continue this analysis until we have computed TE for each activity. We arbitrarily assign TL = TE for the last event and work backward, assigning to each task a value of TL equal to the smallest TL for all subsequent events *minus* the duration of that task.

In our example, we have assigned five days to creating the interview list and two days to scheduling interviews. It will take six days to create the interview schedule and three more to pretest the interview. Interviewing is going to take ten working days. We compute a series of TEs and note that the project will take twenty-eight days (TE for END). Working backward, we compute TLs. The dotted line indicates a *dummy activity*; nothing is actually happening, but we will have to wait after scheduling for the end of the pretest—this event has a duration of zero days.

Our computations finished, we proceed to analyze the numbers. Note that START, CREATE INTERVIEW SCHEDULE, PRETEST INTERVIEW, PERFORM INTER-VIEWS, ANALYZE INTERVIEWS and END have TE = TL. These activities lie on the *critical path*. In other words, slippage of completion dates in any of these seven activities will delay the end of the project. Other activities can slip a bit. How long can TIMETABLE INTERVIEWS slip without affecting the project?

Critical path methods (we have used PERT; CPM differs by labeling the circles as events and calculating times along the arrows) provide the project manager with two decision aids. Knowing which tasks fall on the critical path, the manager can prioritize and assign resources to those tasks critical to meeting schedule goals. Second, the critical path can be "what-iffed" to see what the effects of slippage might be. A number of microcomputer and mainframe packages such as the Super Project Manager™ can assist in creating, updating, and analyzing critical path diagrams.

Manpower Loading Methods

Gantt charts may contain information valuable to computing budgets; they can also be used to compute person-days required for each type of work. In fact, the most reasonable way to begin a Gantt chart is to break down effort into types directly from the task list; which is derived from an **activity structure chart** like the one shown in Figure 15-6. We decompose the interviewing phase of investigation into the constituent activities found in the task list, which subsequently appear on the PERT chart.

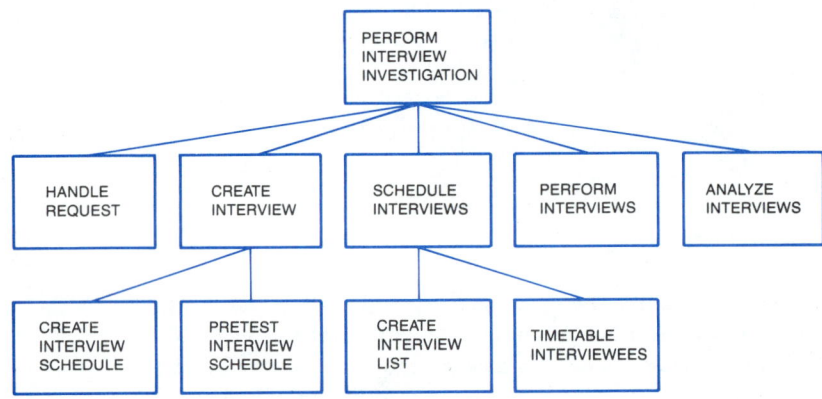

FIGURE 15–6. An Activity Structure Chart

Each task requires labor of specified types. For example, pretesting the interview will require the junior analyst two person-days (four times four hours) and the senior analyst one-half of a person-day, although in terms of calendar time, three days are estimated. We can then break down staffing requirements by type for each activity and then sum across all activities to determine the estimated costs of each activity.

Principles of Work Assignment

Having planned tasks and their structures, determined the critical path, and computed estimated costs, project planning concludes with assignment of work to specific workers with specific goals. Work assignment depends on seven factors: *size* of the project, development *methodology* employed, specific *skills* required, *availability* of specific staff members, availability of *resources*, *timetable* considerations, and *testing and integration* considerations.

Large projects may be more flexible in terms of staff work assignments because they have larger pools of individuals to fall back on. Smaller projects may not have this redundancy—once assigned, a worker may not feel free to leave a project. The particular system development methodology may dictate that certain overhead activities (reporting, conferring, agreeing, negotiating) be performed by specific individuals, thus limiting their availability for other tasks.

Certain skills may be required for specific projects. In some projects, a particular skill, such as interviewing upper management, may be needed for a brief period of time. Those with the credentials and skills may not be available. Working with users requires communication skills not widely available. Thus, the project manager may have to locate or teach these skills.

Project Organization

Projects may be organized in a variety of ways. By far the most common is **matrix management**. In this scheme (Figure 15-7), workers report administratively to a "discipline director" (such as the manager of programming or the manager of systems analysis) and report on a day-to-day basis to a project leader. Matrix management has the advantages of focusing work on projects without burdening administrative managers with details and of handling many projects concurrently. The disadvantage of this management style is that the discipline directors may lose contact with their workers and the work they are doing.

PROJECT	101	203	341	351	372	373	382
PROGRAMMING							
BLAKE	*			*			*
SOROYAN		*					
TODD	*	*			*		
ANALYSIS							
FRANKLIN	PL						
HARRISON		PL		PL			
JUAREZ				PL			
LOMBARDI						PL	
APP. PROGRAM							
ASHBY	*		PL				
BATESON							PL
deGROOT		*		*	*	*	
GRIPTON		*		*			
SANTION			*			*	
VALESQUEZ	*						

FIGURE 15–7. Matrix Management

An alternative is the **team approach**, wherein a semi-permanent team is set up to work on a project over a long term. For instance, an analyst, two programmers, a coder, and a secretary/clerk may work for a project leader through a project office for two years building a system for an operating division. The advantage is total concentration of human resources to a specific, long-term goal. After development is completed, many of the same workers can continue involvement in maintenance and operation of the system. This arrangement provides a long-term life to otherwise time-limited projects. In effect, the team becomes a separate administrative department. The disadvantages are that workers may become stale, disinterested, or unmotivated over a long period; their long-term career interests may suffer, too, especially if the project does not succeed or is canceled.

Summary of Project Planning

The project planning exercise is complete when all the elements of the plan are present and the project is ready to commence. The plan includes the following (Figure 15-8):

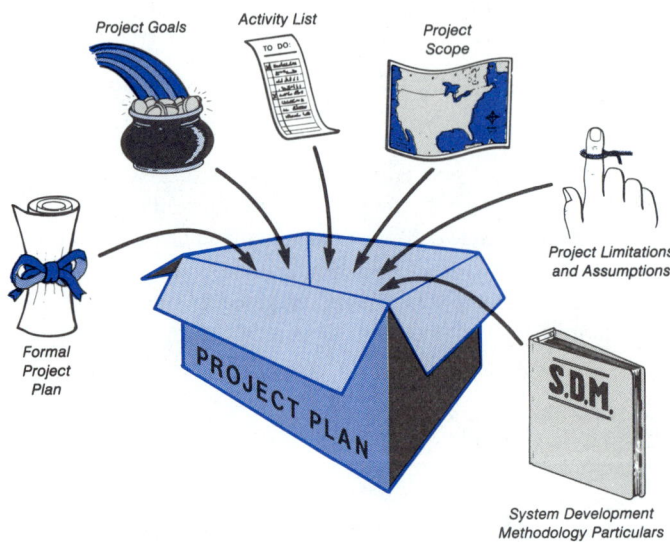

FIGURE 15–8. Components of the Complete Project Plan

- **project scope** (see Chapter 8);
- **project limitations** and assumptions (Chapter 8);
- **project goals**, subgoals, sub-subgoals, etc.: what specific value will be achieved and most important, *how to measure that value* (Chapter 14);
- **project task list**: from the project goals, an activity structure chart (Chapter 4);
- **formal project plan**: summary of the project, schedules, resource requirements, project management responsibilities, assumptions about operations during development, how the system will be constructed, tested, installed, trained, operated, evaluated, and maintained
- **system development methodology** (SDM) details: if no formal system already exists, how projects will be developed, managed, monitored, controlled, and released; who has responsibility for what units; who has authority to delegate these responsibilities.

15.2 PROJECT MONITORING

Once a project has commenced, the project leader's major responsibility is to see that it progresses along the project plan. There are two aspects to this, as illustrated in Figure 15-9. Seen as a cybernetic system with feedback, a project has component parts that sense the project's environment and others that correct project behavior to conform to the plan. The *sense* elements fall under the heading "project monitoring".

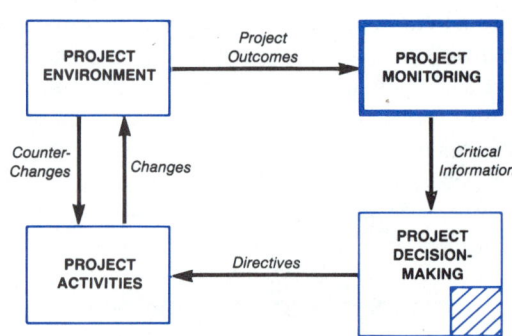

FIGURE 15–9. Project Monitoring Within a Cybernetic Situation

Project monitoring implies overseeing project progress in the critical terms of time, expense, product, and politics. This section discusses reports and meetings which assist in project control.

Monitoring through Reports

Project reports document the progress made toward achieving the product. They compare the current situation with the planned one in the following ways:

- **expenditure tracking**: reports of accumulated expenditure against planned expenditure to date, documenting significant spending variances (Figure 15-2)
- **milestone reports**: indication of expected dates to meet planned milestones and documentation of expected schedule slippage (Figure 15-4)
- **activity reports**: succinct reports on activity in the reporting period; distant early warnings of potential schedule or expense overruns; documentation of staff changes, policy changes, and technical accomplishments.

Project reports are typically addressed to IRU management, since it is IRU costs and staff involved. Often progress briefs are distributed in conjunction with client review meetings (see the next section). It is important to maintain client contact regardless of *formal* reporting techniques.

Diagrams accompanying these reports include: graphs of cumulative expense-to-date against planned expense-to-date, partially completed and recalculated PERT charts, manpower loading charts with details for the reporting period, and summaries. A semi-automated system may provide access only to records and some prompts to enter the data, whereas a fully automated system can produce graphic charts and recalculate critical paths, manpower loadings, and exception reports on expenditures.

Monitoring through Meetings

Several kinds of meetings are held during projects. Figure 15-10 shows the variety of meetings a project manager may call and/or attend:

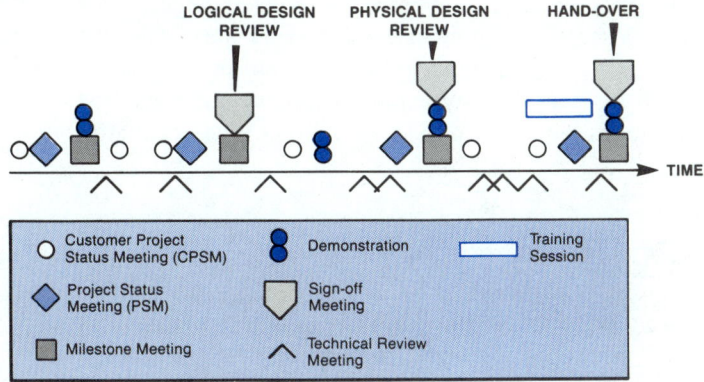

FIGURE 15–10. Varities of Project Monitoring Meetings

- **project status meetings** to prepare for milestone reports
- **client** (or customer) **project status meetings** to update clients on project progress
- **milestone meetings** to conclude phases of work and hand over responsibilities to team or project members
- **demonstrations** of investigation findings, logical and physical design, modules, subsystems and systems
- **formal sign-off meetings** to deliver systems or subsystems to users
- **training sessions**
- **technical reviews** among small groups of staff members to review technical aspects of the project

Project status meetings prepare for writing documentation on the project. At client status meetings, which are often called in response to client fears about overexpenditure or late delivery, decisions on project management or planning should *not* be made, but clients should be informed of major project decisions affecting costs or schedule.

Milestone meetings are technical in nature. They assist the project manager in planning for the next milestone, especially if the original project plan has to be modified. Project status meetings should be called in advance of milestone meetings to find out if there will be difficulties in meeting milestones.

Training "meetings" may be part of a project in the sense that users may have to be trained before full-scale demonstrations can occur. Unfortunately, users (as opposed to clients) are often forgotten in system development, and training sessions become the initial exposure of analysts to users' experience with the information resource being developed.

Finally, there may be many technical reviews during projects. The analysis group may review interview schedules and raw data during investigation phases. The design group may go over and over the physical design while programmers meet in structured walk-throughs. These meetings are often conducted without project management present since they concern only the technical matter under control of technical staff.

Monitoring and Control

The purpose of monitoring project progress is to prepare for changes in the project plan should they become necessary. When expenditures rise beyond estimates, it may be necessary to redeploy staff or, in some cases, make more realistic estimates (with attendent arguments as to why being realistic is more important than being precise in the first place!). The following section illustrates the use of project monitoring to effect control over a project.

15.3 PROJECT CONTROL

Based on the data delivered from project monitoring activities, project management makes decisions to take action such as reassigning work, rechanneling responsibilities, or altering the project plan. Changes may be in response to slowly evolving events, but typically they are in response to crises. Methods of responding include (1) redirection,

(2) ignoring crises, and (3) managing for maintainability. The last method attempt to avoid some of the crises by building in redundancy and anticipating error.

The Project Plan over Time

The project plan is not a rigid document; it changes over time in response to its environment. Because of this, a project is best seen as a *learning system* (Figure 15-11). A second sense cell is added to observe the value of the project plan over time, and another control cell modifies the the plan according to policies set down by IRU management. Thus, while the IRU staff responds to environmental changes according to the project plan, IRU management directs project leaders to alter project plans according to policies.

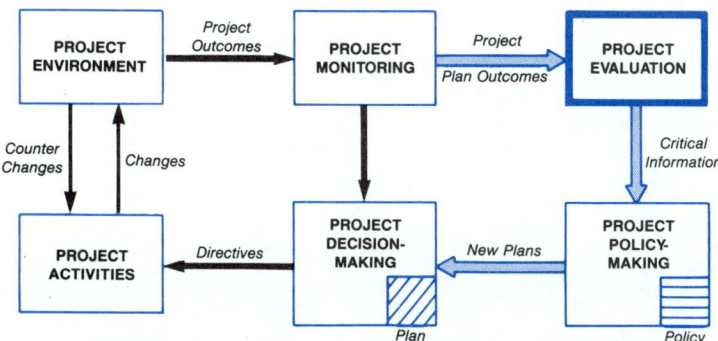

FIGURE 15–11. Project Management as a Learning Situation (cf. Figure 14–9)

In some shops, policies allow project plans to change in response only to long-term needs of the IRU; others may encourage project leaders to respond to short-term staffing requirements. Where planning policies do not exist, project plans may be *over*responsive to the environment. Here are several views of response to the same situation:

> Tony Mullhaven is the leader of the Gasex project for Northwestern Power and Light. Originally the project plan called for a three-week preliminary investigation and a subsequent four-week detailed investigation. But Steve, the junior analyst, quit a week into the preliminary investigation.
>
> *CASE 1:* Tony did the work himself. Because he is also senior systems analyst, he could not do too detailed a job. But the project got the go-ahead, and Tony now finds himself again doing most of the work. The plan calls for sixteen interviews of senior gas marketing managers and analysts. Tony's going to do all sixteen with the help of a raw recruit he hired yesterday.
>
> *CASE 2:* Tony let the deadline slip while he hired another junior analyst. Sarah Stern, the new analyst, has completed the PI, but two of the senior managers she was to interview have gone on vacation, so Tony will delay the detailed investigation until they return in one month. Meanwhile, Sarah has begun writing up and arranging to pretest the interview schedule for the detailed investigation in expectation of approval. This way, they will not lose

time interviewing the sixteen senior gas marketing managers and analysts. Tony is pretty sure he will have to interview in two rounds in case the review meeting turns up something he has forgotten.

In the first case, Tony overcontrolled the project in the hope that he could still meet the milestones, but he sacrificed quality. In the second, Tony underdirected by allowing Sarah to change the plan herself. Neither case had a policy for making changes to project plans. Tony could use some guidelines.

Project Crises

Project crises fall into four categories (Figure 15-12): client-initiated crises, productivity problems, organizational roadblocks, and bad luck.

CLIENT-INITIATED CRISES	ORGANIZATIONAL ROADBLOCKS
Specification Changes	Premature Project Cancellation
Schedule Changes	Lowered Project Priority
Personality Clashes	Loss of Access to Crucial Resources
Failure to Approve	Policy Changes
PRODUCTIVITY PROBLEMS	**LUCK OR BAD PLANNING**
Failure to Alert to Slippage	Late Delivery of Hardware
Defective Products	Employee Absenteeism
Unexpected Turnover	Inadequate Plans
Fatigue, Overwork, Burnout	Fictional Budgets
Insufficient Training	

FIGURE 15–12. Project Crises and Their Sources

Client-initiated crises stem from the essential turbulence of the business environment. Clients may ask for *specification changes*, demand *schedule changes*, or resist schedule changes project managers deem necessary. They may exhibit *personality clashes* with project staff, or *fail to sign off* on perfectly good work. The sources of these crises are varied, and, in truth, not all of them can be anticipated.

Productivity problems often go unnoticed until it is too late and they become crises. They come about because software planning is still relatively primitive. Workers may *fail to alert the project leader to slippage* or *deliver defective products* which have not been tested well enough. There may be *unexpected turnover*, *personality clashes* among staff members, *fatigue*, *overwork*, or *burnout* among staff. It may turn out that a staffer who is supposed to know X does not really understand it at all. Staff productivity problems may be anticipated to some extent with frequent and detailed reporting, but, of course, this overhead activity may by itself impair performance.

Political problems include *premature project cancellation*, *lowered priority of the project*, *denial of access to important resources* such as computer time, or *policy changes* which may overexpose the project or lower its status and flexibility. Usually, political problems have distant early warnings, but IRU management is not always well positioned to receive and interpret these warnings correctly.

Finally, **luck** plays a part. Hardware orders are sometimes late or canceled, employees can get sick, operating systems are sometimes not delivered, and, as a sage professor once said, "The computer always goes down the day before the project is due." Planning principles allow for some turbulence in making estimates. PERT adds the lowest, the highest, and four times the median estimate, and then divides the sum by six. Budgeting is sometimes an exercise in creative writing. And there are crises that are just plain bad luck.

Many project crises can be either avoided or anticipated in the project plan. The next section looks at four types of responses to unanticipated crises.

Project Control Alternatives

There are four general approaches or philosophies of response to crises. The first and easiest is to **ignore** them. This undercontrol philosophy works only in the most placid of environments because it essentially zaps out the control box in the feedback loop. More important, it does not allow the learning system to learn anything about coping with crises.

The next approach, **redirection** or *management by exception*, attempts to correct problems according to fixed policies. As a basic cybernetic approach, it works best in relatively low-turbulence environments. For example, a worker who delivers poor-quality software can be reprimanded and admonished to work better, retrained or coached to do the job better, or reassigned to another job. In some cases, the proper tools need to be made available. Sometimes counselling the worker helps. In all of these cases, the response comes *after* the problem is noted and *before* it becomes overwhelming and can't be corrected.

The problem with redirection is that in very turbulent situations, response may not be quick enough. Redirection depends on good policies and, more strongly, on detecting problems in the early stages. Because of the nature of software development, many problems escape detection in early stages, although top-down development and testing can at least minimize the effects of major failures in low-level modules.

The other problem in a redirection approach is that response may come too quickly. Consider a project leader who suspects that the detailed investigation is not uncovering the right data. He may quickly change the interview schedule and ask other questions. The problem, however, may only have been that the interviewees spoken to so far were relatively uninformed and that the questions were really more appropriate for *later* interviews. The project leader *overreacted* to a fairly benign situation.

The third philosophy stems from the school of software engineering and is called **managing for maintainability**. While the term applies to the design of software, it can be of value to project managers who want the project to be "maintained." This approach requires understanding how projects work and what types of roadblocks and bugs can occur in project leadership. It is beyond the scope of this text to go into this in any detail, but here is an example, based on the previous case:

CASE 3: Tony might have been in a jam had he not been managing to maintain the project. Knowing that staff turnover is one of the most common causes for major project delays, Tony double-teamed the PI. Tony's policy was to put two people to work on all aspects of important projects to build in redundancy in case workers quit or were sick, were on holidays, or at courses. It was expensive, but Tony found that some of the cost could be recovered in the excellent documentation that can be produced when *two* people know what happened. It does not work in all cases, but where continuity is important, Tony planned for it. If Luigi Lombardi experiences any problems lining up interviews, Tony is ready because he lined up twice as many interviews as necessary; he will not cancel any, but the surplus can concentrate on a few detailed areas and be very brief.

Tony's approach is one of *maintaining the project* at production levels in the face of the "usual" problems.

The final approach is to change the project plan in response to every crisis. There *are* times when the plan needs to be changed. If costs are going to be justifiably high, arguments for more funds may fall on deaf ears and estimates of deliverables may have to be scaled down. Staffing problems or opportunities may have to be faced and the plan, changed. On the other hand, the plan is more or less agreed to by clients, and too many changes make the plan look more like an action adventure. "Raiders of the Lost Plan" is not usually a good project management policy.

To summarize the first three sections of this chapter, project management requires a plan, monitoring progress toward the plan, and control techniques to ensure that the plan is carried out. As a *learning system*, a project should have guidelines for changes in the plan, not only to make the project more robust, but to build up documentation of experience. This way, managing projects for maintainability, rather than merely redirecting, altering the project plan, or even ignoring problems, becomes the project philosophy.

15.4 HUMAN RESOURCE MANAGEMENT ON PROJECTS

Often the toughest project problems are those of managing people. Systems analysts rarely receive specific training in management. Most IRU supervisors have never had a human resource management course. Some research indicates that people are attracted to this field by the technology and may even be repelled by personnel problems. Thus, the problems may be exacerbated by the situation and then appear insurmountable.

Issues in Human Resources Management

Four issues dominate human resources management:

1. how to select personnel for and remove personnel from projects
2. how to motivate and train project personnel
3. how to help workers plan careers in a project-dominated environment
4. how to direct and control the work of programmers

Figure 15-13 (opposite), shows the five elements of project management: **planning, organizing, staffing, directing,** and **controlling**. Each has aspects of human resource management. Project leaders *plan* tasks for people to work on and describe them in enough detail to staff them. They then *organize* the tasks into manageable units and *staff* each task, either hiring new workers or assigning existing workers. Next, they *direct* the workers to specific tasks, supplying details of the product expected, listing the work activities, and providing the necessary tools to get the work done. Finally, they *control* the activities of the workers to the extent necessary to keep work on target.

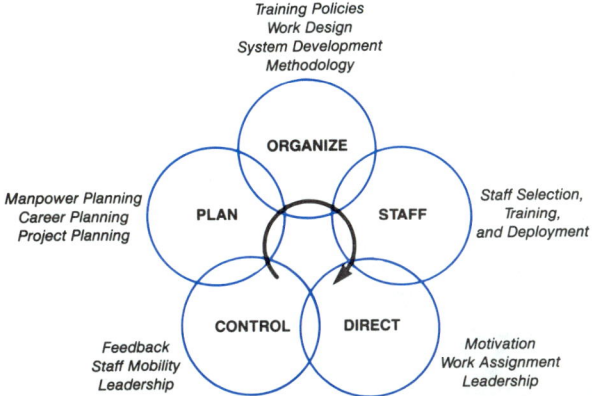

FIGURE 15–13. Issues in Project Management as Related to Human Resources

Selection of Personnel

Project selection is often predicated on the existence of sufficient human resources, so that, to an extent, personnel selection begins before the project commences.

Workers should be available for only those project phases in which they are involved. Skill, aptitudes, and personality factors need to be weighed carefully in putting teams together. The project leader must lead and workers must follow; this means that leadership skills are as important in selecting personnel as they are in directing the people.

A buddy-buddy group may not work; it may just party. Leadership can build team identification without sacrificing work. Thus, personnel selection can be based on whether or not the person can work in the environment the leader intends to build rather than whether or not the team itself is totally and mutually convivial. Needed skills should be there when required. A worker can be trained, but make sure the worker can *learn*.

Because project staffing is often out of the project leader's hands, questions of leadership become very important.

Motivation and Training

Analysts and programmers tend to be more highly motivated by work than by other workers, and more by technology than by their managers' charisma. Usually the most highly motivating aspect of project work is the opportunity to build a valuable product. This implies that a project leader must provide feedback on performance, recognition for accomplishments, opportunity to access the best technology around, achievement-oriented leadership, delegation of responsibility rather than heavy-handed interference, and assignment to other rewarding projects. Managers must pay attention to those who are not typical: overworked, overmotivated, antisocial, stressed workers need immediate attention — their difficulties will *not* go away by themselves.

Training motivates information resource workers because it increases their access to the high technology they crave. Courses in programming languages, new system development techniques, new hardware, and application-specific material are seen as pluses. Courses in technical writing, management, accounting, public speaking, and finance may be more valuable to the firm, but unless technology is involved, workers may not see these as powerful rewards.

On the other hand, training is not merely motivational. A worker who learns X when he needs to know Y may soon seek a job with Z. Also, workers require frequent access to training because technology changes rapidly. There are also valuable sources of training within a firm:

- Consultants should share their knowledge before they are paid.
- Mentoring is a relatively new, but important, source of training, especially concerning the organization as a whole and the IRU in particular.
- Giving a course is a learning experience, too.

In a matrix organization, it is often the project leaders alone who know what training a worker needs. Training should be part of long-term career plans. Until recently, programmers were too mobile (with an average turnover of 30 to 40 percent annually) to plan their careers. Now, project managers help by coordinating training across projects rather than making workers responsibile for their own training.

Career Planning and Projects

One of the least desirable by-products of project orientation in an IRU is a worker's acquiring an unstructured set of skills and experiences, which comes from working on an unmanaged sequence of projects. When projects are matrix-managed, the project leaders — who ought to know what skills a worker has — are too busy with technicalities to *develop* workers; discipline directors have the time and experience, but do not manage workers day to day. This means that training is more likely keyed to projects than to individual predilection or skill.

Staff mobility comes to depend on which projects one has worked on. Experience on a failed project may build character, but it does little for one's reputation. Knowledge of a particularly crucial element of teleprocessing might be important this year, but irrelevant the next. Where projects are long, portfolios tend to be short; where projects are short, experience tends to be shallow. In some shops, people are slotted early to minor sub-sub-specialties, effectively denied any chance to shift to other work because they lack technical credibility.

The short-term nature of projects makes it difficult for workers to find continuity in leadership and limits opportunities for mentoring, counseling, and coaching. Project leaders are often called upon to appraise the performance of their workers. Annual performance appraisal makes little sense in a shop having a lot of short projects. Career counseling then becomes a moot point — who could do it?

These considerations make it especially difficult for workers to view their careers as anything but a series of short-term assignments. In response, project leaders have to remain alert to these factors and practice good career management principles. First, training is for the benefit of the firm, but to retain a worker, training must also fit into the worker's career goals. Thus, project assignment should be based on acknowledged career progression for technical workers. Workers should not be "pigeonholed" permanently, nor should they be required to develop specialties without a regular review of their skills. Project leaders have to know performance appraisal techniques; performance reviews should be keyed to both projects and the calendar, with career planning as part of the performance appraisal exercise.

Managing Programmer Activities

Because systems analysts inevitably lead projects staffed largely by programmers, a review of management principles specially aimed at the supervision of programmers is important. Programmers, like analysts, are technical workers. But while analysts break problems down into constituent elements and propose solutions that are better-designed structures of those elements, programmers compose solutions from the blueprints. Analysts are the designers, and programmers, the builders.

The upshot of this is that programmers need both physical and intellectual access to technology. Couger and Zawacki's work (1980) points out that programmers have an inordinately high desire to **learn and grow** by conquering technology. Since there seems to be a limitless supply of technology, programmers can learn indefinitely. On the other hand, **affiliation** needs seem markedly reduced. One might conclude that the best management of programmers *capitalizes* on this wonderful attraction to technology and avoids making programmers socialize. Alternatively, maybe the best thing managers could do for programmers is to help them develop human relations skills. Neither view, however, is correct. Project management requires teamwork, but it also requires technical expertise. The wise project leader knows what proportions to ask for. Besides, programmers can learn about people, just as managers can learn not to take advantage of programmers.

In addition to managing programmers, project leaders may have user personnel work on their projects; some firms put users *in charge of* projects. Recent work on user involvement in system development points out that users are different from information resource specialists. These two groups will get along best if roles are clearly spelled out. Programmers should not be asked to do analysis work without training. Users should not expect to program without knowing what they are doing. Where prototyping is the specific technique employed, programmers should not do this work unsupervised. On the other hand, it does no harm to provide users with some training in computer concepts so that they can try speaking to programmers.

> *CASE:* A large chemical firm was to automate a major production line. Before beginning automation, they contracted with a local university's information systems department to conduct a series of training sessions on computer concepts with the operators. Most of the attendees were supervisors who would participate in the detailed investigation as interviewees. Many, however, would not see the project again until terminals arrived. The training helped bridge that enormous gap of ignorance and suspicion.

15.5   PROJECT DOCUMENTATION

Projects generate documentation; but to an extent, documentation drives projects. Figure 15-14 illustrates how this happens. As each phase of the project is completed, additional documentation is required as part of the approval process. This documentation, in turn, specifies the remainder of the project plan.

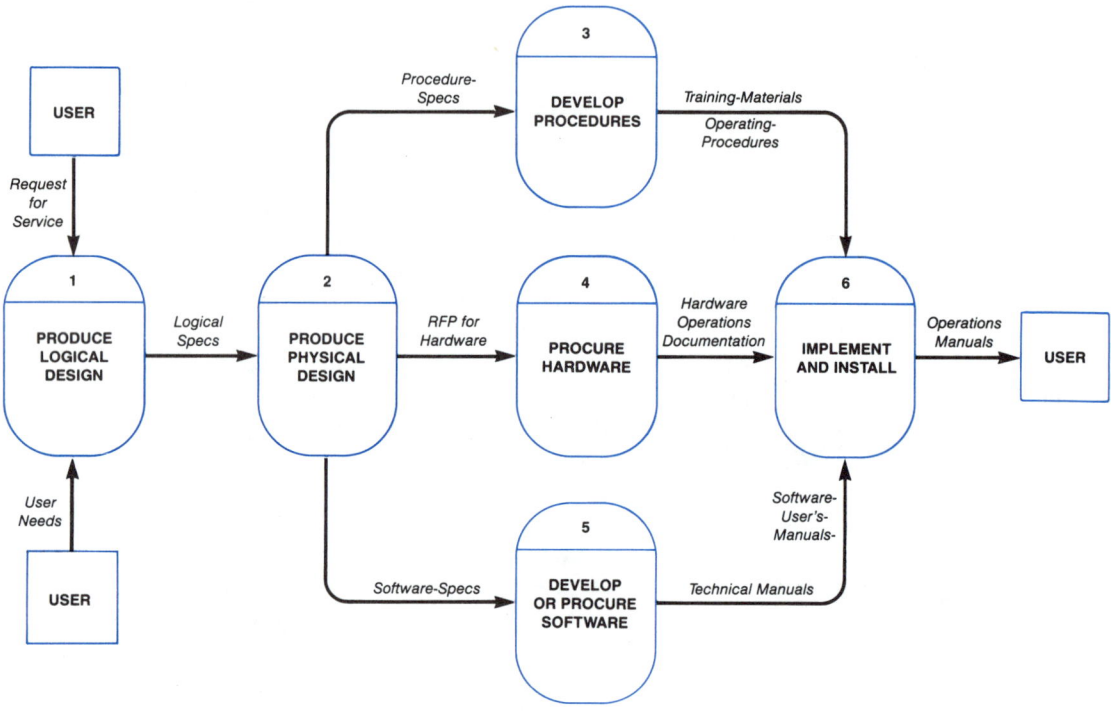

FIGURE 15–14. The Project as a Documentation-Driven Information System

The Role of Documentation

Documentation arises during project activities either as a result of directives from the manager or because a specific system development methodology (SDM) requires documentation to be filed at specific points in the project lifetime. Most organizations either purchase an SDM or develop their own. These SDMs are generally documentation-based procedures. That is, to follow the SDM, workers have to file documentation at specific times during the project.

Documentation serves as a historical record of what was done and what happened as a result. Project documentation serves additional roles (Figure 15-15):

CONTRACTUAL OBLIGATION to CLIENT
PLANS bridging PROJECT STAGES
AGREEMENT between CLIENT and IRU
PACING through S.D.M. STAGES
WORKSHEETS for DESIGN
DIRECTIONS for DESIGN and EVALUATION

FIGURE 15–15. Roles of Project Documentation

1. fulfillment of contractual obligations in providing user guides, system operation and management manuals, and internal documentation for maintenance
2. building plans for each subsequent stage
3. negotiated agreements by client and IRU management
4. going through predetermined steps in an SDM
5. worksheets for design
6. a procedure for design and evaluation processes

Because SDMs are intended to regulate and guide development, they tend to have a kind of legal reality that binds clients and developers. Certain steps may not be deemed completed until specifically mentioned documentation is available. Foremost among these are user, operational, and system maintenance documents (discussed in some detail in the next section).

Design documents — flowcharts, structure charts, preliminary investigations, and logical specifications— are not normally intended to be used after the project is completed, but they act as blueprints. The content of these documents is specific to the system being designed, but the form is strongly dictated by the guidelines built into the SDM. The use of models was discussed in Chapters 6, 9, and 13, and the role of specific reports was discussed in chapters 7 and 10. In the remainder of this section, we will be concerned with less schematic, brief textual or tabular documentation.

Again, because of the contractual nature of the relationship between the client and information resource organizations, project documentation serves as a record of agreements between the groups. These "legal" documents include project briefs, scope and summary statements, approved project plans and alterations, sign-off documents, and any other documents that imply agreement on deliverables and schedules.

Projects move from one item of documentation to another (Figure 15-16). Where there are specific deliverables, such as modules of software or working subsystems, the documentation on the module signals delivery.

Most SDMs include design worksheets that assist designers in working out parameters. For instance, file, screen, and report layout forms aid in both design and documentation. Some worksheets assist in computing volume and traffic within systems, total system size, expected throughput, and the like.

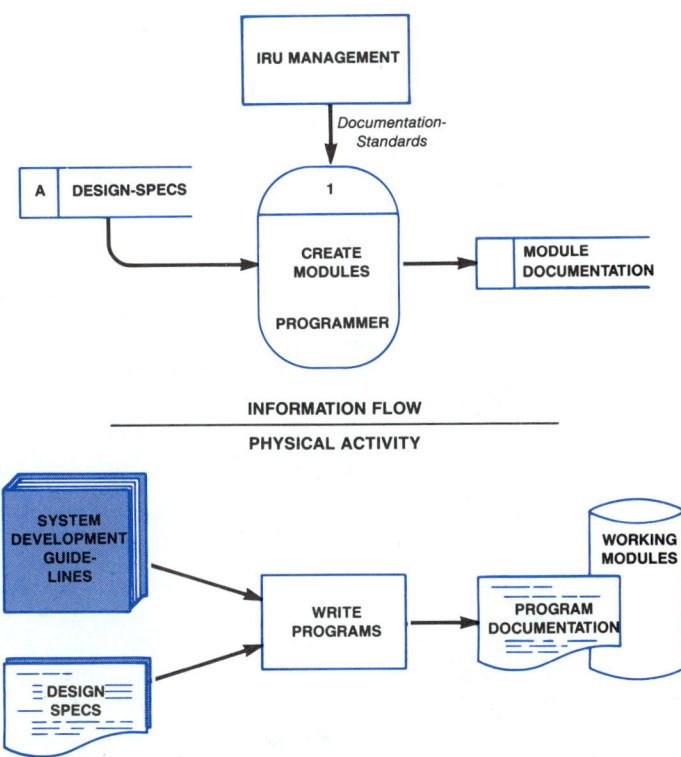

FIGURE 15–16. Documentation Flow Mirrors Physical Activity in a Project

Finally, documentation assists in developing testing and auditing procedures. Knowing *what* the system is supposed to do and *how* it is to accomplish it, how agreement on goals was arrived at, and the remaining historical matrix of the project provides guidelines for developing auditing procedures.

Specific Project Documentation

The following is a list of documentation typically found in projects involving both hardware and software (see also Chapters 5 and 8). Particular SDMs require other documentation. The noted figures illustrate possible examples of this documentation.

1. **request for services** (Figure 5-15)
2. **project summary** or brief (Figure 5-16)
3. **scope and limitations statement**
4. initial and revised **project plans** (Figures 15-2, 15-3, and 15-4) and cost-benefit analysis (Chapter 14)
5. **logical systems design** documentation, including the preliminary investigation (Chapter 8), detailed investigation (Chapters 11 and 12) with supporting models (Chapter 13), sign-off documentation, and logical system specifications (Chapter 10)
6. **physical system design** documentation, including chartwork, pseudocode, and structure charts (Chapter 13)
7. **testing documents**, including test data and testing procedures
8. deliverables, including **users' guide**, **operators' guide** (if different from the user), **system maintenance** or **reference manual**, **installation procedures**, and **operational forms** (Chapter 9)
9. **system-sizing charts** for estimating traffic, throughput, and capacity
10. **layout charts** for screens, input documents, reports, and files
11. additional documentation required for the data administrator defining data requirements (Chapter 7)
12. system evaluation procedures and worksheets
13. project correspondence and memoranda
14. **request for proposal** (RFP) or **request for quotation** (RFQ), if hardware or software subsystems are to be purchased. (These detailed system requirements ask vendors to return with either a proposal to build these subsystems [RFP] or a firm price quote if the system is actually specified by the IRU to the vendor [RFQ].)
15. additional **approval forms** required by the firm for authorizing expenditures

The Final Report

At the end of each phase of development, the project leader writes a report detailing the activities, findings, and results of that phase, and plans for upcoming phases. These reports include preliminary investigation report (Chapter 8), detailed investigation reports or logical design specifications (Chapter 10), detailed designs at the end of physical design, and final project reports. Specific documentation indicates the end of each contributory phase, but the final project report does not contain detailed results, since these are found in the manuals. Instead, it summarizes the results found in other project documentation, details the termination status of the project, and forecasts operational considerations.

The following is a guide to the outline for the final report:

1. Summary of the project to date
    a. expenditures
    b. delivered products
    c. management considerations and concerns
2. Termination status
    a. operational status of the products
    b. work still to be done (but not promised under this project's "contract")
    c. results of testing
    d. need for continuing support by IRU staff
3. Operational forecasts
    a. need for maintenance
    b. initial estimates of efficiency and effectiveness
    c. concerns for operations, recommendations
4. Project termination
    a. special personnel concerns; reassignment
    b. evaluation of project efficiency and effectiveness vis-a-vis the selected SDM
    c. archiving considerations
    d. legal, marketing, vendor-relations considerations

The final project report documents the turnover status of the product, the transition of project personnel from "active" to either "standby" or and "inactive" status, and loose ends that need to be tied up. For instance, hardware may not have been delivered on time or at specifications promised. Whose responsibility is this? Where will project documentation reside? Who will keep it up-to-date?

Consider the efficiency of the delivered product. What problems remain despite the fact that the product has been accepted? What kinds of continuing maintenance will be required and who should do it? Finally, think about how the the project was conducted. Was the SDM adequate? What problems arose that should not happen again? Should the SDM be altered in all cases or only in those that resemble this project? Most IRUs still miss the opportunity to monitor project effectiveness *across* projects and, thus, lose the opportunity to improve their SDMs.

The final report caps the project, but there are usually meetings associated with handing over the product to the users. The next section discusses how those meetings should be conducted. (See also the agenda for a formal sign-off meeting discussed in Chapter 10.)

The Hand-Over Presentation

Handing over the product to the user has three aspects. First, the IRU informs the client that the work has been done, implying that the work has been *done* and that no more project work is necessary. Second, the client approves the work, thereby releasing the IRU from developmental responsibilities, effectively terminating unilateral authority to change the system. These steps complete the contractual nature of the project. The third aspect of the hand-over meeting is social. A relationship has formed between the user group and the IRU. The hand-over meeting is the last formal chance the two groups have to build that relationship. Often, negative impressions formed during the pressure of development can be improved by a good presentation, especially if testing has gone well. On the other hand, a good working relationship can be spoiled by a presentation that shows the IRU to be arrogant, careless, stubborn, or malicious. In other words, the handshake means something.

Systems presentations, unfortunately, tend to focus on technical particulars, whereas a focus on sales aspects would make the audience more comfortable, confident, and convinced. The hand-over presentation is more important than most others since there is a temptation to say "There it is, it's all yours" in fifty different ways. Several important points must be heeded during this presentation:

1. Invite *all* appropriate client organization people.
2. Make sure attendees are comfortable.
3. Avoid going over system operations; concentrate on user confidence aspects.
4. Do not just promise — use data from testing to show how efficient or effective the system is.
5. Refer to the original project summary and use that as a text for the presentation.
6. Avoid a day-by-day retelling of system development activities; instead, concentrate on the future.
7. Make clear what the client is agreeing to — *informed consent is essential.*
8. Indicate the disposition of the project, its human and technical resources. Will these be available to the client after acceptance?
9. Visual aids are important, but also make sure you can be heard; if the project leader is not a great public speaker, find someone else to make the presentation.
10. Distribute the final report before the meeting; if this is not possible, distribute an executive summary no later than the start of the meeting. Prepare a package of materials that summarizes both the approval process and the details of the product presentation.

15.6 CONCLUSION

Chapters 5 through 14 discussed systems analysis more or less as a set of activities that analysts alone engage in. Chapters 15 through 17 look at the context of systems analysis work in more detail. This context is essential for coordinating systems analysis efforts and the work of others.

The **project** is the centerpiece of the **system development life cycle**. Specially trained systems analysts manage the creation, implementation, installation, and operation of information resources. Chapter 15 discusses project management techniques, while Chapter 16 deals with alternate methods of structuring work and management styles. Chapter 17 concerns specific content issues that make the analyst's role in projects and project management different.

This chapter introduces the general structure of a computer-based information resource project. This kind of project follows the SDLC step by step. From a management viewpoint, three aspects of projects are most important: planning, monitoring, and control.

Project planning activities include **critical path planning**, **resource specification**, and **manpower planning**. The critical path methods (CPM, PERT) are mathematical techniques for project scheduling. Resource specification is critical, too, because most projects last a significant length of time and the inherent turbulence in organizational environments almost guarantees that rigid plans will fail. Manpower planning involves the specification of jobs; the recruitment, assessment, and hiring of individuals to fill those jobs; and the continuing motivation and training of job incumbents.

Project monitoring is a reporting function. Project reports consist of interim reports, final reports, and exception reports. In addition, a continuing series of meetings is necessary to remain informed about project progress.

Project control requires redeployment of resources on a routine and crisis basis. This may range from staff redeployment to political hassles involving computer access. Most project control is concerned with financial resources, but recent emphasis on software engineering has expanded the areas of interest to include software reliability, maintainability, and general testing and certification techniques.

Labor costs, rather than technology costs, are paramount. While **project leaders** are often chosen from the systems analysis ranks, half the development work is, on the average, expended in software specification, development, and testing; a similar amount will be spent after the project's goal has been attained to maintain the hardware and software. Thus, a major aspect of the project manager's work is managing programmers. It is in the area of personnel or **human resource management** that the information resources industry has been least successful in the past forty years.

Project documentation is important from the point of view of control, but the client is the ultimate reader of documentation. Finally, projects terminate when their products are considered worthy of use by customers. Two aspects of project termination are the **oral presentation** and the **final written report**.

DISCUSSION QUESTIONS

1. Consider the case of Ed Waller at the beginning of Chapter 15. What are the first steps that should be taken with regard to this project? With what kinds of documentation should the project begin? What agreements need to be in place before any project work begins?

2. Suppose Ed goes ahead with the project and creates a project plan. What would the main headings of this plan be? Consider only the human resource aspect. What needs to be planned for? What are the elements of a financial resources section of the plan?

3. After the preliminary investigation, Ed presented the following structured list of tasks:

    1. Perform detailed investigation (systems analysis staff), 30 md.
    2. Perform detailed design (senior systems analyst), 20 md.
    3. Create RFP for hardware and select vendor (senior systems analyst), 5 md.
    4. Develop detailed software specifications (programmers), 25 md.
    5. Code applications programs (programmers), 80 md.
    6. Code systems software (programmers), 40 md.
    7. Test system modules (programmers), 10 md.
    8. Test integrated system (systems analysis staff), 3 md.
    9. Train user staff (systems analysis staff), 5 md.
    10. Train users (previously trained users), 15 md.
    11. Install system (analysts, programmers, vendor) 10 md. each category plus vendor time.
    12. Hand-over system, 1 md.

    The senior systems analyst is costed at $400 per day, junior analysts at $300, and programmers at $200. Use 1 md. of support staff (clerks, secretaries) for each 3 md. of professional staff time. Vendor time and user time are not included in a budget. Calculate the total manpower cost for this project in each category.

4. The structure of the tasks is indicated below:

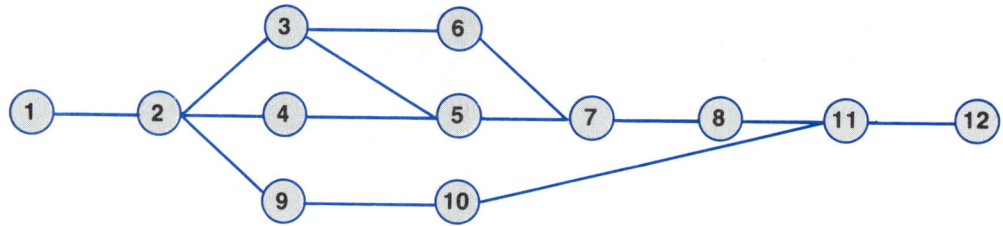

Calendar time estimated for the work is given in this table:

Task	Calendar time (in days)	Task	Calendar time (in days)
1	15	7	20
2	20	8	5
3	45	9	10
4	15	10	30
5	20	11	20
6	15	12	1

Using PERT, derive and describe, in English, the critical path for activities in this project.

5. Given the critical path in the previous project, what effect would a five-day slippage in responding to the RPF have on overall system delivery date? Suppose we could add another programmer who is adept at coding system commands to speed up task 6 and have it delivered in ten days. Would this affect system delivery date? Where is money best spent?

6. Suppose we know that adding workers speeds things up in the following way: doubling the number of workers does not decrease the cost, but it gets the work done in half the calendar time plus a penalty of three calendar days for each added worker. Given the per diem cost of programmers and analysts, how would Ed best spend an additional $5,000 to speed up the project?

7. On what bases could Ed have argued that the project should not proceed beyond the preliminary investigation stage? Flesh out these arguments with (imagined) data. What are the most convincing arguments?

8. What are the pros and cons of organizing this project as an autonomous team? Who would gain? What are the costs to the IRU? What advantages would a matrix organization have over the team organization? Where are the costs here?

9. A project is like a cybernetic system. In such a system there are numerous places things could go wrong. Outline nine different ways in which a project could fail in cybernetic terms and give examples for each.

10. A project is also a learning system. Suppose Ed's management has noted that the project is not going well during the logical design stage. What ways do they have of "changing the program" that governs or regulates this project? What are the costs and the benefits of these actions?

11. Consider the classes of project crises and the four philosophies of crisis management outlined in this chapter. For each class of crisis, indicate how each philosophical approach would dictate a unique approach. Overall, which philosophy has the greatest short-term cost? Which has the greatest long-term benefits?

12. Discuss the reasons why the project form of work organization presents unique human resources management challenges to IRU management. Can you think of ways in which these challenges can be met beyond those mentioned in the text? In Chapter 16, we will look at some alternatives to the project; think about how these challenges might be different as you read Chapter 16.

13. Derive a detailed outline for an oral project report for a project you are working on, or imagine the details for Ed's project as discussed in the previous questions.

14. Derive a detailed outline for the written report of the project used in answering question 13.

CHAPTER 16

SPECIAL PROJECT TYPES

OBJECTIVES

1. To describe the process, advantages, and costs of several alternatives to long-term projects, including prototyping, end-user software development, and the information center
2. To prescribe a prototyping technique for a variety of systems development situations
3. To produce implementation plans for information centers for specific situations, including personnel, technological, financial, and organizational resource requirements
4. To describe the role analysts can play as users
5. To describe the structure and function of projects that involve continuing some and developing other manual systems in conjunction with computerized information resources
6. To complete an exercise in standards writing

SPECIAL PROJECT TYPES

16

CHAPTER

16.0 ALTERNATIVES TO PROJECTS

Retooling at Cornish Tool

Today, noted R. J. Hughes with some displeasure, was the two hundred and twelfth day since he had last spoken with Tom Hutton, manager of software development, and the one hundred fifth day since he had heard about progress on his raw materials control system. Hughes, in charge of manufacturing loss control at Cornish Tool, was a remarkably patient man, but today he lost his cool and marched into Hutton's office. This is what transpired there:

Hughes: Well, Hutton, where's my software?

Hutton: Your software? Well, R. J., why don't you sit down so we can discuss that. You know that we've still got over two months to go on delivery, and we're right on schedule, with only a few minor delays?

Hughes: Yeah, but we need that program *now*. I've got a semiannual waste control review coming up; the VP is on my tail and I've only got estimates, not the actual figures. He wants projections and statistics and stuff like that. You guys aren't really on schedule — the *original* schedule!

Hutton: Now, R. J., we've discussed that several times. When Rolly quit, we lost several weeks bringing on a new project manager. And our unanticipated system change set us back several more weeks, but that was Zilonic's fault that their new operating system didn't work. Anyway, we worked a few new features in there, too, and it's just taken longer for our guys to work on that; besides *you* weren't so available when we needed you to approve those changes.

Hughes: Well, a guy's got to have his vacation, you know. Anyway, I sent McDonald over to approve those changes.

Hutton: Sure, but he really didn't know much, so we decided to wait until you returned to be sure. Listen, R. J., that's not so important. The real issue is when your software is going to be available. I assure you that you'll have it right on schedule, the first of March.

Hughes: Great. I'll have software, but no job.

The problems R. J. is experiencing have become more common in recent years as applications have become more complex, manpower resources more specialized and difficult to utilize, and production demands more exacting. This chapter concerns the ways in which the systems analyst's work has changed in connection with these trends. In particular, we will discuss the alternatives to project organization of work mentioned in Chapter 4, focusing on prototyping and end-user software development.

This chapter also considers two other ways in which systems analysts can work to build or improve systems. An analyst can work within the user group as a user analyst. In addition, analysts are often engaged in writing standards for system development and management. We will look at the work an analyst can put into improving manual systems without computerization. The common adage "If it ain't broke, don't fix it," serves the analyst well if a manual system can be improved without the expense and disruption that computerization can cause.

Chapter 15 covered the basic principles of project planning and management. In this chapter, we will look at alternatives to major projects, alternatives that respond to modern needs, and some long-standing, not-quite-so-modern ones. Whereas the project was appropriate in the early days of mainframe computing, it is less appropriate now because, in many cases, the costs and risks of tooling up for a major project are too high and unnecessary. R. J. and Tom's conversation is not only typical, it is practically a script.

Why Have Alternatives to Projects?

A project is a linear network of activities that attempts to produce a product in a limited amount of time, as defined in Chapter 15. This definition and the way projects are run imply a number of limitations:

1. Projects have limited lifetimes and limited budgets.
2. Because activities are networked, slowdowns and missed deadlines in a few activities can delay the whole project.
3. Projects attempt to deliver a product based on a plan designed many months, if not years, before the product is to be delivered. Because business environments are turbulent, it is unlikely that the product can be delivered to meet the needs that emerge over time.
4. It is quite difficult to forecast resource needs for long periods of time. Many projects exceed the product life cycle of the parent organization by factors of two to ten.
5. Computer technology is changing so rapidly that obsolescence of a project of more than two years' duration is practically guaranteed.
6. Users are far more sophisticated in terms of computer expertise and, more important, the organization's need for data than they were in the past. There is no need to exclude them from the project process.

Thus, a project that excludes users and attempts to produce a product deemed suffic-ient only for an era long past has significantly higher risk and lower value than it promises.

Over the years a number of alternatives to the project structure of activities have been developed. These alternatives attempt to address the limitations of projects in the following ways:

1. Why limit the length of time of the project? Given the ongoing "march of tech-nology," why not have continuous change and development?
2. If the product is only slightly usable at the *end* of the project, why not start with the final product and work *backward*? Why not begin with a crude prototype of the product and refine it rather than build it from scratch through an assembly line network of activities?
3. Can the turbulence of the business environment be countered by drastically shortening or simplifying the planning of projects? In particular, can much of the project work be "canned" in advance with the users responding to their changing needs rather than having IRU staff bear this burden?
4. Resource needs do not have to be predicted if we do not require those resources. More to the point, can much of the human resource requirements for routine work be shifted to the users?
5. If we can shorten the length of project, will we avoid being caught with obsolete technology?
6. How *can* we get these more sophisticated, demanding, and politically astute users involved in project work?

The response to these questions has been the development of rapid prototyping systems and end-user software development tools, primarily through 4GLs and enhanced database managers. These are discussed in detail later in the chapter.

The Economics of Projects

Project costs can be divided into four areas: human, technical, opportunity, and political. The human resource component historically has risen annually while the tech-nical component has dropped. But expectations and opportunity costs have risen, too, as long project development time has pushed benefits back, too. Political costs, while intangible, have also risen as a new cadre of business-school-educated managers has begun demanding increasingly complex, sophisticated systems to plan rather than merely track business, and to operate interactively rather than on a historical or exception basis. On the supply side, business schools and technical colleges have not turned out technical graduates capable of designing these more sophisticated applications on a demand basis without significant (re)learning. Recent research in development of programmers and analysts has demonstrated that while business calls for more business skills in program-mers and analysts, educational institutions are turning out increasingly more technically specialized individuals.

A number of attempts to increase the productivity of the IRU have occurred in the past fifteen years; these are reviewed in the next section.

The History of the SDLC

The system development life cycle (SDLC) is a gift of the engineering heritage of systems analysis. Just as an engineer plans and builds a bridge, the systems analyst/ system designer plans and builds an information resource subsystem. The resemblance is not superficial. Precisely the same planning techniques are employed in the SDLC as in building a bridge, with the exceptions that bridges are seldom built from the top down and a new order-entry system does not need approval by any environmental protection agency.

On the other hand, the products bear hardly any resemblance. Bridges can be built in standard fashion using modular elements. Although cost overruns are common, the costs of individual units can be predicted in advance. And although the bridge may be obsolete when it is opened, it is still useful — few would immediately call for it to be replaced.

The SDLC is a nested series of design and building activities which allows for some adjustment, but essentially requires agreement among parties at various critical points that there will be no turning back. As the project proceeds, retrenchment becomes increasingly expensive, since redesigning interlocking elements may require scrapping entire subsystems (Figure 16-1).

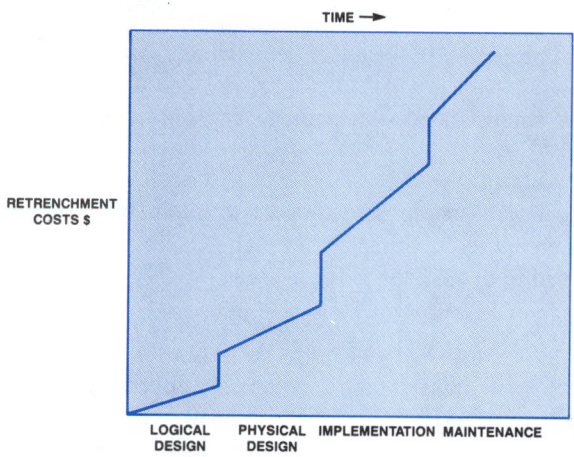

FIGURE 16–1. Retrenchment Costs Escalate over Time

The primary contributor to cost overruns and late deliveries is software, not hardware. Software is labor-intensive; hardware can be bought on a per-unit basis as a capital expense known in advance. However, it was not until the 1960s that this trend became apparent and attempts were made to capitalize on the malleability of software to improve the situation.

The first attempt came as **structured programming**. Before 1970 programmers typically wrote "spaghetti code," which contained uncontrolled branches and connections among subroutines. With the advent of "structured shops" in which code could be written using only the three basic structures of procedures outlined in Chapter 9, programmers began writing neat, indented, and highly modular code. This change is illustrated in Figure 16-2 on the opposite page. There is, however, still a great deal of resistance to legislating programming style, and productivity benefits have been limited. Obviously, more than just programming style had to be limited.

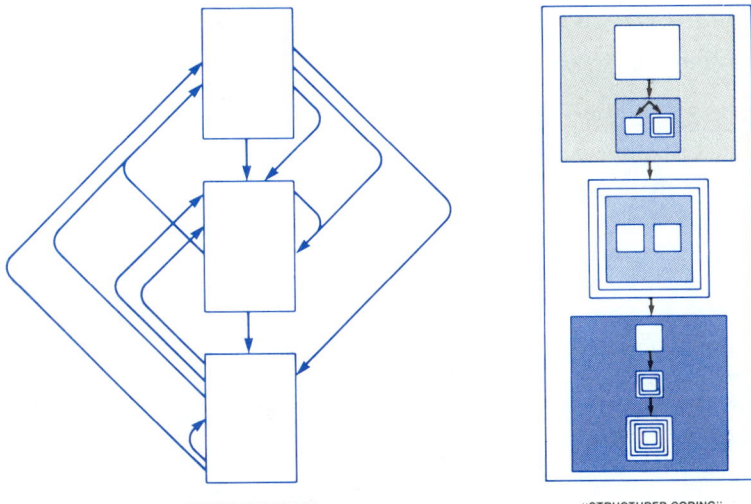

FIGURE 16–2. Spaghetti vs. Structured Coding.

In the early 1970s, the SDLC came under attack. Analyses of expenditures showed that program coding ate up less than 20 percent of all system costs and that maintenance was a far higher contributor to life cycle costs. This is actually an inherent cost in software, one that would not be noticed until major development work was completed in specific shops. Obviously, efforts that decreased the *necessity* for maintenance would be rewarded with productivity increases.

In response to this clearly demonstrated need, a number of individuals, notably Larry Constantine, Chris Gane, and Trish Sarson, developed a consistent philosophical approach to systems analysis aimed at producing better designs. They separated logical and physical design and brought together a number of existing tools to create the practice of **structured systems analysis**, a practice this text is dedicated to. Their basic aims were the following:

1. To develop an information resource model from the *user's* viewpoint to increase communication, thereby reducing the possibility for misunderstanding and significantly speeding up information gathering
2. To facilitate communication between user and analyst by gathering information on functional needs rather than technical needs, as the latter is generally mumbo-jumbo to users
3. To develop a scheme by which detail can be deferred in discussions; more specifically, to develop modeling tools that can be used at a variety of user *levels*, depending on the needs of the discussion

4. to develop a new way of specifying systems so that users can agree to specifications knowledgeably, therefore having less reason to change their minds later, when they think they have been duped, bilked, or misunderstood

5. To develop these specification techniques so that physical designers can build directly from logical/functional specifications

The results were these: making the DFD central in systems analysis, collecting functional needs data in a systematic fashion, diagramming procedures using decision tables and trees, using pseudocode rather than Fortran to outline process logic, specifying data requirements in dictionary form, and analyzing systems in a top-down, output-driven manner, beginning with the data and the system response needs of users. Gane and Sarson (1974) and Yourdon and Constantine (1978) also developed some important tools for system design, refining the structure chart and finally relating the DFD, data dictionary, and structure chart through the HIPO diagram, as covered in Chapter 13.

Structured techniques (which also include structured design, testing, and implementation), however, did not meet the original productivity improvement goals. Martin (1982, pg. 43) estimates an increase in productivity of less than 25 percent on average, which, in the face of the increased need for sophistication, was wiped out.

It was technology, in the end, that came to its own rescue. While programmers tinkered with the processes of coding and data collection and small segments of coding and analysis, the overall problems remained: it took too long to build a complex system for sophisticated but untutored users. The key was that while technologists were becoming increasingly specialized, users and technologists were using computers but, more important, technologists were using them to manage their own work.

Figure 16-3 illustrates these trends. By 1975, it was feasible to prototype systems, and by 1980, it was more than feasible to create tools for users to make their own programs. As we began to understand routine work, technologists were able to build software-generating software for users who could tinker on their own, without having to

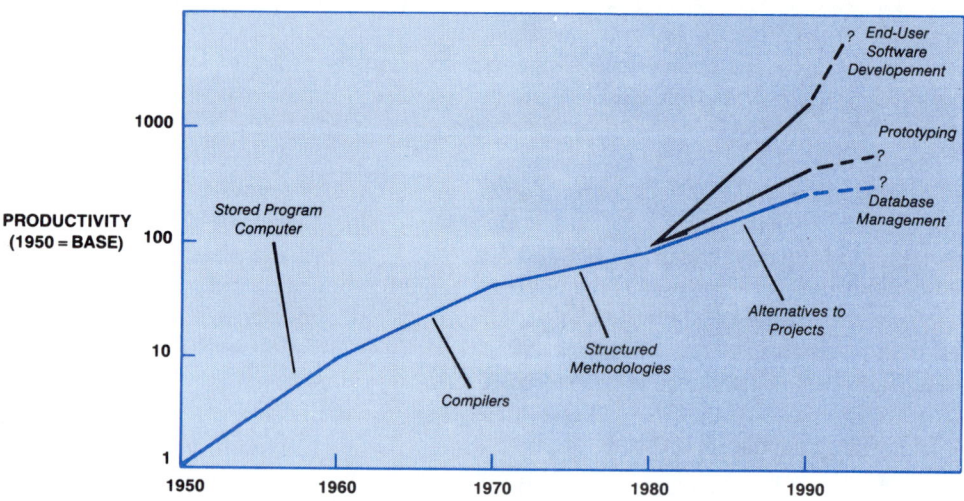

FIGURE 16–3. Technology of System Development Increases Productivity

communicate their tinker to the tailor, so to speak. With the advent of relatively inexpensive, powerful microcomputers in the early 1980s, this trend has accelerated. Today, most users in larger organizations have at least limited access to some sort of end-user-based development techniques.

16.1 PROTOTYPING

Prototyping is a relatively new phenomenon, made possible by a rapid drop in the cost of a CPU cycle with the corresponding rapid rise in the cost of a worker hour. The goal of prototyping is to speed up development significantly by presenting the user with a "working" version of the final product at the start of the development effort. Thereafter, development work is aimed at refining this already-agreed-upon product. This section discusses the advantages and disadvantages of prototyping, a model of prototyping, and techniques of managing prototyping projects.

Definition

Prototyping is the construction of a working version of a final product that is subsequently refined and made more efficient. The process of prototyping is *iterative, nondeterministic*, and *user-driven* (see Figure 16-4):

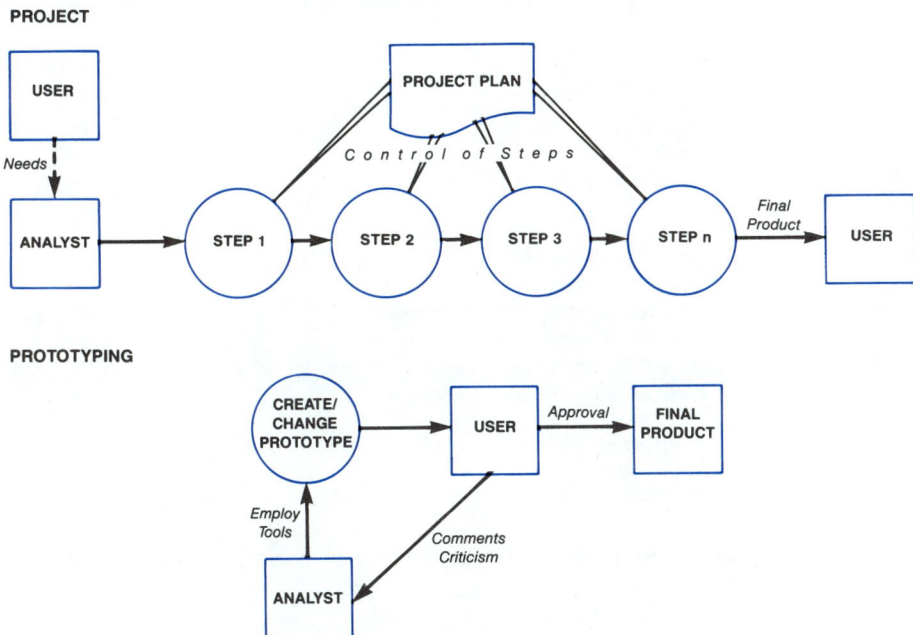

FIGURE 16–4. Comparison of Projects with Prototyping

1. It is iterative because there is repetition of the refinement of the prototype.
2. It is nondeterministic because there is no fixed plan of activities that must be followed to reach the final product.
3. It is user-driven in that the perceptions, needs, and abilities of the user determine what aspects of the prototype are worked on and when these aspects are "approved."

Prototyping pushes turbulence that may arise from changes in user needs and perceptions forward to the beginning of the work. At this early stage, changes to system design are relatively painless and low cost. When the prototype has been approved in all its detail, the costs of producing a technically efficient and valuable system are more controllable.

Advantages and Disadvantages

The major trade-off in using prototyping (Figure 16-5) is putting much of the cost and uncertainty at the start of the project in return for having more control later.

Prototyping has several advantages. Communication between analyst and user from the users' perspective is enhanced when he actually sees and uses the intended product in situations that approach real operating conditions. There is also more chance for informed agreement about logical system functioning between user and analyst. Because of this, prototyping is a more honest approach to system design in extremely turbulent environments.

ADVANTAGE	DISADVANTAGE
Increased Communication in User's Terms	Longer, More Expensive Communication
More Well-Informed Agreement	Agreement may be Delayed, Increasing Costs
More Honest Approach	Honesty Must be Carried through by IRU Management
Better Control Over Costs if System is Finally Implemented	Prototype may Actually be Unimplementable; Promises may be Unrealistic
Higher User Morale, More Confidence	Expectations may be too High
Better Separation Between Logical and Physical Design	Physical Design may be Implied too Strongly by Concrete Prototype Features

FIGURE 16–5. Advantages and Disadvantages of Prototyping

Because the product is seen in early stages, user morale and interest are increased. When user agreement is guaranteed, the IRU can exercise more confident control over costs in later stages. Finally, there is a better separation between logical and physical design, since logical functions are more or less cast in concrete during the early prototyping stages.

There are disadvantages, however. Better communication may also mean longer communication; analyst time is expensive and costs may be very high if there is not some control over the goals of these conversations or if agreement is either not reached or delayed. Prototyping may appear to be honest, but unless IRU and user management approach it honestly and intend to carry through on agreements, disenchantment can result in even worse relations. Expectations are raised along with interest; failure to meet them can negate the advantages of the prototyping process. If the prototype proves difficult to implement effectively, controlling costs in later stages may not be possible. Analysts must not promise more than realistically can be delivered. Finally, prototyping tends to bring elements of physical design back into the logical design process, especially if users cannot generalize from the experience with the prototype but instead concentrate on particular features or procedures.

Another disadvantage is that there may be only a small number of users involved in the actual prototyping. Because not all users are equally informed or responsible, the analyst may choose specific users with whom to work; this can result in an unrepresentative prototype. Analysts may choose to work only with the most articulate, personable, attractive, or politically advantageous users; the result may be a product that inarticulate, unpersonable, unattractive, or politically disadvantaged users cannot employ.

Often, prototyping is used only with certain aspects of information resource maintenance. Most commonly, screen formats, report layouts, and data-entry procedures may be prototyped while other aspects that are completely or partially transparent to users are scoped out and developed in traditional project fashion. Most likely, communication protocols, database management, operating controls, and data security and recovery will *not* be prototyped. Indicators of areas that should not be prototyped are these:

1. Is it expensive to build the first prototype?
2. Will the prototype have to be efficient to function at all?
3. Are sensitive materials or procedures necessary to prototype?
4. Will a large team be required to set up and operate the prototype?

For example, prototyping disaster recovery will require creating a disaster, prototyping some aspects of sensitive financial management may require blocking out transactions or protecting data during the prototyping, and operating a communication network may require either an expensive simulation or the construction of the actual, efficiently operating network.

On the other hand, prototyping *is* valuable where a small number of well-informed, easy-to-work-with individuals can actively inform analysts about how they would like to work with data, especially if they already understand the structure, reliability, and "behavior" of their data.

The Process of Prototyping

After agreeing that prototyping is to be performed (Figure 16-6 on the following page), the **prototyping team** is selected. Members consist of an analyst, one or more informed users, and a programmer. The programmer's activities are performed after the prototype is approved.

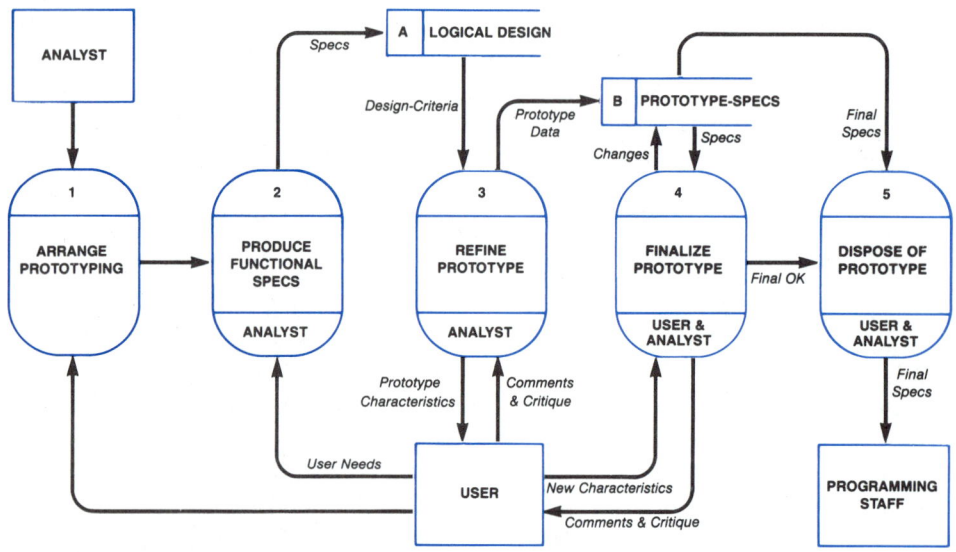

FIGURE 16–6. The Process of Prototyping

The preliminary investigation is aimed at uncovering the major logical functions of the information resource and the structure of the data underlying those functions. Next, a first prototype is constructed. **Prototyping tools** may be programming languages, but more often the analyst will either employ a screen generator, a database manager, and a dialogue manager or build these tools first.

The subsequent steps are demonstration, teaching, discussion, and agreement. The analyst demonstrates the prototype and teaches the user how to use it. The team then discusses user's experiences. The analyst offers advice about feasibility, even in this early stage of design. The user indicates which features were liked or disliked or may recommend an entirely new approach. The conversation takes place entirely in **user language** augmented by "techy-talk" only as necessary to navigate through the demonstration ("Press the function key," "Notice the blinking area of the screen," "Is what you want to see on the screen or off to the right?").

The analyst then refines the prototype and schedules another session. Finally, analyst and user agree that a prototype resembles the desired product closely enough so that continued prototyping will be unnecessary. In this case, the programmer and analyst meet to determine the next steps, which may be immediate implementation, actual employment of the existing prototype without modifications (especially if the final product is to be used only a small number of times), or extensive development after more data gathering.

Where a prototype is to be used only a small number of times, inefficiency can be tolerated to the benefit of a limit on costs. In the 1960s, Fred Brooks admonished the readers of *The Mythical Man-Month* to be prepared to "throw the first one away." The prototyping philosophy is to throw all but the last away. Commonly, a user may discover the results desired during the prototyping process and not require a final, working version after all. In this case the analyst has thrown *all* of them away!

Because users are involved at early stages in the development of their own products, they often become motivated to employ prototyping tools acquired or built to construct the products they wanted. Ultimately, they may be interested in developing and maintaining their own information resource using a version of the prototyping tool. For example, Lotus 1-2-3 is a good prototyping tool for development of interactive spreadsheets. It is also a fine tool for users to employ.

Managing Prototyping as a Project Leader

A systems analyst who chooses, as a project leader, to employ prototyping must address three issues:

1. What aspects of the information resource maintenance work are most amenable and most profitably done by prototyping?
2. What techniques of prototyping should be employed?
3. How will prototyping be carried through? What happens to the prototype(s)?

In addition, the project leader has to select individuals to staff the prototyping effort and choose appropriate prototyping tools.

The user interface is most likely the first candidate for prototyping. Certain system functions can then be ignored. In particular, error routines can be effectively delayed. Also, prototyping can help locate just where error detection and recovery, backup, and efficiency are required. It is possible to select some aspects of the information resource for prototyping and use more traditional methods for others; this naturally leads to the concern the project leader has for integrating an iterative, nondeterministic process with other, more completely planned processes.

Three major prototyping methodologies are in use. **Scenario-based design** (Mason and Carey, 1983) determines the users' requirements in terms of scenarios of production. This technique is advantageous when specific output is required for specific input. Scenario-based designs tend to be driven by screen formats and are sometimes called "screen generators" or "dialogue managers."

Special-purpose languages are used to generate applications quickly based on user recommendations. Such languages are often called "applications generators." This writer has produced an office automation applications generator called LAMP to create specific systems. IBM's ADF and CICS are examples of programmer-oriented applications generators. Spreadsheet processors can be considered application generators, too.

The third approach, **reusable software**, relies on recognition of a new application as merely a unique structuring of other, smaller problems already solved as a set of routines in software. Some applications generators work this way, especially those based on menus. Each menu choice refers to a subroutine.

Two other techniques are being developed. One technique, called **simplifying assumptions**, documents simplifying assumptions that the programmer/analyst makes in coding. These assumptions can be referred to later, after the initial design is approved by the user. Such assumptions can account for a significant part, if not the bulk, of difficult coding work, making rapid development possible. Speeding up the development cycle makes it effective to bring the user directly into the software approval process, despite the fact that traditional "coding" is taking place. Assumptions such as error handling and recovery, to name one example, may account for 80 percent of all input software coding. This proportion of 20 percent of the situations accounting for 80 percent of the coding work is referred to as **Pareto's Law**. Simplifying assumptions takes advantage of Pareto's Law to cut initial prototype development time by a factor of 5 (from 100 to 20 percent) while still allowing effective discussion of four fifths of all cases in an early stage of system development.

A second documentation-based technique relies on **executable specifications**. An executable specification language documents what is needed without programming the procedure and telling how it is to be carried out. Several examples include GIST, RSP, OBJ, PSLAIR1, and USE. These build models of processes that can then be directly interpreted by software to "implement" the process.

After prototyping is completed, there are a number of possible uses for the prototype. One view is that the prototype can be "thrown away," its value derived from the learning/teaching process the analyst and user have engaged in to understand the user's application. This view sees prototyping as a technique of needs analysis (Chapter 7). On the other hand, the sequence of prototypes may serve as documentation of the needs analysis process. In some cases, the last version of the prototype may *be* the delivered product, assuming that questions of efficiency, error handling, security, portability (the ability to move the product from machine to machine), and maintainability are either unimportant or have already been addressed.

Sometimes the prototype can be mechanically translated into a working product, either manually or automatically. Where the prototype is created in an executable specifications language or as a product of a special-purpose language or applications generator, the translation is done automatically during prototype development. As an alternative, the prototype may have to be tediously hand-translated by a master programmer. In this case, costs may rise, negating the initial cost advantage of prototyping.

Case: A Client-Tracking System

Darlene Hicks is district director for a state welfare agency. She manages thirteen district offices primarily concerned with child welfare and public assistance. Her most recent concern has been with tracking children who have been placed in foster homes. Because of the human and financial costs of foster placements and because of the political visibility of mistakes in placement, it is important to know at all times which children are in which homes. Because the status of children may change, these records have to be reviewed periodically. Also, it is important that children not be placed into unsafe homes.

Darlene has been using a batch system to maintain these records. Each time a child's status or placement changes, one of a number of forms is filled out by a social worker and sent to the department data center for processing. Often there are significant delays in processing because of errors in filling out the forms or in transcribing them. Currently reports are available on a weekly or monthly basis; the weekly reports are about four days late and the monthly ones arrive about ten days into the next month. Where crises develop that are politically sensitive (especially where a child is hurt while in foster care or a foster parent is accused of unfair billing for child care expenses), the lateness of these reports can be politically damaging.

Terry McNeil is an analyst assigned to work with Darlene to prototype a new on-line data-entry system. Terry is working with a microcomputer forms-generation program to prototype entry forms. She uses a report generator to create report formats for Darlene, Stu Bessinger, a clerical supervisor, and Phyllis Andover, a child welfare supervisor, to look at. Figure 16-7 illustrates a sample input form and Figure 16-8 a sample report format prototyped by Terry.

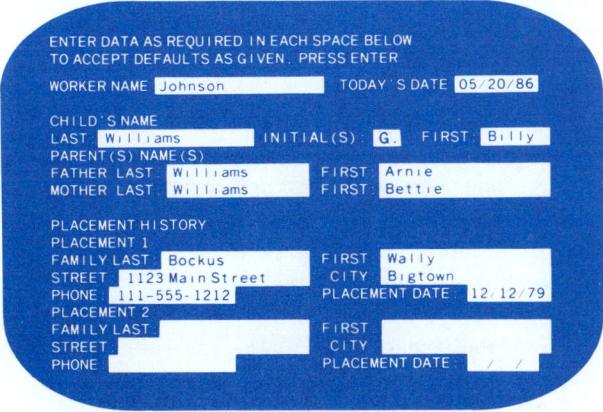

FIGURE 16–7. Prototyping of an Input Form (uses dBASE III)

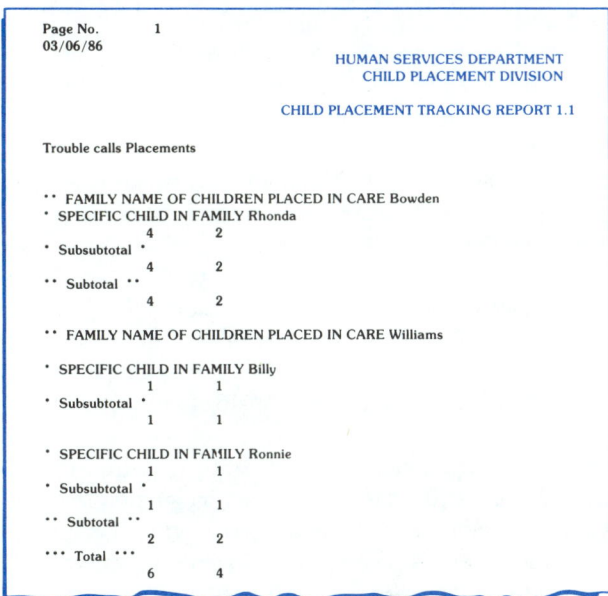

FIGURE 16–8. Prototyping a Report Form (using dBASE III)

After settling on the report formats, Terry is going to design the database query "conversation." With on-line data entry, on-line query is possible, but to keep costs down, the system design goal is to supply a number of "canned" queries that can be combined in limited ways. Here is part of a conversation between Terry and Phyllis:

Terry: OK, now you're at the terminal and you want some information...

Phyllis: Yes. The most important thing is knowing which children have been in foster homes for a certain long period of time...say over a year...and how many homes they've been in...

Terry: Is this one-year limit fixed or is that just a common value?

Phyllis: No, it's not fixed, but we will want to list placement in order of the length of time the child's been in care or in order of the number of homes the child's been in.

Terry: So you'd like a list of children sorted either by length of time in care or by number of placements. But only for those children in care a year or more?

Phyllis: Yes. Though 85 percent of our children are in care for less than one year, the one-year limit will still list hundreds of children. Let's say two years minimum. And the number of placements should be at least two—we can find out easily those children with only one placement another way.

Terry: Now, that takes care of time and number...[enters something on the screen]...how about this; we'll have these menu choices [Figure 16-9 on the opposite page]...but what about combinations? Will you ever want to look at really *bad* cases, say a long time *and* any placements?

Phyllis: Well, now you're really into it! Yes, we'd really like the computer to pick out trouble spots before they blow up. Maybe we could cook up an index of possible problems and shoot it through all the placements.

Terry: [Enters something else on the screen—see Figure 16-10 on the opposite page.] Yes, I see. Here's another menu entry for "Trouble" or whatever you want to name it. Do you have such an index available now?

Phyllis: Well, not exactly, but it's pretty easy to make it up.

Terry: What you really need is some kind of "what if" thing to see which cases are "bad" if certain things are true and then finalize your formula. Let's see if we can work out some dialogue to get that out.

At this point Terry and Phyllis work out a scheme for finding out what factors should go into a trouble-warning program. Phyllis is not really certain of what factors are important — she has to go back to her literature — but Terry has already outlined on paper how to edit such a function and she is prepared to bring a sample dialogue to the next meeting.

When the prototyping is completed, Terry will hand the screen, report, and query requirements to the project's senior systems analyst for hard-coding and integration into the on-line system. The final prototypes of screens, reports, and queries will form an important part of the project documentation. There is some talk of integrating micro-computer packages into the network. Because the network is mainframe-based, integration of microcomputers was not planned for. However, since a number of local

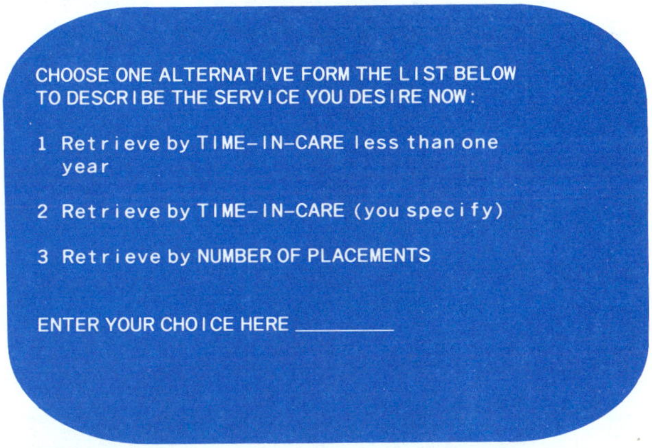

```
CHOOSE ONE ALTERNATIVE FORM THE LIST BELOW
TO DESCRIBE THE SERVICE YOU DESIRE NOW:

1  Retrieve by TIME–IN–CARE less than one
   year

2  Retrieve by TIME–IN–CARE (you specify)

3  Retrieve by NUMBER OF PLACEMENTS


ENTER YOUR CHOICE HERE _____
```

FIGURE 16–9. Prototyping a Menu

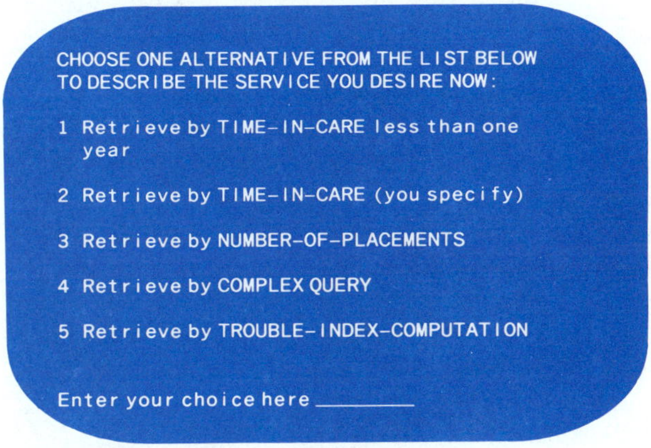

```
CHOOSE ONE ALTERNATIVE FROM THE LIST BELOW
TO DESCRIBE THE SERVICE YOU DESIRE NOW:

1  Retrieve by TIME–IN–CARE less than one
   year

2  Retrieve by TIME–IN–CARE (you specify)

3  Retrieve by NUMBER–OF–PLACEMENTS

4  Retrieve by COMPLEX QUERY

5  Retrieve by TROUBLE–INDEX–COMPUTATION


Enter your choice here _____
```

FIGURE 16–10. The Next Stage in Prototyping a Menu

offices have acquired micros, Phyllis's prototype may yet see major use as an operational item.

16.2   END-USER SOFTWARE AND THE INFORMATION CENTER

There are two critical factors active in prototyping that are not active in the standard SDLC: user involvement and rapid feedback of results. The prototype is built with the user's active assistance, and results of the prototype are fed back rapidly to both user and analyst. In the SDLC, users are involved only as information sources and feedback may be delayed by years.

Figure 16-11 illustrates a number of development alternatives in terms of these two dimensions. Note that end-user software development lies at the extreme value of high user involvement and rapid feedback. Of course this is possible only where users succeed in developing their software!

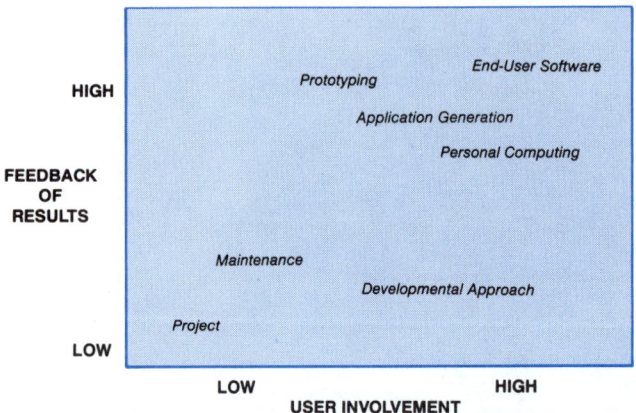

FIGURE 16–11. User involvement and Feedback in Six Development Alternatives

This section discusses the recent phenomenon of users developing their own software, the mechanics of application generation, and the revolution of the information center. We will examine how an information center can be created and run and the role the analyst plays. We then will move on to some difficult issues that the information center raises for computing in general and traditional systems analysis in particular. We will end this section with a case study involving administrators developing their own software at Southern State University.

The Revolution in Software Development

Many people forget that in the earliest days of electronic data processing (EDP), users developed their own software because no one else could! With the advent of the professional programmer, assisted by compilers and guided by operating systems, such user-programmers faded from the scene, although many, if not most, engineering and science applications are written by the engineers and scientists themselves.

Between 1975 and 1985, the appearance of microcomputers changed that. Most 16- and 32-bit microcomputers have software available to assist users in generating their own applications in very high level **fourth-generation languages (4GLs)**. Like NOMAD™, Sperry's MAPPER™, database user languages such as ON-LINE ENGLISH™ and QUERY-BY-EXAMPLE™, these retain some characteristics of programming languages but are incremental improvements over procedural languages like COBOL and Fortran.

The End-User Software Development Concept

Figure 16-12 illustrates the end-user development process. Beginning with an idea of an application, the user first chooses to use a traditional development technique or to develop or her own software. This is guided by available development environments, user skills, and economic considerations (most important, time available).

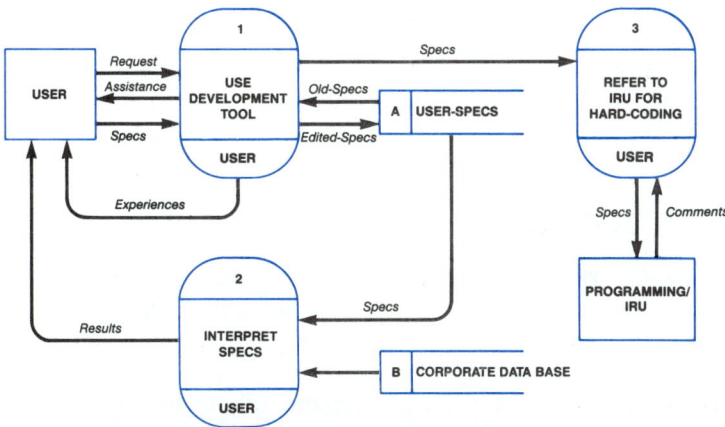

FIGURE 16–12. End-User Software Development

Next, the user selects a **development environment**. This can be an information center, a personal computer, or a terminal to a mainframe. In the first case, a variety of tools and personnel will be available. A personal computer may have a limited set of development tools. The most common tools are a **database manager**, a **spreadsheet calculator**, a **report generator**, and a **word processor**, the latter for documentation. **Integrated packages** assist in development by decreasing learning time with some sacrifice in facility. A variety of sophisticated decision-support systems based on artificial intelligence (AI) concepts are now available; calling these "programming tools," however, may be stretching the definition of programming. IFPS™ is an example of another approach to decision-support systems using a programming-like language.

The third choice is to utilize a **package** such as MAPPER through a terminal into a mainframe computer (it is also available on microcomputers). The well-known packages SASS and SPSS for statistical analysis resemble programming languages, and they, too, are being ported to microcomputers in limited versions. **Database query languages**, like NATURAL™, working in conjunction with ADABAS™, make the power of the mainframe available through terminals to users sophisticated enough to learn them. The database approach requires a **data administrator** (DA) to mediate, however, which complicates development efforts. Given the rather steep learning curve and the need to safeguard corporate data accessed through terminals, many firms limit access to the information center environment.

The user then develops an application. In the information center environment, training and consultation are integral to use of the machinery; in other environments, users may be on their own.

When the application is "completed," users have the same choice that analysts have in prototyping. Often they end up with suboptimal, error-prone, undocumented programs that may be lost without proper archiving. The information center tends to standardize disposition of software through standard archiving techniques, although the political questions are still paramount.

In summary, when end users control software through all stages of development, they either learn the tools on their own or get help from IRU staff or outsiders. The most typical aspects of a larger application programmed by users are data entry, report generation, query, decision-support and modeling, and statistical analysis. End users decrease direct IRU expenses in IR maintenance, but it is not clear that long-term cost savings will prove significant. Obviously, though, end-user computing, pushed forward by microcomputing, is here to stay.

The Information Center as a Workshop

The idea of end users creating and using their own software like home handymen was so foreign that, until recently, few analysts and programmers set their sights on developing tools for the "handy" marketing analyst, the "handy" banker, or the "handy" executive.

We can consider the **information center** as a workshop (Figure 16-13) in which semi-amateur programmers can labor in the vicinity of master programmers. With a few standard software tools, some training, ready consultation, *and an IRU willing to support the idea*, the programming workshop concept can work. In many firms with information centers, the terminals and microcomputers are busy all day increasing the information-related productivity of nonprogrammers significantly.

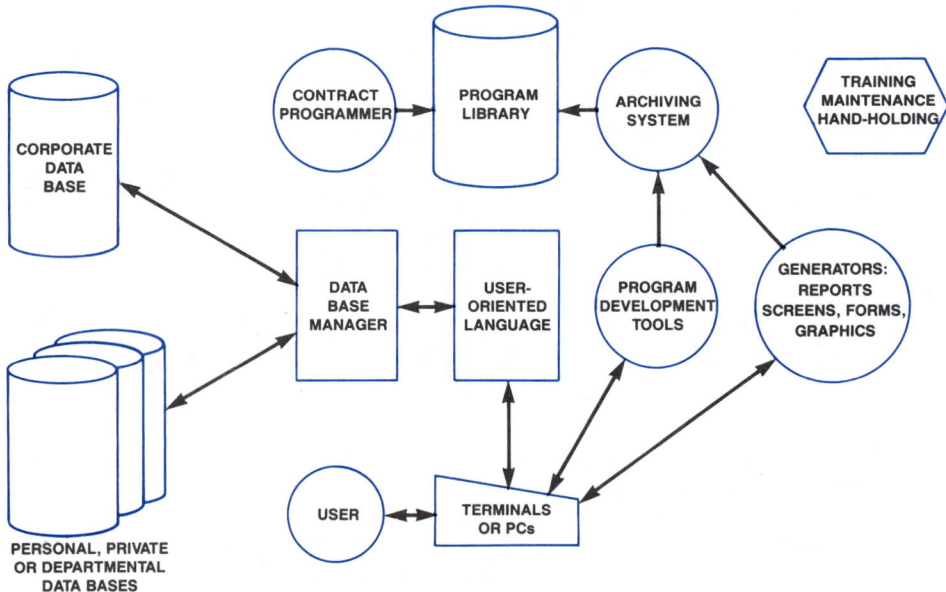

FIGURE 16–13. Components of the Information Center

The programming workshop consists of the following components:

1. access to the **corporate database** through a single standard database-management system dialogued through a simple, user-oriented language, for example the ADABAS-NATURAL combination
2. access to private or **departmental databases** through a similar interface
3. **contract programming** with professional programmers and/or analysts to design, create, populate, and maintain these databases
4. **report, screen, form, and graphics generators** and transparent access to the required hardware (plotters, color monitors, and printers)
5. **archival and library systems** to allow natural access to software created by nonprofessionals, using their own tools
6. a set of program and quasi-program **development tools**
7. **training**, **maintenance**, and **hand-holding** help available on demand at reasonable emotional, political, and monetary costs

Other less common and optional facilities include **networking** to remote locations, specialized **decision-support software** other than commercial packages, and "hard" **programming languages** like COBOL, Fortran, and BASIC.

The **database** is the key element in the information center. Originally developed to assist programmers in generating programs, the database concept has grown into a management system to ensure integrity, efficiency, and documentation. These attributes make a good database manager a fine tool for users since it ensures that none of the code need refer to the location, structure (scalar, array, outline), type (logical, numeric, character, switch, etc.), or value of a data item. The data administrator approves data designs and controls vocabulary, thereby guaranteeing that reference to a variable by name automatically takes into account its various attributes.

This is the "giant step for users" into programming freedom. Users refer to data by name in their work: "The Smith report," "Joe's data," "Tom's work history," "sales for January in Region III." With this **data independence**, users continue to refer to data by name only.

The need for private or departmental databases, however, brings IRU professionals back onto the scene. Criteria such as efficiency, security, audit, cost, training, and documentation are rarely considered by users rushing to meet deadlines to develop "personal" software. Professionals can mediate between users and the database facility to provide a low-cost, efficient, documented, and relatively "safe" private database design. Considering that the database will span employees, applications, and probably years, such concerns are not unreasonable. Figure 16-14, on the following page, shows how IRU professionals provide this mediation.

A variety of generators help users with actual software creation. **Screen generators** create temporary report formats for query results or signaling exceptional circumstances. Forms generators assist amateur programmers with creating input screens for data entry. Report generators and word processors create layouts for item-by-item and summary reports.

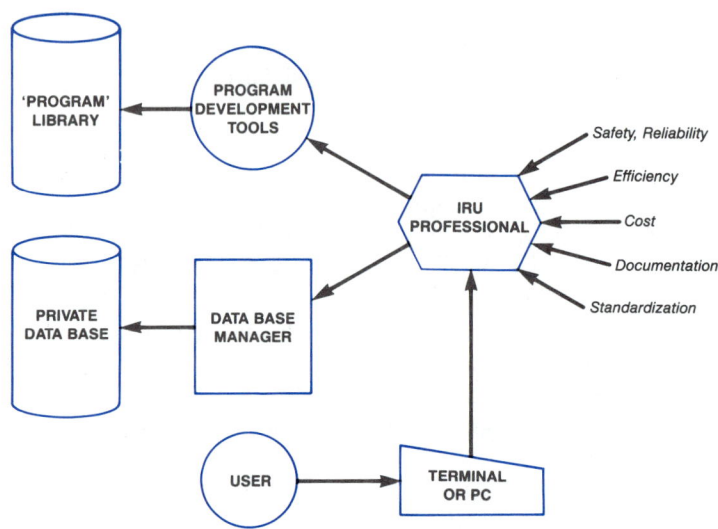

FIGURE 16–14. The Role of the IRU Professional in Information Center Activities

Users build charts, graphs, and displays from their data through **graphics genera-tors**. Simple generators can produce simple line graphs; complex ones can generate three-dimensional graphs in many colors from any angle of view. Most spreadsheet calculators have at least some graphics capabilities. Machinery to print the graphics is often expensive and must be centralized; transparent access to the color plotter or the laser printer must be available with a few keystrokes.

Because users want to save the software they generate, and in some instances wish others to use it later, **archiving and documentation facilities** should be available. Pro-cedures to save, update, document, and distribute software must be integral parts of the programs that create them; otherwise, IRU professionals will have to be involved. A natural adjunct to the information center is software to create training packages, like COURSEWRITER III™.

Tools can be **applications generators** like LOTUS 1-2-3; **fourth-generation lan-guages** (4GLs) like FOCUS™, RAMIS™, or NOMAD; **enhanced DBMS** like MAP-PER or NATURAL; or **programming-language end-user tools** like the macro facility in LOTUS, SYMPHONY, and FRAMEWORK™; or the real programming languages in IFPS, SPSS, dBASE III, APL, or SASS. These nonprocedural **pseudo-programming languages** specify results rather than processes. Spreadsheet calculators work in a net-work rather than linear fashion to calculate, thereby qualifying as pseudo-programming languages. Their macro facilities, however, are inherently linear and are thus true pro-gramming languages.

Underlying the information center is the human resource. James Martin (1982, pp. 305-307) divides information center responsibilities between **consultants**, who provide user assistance, and **technical specialists**, who keep the place running. The consultants train, encourage, sell, hand-hold, spoon-feed, prototype, consult, help users decide if something is worth doing, try to maintain standards, and alert the data administrator regarding problems in the common set of tools and data. Technical specialists set up the system, support it technically, maintain its tools, provide backup and recovery, calculate costs, and choose hardware and software. In other words, technical specialists operate the information center as another major function of the IRU.

The workshop aspect of the IRU is best illustrated by example. Consider a marketing analyst who wishes to build some software to locate strong and weak markets for particular products. Figure 16-15 shows how the analyst would interact with the information center.

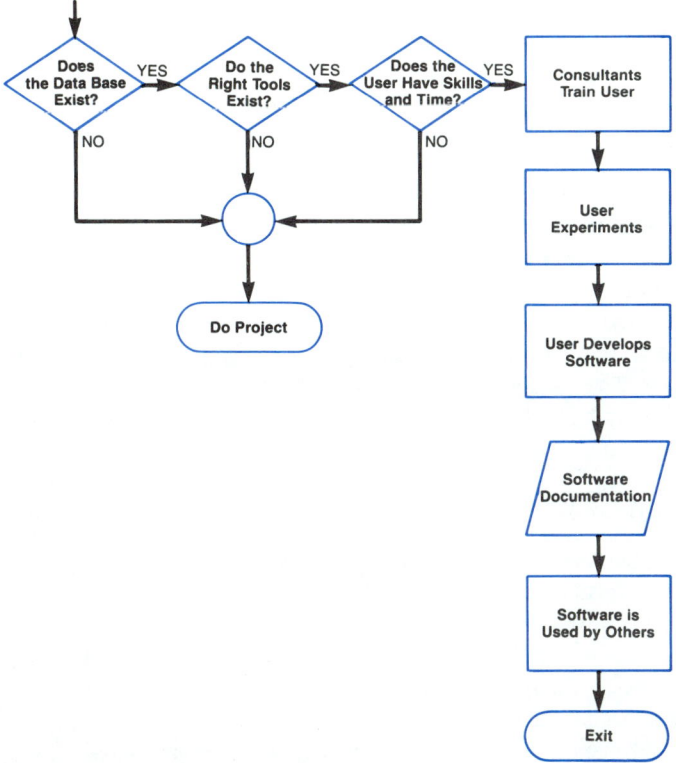

FIGURE 16–15. Using an Information Center

1. The analyst approaches the information center management with a problem.
2. Information center management works with the analyst to ensure that there is an appropriate database, that there are appropriate tools, and that the analyst has the skills and time necessary.
3. Information center staffers show the analyst how to use the information center, its databases, and its tools.
4. The analyst spends a while learning, experimenting, and becoming familiar with the information center facility.
5. The analyst develops software using a specific tool.
6. Helped by a consultant, the analyst archives and documents the software and sets up for later access to it.
7. Others use the software to produce reports.

The Analyst and the Information Center

The information center spells out a new role for the analyst, and as a person who may work either in the information center or as a consultant to it.

The analyst can, of course, help design the information center. Database designer, selector of tools and languages, and trainer of information center staff are natural roles for the analyst. Once the information center is designed and implemented, alterations and enhancements can again be made in traditional fashion.

The analyst might work as a consultant in the positions, learn about the firm's applications and tools, thus keeping in touch with the operating divisions of the firm. Senior analysts often manage the information center.

In addition to performing traditional systems analysis tasks, analysts may provide consulting services to information center management. Firms that implement successful information centers may wish to share their experience and help other firms set up information centers, especially when these firms are in the same corporate family.

Finally, the analyst may continue to function in a traditional manner, but utilize information center services for user-developed applications. This means sending users to the information center at the start of a larger project. For example, the analyst may be building a complex decision-support system for upper management. Routine reports on the database need not be from scratch, but can be created either by or for upper management through the information center. The "software behind the software" (the complex models that need to be constructed) may be built by technical specialists using a program generator like IFPS while the analyst coordinates these efforts.

Difficult Issues in Information Center Use

The information center will assist in some ways, but it may raise some difficult questions. IRU staffers may see the information center as "skimming" easy applications and increasing productivity more than traditional IRU functions. The availability of private databases brings back the issue that DBMS was supposed to address in the first place, thus putting additional burdens on the data administrator and information center consultants to keep data integral, nonredundant, and well documented. There are also potential problems with security, efficiency and maintainability of the user-generated applications.

Despite these issues, the benefits of downloading programming to users continue to outweigh the costs. When coupled with an intelligent and sensitively run screening process, users can begin to integrate the information resource into their daily labors by creating and using their own software.

Case: Downloading System Maintenance at Southern State University

Jim Taylor was dean of Fine Arts, but he had not mastered the fine art of computerization. When he began to notice that his student body was getting older — or so he thought — he started to wonder how the computer might help him plan courses and programs.

Jim asked Rita Dalgleish, his administrative assistant, to go through registrations for the coming fall term and find the average age of students coming into first-year fine arts and those declaring majors. He had a hunch that students were older than ever and, therefore, were somehow different from previous groups.

Rita threw her hands up in disgust and said, "Jim, now you can stand there and *ask* me to total up hundreds of numbers on my calculator, but there's no way I'm going to get that done today. Besides, all that stuff's already on the computer, at least the birthdates are. So why don't you send Sandy over to administrative computing services to ask them to write you some programs?"

Jim was perplexed. "Rita, you know that it takes six months just to get those computer folks to write their names. All I want is to find out something today about students' ages. Can't we do something?"

"Well, I've heard about the information center, where you write your own programs, but heck, I'm no programmer. Maybe Sandy can find out."

Sandy McLaren, a fourth-year music student, was working in the office for the summer. She went over to ACS and inquired about their information center. After being shown where the office was, Sandy spoke with Deanne Miller, the information center manager.

"Sandy, I'm glad you came by. We don't have anybody from Fine Arts using our service, and we're supposed to be university-wide. What do you want to do?"

"Dean Taylor wants to know how old our students are and whether or not they've been getting older over the past several years. Mrs. Dalgleish says that data's on the computer but that it will take you six months to write the program."

Deanne smiled. "Well, not *six* months, maybe only five. But Mrs. Dalgleish is right, the data's all there, and if you have an hour to spare, we can show you how to get the answer to the dean's question and many others."

"OK, I've got an hour. But the only thing I've ever programmed was a Moog synthesizer."

Deanne introduced Sandy to Byron Ford, a consultant in the five-person information center. Byron showed Sandy the networked microcomputers and how they could tap into pieces of the University administration's database with a few simple commands. Sandy quickly learned the basics of micro MAPPER and was performing simple queries within an hour. It was already noon, and students being the age (older though they may be) they are, Byron asked Sandy to lunch.

"By, I've already got the answer to the dean's question. But how about surprising him?" she asked between mouthfuls of Crackerburger. "Maybe I could let him look at trends by department or sex or even by concentration, say piano or dance?"

"Sure, Sandy. That's not hard. Micro MAPPER's got a way to break down requests. It even uses words like "by department" or "by concentration," as long as you spell the words right. Anyway, you sound like a programmer now. Why is the Dean so interested in how old his students are?"

"I'm not sure, maybe the computer can tell me *that*? Would I ask for a breakdown of the dean by information overload?"

16.3 THE ROLE OF THE USER-ANALYST

Our emphasis so far has been on explaining how an analyst can work within the constraints of the IRU as either an employee or a consultant. A recent trend, accelerated by user interest in microcomputers, has been for user groups to employ analysts to look at systems in the information resource from the user's point of view.

Ein-Dor and Segev (1982) examined the evolution of the "systems" group in over forty major organizations. They found a natural progression from **subordinate** IRUs working in functional areas or accounting, through **coordinate** IRUs functioning within finance or controller organizations, to **autonomous** IRUs reporting at a vice-presidential level. This spells out the increasing tendency of the IRU to accumulate a mission external to that of all operating divisions or administration in general.

On the other hand, the diffusion of microcomputer-based skills and the increasing sophistication of user-managers has increased the likelihood that user groups, rather than an autonomous IRU, will initiate projects and perform at least some initial needs analysis. Where information centers are not available, user groups may acquire their own development tools and put together their own software. Obviously, manpower is needed to coordinate this effort, and the position of "user-analyst" has evolved to meet that need.

The **user-analyst** is a skilled user who understands and probably uses the information resource of an organization as a user rather than as an analyst or programmer. For instance, the user-analyst may be a clerk or professional worker. Consider a marketing analyst skilled in the tools (manual and computerized) of market research. If the marketing group decides it should acquire a marketing package, the marketing analyst may begin looking at market research packages for the firm's computer system. In addition, the analyst (let us now call this person a user-analyst) may begin an audit of existing packages useful to market researchers.

The result of this effort is a report detailing existing capabilities, some possibilities for purchase, and existing needs. At this point, the user-analyst may not have contacted the IRU at all. Instead, there is a lengthy document that spells out options and important questions or needs.

Systems analysts, of course, can function as user analysts in their daily work. It is doubtful, however, that any harried analyst can afford to put the time into becoming intimately familiar with detailed operating division procedures and goals. Instead, the analyst has to infer these from lengthy investigation.

As users become more familiar with information resource options and the methodology of analysis, the user-analyst will become a viable companion and — in some cases — alternative to the systems analyst.

16.4 WORKING ON MANUAL SYSTEMS

A large proportion, maybe the bulk, of analysis work in North America is the investigation and streamlining of existing manual systems (see Figure 16-16 opposite). Much of this work is *not* performed by people bearing the title of "systems analyst."

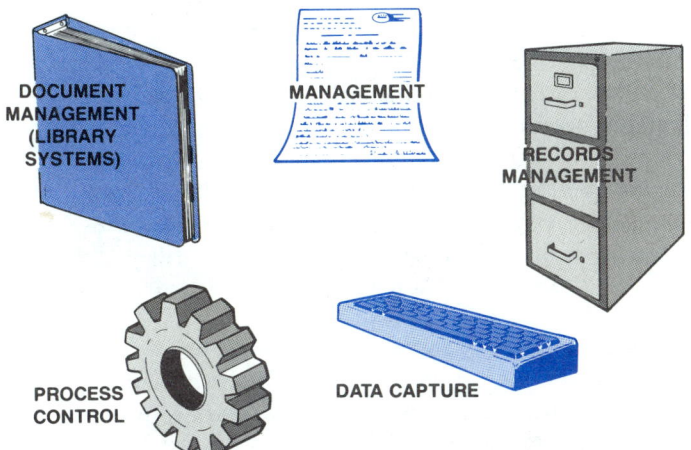

FIGURE 16–16. Areas with Heavy Manual System Contribution

Instead, they may be forms analysts, policy and procedures analysts, or records management consultants or vendors.

They concentrate on manual or mechanical systems, some of which are quite sophisticated. With the blending of computer and video technology in video disks and CAD/CAM (computer-aided design/computer-aided manufacturing) systems and with increased attention to automated manufacturing tied to inventory and order processing, the boundary between the purely manual and the totally computerized is blurred.

There are five areas in which at least some measure of manual systems are often or sometimes maintained even while computerizing: records, forms, and document management, data capture and process control.

In a **records management** application, the emphasis is on efficient storage and retrieval of paper records. Analysts may be called in to examine the need for specific records, their internal organization, and how they are distributed. Recommendations may include the purchase of an integrated records management system, mechanizing a microfilm or microfiche documentation process, or automating microrecording and retrieval.

Forms management includes form design. Often, forms are poorly designed and hard to fill out, read, or correct. Many organizations have dozens of obsolete or seldom-used forms; some may be hard to get and others, simply unavailable. If, after a time, user organizations evolve away from most of the content of a form, those contending with the form may find most fields irrelevant or confusing. These are matters of forms control. The forms analyst controls both the content and the use of forms by investigating usage and problems.

Document management may be a library application or a typing concern. In the first case, the analyst works with problems of acquisition, cataloging, retrieval, circulation, and sometimes transcription. For instance, a small technical library in a natural resources firm may subscribe to a technical report series. As each report is received, it must be logged, cataloged, and prepared for circulation. Where the original copy is precious, there may be a transcription (photocopying or microfilming) procedure. As the report circulates, it should go either to those who have requested it or to everyone who might have an interest in it. After an initial closed circulation of this type, it should be prepared for general circulation, which implies putting it somewhere accessible.

The other document management application involves the origination of documentation, which may range from designing procedures to accept, handle, and return typing within a pool to procedures for drafting copy, artwork, printing, and binding of written documents. Other specialists in printing may already have this responsibility, but if these systems are candidates for computerization, an analyst is consulted. An effort to automate typing, for instance, is handled outside the IRU. Many authorities have written on the subject of the "Data Processing/Word Processing Confrontation" regarding who has responsibility for these systems.

Data-capture applications involve the creation of forms and the building of procedures to handle them, including archiving and correction. Since much transaction processing begins with handwritten documentation, the design of these data-entry forms is critical. Because it is often customers rather than employees who fill them out, extensive pretesting is necessary. Here, an analyst who is aware of important human factors considerations is valuable. Consider the forms you fill out in banks and government offices, and even for contests, and you will understand how critical the design of data-entry forms is.

Finally, the burgeoning field of automated **process control** requires the design of many **manual procedures** for reading and recording data, resetting equipment, taking it off-line for repair or replacement, and ol procedures. A number of books and articles have been written about the events at the Three Mile Island reactor in Harrisburg, Pennsylvania, citing the problems involved in manual procedures.

The problem at Three Mile Island began when two valves that should have been open were shut. Apparently, operators could not see the tags indicating the valves were shut because they were hidden by the handles on the valves themselves. The closed valves made the water that should have been sent to cool the reactor unavailable. So many things went wrong within milliseconds (termed "common fault failure") that pinpointing the problem was impossible.

In the absence of an indication of how *much* water was in the reactor (water is used to cool the reactor and move heat to the turbines to generate electricity), operators relied on a *pressure* reading, unaware that high pressure can result from either a lot of water in liquid state or very highly pressurized vapor. This, unfortunately, led operators to choose to override the computerized system, which in fact was doing quite well, and turn off an auxiliary water supply, with the mistaken impression that there was *too much* water in the system. In fact, the reactor core was uncovered and was in danger of overheating and melting the reactor (the so-called "China syndrome"). This did not happen and, although damage was extensive, steps were taken to stabilize the reactor without endangering central Pennsylvania.

Operators had been trained only in *normal* procedures, not in *emergency* ones, and had made their decision to override the computer based on poor information. In addition, there were no *management* procedures in place to keep upper management informed of progress. To summarize the experience, there were failures in carrying out procedures, the management and public relations systems, and in training operators. The computer system, based on evidence to date, worked fine.

The Necessity for Computers

Computers were not necessary until the 1950s. But of course, this is a chicken-and-egg analysis. Computers begat computer applications, which begat computers. Are computers necessary now? Should all business information applications be computerized?

Figure 16-17 shows the advantages of computer systems and manual systems. Computers have the advantage of high-speed computation, inexpensive and rapid backup, tireless repetition of operations, and sharing of data without redundancy. Manual systems are more capable of handling special, private, and familiar cases.

COMPUTERIZATION	MANUAL IMPLEMENTATION
Much Computation	Little Computation
Speed is Important	Speed is Unimportant
Repetition of Operations	Special Handling
Data Sharing Without Redundancy	Privacy and Security Important
Little Discretion	Backup Unimportant
Few Exceptions No General Case	Lack of Procedures

FIGURE 16–17. Comparing Computerization and Manual Implementation Criteria

How are these two sets to be compared? We tend to computerize where a procedure exists, where exceptions to the rule are rare and *easily noted*, where many computations and copying operations are needed in each procedure, and where time is so much of the essence that learning time is unimportant. Where there are many exceptions to a relatively simple computational procedure, where backup copies are less important, and where data do not have to be shared (in fact, where privacy and security are paramount), then a manual operation seems indicated. In between? Call a judge!

Computers are not always necessary. The old adage "If it ain't broke, don't fix it" is important. A well-designed manual system is *not* a candidate for computerization. Users will revolt at having to learn new procedures, and analysts will be under a lot of pressure to simulate the manual system on the computer. This is not always a good idea. What a person can do with a pencil in a random inspection of a printed document is not necessarily analogous to what the same person can do with a cursor on a 24-line screen.

Resistance to computerization is often dictated by users' perceptions that analysts desire computerization of everything. Unfortunately, systems analyst training strongly stresses computer applications while ignoring working manual systems.

Differences for Manual System Development

There is little modern research specifically on the topic of how to proceed with manual system development. In fact, the SDLC is not system-specific; logical system design, for instance, does not mention computers. Within this context, there are few real differences.

There are some feasibility considerations that are different. First, manual system development almost always means **purchasing equipment** for which computer-trained systems analysts have no design experience. Where physically large systems must be installed in user areas (such as filing cabinets), the skills of an interior designer may be required.

Readability, layout, color, format, paper stock, type face, and durability are usually ignored in building most computer systems; they are critical in physical design of manual systems. Durability, weight, and color may also influence logical design in that certain information may be physically hard to get together. Manual forms cannot be "collated" field by field as they can be on a computer; thick documents cannot easily be bound together for access; if data can be smudged through handling, then it should be replicated elsewhere.

Users expect manual systems to evolve in small steps with the addition of new forms, revision of fields, and gradual change of procedures. The degree of change an analyst can assume acceptable in the logical design may be limited by these considerations. In terms of a firm's image, manual systems development is usually financially and technically undersupported. Analysts cannot expect to spend months redesigning a paper form, but they might get away with a lengthy screen design exercise. Firms like to be seen computerizing; they do not like being seen developing manual systems.

It is politically and often economically infeasible to "throw out the computer." The costs of decomputerizing can be prohibitive. Some small firms may experiment with computerizing through microcomputers and then discover that their customers expect to see computerized bills.

These considerations influence an analyst's approach to manual systems development in the context of a change to the way an organization functions as a whole. As time goes on, there will be increasing pressure to dispense with manual systems, since analysts are increasingly trained to computerize.

The Future of Manual Systems

There *is* a future for manual systems, but it seems extremely limited. Where the funds are available for computerization, only procedures requiring extreme privacy or individual attention will be implemented through manual systems. Although I still find it convenient to total up a column of figures with a pencil rather than using the SIDE-KICK™ calculator on a microcomputer, I would not want to do that sort of repetitive calculation for a living. An increasingly important area is CAD/CAM, in which semi-automated systems are used to collect information from operators to assist, rather than create, designs and products. Such symbiosis is, in this writer's opinion, the future of any manual systems development.

Case: Streamlining the File Room at BelCo

BelCo is a medium-size manufacturer and retailer of geegaws and gimcracks, otherwise known as balloons, party favors, party hats, pennants, and souvenirs, many of which it imports. BelCo has a manufacturing division and a retail division; each produces voluminous written and typed records of its operations. Started ten years ago by Richard Belinsky using $3,000 borrowed from a local bank to import and resell souvenirs at county fairs, BelCo has grown rapidly. The information resource, however, has not fared as well. Most records are handwritten, and although payroll and much of the accounting is handled through a computer service bureau, sales records are still compiled manually. This is due partly to large numbers of suppliers and customers and partly to rapid expansion. The result: the file room is a mess.

Leslie Wasserman is a records analyst hired for six months by Belinsky to overhaul records handling, beginning with the file room, where everything ends up. Noting that over eighty different forms (including copies of forms from suppliers) are filed in several different ways — alphabetically by supplier or customer; by salesperson; and numerically by part number, by date ordered, and even by order size (large vs. small) — Leslie proposed a unified form consisting of a manilla folder with index tabs of different colors to indicate internal part number. She reorganized the paper handling, cut down on the number of copies made on photocopiers (the cost was getting out of hand), and organized access to records only through this internal part number. Leslie prepared a set of easily maintained indexes to cross-reference external part numbers, suppliers, customers, and sales people. She developed a log book for sales people to eliminate references by date of ordering. Finally, she helped Richard specify individual responsibilities within the file room and among staff originating data. She built a procedure to request information from the file room in library style and set up a circulation control list to make sure information was never unaccounted for.

Leslie noted that documentation more than six months old was almost never used, so she proposed that six-month-old documentation be archived in boxes and moved out; financial summaries were made automatically so that the original data normally would not be needed. The new file room worked efficiently and rapidly. Those who actually needed the data were pleased; those who had had fun browsing in old records complained, but an important security loophole had been plugged. The word "computer" was never mentioned. The total cost of the project was $8,500 and Sandy finished it two months early.

16.5 WRITING STANDARDS

Another major responsibility that falls to analysts is **standards writing**. A standard is a guideline or measuring instrument through which other things are guided or measured. Standards for behavior at senior proms, for instance, dictate that gentlemen wear formal garb, escort ladies on their right arms, and lead in dancing. Programming standards might be that all programs are self-documenting or that all numeric variables begin with letters between I and N. A system design standard might be that all designs include data flow diagrams.

What is a "Standard?"

A **standard** is a fixed post. It either guides someone in a particular direction or it provides a common measurement. Standards are necessary any time two or more people work on the same product or engage in a cooperative venture. For instance, standards would help if one programmer wrote a calling routine expecting variables to be *external* and another programmer wrote the subroutine expecting the same variables to be present in the calling sequence as parameters.

We tend to call something a standard if it has either legislative or authoritative power behind it. For instance, the "standard" communications interface to a modem is the RS-232 electrical connection, agreed to by a "legislative" standards body in the EIA (Electronics Industry Association). On the other hand, the "standard" transmission on automobiles (i.e., the manual one, not that so-called "automatic" one) is merely a convention among automobile makers. The MS-DOS™ operating system is the "standard" operating system for 16-bit IBM-compatible machines because of the commanding authority of IBM in the marketplace. However standards arise, they create common communication, make for natural agreement, and give a reference in resolving disputes.

How Standards Are Developed, Publicized, and Maintained

Standards are usually produced by a **working committee** of practitioners. The committee is struck by legitimate authorities who vow to put their weight behind the standard if it is reasonable and practical. Thus, international standards bodies typically develop standards that members swear, in advance, to follow. Such standards as X.10 for modem connection and X.25 for packet switching come from bodies associated with the ISO (International Standards Organization). Within an organization, a standards group for information resource management will usually be given its mandate by the director or vice president and will contain analysts, programmers, and managers. Users may be present if standards for user interfaces, for instance, are to be developed.

The resulting standard regulates the behavior of system developers and thus limits the possibilities, making it easier for them to communicate with each other and reach agreement. General standards are reached by general standards committees; individual projects may derive *ad-hoc* standards that exist for the life of a project. Data administrators usually build policies that generate standards that they follow — and, of course, indirectly force others to follow, since they administer the data.

Standards have to be publicized. Two techniques are to integrate them into training and to have regular standards publications or newsletters. This implies that standards are well written in the first place. A standard such as "All calling sequences will be documented" is not as good as "Each data flow appearing in a DFD will be documented in the XLD skeleton provided by Excelerator."

Standards are maintained by awareness of how well they function. Analysts try out and use system development methodologies (SDM); programmers use programming conventions; coders use coding conventions. Formal systems of standards maintenance are not common, but analysts become aware of problems during system development and should forward these concerns to standards committees.

Case: Grow Your Own SDM at Major Foods, Inc.

The systems group at Major Foods, Inc. has recently initiated a project to produce a system development methodology (SDM). A task force (the manager of programmers, the chief systems analyst, and the chief operator) has looked into the acquisition of a commercial SDM and has assembled a number of documents. It has also received sales visits from two vendors pushing established but expensive manual systems that ensure that system development proceeds in a logical, orderly, and well-documented fashion. However, the expense of acquiring these systems and the sheer volume of documentation seem unnecessary for a group consisting only of three programmers, two analysts, and a small set of operators of an enhanced minicomputer.

Instead, the standards group expanded by bringing in an outside consultant from a consulting firm, one programmer, and the other analyst. They meet every week for three hours to discuss the issues involved in standardization of system development methodology. To date, they have settled on some coding standards (all in COBOL), quite a few flowcharting and documentation standards for programming, and a number of standard documents to be produced in new system development and maintenance. They also have decided to purchase a number of project management and system development aids, including Excelerator and Super Project Manager™.

As they go along, they document standards and discuss characteristics of good standards. They have noted that a standard should be implementable, documentable, teachable, enforceable, and, of course, valuable. There is little sense in enforcing a bad standard. They have plans to produce a regular newsletter for their group of ten, especially since there is talk of expanding the group to seventeen within a year by adding two programmers, four operators, and a tape/disk/diskette librarian. They feel it is important to have good standards in place by the time these new recruits are brought into the group.

Standards for Standards

Analysts are often involved in an exercise of deriving a system development methodology for their IRU. Figure 16-18, on the following page, illustrates the process of standards development. In this process, data are collected in a variety of ways: through standards committees such as at Major Foods, through a standards project, or from a *quality circle* (Couger, 1984). This last format is a committee struck by management, composed of workers, with the task of improving the quality of a product, process, or job. The quality circle is trained in problem-solving techniques and then works to come up with solutions. In the case of standards development, the solutions will be standards.

The initial standards document is promulgated through training, newsletters, or, less effectively, memos. As standards are used, they should be regularly reviewed and changed; a standard that changes often, however, is not a standard.

Areas in which standard development is usually concentrated are the following: language choice, documentation format and content, coding conventions, manpower deployment and responsibilities, order of phases in the SDM and procedures for signing off, testing procedures and responsibilities, and modeling tools.

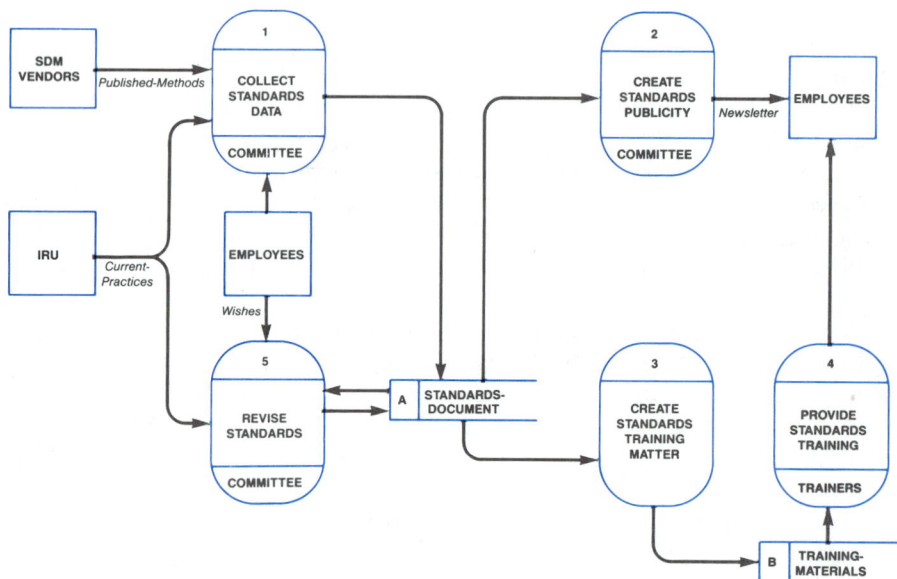

FIGURE 16–18. Creating and Using Standards

Responsibility for enforcement of standards often falls to the data administrator, who already enforces data naming and format conventions. Programming standards are best enforced within the programming group.

Four guidelines for enforcement of standards are (Licker, 1985):

1. Standards are not a substitute for good project management; they are a response to a need for rational coordination of activities.
2. If standards are imposed, they are more likely to be ignored; if standards are developed within a unit, they are more likely to be followed.
3. Enforcement of standards is an educational function, not a disciplinary one.
4. Standards *do* change; poor standards should be reworked.

16.6   CONCLUSION

There are several alternatives to project structures; there are also special kinds of projects an analyst might be involved in. The alternatives include **prototyping**, **end-user software development**, and the **information center**. These alternative structures are changing the way analysts work and the IRU.

Analysts can and should handle **manual systems development** in an era of burgeoning computerization. Another area of interest for analysts is the **writing of standards**. Writing standards is important at two stages of an IRU's growth: prior to building an information resource and when the resource is functioning.

Following a discussion of project management in Chapter 15, this chapter pointed out alternatives that may be less expensive, more easily managed, and more politically viable. Chapter 17 goes into the content of specific projects in more detail.

DISCUSSION QUESTIONS

1. I have to get a piece of software developed quickly. I am a marketing manager in a retail drugstore chain, and we're seeing an apparent change in buying habits among our customers. I'd really like to know what's moving the fastest. Our inventory system is already built, but it gives only monthly reports and only for product groups. Consider the traditional project, prototyping, and end-user-software development (using an application generator). Discuss the relative advantages and costs of each alternative, specifying both tangible and intangible costs. Assume that an application generator *is* available for me to use.

2. The current batch payroll system provides adequate reports to M. Monroe, director of finance, but he is not sure that he's really getting all the information he needs. Describe how an analyst might work with Monroe to prototype an on-line query system that works through a DBMS and a terminal.

3. Consider M. Monroe and others like him. Given that the firm already has in place at least one prototyping tool and several executives like Monroe, what steps would you outline to build and operate an information center in their firm? What are the "up-front" costs and the hidden ones? How would you measure the benefit of an information center? Whom would you expect to use its services? What personnel would staff the center; which computer resources of which type might you look into? What *organizational* resources would have to be available?

4. You are a manager in the sales office of a very large construction firm, part of a conglomerate with an extensive information resources department. You are not happy with the speed with which the IRU responds to your requests for reports on project progress and expenses. Outline a plan by which you could locate, hire, train, and utilize the services of a user-analyst. Who would be likely candidates? What skills would you require? How would this person work with the IRU?

5. Consider how your user-analyst (Question 4) might function within the context of an information center. Would there still be benefit to having a user-analyst on staff?

6. Jones Warehousing and Transfer, Inc. has a complex forms processing mechanism that involves over a dozen people stamping, collating, bursting, completing, and routing the forms required to provide estimates on moving and maintain appropriate records as households and industrial goods move across North America. As a consultant brought in to look at computerization, what might you consider before recommending computerization? List several criteria for computerization and detail how you would weigh them in making this decision.

7. Write a "standard" for producing a data dictionary entry. Use English to compose it; do not resort to computerese. Assume that systems analysts will be composing the DDEs for others to use in writing software specifications.

CHAPTER 17

SPECIAL PROJECT CONTENT

OBJECTIVES

1. To describe and implement a process by which decision-support systems are constructed and used
2. To describe, implement, and indicate the methods of control for a project to construct a management information system
3. To compare and contrast the construction of an office information system (an electronic office) with a project to created an MIS or DSS
4. To indicate how building systems on microcomputers differs from projects aimed at mainframe or mini-computer implementation
5. To discuss the major aspects of records management projects
6. To show how the construction of an information retrieval system differs from that of an information processing system
7. To discuss the peculiarities of developing consumer-oriented computer systems such as games and training packages

SPECIAL PROJECT CONTENT

17

CHAPTER

17.0 INTRODUCTION

This chapter looks at the wide variety of *content* of systems analysis work. Seven content areas are briefly covered, although they do not form an exhaustive list: (1) decision-support systems development, (2) management information systems (3) office automation, (4) microcomputer systems development (5) records and forms management, (6) document management, and (7) consumer products.

How Content Influences Technique

Content factors that influence technique (Figure 17-1) are the **nature of the client**, the **proportions** of hardware and software on the project, the role of the **vendor**, the **computer/manual** mix, the nature of the **user**, the technological **positioning** of the work, the need for **security or secrecy**, and the **breadth** of the work.

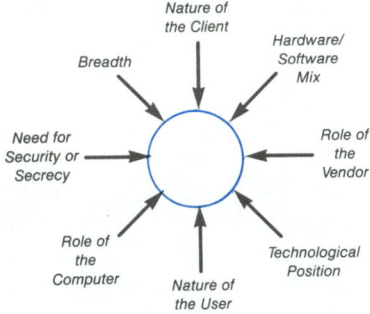

FIGURE 17–1. Content Influences on Choice of Development Technique

Sophisticated clients may require more investigative information before proceeding while less sophisticated clients may need instruction in the SDM and its implications. Where a lot of hardware is on order, more work in developing RFPs and RFQs is required and when vendors get involved in that work, vendor relations become more important. Developing manual or computer/manual systems means that analysts have to learn about the manual procedures and orient their thoughts more toward the human factor. If users are sophisticated, "user-friendliness" is less important, but flexibility must prevail; on the other hand, first-time users may have fears and interviews may be trickier.

State-of-the-art and pioneering projects mean more exploration, more risk, and correspondingly greater degrees of project monitoring. Learning curves may be steep for staff members. Where a particular kind of project (say prototyping) is new to an organization, accountability procedures may prove difficult and the politics of the project may overwhelm the technical work. Well-trodden developmental paths may be unattractive to high Growth Need Strength technical workers, and poor user relations may result if users detect arrogance or boredom on the part of workers.

If security must be tight, then development has to be more strictly supervised to keep secrets in and prying eyes out. Screening and training of staff are security considerations. Finally, when a project is broad in scope, large numbers of organizational considerations must be handled. Many people may be involved, some of whom are not connected with the IRU. These people have a political interest in the project, as Ed discovered in Chapter 15. Broad-scope projects require a broad list of staff skills and may present an opportunity to expand the staff; however, such broad groups may be difficult to manage.

We now turn our attention to the seven content areas, each of which has unique implications for project management and systems analysis technique.

17.1   THE DSS "DEVELOPMENT" PROJECT

What is a Decision-Support System?

A **decision-support system** (DSS), such as the one diagrammed in Figure 17-2, supports management decision making by providing data, graphics, and decision-making tools to a relatively sophisticated manager. The components of a DSS are a corporate **database**, a **model base**, and a **dialogue manager**. The dialogue manager allows the manager to converse with and access models to be run against the database in an attempt to answer "what-if?" questions. The what-ifs are the potential courses of action. These questions are answered by running mathematical models against corporate data.

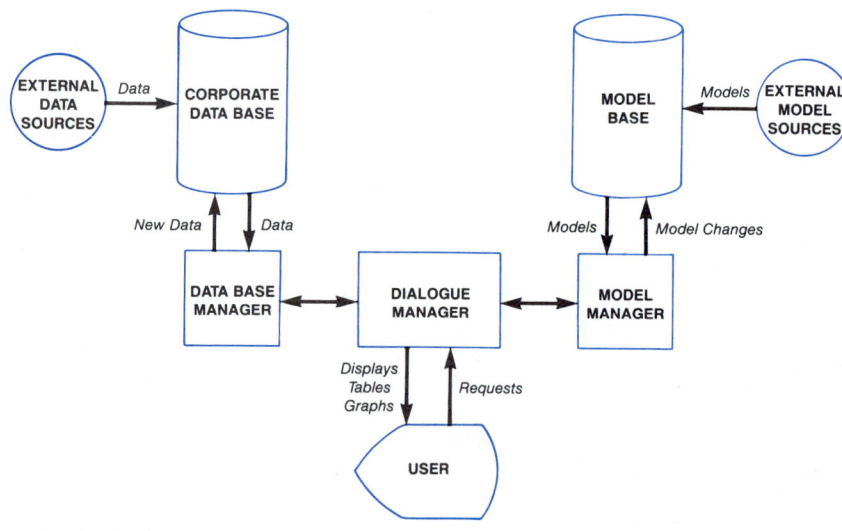

FIGURE 17–2. Components of a DSS

Results can be displayed either graphically or in tabular format.

Sprague and Carlson (1982) created the terminology of DSS development. A **DSS generator** (DSSG) is a tool used to create a **specific DSS** (SDSS), which would be appropriate in a given DSS circumstance. For instance, an analyst may create a spreadsheet (SDSS) using a spreadsheet generator (DSSG); another may build a model using a language like IFPS (DSSG) to create a specific program (SDSS) for a manager to do financial planning.

Thus, a DSS has both a potential (DSSG) and an actual (SDSS) form.

The Major Players in DSS Development

Sprague (1980) described DSS development as an iterative approach involving a number of people: the **user**, the **analyst**, and the **toolsmith**. The analyst, acting as a go-between, talks to the the user, who knows the subject matter, and the toolsmith, who constructs the DSS generator. The analyst works with the user in a prototyping mode, but the tools keep changing to meet the user's needs. The analyst has an interpretive and educational role here as the link between the user and the toolsmith.

The Special Role of the Analyst

DSS development is a kind of prototyping, with the analyst working closely with the toolsmiths, who craft the prototyping tools and may actually build the model. Toolsmiths, however, may have little knowledge of system development techniques, the crafting of dialogues, or the organization's information resource needs.

The analyst interprets system needs for the toolsmith much as he interprets user needs for system designers. In this way, system needs for compatibility, transferability among staff, preservation of data integrity, and efficiency are maintained. Users have little opportunity to communicate directly with toolsmiths and vice versa. The analyst plays a crucial role in interpreting between the two groups.

In addition, the analyst plays an *educational* role, teaching the user about system capabilities and helping the user become more sophisticated regarding his own needs.

Case Study: DSS for Portfolio Management

Baldwin, Chase, Scanlon and DeWitt is a well-respected brokerage firm in Bigtown. With over one thousand clients and several hundred million dollars in investments, the eleven brokers at BCS&D have a lot of responsibility. Recently, in an effort to help the brokers, Stuart Baldwin, the senior partner responsible for institutional investments, approached Danny Mallory, director of data operations, about building a system for portfolio management. Stu wants to use the same terminals that stock quotes appear on to manage portfolios for clients, to tailor portfolios to client wishes, and to take advantage of capital gains tax provisions.

Danny put Stu in touch with Rita Palmetto, a systems analyst. Rita spent a morning with Stu discussing his needs and wrote up a preliminary investigation report. Rita could do this quickly because she was already familiar with Stu's terminology and operations.

In her report, she recommended the developmental approach to building a DSS and thought a team composed of herself, a senior broker, and a programmer would be ideal. Stu looked at the budget (almost $15,000) and the time frame (four months) and approved the project.

Rita worked with Arnie Churloff for a month and a half, using IFPS available on a mainframe, and built a portfolio management DSS. Arnie provided the information about balancing, Sally Bohannan programmed the models, and Rita provided the links to the on-line quote system. Within the four-month period, the DSS was ready. Rita and Arnie tried it out on some of the other brokers and, after obtaining their feedback, made some final modifications.

The new portfolio management system is not complete; it does not yet interface with the main information system that holds actual current portfolio information, but it does provide quick and accurate assistance in answering clients' "what-if" questions.

17.2   THE MIS PROJECT

The components of a management information systems (MIS) (Figure 17-3) are:

1. a permanent **database**
2. a database of "transient" records called **transactions**
3. a mechanism for **merging** transactions with permanent database records, providing location, update, summary, and exception reporting
4. a set of human-machine interfaces to **enter transaction data**, call for specific, predefined **reports**, and schedule reports

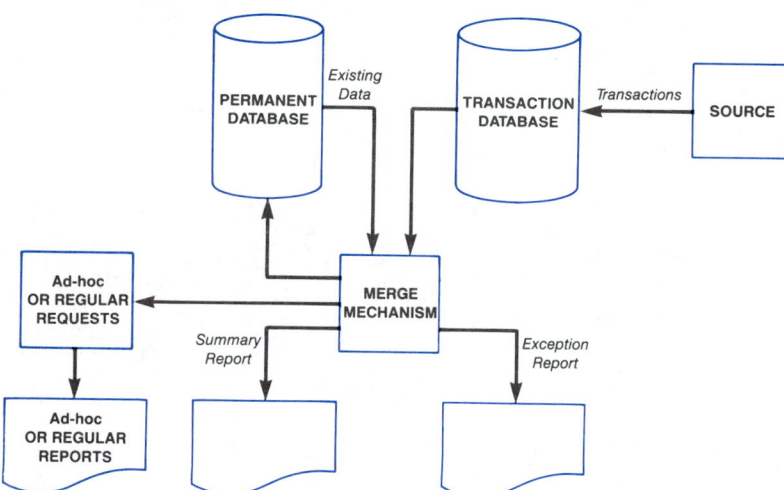

FIGURE 17–3. The Components of a Management Information System

Both the permanent and transaction databases have predetermined contents, although these may change in a database management environment. Transaction processing is normally done in either a **batch** or **on-line** mode. The batch mode requires the collection of transaction records for simultaneous processing at specified times (daily, weekly, monthly, quarterly, or yearly). The on-line mode processes each transaction as it is entered into the system. In the batch mode, reports are generated with each batch; it makes little sense to produce reports between batches since the system's status has not changed. In effect, the system runs at the rate of the fastest batch and data is no more recent than the latest batch. The on-line mode makes the concept of batch irrelevant. System status reports can be made at any time and reflect the current status.

Transaction processing consists of matching transactions against master records and merging, accumulating, replacing, or deleting information from the master record. Figure 17-4 illustrates the alternatives in transaction processing. The goals here are (1) to maintain **master records** in the most current status and (2) to retain **historical information** on transactions.

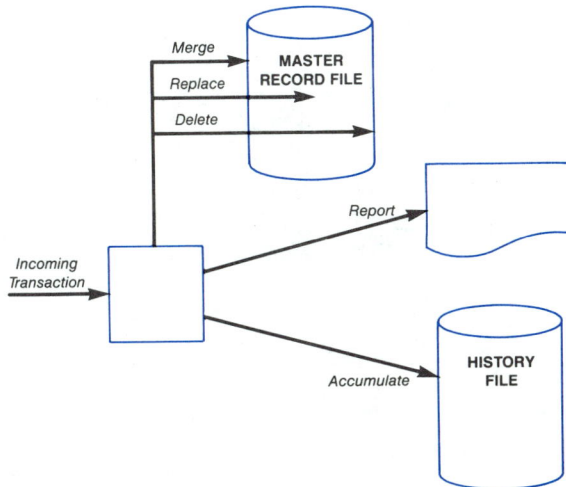

FIGURE 17–4. Alternatives in MIS Record Processing

In addition to transaction processing, master records and transaction histories are routinely analyzed for trends. For instance, Stu Baldwin's firm keeps track of client purchases and processes transactions on-line. Client master records are updated, and there is a history file containing all transactions. Each quarter, tax-related information is pulled off the master client file and prepared for clients. Each month, specific reports are prepared for Stu on institutional sales. To obtain this information, the entire month's transaction file is searched for institutional sales. Stu can also sort the accumulated transaction file by client type, age, gender, institutional affiliation, broker name, total portfolio value, total monthly transaction value, or number of transactions.

Report-Based Design

The thought given to Stu's needs has motivated a number of reports, since the design philosophy of the director of data operations has always been report-based. In **report-based design**, managers' report needs are the driving mechanism for design work. Information necessary for reports is traced in form and substance to inputs available within the BCS&D transaction processing system. This limits reports to available data, but as new data needs arise, the system has the flexibility to generate reports from *all* available data.

Report-based design has the advantage of working from the known (what managers say they want to see) to the unknown (the mass of data in the system). If managers uncover a need that cannot be satisfied, this provides ample evidence that the input side of the system needs redesigning. The disadvantage of report-based design is that redesign can be difficult and expensive. Existing data are incomplete from the moment new data are collected, which indicates a significant phase-in period.

The alternatives to report-based design are **input-driven design** (what reports can I generate from this data?) and **process-driven design** (what can I do to the data?). Neither is appropriate in an organization of sophisticated managers who know what they need to know.

Operational Considerations

The MIS works as indicated in Figure 17-5. Data are input in either the batch or on-line mode. While data are being input, they are **verified** and, if the mode is on-line, checked against master records. Verified transactions are then **merged** with master records and **reduced** by removing pertinent information from the transactions to the master file; transactions are then filed in a **history database**. **Reports** are then generated and distributed to the appropriate audience.

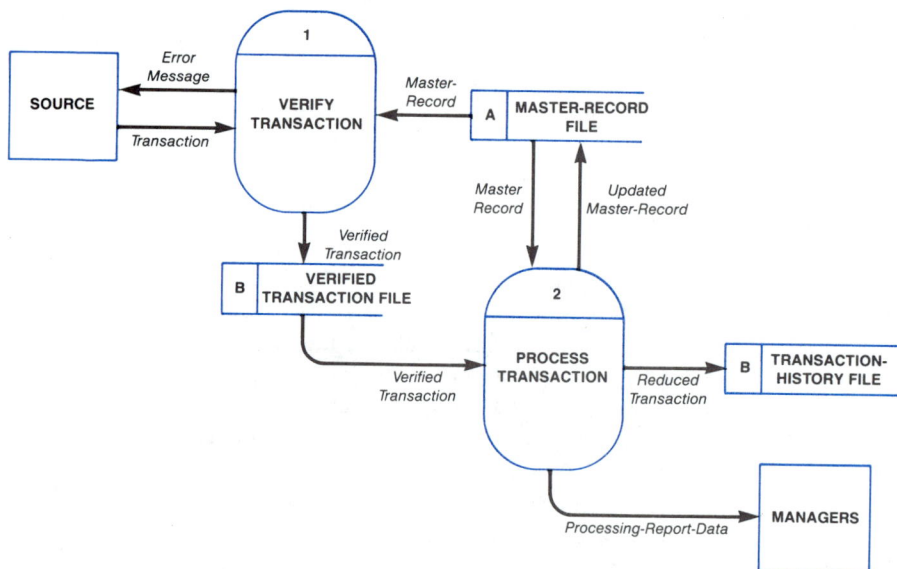

FIGURE 17–5. The MIS Model

Data entry is tedious and error-prone. Often, errors introduced at this stage cannot be removed effectively. To prevent errors, checks on ranges, reasonableness, and rules are often used to keep the data **clean**.

Master records can become "corrupted" through error, electronic accident, or subversion. Protecting against unavoidable or inadvertent loss or corruption of data is big business today for security experts. The MIS system requires significant degrees of redundancy because of the wide exposure of the master record file to hazard and the large volume of transactions.

Users of an MIS often complain about information overload and ignore thick reports, preferring to look at summaries. The temptation to overlook signals of disaster is high. Periodic and frequent data audits are necessary in complex systems.

A report can either appear on a screen or it can be printed. The moment a printed report is distributed, it is out of date. Having the "most recent data" is important, but wide distribution of reports may negate the system's capabilities in this regard by putting copies of dated material into the hands of many people who may rely on it too heavily.

An MIS tends to grow in complex ways via the addition of reports, data fields, and users. Not only are data audits important, but human resource audits may prove necessary to find out who is actually using the system in the ways it was intended to be used. This author has experience with a system that worked well for over a year until a customer switched operators. The new operator had decided he did not have time to reconcile the 1,000 bills which were being filed weekly. Not surprisingly, the system slowed down under the weight of unreconciled records.

Because new transactions arrive continuously and because organizations come to depend on the MIS for their normal cash-producing operations, the effects of these problems are exacerbated.

The Corporate Database

The centerpiece of the MIS is the **corporate database** (Figure 17-6), which contains all operational information about an organization in the form of master, historical, and transaction records. In a well-built MIS, one could run the history of the firm backward and produce adequate and accurate reports for any period.

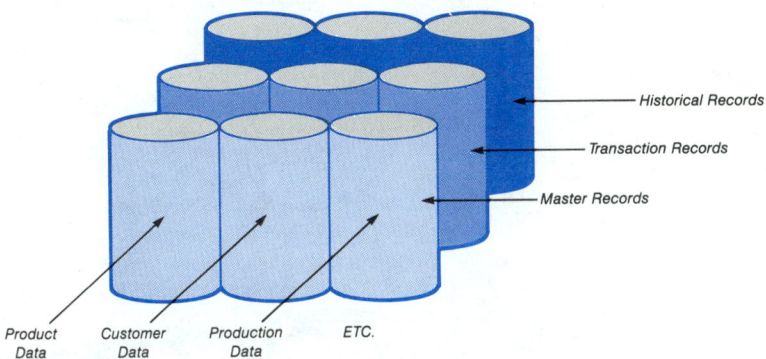

FIGURE 17–6. The Corporate Data Base

As the database grows in size and complexity, as the user group grows in number and sophistication, and as applications are added to the system, maintaining the MIS requires larger fractions of the total budget allocated to IRU work. Most IRMaint work becomes true maintenance of existing applications, a spur to end-user software development and application generation.

Finally, the corporate database fits in well with decision-support systems and office automation. Because the data are about the organization, decision makers can use the data (as opposed to the reports) to make decisions about the organization. Because the data concern everyone in the organization, the communication facilities of the automated office naturally assist workers in accessing the corporate database and incorporating it into their daily office work.

17.3   THE AUTOMATED OFFICE

Office automation, a recent phenomenon, brings computers and office work together. Although the term "office automation" has been around since about 1957 (when it referred to the use of tabulating equipment in support of office functions), modern use of the term began in the early 1970s. Because office automation combines computers and relatively untrained operators to accomplish varied tasks, designing an office automation system requires a wide range of skills. Development may not proceed in a linear fashion from start to finish, and the final product may not resemble original plans. Microcomputers often come to the office without much planning, which complicates their rational introduction to office functions as an integrated whole. In addition, the political situation brought about by computerizing the work of large numbers of professional workers, as opposed to clerical workers, makes the development work even more interesting.

What is an Automated Office?

Definitions of office automation vary. Because so many different products with different emphases are marketed, simple definitions can no longer describe the "phenomenon." Instead, an ostensive definition has grown describing office automation as a package of features or subsystems. Our definition will follow that model (see Figure 17-7):

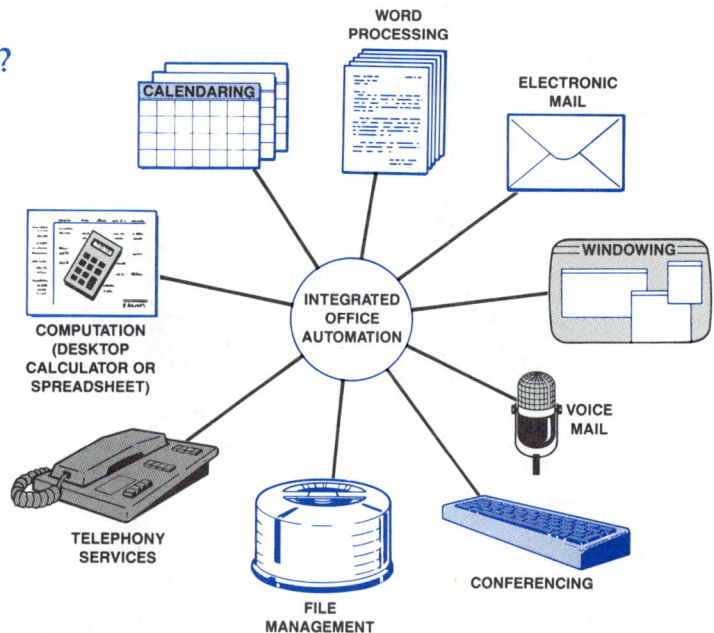

FIGURE 17–7. Components of the Integrated Automated Office

Office automation is an approach to introducing computers to business offices that attempts to integrate computation, communication, and information handling into single **windows** available from workers' desks. Office automation systems supply at least some of the following functions:

- electronic mail and messages: entry of typed messages intended for individuals or groups (Figure 17-8)
- voice mail: entry of spoken messages for later retrieval by the recipient(s)
- scheduling, reminders, calendars
- simple (calculator-type) or complex (spreadsheet-type) computational facilities
- integration of telephony with computation: textual information available on incoming and prescheduling of outgoing calls, generally through a CBX (computer-based branch exchange)
- file management: personal and group files, note files, saved electronic mail
- data and voice conferencing: audio teleconferencing (several persons in a single conversation), data conferencing (an extension of electronic mail), and combined data and voice conversations linking screens together
- windowing: rapid transfer between functions without losing one's place
- word processing

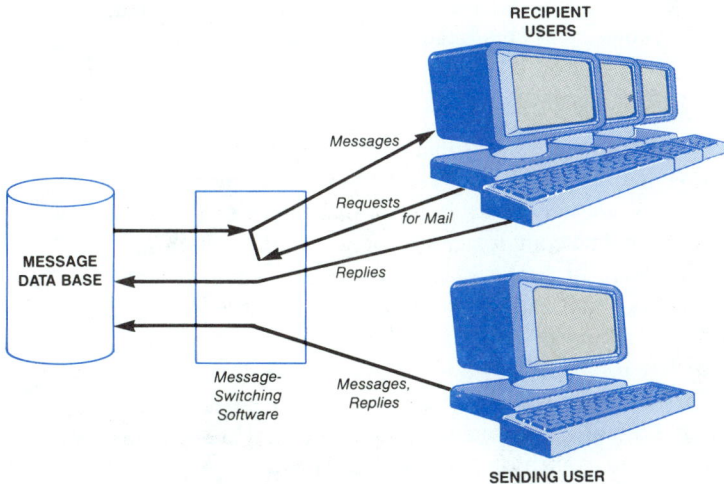

FIGURE 17–8. Electronic Mail

Word processing packages and their relatives (computerized typesetting and document production, graphics, mail-merging of documents with mailing lists, and spelling checkers) have been the most successful office automation products, especially on microcomputers. Each manufacturer of minicomputers and mainframes has its own word processing package, some of which (like Displaywrite III™) have trickled down to the micros. There is little doubt of the continuing viability of word processing. A major issue, however, is the *management* of word processing — is it an IRM function or should it be handled elsewhere in the organization?

Electronic mail and **computer conferencing** have had some measure of success. Bair (1978) pointed out that the best productivity gains were in the area of *executive* functions. Predicted gains of up to 15 percent in executive productivity (Booz, Allen & Hamilton, Inc., 1980) have not been realized, but electronic mail has been successfully integrated into executive communication systems.

Other communication facilities have had less-than-thunderous receptions. **Schedules, reminders,** and **calendars** are generally clumsy on a computer. **Voice mail** is interesting but expensive. **Shared filing** may not be a natural mode of operation for professionals. Integrating **telephony** and **computation** seems valuable, but it is still unclear *how* people would use these facilities other than to manage missed phone calls or as simple extensions to existing phone features (such as call forwarding, speed dialing, and the telephone directory). The fact that half of all attempted business calls are unsuccessful — ending in busy signals, no answer, or called party unavailable — is enticing to office system designers. It is the *team* approach (the professional knowledge worker plus professional clerical/secretarial worker) through integrated communication that could provide the productivity edge.

The unadulterated computing facilities (spreadsheeting, filing, and word processing) are probably going to be the most successful in the short run. Most business professionals have had at least some exposure to computers and computation. Accountants, engineers, analysts of all types, bookkeepers, and others used to working with numbers welcome the easy access to spreadsheeting, very smart calculators, and electronic paper.

Investigating Office Processes

Office processes differ from information processes in a number of ways. Thus, they must be investigated in a different manner. Most professionals are not sophisticated information manipulators; usually, they are not clerically adept or as used to following procedures as clerks and secretaries are. The communicative aspect of information sharing in an office is as important, if not more so, as the information aspect. Whereas clerks have a great deal of experience with computerization, professionals have little history to refer to; they may be either mistrustful or overly optimistic.

Office automation systems are often bought as whole units and then customized; a complete logical design may be fruitless in light of what actually can be purchased. There is little research available on how to investigate office processes; there are conflicting views as to the importance or relevance of traditional report-based, top-down logical analysis with regard to communication processes in an office. Finally, there is a high degree of concurrency, shared activity, and indeterminacy in office processes; these run counter to the assumptions of structured programming and may make the design exercise quite frustrating.

Investigating office processes is more art than science, and conducting a detailed investigation may require skills not normally tapped by analysts. Consider the following transcript of an interview between analyst Colin Griffin and marketing manager Wendy Bullard:

CG: Wendy, I'd like to know something of how you operate here. What kinds of communication do you do?

WB: Colin, that's a difficult question. Yesterday, I had four hours of meetings. Around those meetings, I spent over half the rest of the day on the phone calling my agents around the state. Half of these calls were wasted because they weren't in. Then some tried calling me back and I had four calls on hold on four lines at one point. Anyway, what do you mean "what kinds?"

CG: Well, what I mean is, what kinds of data do these messages or conversations involve? Do you need printed matter, for instance?

WB: Sure, but it's hard to say how much. For instance, one call went on for fifteen minutes because Edna Gross and I couldn't agree on the wording of a sales contract for a guy — he'd changed the standard contract and Edna wanted it approved, but I didn't have other examples available; anyway, Edna had to read whole paragraphs to me that just didn't make sense.

CG: So you'd like to see what she's seeing when she's reading material?

SB: Sure, that's right. But she needs to see my stuff here, too, because we've got some standard forms for unusual contracts, you know, promise of delivery, large lots, returns, and so forth. Really the problem was that she needed to speak with our contracts guy, but he wasn't here and I didn't have the key to his cabinet.

CG: So there's a need to get a third person involved?

WB: And a fourth — his secretary — and a fifth — mine! Really, do you guys think you can help?

Colin is having problems with this interview. Wendy has such a variety of complex situations, handles such varied material, and has so little understanding of the dynamics of the interaction that Colin, who normally lets clients lead through the information analysis, is at a loss to begin to uncover Wendy's "kind of communication."

Certain things must be covered in these interviews:

1. the need for **serially shared documents** passed from person to person for approval (Figure 17-9 on the following page)
2. the need for **store-and-forward messaging** (messages which are later saved for reading by the recipient)
3. the need for **personal filing** at one's fingertips: reminders, phone notes, short memos, charts, etc.
4. the role that computation currently plays in the interviewee's job
5. the current use of **formal filing systems**, calculators, and telephone features
6. the existence of a **compunication** unit, typically composed of a professional, a secretary, and possibly a clerk
7. stepping through several typical "compunication" events such as recording a phone call, taking notes during a phone call, and mail handling

The term **compunication** is nonstandard; it refers to an event that is communicative in nature but involves significant computation or data handling. A term that describes this field of study is **telematics**. Regardless of the terms, finding out about integrated communication/computation events rather than routine data handling is the focus of the office automation investigation interview.

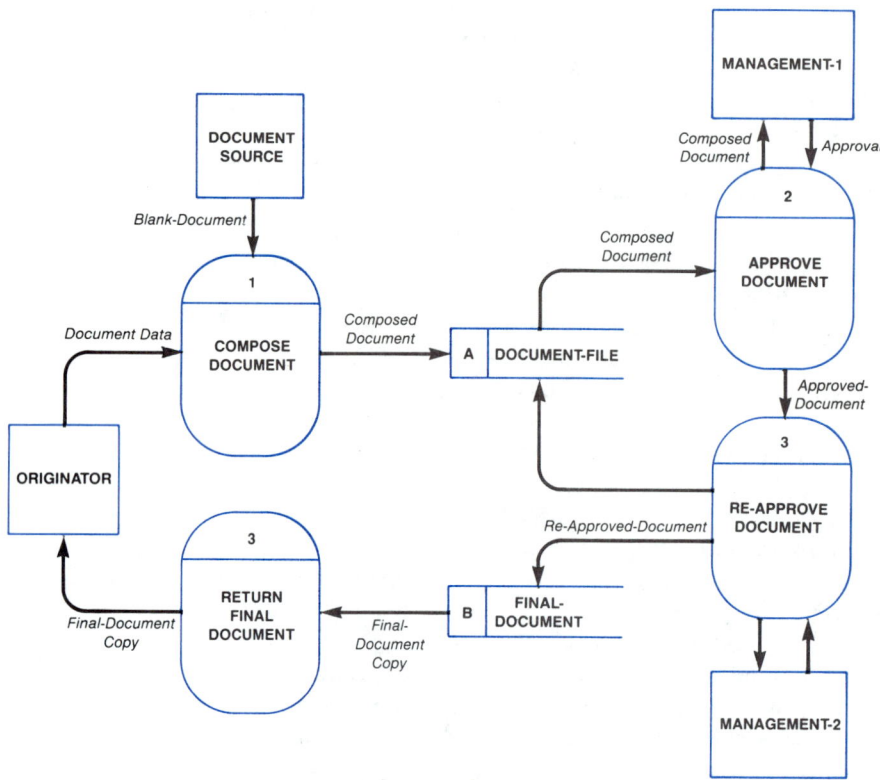

FIGURE 17–9. Serially-Shared Documentation

The Role of Bundled Hardware-Software Systems

There are now myriad vendors of integrated office systems. The two major approaches are (1) the specialized processor and (2) the local area network. Each is built around a computer and some communication hardware. The first adds terminals to a communication processor that is specially enhanced to handle files; the second adds a communication network to a computer.

The specialized processor is often a C(P)BX (**computer-based (private) branch exchange**), the successor to the older electrical or mechanical PBX (**private branch exchange**). Most office phone systems have hold features and communication equipment to handle multiple incoming office lines. Over the years, PBXs have become quite sophisticated, adding many features such as **call forwarding** and **speed dialing**. Since the latest PBXs are really computer-based and since the telephone terminal, with its ten or twelve pushbuttons, has a keyboard-like capability, it was natural to attempt to turn the PBX into a computer (see Figure 17-10, opposite).

Many communication firms have been upgraded their PBXs to provide sophisticated communication capability enhanced by some computation, file sharing, electronic mail, electronic directories, and even spreadsheets. Manufacturers are rapidly converting their mechanical switchers into sophisticated computers.

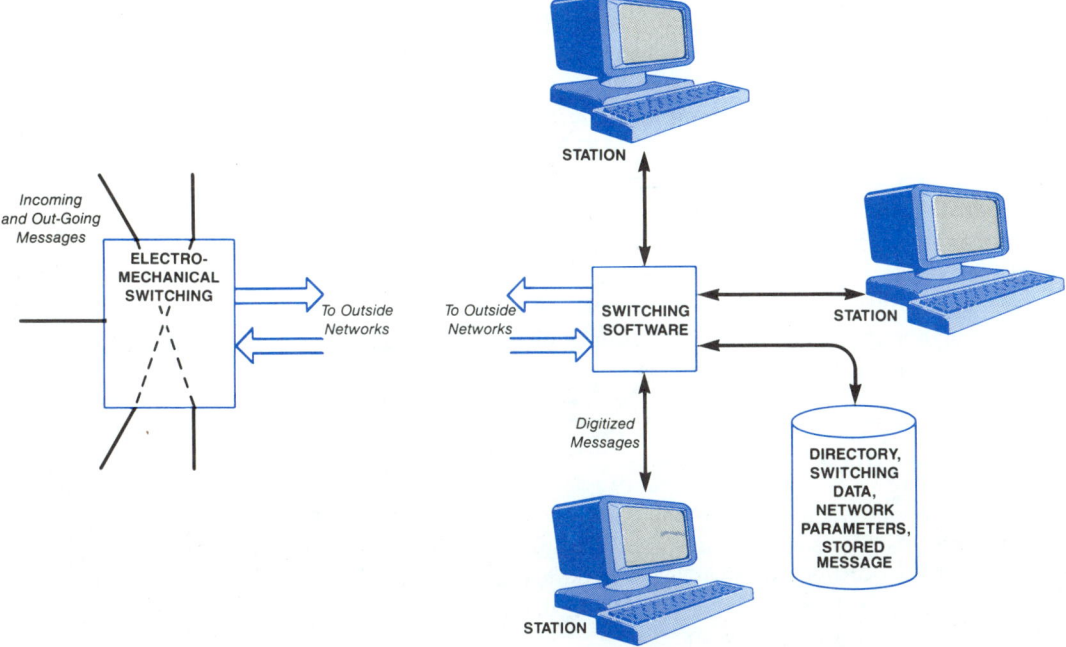

FIGURE 17–10. PBX vs. CBX

The other approach is to take an existing computer installation and add networking hardware and software. The most available form is the **local area network** (LAN), which ties a number of computer terminals, perhaps augmented by voice equipment, into a central computer or series of computers called "servers."

In recent years, most computer vendors have been developing LANs around microcomputer-based intelligent terminals, although the concept does not require micro-computer intelligence. Many successful networks are built around mainframes and dumb terminals that simply act as twenty-five-line screens. Because these systems are computer-based, communication facilities tend to be limited to electronic mail and computer conferencing, with telephony an add-on. Manufacturers like IBM and DEC and equipment manufacturers like Corvus, Nestar, and Davong have either marketed or supported LANs in recent years. The recent AT&T divestiture has enabled AT&T to enter the computer field aggressively. Industry watchers have a few interesting years ahead as they wait to see which approach, if either, wins out.

The effect of the variety of approaches is that hardware and software are effectively **bundled** for the majority of the system's features. Specific elements for the customer's installation may be custom-ordered, customer-developed, or custom-installed by the vendor, making the IRU a support unit for a system developed elsewhere. This implies an overhead for staff training and a need for policies governing enhancements to the installed base of equipment. As stated previously, this means that investigation may lead only to selection of a particular system with its own proprietary design, reducing the need for a full design exercise.

The bundling of hardware and software occurs in another instance in the automated office. Word processing may be purchased either as a **dedicated processor** or as **software** intended to run on a general-purpose computer. The specialized processor of the stand-alone word-processing system is optimized for and dedicated to functions such as text manipulation, document file management, and keystroke capture. Some processors allow for custom tailoring through the addition of functions in special programming languages. The drawback of this approach is incompatibility among processors.

The other approach, to build a software-based word processor on a general-purpose microcomputer or minicomputer, has the advantages of integration of word processing into other packages (such as graphics, spreadsheet, or database management); device independence (to some extent), allowing porting of the documents between computers; and some occasional opportunities for easier custom tailoring of packages. Software costs may be high when many copies of the same software have to be purchased or licensed.

Prototyping in the Office

Bundling makes prototyping an obvious approach. Most systems arrive with a large number of features disguised as complex keystroke configurations that have to be taught to users. Some homegrown systems are installed with only a few programmed functions; later they are developed and released in more complex fashion as features are prototyped **after** users become accustomed to the system.

This writer has developed a prototyping language called LAMP (Language for Active Message Processing), in which either users or analysts can prototype combinations of electronic mail, reminder, messaging, and filing functions into packages for later use. Figure 17-11 illustrates some LAMP programs that can be filed and combined.

```
DICTATION: ### A voice message is placed
    in here between the hash marks ###;
    RELEASE TO MY-SECRETARY;
    IF IT IS TOMORROW THEN HOLD UNTIL
       NOON AND ALERT ME;
    QUERY ### If you can get this done
       before noon, type in Y E S. If
       you can't type N O and I'll find
       some other way to get it done. ###
    IF REPLY IS "NO" THEN (CONNECT US
       OR LINK SEND-TO-POOL) AND RELEASE
       PARAGRAPH-OF-THANKS-ANYWAY TO
       MY-SECRETARY;
    PARAGRAPH-OF-THANKS-ANYWAY;
       ### THANKS ANYWAY ###; END
```

```
VOICE-TELECONFERENCE: "WELCOME TO THE
    TELECONFERENCE.
    EVERYTHING IS UNDER SOFTWARE CONTROL
    AND YOU NEED DO NOTHING BUT SPEAK
    OR TYPE
    AS YOU WISH." INSTRUCTIONS: "...."
    AGENDA: "...." RELEASE TO CONF-ONE-LIST;
    CALL PRESENTATION; LISTEN TO
    CONF-ONE-LIST;
    IF SENDER IS TOM SPEAK TO CONF-ONE-LIST;
    IF SENDER IS HARRY THEN CENSOR ALL
       AND SPEAK TO HARRY;
       MONITOR ALL; IF IT IS AFTER NEXT MONTH
       THEN DISCONNECT ALL AND ALERT ME: END;
```

```
TEXT-MESSAGE: "HI PLEASE PHONE BEFORE
    MIDNIGHT.";
    IF IT IS BEFORE MIDNIGHT TODAY THEN
       RELEASE TO 555-1212 AND (CONNECT
       US OR ALERT ME): ELSE DESTROY AND
       ALERT ME AND MY-PERMANENT-FILE; END;
```

```
MEMORANDUM: "HOORAY, SALES ARE UP"
    RELEASE TO TOM OR DICK OR HARRY;
    PASS TO BOB AND JOE IN ORDER;
    QUERY "TYPE IN OR SPEAK YOUR
       COMMENTS"; MONITOR REPLY;
    IF RECEIVER IS JOE THEN DESTROY
       AND CONNECT (ME OR MY-BOSS);
    ELSE IF RECEIVER IS TOM OR DICK
       THEN ALERT ME; END;
```

FIGURE 17–11. LAMP — A specialized Office Automation Generation Language

Based loosely on PL/1, LAMP can be further layered by combining with a higher-level query-type language.

Ergonomics, Interior Design and Task Analysis are by Components of Office Automation Investigation

Ergonomics, Office Landscaping, and Decorating

If the situation were not already complicated enough, placing computers in offices, rather than in clerical work areas or data-entry sections, leads to a concern for interior decorating that has a real physiological basis. Designing equipment and procedures for normal mortals in an office situation is a difficult task and one that becomes as important as the software and hardware controlling the telematics. These efforts fall into a field called **ergonomics**.

Consider office lighting, for instance. Normal office lighting is set up for paperwork, which is usually dark letters on light paper. To strongly contrast the paper and the letters, lighting levels are high. Computer displays (CRT, VDT, VDU), however, are usually bright letters displayed on dark backgrounds. Even the reversed image (dark letters on a bright background) is light supplying, not light reflecting. Average office lighting levels are too high for comfortable work at terminals.

Other concerns in ergonomics are (Figure 17-12 on the following page):

1. **lighting:** glare from the screen, distraction, distance from the eyes, neck strain from having to squint for long periods of time, and improper eyeglasses
2. **sound:** some terminals now talk, many beep, and quite a few sing
3. **seating:** poor posture leads to neck and eye strain, back problems, and circulation problems in legs and arms
4. **work area:** papers and documents must be placed nearby; beverages can spill into the keyboard if space is crowded

5. **traffic patterns, cable runs:** exposed cable is a safety hazard, and traffic moving nearby may be distracting

6. **landscaping:** because *whole* offices are redone with computers, there is often an opportunity to redesign office layouts. The open-office concept (offices without walls) makes running cables easier, but runs smack against a social system that rewards longevity or success with privacy. Also, open offices are harder to light and make it more difficult to baffle sounds of beeping computers, too.

7. **Work stations:** integrating phone, computer, notepad, and books into a single station may sound good, but putting it all on a manager's desk may merely clutter it. Good work station furniture and seating is expensive.

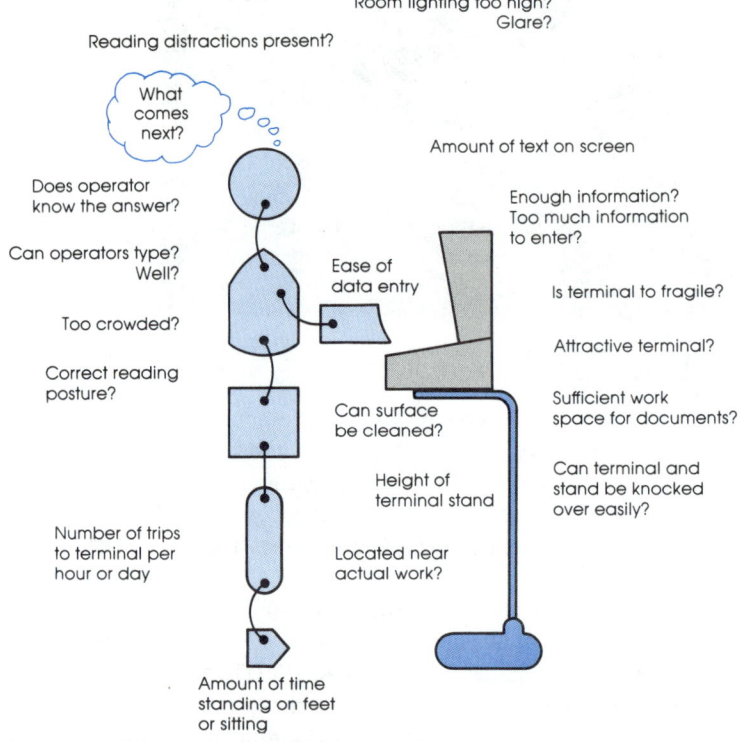

FIGURE 17–12. Human Factors Design Considerations

Case Study: E-Mail at the University

Thadeus McGovern III is Public Affairs VP at Bigtown U. For the past four years, he has handled relations with alumni groups, business organizations, benefactors, governments, angry parents, angrier students, and neighborhood action committees protesting just about everything. He has escorted visiting engineers from Thailand, royal dignitaries from various emirates, legislators checking the buildings they funded, and millionaires checking the buildings named after themselves. For the past four months, he has been using BUGD, Bigtown University General Delivery, an electronic mail system built around the administrative computer system. Thad is pleased with BUGD and how he came to use it.

BUGD is the brainchild of Marian Brenner, the president of the University. Tired of getting busy signals, speaking with secretaries, and having to remind her VPs and deans to respond to memos, Marian approached Brian Depuis, the administrative systems director, to see if the existing administrative computer could help out. She had read about electronic mail in both general press and technical accounting publications and was eager to try it out.

Brian felt that the existing system could handle electronic mail and set a team of analysts to work looking for existing packages for their mainframe. Two seemed appropriate. Brian sent the team out to observe and use these systems at customers' sites; they brought along Marian's executive secretary, Viola Cates, to size it up for Marian. Neither package appealed much to the team or to Viola. "There's too much keyboarding," said Viola, "and neither system can get Marian the reminder facility she wants. Besides, it's too easy in both systems to ignore messages and not respond to them."

The team next went to several vice presidents and deans and spent a month interviewing them specifically on their communication with the president and with each other. They discovered that strong units formed of a vice president or dean plus a secretary plus an assistant (or two) accounted for over 80 percent of all work-related communication, mostly approvals or requests to distribute incoming material to appropriate persons. The other 20 percent consisted of informational memos, directives from the President, and requests for meetings and meeting minutes from committee conclaves.

Marian selected Thad and several other VPs to prototype a system to be constructed by Brian's group over a six-month period. First the group constructed an E-mail facility that was nothing more than a message-switcher through the operating system. This was rejected immediately by Thad and the others because of the difficulty they had in accessing messages. Then the group overlaid this with a program to smooth over the user interface, create personalized access sequences, notify the sender of the receipt of messages, and enable everyone to file messages into bring-forward files. Thad especially appreciated this last feature because his busy schedule made it difficult for him to remember when to respond to what.

At the same time, Brian's group was looking at the human factor. University executives can type, but not well, and their offices are not set up to make working at a CRT comfortable. The group designed a work station that could be wheeled over to a desk and that could easily be slaved to the secretaries' terminals. The presumption was that the secretaries would do the bulk of keyboarding, message reception, and screening — the executives would create texts the secretaries would flesh out. Because each VP and most deans had assistants who needed to look at the messages, too, Brian decided to begin with an inexpensive printer attachment and, ultimately, to put terminals on the assistants' desks, also slaved to the secretarial screen.

Four months into the prototyping phase, Tad is satisfied with the results. Not only is Marian rarely put out because Thad forgets to handle a memo, but he finds her more accessible now. Meanwhile, he has lots of ideas about word processing and somehow putting the phone (which provides him with most of his contact with the outside world) together with his keyboard-based messaging system. He remembers reading something about integrated voice-and-data systems...now just where was that article? Maybe a little personal filing system might not be bad....

17.4 MICROCOMPUTER SOFTWARE DEVELOPMENT

The microcomputer revolution, which began in the late 1970s with the introduction of the 8-bit personal computer and is continuing relatively unabated with 16- and 32-bit business and super-microcomputers, has had a significant impact on information resource maintenance. Most important has been the **democratization** of computation. In effect, the personal computer has broken the monopoly that the IRU has held on computer systems development, making computers and programming available to a wider segment of the organization. This, in turn, has caused the IRU to take a good look at itself as proprietor of computation and has enabled the information center to come forward as a means of employing microcomputers more effectively within a larger IRU strategy.

Another effect of the microcomputer revolution has been the growth of a lively and viable software market. This market offers powerful software for management decision making and clerical operations for only hundreds (often tens) of dollars, well within the discretionary limits of most managers, yet again, outside the control of the IRU. This often results in users having access to more powerful software at far lower costs than IRU programmers.

An increase in the sophistication and expectations of users caught up in "micromania" is also a dark side of the micro revolution. Because micro software can be purchased for a fraction of development costs, the IRU somehow loses face when it cannot offer, say, spreadsheeting. Users may be surprised to discover that their carefully crafted spreadsheets cannot be uploaded to the mainframe and saved, and they may be horrified to discover that no one at the IRU wants to maintain their software for them (Alavi 1985; Davis 1982).

How Microcomputers are Different

In addition to organizational impacts, consider the IRU that undertakes a significant microcomputer software product effort. There are a number of important differences in how microcomputers work and are employed that make developing systems on microcomputers different.

Unnetworked microcomputers are independent. Coordinating efforts in which the "latest version" may be on some diskette somewhere is a difficult goal to meet. System development tools are still quite limited, although this is changing. Ranging from relatively unsophisticated text editors to complex file management subsystems, microcomputer system development environments still cannot compete with the well-developed library systems found on mainframe computers.

The information center makes it easy for users to build bad software that the IRU may have to account for, either by repairing, integrating, or replacing in production. Davis (1982) pointed out four **quality risks** of end-user-developed systems: lack of extensive testing, inadequate design for validation and checking, re-inventing the wheel, and threats to security and integrity of the corporate database. Not only may the IRU have to rescue poor software, but the poor experience users may have with their own software may seriously inhibit future development; good experiences may simply raise expectations.

Microcomputer hardware is far more reliable now than it used to be, although it is not as reliable or as powerful as mainframe hardware; disaster planning and recovery are trickier. Memory is limited; multitasking, where available, is trickier; on-line access to data may be restricted to on-hand floppies; and archiving may be difficult.

User interest in microcomputers has fragmented computer budgets. With business microcomputer purchases of hundreds of millions of dollars annually and growing, the clout that user departments have must be reckoned with. In fact, a *PC Week* survey showed that only 8 percent of corporate microcomputer purchases in 1985 started with the IRU and about a quarter do not involve the IRU at all. The result of this is that users may present the IRU with a *fait accompli* in hardware and operating systems terms, while demanding maximum flexibility from the IRU during a system investigation — a flat contradiction.

These factors change programmer and analyst responsibilities (these days, there is a greater chance that the programmer and the analyst will be the same person).

The Responsibilities of the Programmer and the Analyst

Given the lack of development tools available, systems tend to be smaller and development time, shorter. Because hardware and software change rapidly, changes to the ultimate operating environment may be dictated from outside during any stage of development. This makes programmer-analyst communication crucial. Because system documentation is also more limited, analysts cannot expect programmers to make important changes routinely — the documentation may just not be available.

Microcomputer systems tend to be more limited in terms of clientele than others, too. For this reason, prototyping and end-user software development tend to be favored development modes. This may require programmers who can speak to users, users who can speak to programmers, or analysts who are adept at programming.

Formal investigations may be either brief or unnecessary in the face of known, limited capabilities in a pre-existing database manager. On the other hand, since communication with users is more important, the selection of modeling techniques that clarify user operational concerns is vital. A DFD is far easier to read and understand than a system flowchart. Logical design may be even more important where user approval is required quickly.

Finally, analysts have to know microcomputer hardware better and be more aware of announcements, rumors, and advertisements than analysts who work solely with mainframes. Users read the same ads and articles as analysts and are full of questions and opinions. Knowing that release 4.1 was reviewed in *Byte* and is "a disaster," users may question an analyst's decision to stay with release 4.1. It may prove difficult for analysts to stay on the topic of logical design when users are asking for multitasking, color graphics, and 4800-baud modems.

Case: Developing a Case Management System

In the fall of 1983, a demonstration project with the goal of introducing microcomputers into the practices of the family therapy team began at Bigtown Hospital for Children (BHC).

For the first two years of the project, the team director and two professors from Bigtown University worked to specify the systems to be employed. In the earliest stages, basic hardware acquisition decisions were made to satisfy the government funders providing the seed money and to provide a stable environment for development. Sally Kelly was hired as a programmer/analyst for her ability to get along with users; she had a technical school programming degree and a stint training users for a local software vendor. A team of graduate students was to research potential microcomputer applications to social work in general, and family therapy in particular.

Over the first year of the project, a number of decisions were made: the system would be built around a commercially available database manager; additional software would be written in BASIC; a single line of computers would be purchased; the system would consist of a loosely connected set of modules available from a single menu; graphics and color would be employed where possible to make the system attractive; and usage statistics would be automatically gathered and analyzed to facilitate reporting.

Within weeks of making each hardware decision, newer, faster, and cheaper hardware was announced. Within weeks of purchasing each software package, others with more capabilities became available. Upgrading became almost a full-time occupation of the technical staff, and learning new packages proved to be a burden.

Whereas mainframe and minicomputer software is voluminously documented, most packages come with brief "user" documents which are aimed at the casual user and do not contain technical detail. The BHC project aimed for building a product easily transferrable to other family therapists and social workers. It was important to produce a flexible, efficient system, but the documentation was always incomplete. Unable to get clear documentation about packages quickly enough, the project ultimately purchased and learned no fewer than four database managers, two word processors, and two major spreadsheet programs.

The limited capabilities of microcomputers quickly became a factor. Grandiose ideas of instant retrieval became, in practice, grindingly slow disk searches. Daily hardware failures pushed deadlines back, and undocumented software bugs bit the staff without warning. On one particularly disastrous day, both hard-disk-based micros became inoperative, one because of a major disk failure, the other because a student accidentally installed an incompatible operating system. These problems normally would not have been so bad, but the project had scheduled a major paid-admission seminar the next day! Once, two portable microcomputers both failed during a conference seminar, leaving the software undemonstratable. These were "compatible" microcomputers that turned out to be not quite so compatible. Another time, a student accidentally wiped the hard disk clean.

Another difference was in the system development methodology. Normally, a system such as this would have been carefully specified during a preliminary and detailed analysis. Most applications within the system were prototyped with selected therapy team members. Curiously, a spreadsheet-based system to provide evaluation of the team members' practice techniques in Lotus 1-2-3 was *not* prototyped, but instead built from specifications without user involvement. This decision was made because the subsystem was essentially a computerization of a well-known techniques book available to and endorsed by the staff.

Because the hospital had had no previous computer experience, an advisory committee of interested local social workers, hospital administrators, academics, and vendors was put together to counsel the project. This committee proved invaluable in anticipating problems, pointing out need for improvement, and suggesting ways around administrative problems.

Beginning in the nineteenth month of the project, live data were collected and reports were generated. It became clear that the project would succeed, if only because facilities such as a consulting system to peruse past cases and provide advice had been infeasible before. A report writer produced easy-to-read reports from raw data which were fleshed out in pencil by staff members and then given to a secretary to enter using a commercial word processor. Rapid query facility helped staffers understand their own and others' cases. There was even a game to assist therapists in learning about their families.

17.5 RECORDS AND FORMS MANAGEMENT

Records and **forms** are general **documents**, and documents are collections of information concerning a case, which is a continuing record of events. For example, a bill is a document of a month's purchases; a registration form is a document of identifying events concerning an automobile; a paycheck stub is a document recording a single payment event.

Documents create a unique problem in information resource management because they refer to a multitude of data items whose relationship is mostly historical (i.e., accidental). It may be difficult to specify in advance precisely what data items are necessary, what form(s) they should be in, and how they should be structured.

A document may be a free-format collection of information called a **record** (from the Latin "to remember") or a rigid-format fill-in-the-blanks **form**. The term "record" also means a fixed-format electronic data set. Because both records and forms document events, collections of them become important; therefore, methods of collecting and storing records and forms is the major aspect of records management. The term **forms management** relates more to the design of forms and the control of their usage. This gives the odd situation of assigning managing record forms to forms management and forms recording to records management.

Figure 17-13, on the following page, illustrates the **document management cycle**. Documents first must be designed, then composed (filled in), filed, retrieved, edited, distributed, refiled, and so forth. The design of forms follows some well-known principles. Factors known to affect the error rate in forms completion are color, legends, order of reading, and size and shape of spaces to be completed.

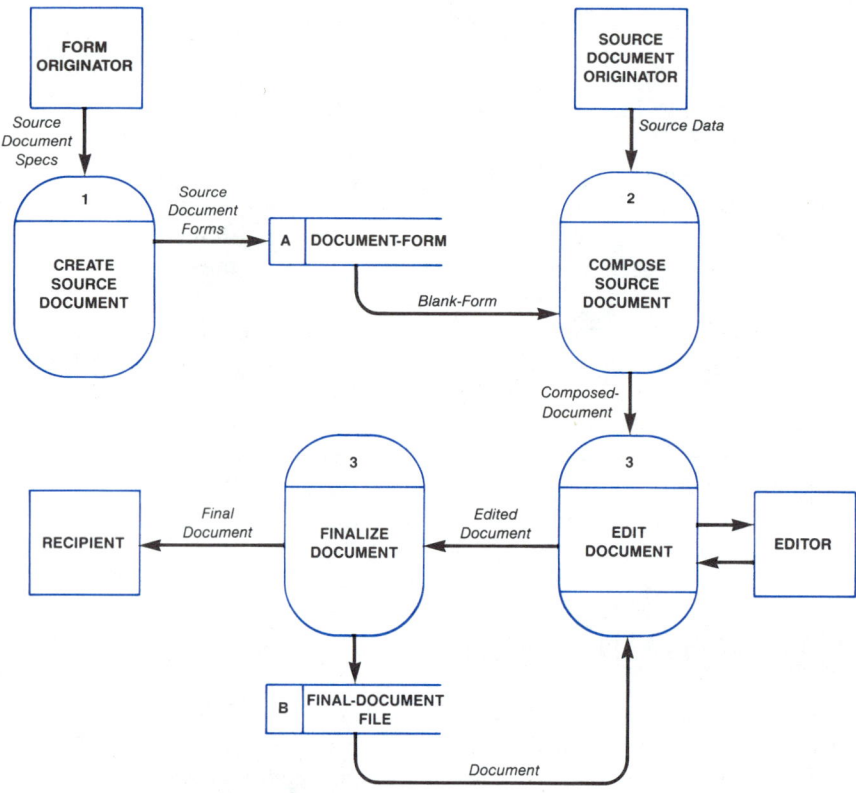

FIGURE 17–13. The Document Management Cycle

Next, documents are composed. Guidelines for completion of documents may be simple or complex, but the analyst's major contribution here is providing clear instructions. For example, there may be 39 codes for occupation in a particular form; finding this list and accurately translating a job into a two-digit number is a non-trivial task. Following a step-by-step guide to a form (such as an income tax form) is not only *not* easy but sometimes unpleasant. Writing clear, concise instructions is crucial to obtaining valid data.

After editing and finalizing, documents can be filed by any of these techniques:

1. alphabetically by content, author, etc.
2. numerically by acquisition or serial number
3. chronologically by date composed
4. categorically by a system of types and subtypes
5. cross-indexed through an index list
6. in single copies or multiple copies
7. in controlled (closed) access or uncontrolled access
8. in originals, copies, or both

Several of these schemes can be applied simultaneously. Where documents are of many types, it is not uncommon to find memoranda filed by date; reports by a library index or acquisition number; correspondence by sender or recipient; and forms by customer identifier, with copies of these filed by some other scheme. Where statute requires original data filed by specific category, large files may be multiply indexed, which makes maintaining them a superhuman feat.

Machinery for a Records Management System

Paper records, unlike the electronic variety, acquire bulk and place demands on physical space, and space has to be managed: where will the documents be placed? How much walking room between shelves is necessary to be able to find documents? Where will documents taken from shelves be placed? Navigation among documents becomes serious, time-consuming business, often physically taxing or, at best, clumsy: where, physically, is a certain document? Is it worthwhile climbing up there to get it? Where can I put this armful of paper?

The issue of physical space management has led to the development of a number of clever **filing systems**. These often consist of movable shelves, drawers, or cabinets, often massive outfits on rollers. The trade-off between efficiency of storage and efficiency of search generates either movable cabinets with a single aisle or a grand series of fixed cabinets and aisles, using a lot of floor space.

Navigating among the documents presents serious dilemmas to would-be information sailors. A good set of indexes reduces walking-around time dramatically. Placing commonly used documents nearby makes it easy to find the 20 percent with the 80 percent of the use, but then the remaining 80 percent may still be hard to find. Users may resort to tabbing or tagging documents with numbers, colors, or other visual devices to make specific items more visible at search time. These are mostly elaborations on the file-card system found in any 3x5 file box with index tabs.

Another approach to solving physical problems has been to photo-reduce the original documents (which are rarely required and may be subject to wear or decomposition) to microforms. The most common versions of microforms are **microfilm** (a role of 35mm film which can store the equivalent of a page of newspaper on each frame), **microfiche** (roughly the size of a computer-punched card or smaller, storing 20 to 60 A2 pages), and **optical disk** (which normally stores about 30 minutes of video images that translate to about 54,000 poorly resolved A2 pages or 54,000 half-A2 pages). In some instances, where records are to be archived only, **videotape** may suffice.

Writing Procedures

Because paper records have bulk, weight, and a physical location at all times, because paper can (and sometimes should not) be copied and the copies tend to proliferate, and because a physical records management system hands over the document to a user and hopes to get it back, procedure writing for records management systems is critical. Procedures can reduce the physical labor involved in locating documents, the chance for loss or damage, the proliferation of copies, and the time required to handle documents.

Specific attention needs to be paid to the following procedures. How will documents be located? What indexes need to be created? How will circulation be maintained? What are the backup and recovery procedures for paper records that might be damaged, lost, or destroyed? How will documents be entered into or permanently removed from the system? What kind of access controls exist for the circulation of documents? How will an inventory of the system be conducted? What are the security procedures for protecting specific records from specific people or for allowing only restricted access to specific records? Finally, how is missing information to be handled?

These procedural issues show up in computerized systems, too, but they are not as serious because the computer handles them more reliably. People, however, are not so easily programmed and procedures can be forgotten, misperformed, or circumvented. Checks on procedures become even more important. Consider the simple access controls on safety deposit boxes in banks. These can afford to be simple because customers need to have their keys — without keys, no one can get into any box. In other words, procedural controls are simple because physical controls are highly restricting. On the other hand, because anyone can request general delivery mail, identification procedures are necessary to prevent postal pilferage.

Case: Finding a Door at Part With It

Part With It is a unique car junkyard. Unlike the traditional junkyard, which has a jumble of crushed, mangled, and deserted hulks, Part With It has a relatively neat arrangement of automobile carcasses displayed on its ten-acre lot just south of Bigtown. There the amateur auto-repair expert can come and try to locate a door for a 1976 Datsun at a fraction of the cost of purchasing a new door and a fraction of the time he would have spent waiting for his neighbor to junk *his* 1976 Datsun.

When a customer arrives at Part With It, he approaches the sales desk and describes the part he wants, say, a right front door for a 1976 Datsun. The clerk consults a microfiche listing of available parts, broken down by type of car. In this case he has found the Datsun card:

> *Clerk*: Yeah, we've got the car. I think it's still got all four doors on it, too. It says here that its body's intact. You don't want a radio, do you? It's got one of those too...
>
> *Customer*: No, just the right front door. OK, what do I do?
>
> *Clerk*: Well, we could go get the door for you, for $20. Or you could take that wrench set over by the door there and go take the door off yourself. You leave us a $25 deposit for the tools, though.
>
> *Customer*: OK, I'll take the tools (removes $25 from his wallet and gives it to the clerk). When I get the door, what do I do?
>
> *Clerk*: Well, you bring it down here and I'll write up the bill for you. We gotta keep track of the cars and I'll write up a tag for the inventory manager. Next week when somebody comes in for a right-front Datsun door, he won't have to walk out to the car and see that there's no door.

Customer: Oh, yeah, I forgot. Just where *is* the car?

Clerk: Well, you turn right outside the door, go down about forty feet (consults the microfiche again), no, 80 feet, turn right again, and look for the pole with C4 on it. Now the Datsun's about two cars north of that pole. You can't miss it, it's got a Pinto on top of it.

17.6 INFORMATION RETRIEVAL

The topic of records and forms management leads directly to the concept of **information management** in general and **information retrieval** systems in particular. Most systems have aspects of information management, particularly where the volume of records (electronic or otherwise) and activity is high.

A familiar system is the 411 service we all use to find telephone numbers that are not in the directory. We ask questions like "What's the number for John Smith at 111 Main Street?" The directory assistance operator uses a computerized system to find John Smiths at 111 Main Street or some close match. "We have a John Smith on Main Street. Is this your party?" might be the response.

This chapter concerns large-scale information retrieval systems and focuses on libraries rather than item- or record-oriented database management systems. Library systems differ because the items retrieved actually "circulate" and are expected back, users often have very poor navigation skills, physical location is extremely important, and circulation and access control are the most critical elements.

Information Management Concepts

Information can be considered to be the content of documents. A document is identifed by a unique name or **code** and a series of non-unique descriptions or **indexes**. A single code therefore corresponds to a number of indexes; a single index may correspond to a number of codes. Codes themselves may be chosen to signify something (for instance, the P codes in the Library of Congress classification system refer to mathematics), or they may be only serial identifiers. The indexes, on the other hand, must signify known values, because they are used to track down documents.

Information management is concerned with managing information about documents as opposed to the information *in* them (which is of some concern in records and forms management). Thus, maintaining information on circulation, acquisition, indexes, series information, and location are the major concerns in information management.

How a Library Differs

A library is a system for complete information management. It differs from a filing cabinet because of the importance placed on **indexing** and **circulation control**. It differs from a database query system in its need to know where an item is *outside* the system and lack of concern about the content of documents. It differs from an electronic mail system in that documents are *permanent* rather than transient messages, although they are sometimes on loan. It differs from a management information system in that information in documents is not processed. Finally, libraries differ because they have traditions and histories far predating computers, large bodies of research literature, and sets of publications quite separate from the computer science and MIS bodies of knowledge.

Libraries also differ in the sheer volume of information they manage. A typical 300-page book contains about 3,000,000 characters. A small library, containing 20,000 volumes (exclusive of periodicals) would then require about 60 billion characters of storage. Each of the 20,000 volumes may have another 1,000 characters of information about it to be manipulated (another 20 million characters), and a daily circulation of 200 generates (assuming 100 characters for each transaction) 5,000,000 characters a year of circulation information alone. This computation does not include acquisition and disposal paperwork amounting to an additional several million characters each year. The management of this information, most of which is usually on paper, requires significant effort even for a small library. A large university library with a few million volumes and daily circulation of thousands of volumes presents interesting problems to the system designer.

Aspects of a Library System

Space in this text does not permit more than a superficial exposure to library systems. Figure 17-14 illustrates how a library system manages information about documents.

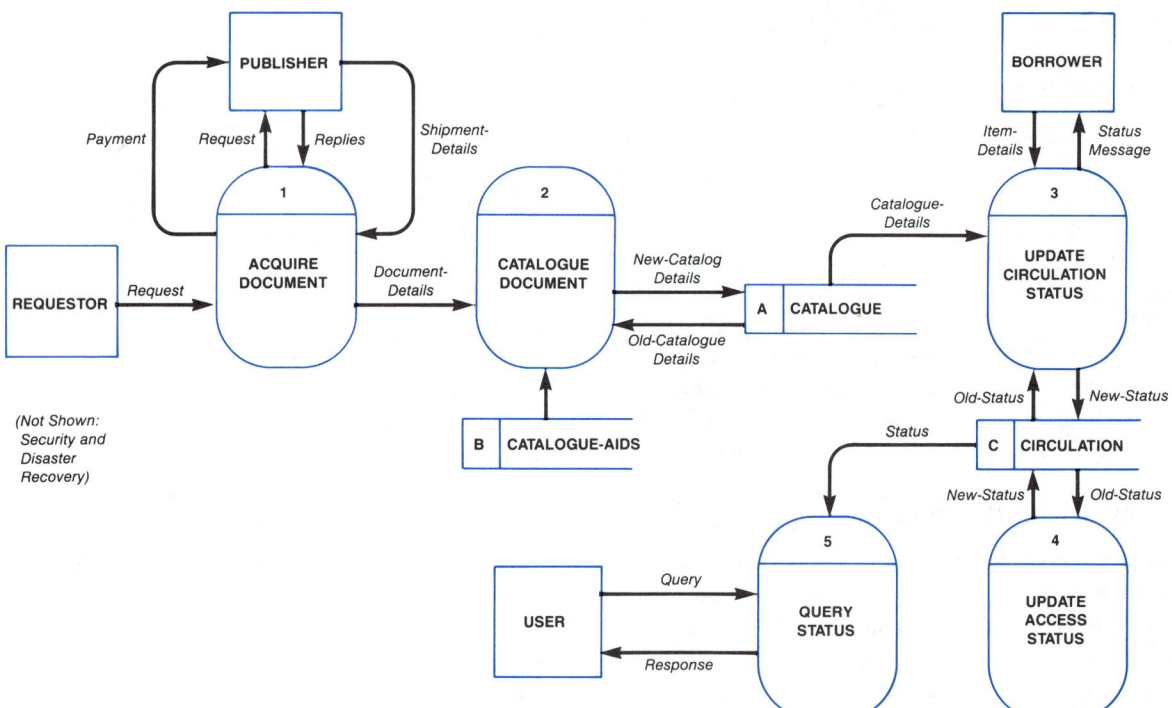

FIGURE 17–14. Library System Information System Documentation Cycle

First, a document is acquired. Prior to **acquisition**, a great deal of paperwork may have to be done to locate and order a document, approve purchase, and transfer funds.

The document is then **catalogued**, which may be as simple as assigning a serial number (1,2,3, etc.) or as complex as completely indexing it within a computerized bibliographic system. Generally, libraries adopt one of two **classification systems**: either Dewey Decimal (555.03 denotes a book on dinosaurs) or Library of Congress (TJ 211 C63 1983 catalogues a book on robot technology written in 1983 by Philippe Coiffet). In addition, to anticipate queries from readers, libraries provide series of indexes to some documents, particularly in-house research reports, memoranda, and the like, especially where these documents are not normally classified in advance (as most are today) in the Library of Congress (LC) system. Cataloguing also requires maintaining a card catalogue, although the cards are being replaced increasingly by electronic files.

Documents then **circulate**. They must be formally registered when their status changes. As a book enters the system, it is catalogued and filed. If a research report is removed, a note of who checked it out and when is made. When the report is returned, another note is made. Occasionally or automatically, notices are sent to those who have not returned needed or overdue materials. Readers can also request material on loan. All of this information is indexed by borrower name and unique document code; it may also be filed by date due. Where transfer is internal (from open stacks to reserve reading, where circulation is more highly controlled), a note needs to be made, too.

Only certain individuals are allowed to access documents. In addition, certain documents may be further restricted. Normally, library cards identify valid borrowers; cards must be presented to check out materials and, in some cases, even to enter a library. Library cards may also identify specific borrowing privileges. Because library cards, like credit cards, are deeds of trust, a separate administrative mechanism needs to be set up to issue, revoke, or temporarily replace them, and enforce legal sanctions against misuse.

What would a library be without people asking questions? Librarians are prepared to answer questions about the status of a document about x, the best source for information on x, and even specific facts about x. Libraries are well-known sources of information on other library materials and reference matter: dictionaries, encyclopedias, bibliographies, who's whos, and business indexes are the most commonly used materials. Periodical indexes, too, can answer queries and maintaining these indexes in series is another librarianship task, as is maintaining relatively up-to-date information on the contents of those indexes. With the advent of the on-line card catalogue, facilitating reader queries regarding document status has become a concern second only to circulation control.

Security, **disaster planning**, and **recovery** are other concerns. Physical security and safety are important to prevent theft, vandalism, and accidents. Where threats such as fire or water may occur, a plan needs to be in place for handling potential disasters. The recovery of records indicating status of materials is extremely important; many libraries keep duplicate paper or microform records in safes or in off-site locations. In addition to coping with disasters, libraries have to dispose of materials that are worn out or deemed useless for archival purposes.

Thus, a library system attempts to control the flow of documents through it, trying to balance the need readers have to access and borrow materials with the library's need to maintain its inventory. The following case study discusses the development of an on-line system for Milner Machinery's small technical library.

Case: Milner Machinery's Library

Located in an industrial park on the northern edge of Bigtown, Milner Machinery is a leading world-wide supplier of high-technology computer-controlled tools to industry. Trevor (Tank) Milner started the firm fourteen years ago after a stint in the tank corps got him interested in machine tools. Armed with a degree in electrical engineering and a keen interest in robotics, Tank, along with some financial backers, started Milner Machinery to fill the gap between mechanical tools and robots. With the introduction of industrial robotics ten years ago, Tank began manufacturing machinery to produce close-spec tools using robot technology. This required hiring some high-powered engineers and researchers; this, in turn, necessitated building a library.

Tank felt that a central source of documents could keep his engineers up-to-date, but he wanted his library to be computerized from the start, but this proved infeasible in the short term. However, three years ago, Tank contacted Infor, Inc. to conduct a study aimed at putting together a first-class computerized library system. He hired Nancy Chow, a recent library school grad with some technical library experience, to head the library and work with Infor, Inc.

After several weeks of study, Infor recommended that Milner Machinery acquire LOLS (Library On-Line System), a commercial library package, and tailor it specifically to the kinds of circulation and queries that Milner's engineers needed. Nancy and Tank agreed with this recommendation, and both felt that Infor could also do the tailoring.

Infor, Inc. worked to design library procedures that would operate under LOLS. Because the Milner library was small (15,000 technical reports, about 4,000 books, and a specifications file amounting to about 82,000 pages of past work that had some commercial and research value) but growing (at about 25% annually), the acquisition and cataloguing functions needed the most attention, while LOLS circulation control would be adequate. Access control was not difficult yet, since the librarians could easily recognize the engineers. It soon became apparent, however, that "on-line" also meant "off-campus" and since there were plans to make some materials available through the computer, access control was going to have to be tightened.

Another critical concern was knowing at each moment where a particular technical report was, since an engineer might need it to complete a circuit or device on short notice. LOLS had no facility for framing complex queries about the location of specific items (you could locate an item by acquisition number only), so Infor added a conversational front-end to assist engineers in locating material without going through the librarians. Of course, this complicated circulation control since an engineer might walk over to another's desk and "borrow" an already-borrowed report. This prompted Nancy to request an electronic mail facility through which messages about borrowed materials could quickly be sent to the CAD/CAM terminals on engineers' desks. In other words, a simple library system quickly grew into a complex multi-faceted information management and communication system. In fact, now, three years later, Nancy is seriously considering completely overhauling the system as acquisitions have plateaued, and concentrating on a single interface to the library's information, including the project management software's automatically-generated tool reports.

17.7 CONSUMER PRODUCTS

The previous areas have all concerned business applications. Now we will turn our attention away from the use of computers and information by professionals to a more general audience, namely consumers. This diverse audience differs from business groups in a number of important ways:

1. Business workers have well-known procedures to follow, often documented in procedures manuals.
2. Workers are *extrinsically motivated* to follow these procedures by money and other perks and are directed by managers, whereas consumers use software to achieve their own *intrinsic* aims.
3. Consumer purchase of software is responsive to a wide variety of influences, whereas business-oriented software is often sold on technical merits (although advertising and packaging play a large part).
4. Business software users, although their backgrounds are often nontechnical, usually have good performance standards by which to measure their software, whereas consumers may judge software by nonperformance factors such as appearance.

In other words, consumer software (and hardware) products have to be developed, manufactured, marketed, displayed, and serviced like other consumer products. This may mean foregoing technical efficiency for colorful graphics, complex processing alternatives for straightforward canned performances, and performance-oriented features for ease of advertising. If the current experience of marketers of technically oriented business software is any guide, advertising will play an increasing larger role as a variety of standard software items (spreadsheets, database managers, desk-top simulators, communications packages, terminal emulators, graphics packages, and project managers) arise to "define" the market from the producer's viewpoint.

The Growing Consumer Marketplace

Until 1978, the only consumers of information resource systems products were technical people or managers in business-oriented or scientific areas. The idea of selling software or hardware to Joe Buyer and Jane Consumer was nonsense because there simply were no "consumer" computers available. The development of the home and personal computer changed this.

With IBM's entry into the personal computer market, a great deal of standardization occurred. Whereas hobbyist computation can appeal to the bulk of technically minded experimenters who enjoy playing games without looking beyond to general information resources in the home or office, the power of the computer is being made available in certain standard ways through relatively standardized 16- and 32-bit machines that offer a limited number of "concepts" such as word processing and database management. (The market for games has diminished in importance and is now concentrated in the nonprofessional market, aimed mostly at children and including entries by Coleco and Mattel. This marketplace can be considered an extension of the handheld- and board-game market.)

On the other hand, after three years of chaotic activity, the demand for sophisticated *business* software has resulted in a few giants leading the generic software pack, with literally thousands following, offering *consumer* software for everything from match-making to divorce counseling, and tax planning to group dynamics. While the market for relatively simple applications has not materialized, it is apparent that there are a number of niches that can be filled by individuals willing to put the time into marketing a working product.

Software publishing is moving quickly to resemble the other kind of publishing, namely the printed word. The following types of software will probably arise:

- general-purpose, "tool" software for the professional software developer and technically minded business user and entrepreneur
- items aimed at specific marketplaces, as most of today's magazines are aimed at specialized markets (fitness, cycling, home improvement, and the like)
- games and entertainment
- software controllers for consumer products (such as talking microwave ovens and cars) and arcade games

Into the first category fall tools used in the business market to create software or solve business problems. Spreadsheets and word processors were the first major products in this marketplace. Also in this category we find communication software and hardware, business data services, specific business (usually accounting or accounting-related) software, and desktop publishing modules.

The second category includes engineering and scientific applications, educational software, training packages for other software, knowledge-based (expert) systems, property management and tax packages, and a host of very specialized software ranging from how to manage to how to diagnose illness. The third category encompasses interactive games, board games, and puzzles.

Finally, consumer products are getting, if not smarter, at least more talkative. Ovens, automobiles, elevators, and alarm systems are acquiring voices, and the micro-circuitry built into a variety of products such as VCRs requires software. Some parents note with alarm that arcade amusement games now go beyond simulating life-like battle situations and actually create them based on player input. However, since these products are not marketed directly to consumers as individual items for purchase, we will not further elaborate on their development.

How an Analyst Might Work with Consumers

The major difference between building a specific system for a specific user group (say, a department in an organization) and putting together a product to be mass marketed is the accessibility of the target group. A consumer group is larger, less controlled, and has more discretion regarding use. Furthermore, the purchasers of business systems tend not to be the ultimate users; more to the point, the users do not make the initial decision to implement (although they can bring pressure to bear during and after implementation), whereas consumers make purchases usually for their own use (or as gifts). Once an item is acquired in a mass market, its continued use depends on its attractiveness more than direct measures of its productivity. The final penetration of an item into the marketplace depends on advertising and, to a great extent, word of mouth. Executive and managerial edict are also important in maintaining use of a developed information resource. Finally, the analyst (i.e., developer) of a consumer-oriented system may conduct the initial investigation with persons who may ultimately never purchase or use the product.

Thus, many aspects of the initial and detailed investigation become suspect in the mass-marketing situation. First, **determining the problem** to be solved may not be easily done. Whereas, typically, users bring concerns to the IRU for solution, in the consumer-marketing situation, a budding entrepreneur may bring a perception of some nagging problem or a conception of an opportunity to the development situation without any idea that specific consumers may actually want the product. This stands the usual SDLC on its head as it represents a solution looking for a problem.

Second, **selecting investigation participants** is a matter of sampling. Whereas most information resource maintenance efforts work within a narrowly circumscribed unit of individuals, developing a product for a mass market implies that representative potential users will be investigated. Statistical considerations become important. After speaking with nonrepresentative individuals, the developer may create a system to satisfy persons who may be completely different from everyone else who might buy the product.

Third, **installation** of a consumer-purchased package is quite unlike installation in a rigidly controlled environment. The consumer picks the package off the computer-shop shelf, slips off the shrink-wrap, maybe reads the installation instructions, and sometimes installs the package correctly. As we have discussed, installation of an information resource is part of the tightly controlled SDLC, bolstered by testing, phased-in or parallel installation, and post-installation audits. Consumers can be forgiven for not wanting to do all that.

Finally, one has to allow for a far greater diversity in skills, attitudes, and opportunities in the general public than in trained (or trainable) operators in business environments. Fewer assumptions can be made about reading speed, preference for various colors, typing ability, ability to follow instructions, patience, motivation, and fear. Granted, there is some variance in the business environment, but sampling potential buyers is nothing like interviewing all prospective users; having an idea about what prospective users may be thinking bears no resemblance to knowing what procedures workers *will* follow in requesting a report.

Principles of Human Factors

When building systems to be used by the general public, one must take into account the variability of human skills and capabilities.

1. Reading skills are paramount since you will sell and document use and installation of your product with words. The average reading level in the United States is about eighth grade.
2. Typing skills in the general population are poor. Menus (pop up or pull down) are popular. Mice and touch-sensitive screens are ways around this limitation but they may limit your market.
3. A "session" may prove a grueling experience for many users and eye and back problems may result. Keep eyes moving, but reserve specific parts of the screen for status and warning messages. You cannot *make* people take rests, but remember: people can become fatigued after an hour at the keyboard and make errors you would never (?) make. Murphy was right!
4. Color helps. Too much color distracts.
5. Users will regularly forget filenames, specific commands or keys functions, how to quit, and how to re-install their systems; they will forget what is where and in which manual. They need to be reminded of these with appropriately labeled and explained prompts from time to time.
6. When something goes wrong, a business employee will "know" to go to a manual or a supervisor. Your consumer may not have access to either. Assistance must be provided on the screen at the appropriate times.
7. The program has to work precisely as documented, advertised, and sold. You cannot send out memos and you cannot expect people to think like programmers.
8. You must expect that users will provide input different from what the documentation says "absolutely must be entered." Idiot-proofing is required.
9. Tutorial programs are fashionable and valuable. The more unobtrusive the training you provide, the better users your users will be. Avoid computerese and cyberbabble. Try to be nice without being condescending, as this sentence is, OK?

Case: For Managers Only

Tammy Johannson is a free-lance organizational development consultant. Over the past four years she has worked a lot with undermanaged organizations. Symptoms of undermanagement are lack of effective controls over employee behavior and productivity, low morale, feelings of desertion and/or omnipotence among workers, and the rise of "self-directed" groups in which certain charismatic, but not necessarily competent, individuals "take over" the group management functions. Earlier this month, Tammy consulted with an engineering firm whose major problem was illness of both the president and vice president for eight months. Because the employees started to wonder about leadership, they turned to the most vocal of the senior engineers who, unfortunately, had some axes to grind. When the president returned to his job, he found he could no longer manage effectively.

Tammy built an index of thirty "indicators of undermanagement" which she applied informally to derive organizational remedies in her own consulting practice. Now she is building a microcomputer-based product to be distributed nationally through a software publisher. This product will collect management and leadership perceptions from individuals in a group and produce a detailed action report on the group to be used either by clients for self-study or as part of Tammy's practice.

Tammy began her development work by speaking about her idea with Peter Husak, a consulting systems analyst. Peter located some software that created and ran questionnaires and showed it to Tammy. They talked about who might purchase the package and how it could be used in a corporate environment. The need for a facilitator became apparent. The facilitator would be a person who would coordinate data collection and brief management on the results. Although the indicators Tammy had assembled seemed to her easy to understand, when they tried it out as a questionnaire on three of her former clients, several questions seemed ambiguous and two of them could obviously be faked. There was a lot of work to do on the instructions, too; it became apparent that a detailed step-by-step manual would be necessary.

Next, Tammy put together a focus-group interview with several executives on the topic of "undermanagement and staff development." She focused on ways of improving poor situations by developing staff to take on leadership roles. In this discussion she elicited information about various formats, including use of a microcomputer to collect data. She had Peter put together a prototype of the questionnaire and some preprinted sample reports and passed them around for comments. While she expected her ideas to be received favorably, she was amazed at how fascinated these executives were with the idea of computerized "testing." They saw opportunities for a whole package of leadership training and development that she had overlooked.

Back with Peter, Tammy decided to expand the package to include leadership training for both workers and executives based on questionnaire results from the group and the individual. A detailed data flow diagram showed where various sets of data could be collected, merged, analyzed, and reported on. A data dictionary was developed, partly to facilitate system development and partly to control the vocabulary Peter and Tammy were creating to talk about the product.

When modeling was complete, Tammy went to another group with the models, prototypes, and mockups and elicited comments. Reaction was positive, and several executives asked her when the package would be available. Tammy promised three months. Peter scolded her for rash promises, but three months later, to the week, Tammy had versions of Leader of the Pack available for members of both focus groups. She had already tried out the package on a client with positive results and had contacted both a technical writer and a software publisher to get the package ready for marketing. At $365 a package, Tammy knew that she would make money, but more important, she had built an enhancement to her own consulting practice.

17.8 CONCLUSION

Seven specific content areas were discussed in this chapter: One-user decision-support systems, large MIS projects, office automation implementation, microcomputer software development, records management, information retrieval, and consumer products. These form a major part of the work that systems analysts are involved in, but each requires a different blend of the techniques that analysts employ.

Decision-support systems are constructed in a "developmental" mode, with cooperation among a single **user/client**, an **analyst**, a **toolmaker**, and one or more programmers. Because mathematical models are built and tuned over time, the system "develops" from a clumsy tool to a refined system under the guidance of the analyst. Many DSSs are constructed using a prototyping tool, particularly on microcomputers where integrated packages are the rule.

The **MIS project** involves the full range of analyst skills because the resulting system has to reflect the needs, expectations, and skills of a varied user group composed of many different kinds of workers. The MIS project involves all of the following: **data entry** with **checking** and **validation**, **data reduction**, **report generation**, and **report distribution**. Internally, **data integrity**, **data security**, **file maintenance**, and **resource sharing** are paramount considerations.

The **automated office** is a complex system of machinery, software, and work procedures; its goal is the simplification, rationalization, and streamlining of office tasks such as memo typing, file sharing, scheduling, bring-forward filing, graphics, and voice communication. Recent advances in computers and communication equipment have made integrated voice/data systems economical. Analysts need to consider the organizational communication implications of shared data and integrated communications systems. Therefore, prototyping may be appropriate. In many cases, specific stand-alone systems intended for word processing may be mandated before the study begins. An office study can concentrate on either secretarial/clerical functions, executive/managerial functions, or both. Finally, the electronic office can be deemed either part of an overall MIS strategy or it can stand alone in organizational plans — which also affects the analyst's activities.

The microcomputer revolution has had a major effect on how management sees computers. **Microcomputer software projects** are similar to those involving larger computers. The existence of end-user-oriented software tools makes the task of analysis even more complex, however, since users create, use, and share software.

Records and forms management, a major part of information resource management, is often ignored in systems analysis because it does not always involve computers. Controlling the content of filing cabinets and forms shelves may not be as romantic as building a nanosecond-speed system, but for most firms, this aspect of the information resource is by far the most pressing and restricting.

Information retrieval, in the form of a library system, is another aspect of systems analysis that reports-oriented analysts often ignore. Library systems are complex because users come in two varieties: librarians (or information specialists) and patrons. Automated systems allow the following components: **ordering**, **cataloguing**, **circulation control**, and **query/access**. While the term "library" may bring to mind an image of dusty books, most organizations possess large numbers of volumes that are haphazardly filed and indexed; information management (as opposed to management information) requires a systematic approach to maintaining document files so that they can be kept up-to-date, available, and accessible.

Today, many **consumer products** involve computers. While games may be the most visible aspect, several other kinds of consumer products — including microprocessor-controlled appliances and vehicles, calculators, and home information-retrieval systems, especially when connected with cable televisions — are growing in importance, too. Several types of communication networks, in particular cellular mobile radio-telephones, depend heavily on computers and the maintenance of a complex of information.

With this discussion of project work details, the text has ended. Those readers planning to apply skills and attitudes discussed in this book are urged to pursue content areas, as they will in any event on their first project. This book has stressed *management* and *organizational* skills as much as the techniques and technicalities of computerized information resources. The reliance that most technical people in this area place on expert power is often seen as a stumbling block to development and good relations with users. Shared values and empathy play just as important a role. I hope that some of that has come through in this text.

DISCUSSION QUESTIONS

1. The text mentions eight factors that directly influence how systems analysis proceeds on a project. Three others were mentioned in passing: technology employed in development, intellectual atmosphere, and organizational context. Discuss how each of these three might influence an analyst's work using the following three situations as examples:
   a. A system is to be developed on one computer for use on another after that second computer, which is on order, has arrived.
   b. Prospective users of the system have all previously used a system that was so bad that it had to be removed and the manual system, reinstalled.
   c. The organization's major product is computer software.

2. *M* is a manager at a municipal water works where a computer to monitor production is already in place. *M* would like a system that enables her to make quick decisions and do longer-term planning regarding water-use restrictions during drought periods. You are an analyst with the central municipal IRU who has been called in to work with *M*. Describe how you would set up a project to develop a decision-support system for *M*.

3. "Richards, get in here... *now*!" bellowed Phil McCorkle over his phone to his assistant, Randy Richards. "Just what does all this stuff mean? Look here. These sales figures can't be right. Eight *hundred* returns in one month? And where are the figures for Bigtown district? How come they're not in this report? It makes the gross too small. How come I get only summaries on a *quarterly* basis by store — where are the details for each week? Anyway, look at the date on this report! It's two weeks old already; I thought this computer system was going to give me data every day. And look at this! There must be

forty different reports in here, but there's no index, no order to them, no legends, no explanation of what these fields are...I mean, what's EXTOT supposed to be, Randy? What's going on?'' Try to tell Phil what is happening to his MIS and what steps might be taken to alleviate the situation. Who do you feel might be at fault here?

4. How does communicating differ from informing? What happens during communication that is *more* than merely passing information? What factors have to be taken into account in designing an office communication/information system that are less important in a pure information system? Also, what do we know about how people use an information resource that brings communication back into the design? Which logical design techniques and tools are most appropriate for representing office processes and which are less appropriate?

5. Microcomputer system development has taken a number of turns that make it different from mainframe development. In a sense, the microcomputer has brought user and developer closer together (creating what Alan Toffler, in his book *The Third Wave*, has called the ''prosumer''). In another sense, however, this closeness is artificial, since microcomputer capabilities are often greatly reduced or simplified. Discuss the pros and cons of this movement toward unifying user and developer. What effect has this had on the role and effectiveness of the systems analyst?

6. What tendencies might an analyst who is used to working on computer-based information resources demonstrate when developing a partially manual system? Can you offer some advice to this analyst toward eliminating any negative tendencies? Is working on manual systems merely ''humbling,'' or can important skills be learned and transferred to the development of a computer-based system?

7. Library systems are increasingly computerized for obvious reasons of efficiency. What are the gains from a user viewpoint? What are the losses? Given these gains and losses, what should the analyst watch out for when developing computer-based library systems?

8. The concept of information management is clearly not the same as that of records and of database management. Compare and contrast these three concepts, especially with regard to the kinds of questions an analyst might ask in investigating applications in these areas.

9. Alan Bornoy is sales manager for Bigtown Cablevision, Inc. BCI distributes 30-channel cable television signals to over 60,000 homes and offices in Bigtown under municipal license. Alan has been approached recently by a team of student entrepreneurs who want to use his network to distribute a graphics-based computer bulletin-board service to consumers who would use their telephone touch-pads to request a variety of information services. What advice would you give Alan about requesting a formal proposal from the students? What should appear in the proposal? What kinds of skilled people should be on this team? Regardless of the *economic* implications of this scheme, what plans should be in the proposal?

AN EXAMPLE OF A USERS' GUIDE

A

APPENDIX

PHYSICALC™ USERS' GUIDE

Contents

Chapter 4: Patient Visit Data

This chapter is concerned with handling patient visits prior to billing. PHYSICALC calls this function "reception." Normally a patient's visit is scheduled in advance; sometimes patients are handled on an emergency basis and are seen when they come in. In some circumstances, the physician may see a patient outside the office and the receptionist will never really "receive" the patient. PHYSICALC provides the receptionist with the capability of pre-scheduling, slotting, or post-scheduling patient visits.

From the point of view of later billing, each visit consists of a patient (with a *patient name* and medical insurance *AHC* number) being seen at a *time* on a *date* for a reason identified by *dxg*, a diagnostic code. Each visit may result in one or more services, each identified by a *feecode*. Part of each service is covered by insurance (*$AHC*) and part by the patient (*$private*). A number of other codes, such as worker's compensation (*w*), may also be attached to each service. These data constitute each line of billing. The visits file consists of these lines of billing. It is organized into 366 segments, one for each day of the year. PHYSICALC allows you to access a day's billing quickly and efficiently for examination, printing, receipting, or billing.

PHYSICALC also stores some information on each patient in a file organized alphabetically, like a file box, separated by 234 INDEX CARDS, 9 per letter of the alphabet (for example: A, AD, AF, AH, AL, AO, AR, AU, and AX) for quick access to a patient. A patient has information stored as follows: name, address, home phone, work phone, AHC number, birthdate, date of last visit, feecode of last visit, and comments. Version 3 of PHYSICALC will store a significant volume of treatment data. This last feature is not available in version 2.

Reception functions provided by PHYSICALC include the following: Find patient, print patient data, enter a visit, compute fees, check AHCIP number, maintain fee-code price list, maintain referring physician list, modify patient data, print patient recall list, print daily billing list, as well as retrieval of patient or visit information by any data item. All of these are accomplished with a small number of keystrokes based on the function keys located at the left side of the keyboard:

F1:   Access secondary menu (then press another function key):
      F1:   Bring current record to top of screen
      F2:   Scroll until F1 is pressed
      F3:   Locate records (see below) until F1 is pressed
      F4:   Same as F3, but scanning file backward
      F5:   Refile record into order by key field (*name* in patient file, *date* in visits file)
      F6:   Undelete record previously deleted
      F7:   Duplicate the record at the cursor below it
      F8:   Create records based on a pattern
      F9:   Find patient record for current visit record
      F10: Unassigned
      nnnnn: Locate a specific record by record number
      BKSP: Bring up record previously examined (unless deleted)
F2:   Scroll next ten records forward
F3:   Find the next record with a specific value on the field under the cursor
F4:   Same as F3, but scanning backward
F5:   Enter data into record (keyboard normally is locked)
F6:   Delete the record under the cursor
F7:   Add a blank record under the cursor
F8:   Print a recall list of patients not seen in one year
F9:   Toggle between patient file and visit file
F10: Help screens or exit from PHYSICALC to PHYSICALC startup program

AN EXAMPLE OF AN OPERATOR'S GUIDE

B

APPENDIX

PHYSICALC OPERATIONS MANUAL

A1 The first screen you see when PHYSICALC is loaded prompts you for the eight-digit PHYSICALC key. This key is entered during system generation (see A5).

A2 The next screen prompts for either a redefinition of system configuration options or a display of the current system status. Enter *1* to redefine the system configuration or *2* to look at the status (See A3). Enter *!* to proceed to PHYSICALC's main menu.

A3 The status consists of this information: the number of the last user of the system, whether or not practice history records are automatically filed on reconciliation of bills, whether or not prompting for filing of bills to a disk file is default, which sort of printer (IBM or Okidata) is to be used, whether or not a unified General Ledger is to be created (and, if so, under which physician's identifying number), and what the files will be named. After reviewing the status, press any key to bring up the A2 screen.

A4 If you request redefinition in A2 by pressing *1*, you will see this screen. Each screen allows you to change an option in the system configuration. *Most never need to be redefined.* These options are described as follows: practice history recording (A6), default billing file (A7), printer driver (A8), blended general ledger (A5), and file name spec (A9).

A5 Enter *!* to cancel practice history recording. Any other key will allow such recording. The next screen you will see is A3.

A6 Enter *!* to make printing of bills on-line to the printer the default. Any other key will default to saving bills to be printed in a file called BILLING.LST. The next screen you will see is A3.

A7 Enter a digit identifying the user having the General Ledger (G/L) file. Usually this will be a *1*. If you enter *!*, each user will have a unique G/L. A CR will accept the currently specified situation and return you to the A3 menu to show the results.

A8 The letter you enter will be used as a base for naming files; we recommend *t* or *a* (lowercase letters). Files that pertain to the first user will be *.DFT if you specify *t*; those of the second user, *.DFU; those of the third, *.DFV; and so forth. Note that with *t*, you can have only seven users (T, U, V, W, X, Y, and Z), while with *a* you can have 26. The next screen you see is A3.

A9 The choice here is between using *I* or *i* for an IBM and either *O* or *o* for an Okidata printer. The next screen is A3.

A10 You can change the specifications for the screen by indicating that the screen is an IBM color monitor, an IBM monochrome monitor, or an Olivetti monochrome monitor. If you reply with a CR, you will cause the screen to become "Unspecified," which means, in effect, an IBM monochrome monitor. This specification is *necessary* before you can run any programs.

A11 Certain conditions require locating a set of programs on another disk drive. *Normally you will not need to make any specification*, other than at the initial system set-up. Be careful. Usually programs are located on the C: drive for an IBM or Olivetti *hard-disk* system. Note that this option is *not* specified on the REDEFINE OPTIONS menu, but it is #7.

NOTE: The *first* time you bring the system up, specify each option. Pay special attention to Practice History Recording, setting a unified General Ledger (if one is required), and the file base. We recommend the following set as "standard":

    Practice History Recording: ON
    Default Bill Print Prompt: OFF
    User Having G/L: none (use *!* for separate G/Ls)
    File Base: *t*
    Printer Type: *i*
    Program Location: *c* required for IBM hard-disk systems

AN EXAMPLE OF A TECHNICAL MANUAL

C

APPENDIX

PHYSICALC TECHNICAL MANUAL

System Parameters

All files in PHYSICALC that can be edited on-line are limited to 10,000 records (actually that number is usually less than that by 28 plus the number of index cards, which may vary between 100 and 366).

A PHYSICALC data file other than STATUS.DFT (which stores system status), transactions lists (which are sequential files opened for "append" to copy all activity in the G/Ls), and PRINTER.DFT (which configures the system for specific printers) has the following structure:

Record	Offset	Contents (ASCII Characters)
1	1-16	Informal file name that appears on the file description line
	17-32	Owner, the name of the physician who actually "owns" the data in the file
	33-36	Number of records (including index cards)
	37-40	Record number of first free element
	41-44	Record number of first data record
	45-48	Record number of last data record
	49-52	Number of fields in this file
	53-56	Field number for index field
	57-60	Record number of first non-index record
	61-64	Record number of last written record

2–27		One record for each field
	1-16	Field name to appear on screen
	17-20	Field type (0 = character, 2 = dollar value)
	21-24	Data offset in record (minimum of 9)
	25-28	Number of characters in data field
	29-32	High-order digits (type 2 only)
	33-36	Low-order digits (type 2 only)
	37-40	Offset on screen (between 9 and 79)
	41-44	Length of display on screen (2 or more)
	45-48	Display number (1 or 2)
28–x		Index cards
		Alphabetic: 234 (nine for each letter A - Z)
		Accounts.DF?: 100
		Visits.DF?: 366 (one for each day of the year)
	1-4	Forward pointer, record number in ASCII
	5-8	Backward pointer, record number in ASCII
	9ff	Data in fields as specified in records 2-27

The BASIC VAL() function translates ASCII values to numerical values; the user-defined function FNFR$() converts a number to a four-character string.

Files are currently limited to a record length of 128. Files may have no more than 26 fields. All files in the standard PHYSICALC must be located on the same drive. Exceptions are for DTA, LST, RPT, FMT, and QRY files used in the database management program. These are not PHYSICALC data files but are used to create and store queries and reports or, in one case, to save bills to be printed. Here is a list of reserved file extensions:

Extension	Meanings
.DF?	PHYSICALC data file; the last letter identifies the user, depending on the base letter set in system generation.
.LST	Target files for billing and records retrieved using F3 in the database management program
.QRY	Files that store queries for reuse in the database management program
.RPT	Files that save the results of reports in the database management program
.FMT	Files that save the formats for reports in the database management program
.WP*	Form letters created by EDLIN that are used in generating reports in the database management and for recalls in reception. The * stands for the sequence number of the physician (1, 2, 3, and so on).

CASE STUDIES

APPENDIX

CASE 1: HAIRLOOM STYLISTS, INC.

Background

You have already worked extensively on **Hairloom Stylists** information resource. Now it is time to complete the analysis and design by producing the logical design package, the functional models, the user's and operator's guides, a technical manual, and a working system in dBase III.

What Doug really needs is an application generator that he can use to create *ad-hoc* queries of his files and produce analyses on the screen of his business. Based on the interviews, it looks like he needs the following kinds of information:

1. Stylist profile: all the work a particular stylist has done
2. Client profile: all the work a particular client has had done
3. Productivity reports on stylists
4. Sales analyses for services and products
5. Inventory reports on specific products
6. Supplier analyses for profitability of product lines
7. Training and skills analyses for stylists

These reports have **parameters**: names of stylists, ranges for products or services, names of clients, skill codes, and so forth; time periods (weekly, monthly, annually); and others such as profitability criteria.

There are already several query facilities built into the program, but Doug has to go through each query, laboriously tracing or cross-referencing items. He needs a way to create and store queries that he uses on a regular basis and be able to change them is necessary.

Application generators work by asking for application parameters and storing them in a file for later use. At this later time, the application generator asks the user for the name of the application to be accessed, finds the parameters, generates the required application, and produces the report. The user can create, access, and change applications merely by changing the parameters in the file.

Build an application generator for Doug to query his files in a more coherent and efficient way than HAIR now does. Doug would also like to be able to enter data to files from screens that check the data rather than using the APPEND statement and the cold "form" that appears.

Statement of Scope and Objectives

> **Problem:** Many *ad-hoc* queries have to be rereatedly re-entered; data entry is cold and uncontrolled.
>
> **Objectives:** An application generator to create, access, and change query applications concerning clients, stylists, products, services, suppliers, and skills; a set of data entry screens.
>
> **Criteria:** Must work around and with current HAIR system (which is to be improved) without changing existing file structures or operating procedures.
>
> **Scope:** Data entry and query facilities. System maintenance is not to be changed.

Additional dBASE III Commands

To add data (other than through APPEND entered at the dot prompt), you should use the following sequence:

APPEND BLANK	Add a blank record
@...SAY...;GET x	Read in some values
REPLACE a WITH y	Replace the value of a field with the value read in or computed.

The REPLACE command replaces a field with a value. You can also use REPLACE to edit existing records. Note that the @...SAY...;GET sequence can directly replace values in fields.

Here are some examples:

```
@ 1,1 SAY "Enter the Stylist's NAME";
  GET N
READ                                    Get the name of the client
FIND &N                                 Already on file?
IF EOF()                                If not...
    APPEND BLANK                        Add a blank record
    REPLACE STYLNAME WITH N             Replace the name
    @ 2,1 SAY ...                       Add in the information
    ....
ENDIF
...                                     Continue processing record
INDEX ON STYLNAME TO ....               Be sure to recreate the index
                                        file with the new record.
```

Whenever records are added to an indexed file, the file has to be **re-indexed**. dBase III will normally prompt you with the message

>   *x* already exists. Overwrite (Y/N)?

and you should respond appropriately. To *avoid* this message and the need to respond, you can SET this message OFF:

>   SET SAFETY OFF   Do not prompt for safety.

>   SET SAFETY ON   Prompt for safety.

To **check the validity** of data entered in response to @ commands, dBase III has a RANGE option (numerical data only):

@ 3,1 SAY "Enter Activity ID";	Read in an ID as a
GET NUM RANGE 1000,1999	number between 1000 and
READ	1999. The user will be
	told if the value is not
	in this range; then
ACT = STR(NUM,4)	convert it to a four-
	character string.

There are a number of additional commands that are useful in editing .PRG files. These include the following (upper or lower case):

^A	Go left one word.
^B	Go to right end of line.
^F	Go right one word.
^G	Delete the next character right.
^I	Tab five spaces (or insert if in insert mode).
^M (or ^N)	Insert a blank line below this one.
^Q	Cancel editing of this file.
^T	Erase the next word.
^V	Toggles insert mode (Ins key does the same).
^Y	Erase this line.
^Z	Go to left end of line.
Esc	Equivalent to ^Q.

(Note: ^ means hold the Ctrl Key down while pressing the next key.)

Here is a **template** for the keyboard indicating which keys do what:

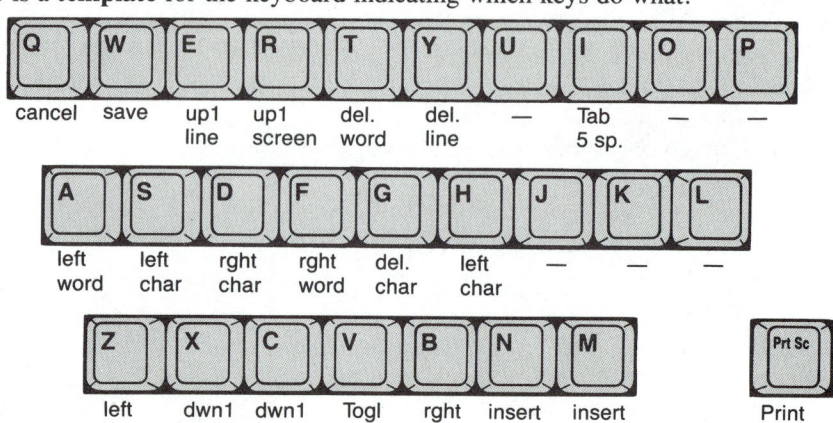

Q	W	E	R	T	Y	U	I	O	P
cancel	save	up1 line	up1 screen	del. word	del. line	—	Tab 5 sp.	—	—

A	S	D	F	G	H	J	K	L
left word	left char	rght char	rght word	del. char	left char	—	—	—

Z	X	C	V	B	N	M		Prt Sc
left line	dwn1 line	dwn1 screen	Togl INS	rght line	insert line	insert line		Print Screen

Note: Backspace deletes the character at the cursor on most machines, but on the IBM PC/AT this is not the case. Backspace merely moves the cursor one character to the left. Use the INS key to delete the character just to the left of the cursor.

CASE 2: MCNAMARA EXECUTIVE CONSULTANTS, INC.

Background

McNamara Executive Consultants was founded in 1968 by Barbara McNamara. Its primary business is placement of executives in sales, marketing, accounting, data processing, and technical management. Its clients are organizations with executive and professional personnel needs. McNamara's competent, experienced consultants are concerned with rapid placement of candidates, as clients require immediate responses.

Barbara McNamara is the President of the firm, but Grant Byers, the general manager, is responsible for day-to-day operations. There are six consultants on staff, and over 3000 candidates being considered for 150 to 250 positions. Three secretaries and two clerks assist.

Barbara has noticed that as clients' needs become more complex, the ability of McNamara Executive Consultants (MEC) to respond in a timely fashion has decreased. A great deal of time is spent searching through files and several client contracts' time constraints have not been met. Barbara has been interviewed and these concerns have come out:

1. The candidate files contain redundancies that are hard to update.
2. Clients' needs are often expressed in too vague and general a manner.
3. Candidate information is too general and there seems to be too much of it.
4. None of the people who consult the files do so consistently or efficiently.

Specific findings made during the preliminary and detailed investigations are these:

1. The job order form consists of general categories while the information required is quite detailed and specific, resulting in incomplete information.
2. The candidate data sheet is used for too many purposes, resulting in confusion, redundancy, and lost data; in addition, the data is not very specific and not useful for making complex decisions.
3. When the candidate data sheet is filled out, the candidate usually has experience in three or more different occupations; these forms are photocopied and filed by occupation; with 3000 candidates, this means up to 10,000 sheets of paper have to be examined periodically.
4. Occupational categories are poorly defined.
5. Candidate files are updated only when a candidate is contacted for a possible placement; five-year-old data sheets can still be in the system in their original form with obsolete data.
6. Most data sheets are incomplete (from mail or phone information) and since the forms were redesigned three years ago, the sheets are in at least two formats.
7. Given these problems, several of the consultants usually bypass the file system and recruit directly.
8. Placed candidates' data sheets become inactive but are usually placed back in with the active files.
9. The consultants and the GM each have their own personal files that duplicate or extend the information in the common files.
10. Candidate files more than one year old should either be purged or the candidates contacted to renew their entries.
11. Barbara has little idea of overall efficiency and effectiveness of the existing system and little control of activities. She does not know, for example, how long the average placement takes, how many files are really dead, how many new candidates are filed each month, or how many interviews it takes to achieve a placement for a specific job category.

Statement of Scope and Objectives

Problem: The current system is untimely; search time is too long and files are difficult to use and maintain.

Objectives: Reduce redundancy; reduce search time; improve relevance of the information; improve reliability and maintainability of the files.

Criteria: No increase in personnel; minimize time for change.

Scope: Placement activities only; business aspects of MEC are to be left alone.

High-Level Analysis

1. Enter Candidate Data (from candidate and consultant via interviews, phone surveys)
2. Update Data Sheets (from candidates' phoned-in or mailed-in information)
3. Enter Client Needs on Job Orders (from clients)
4. Determine Appropriate Candidate (from all information)

Current Data Forms

1. Coding Sheets
   applicant rating scales, occupation codes
2. Candidate Data Form
   name, address, birthdate, residence and business telephone number, marital status, current employer, driver information, highest educational degree, position desired, salary desired, geographical preference, travel preference (Y/N), current and three previous positions with salary, employer, tenure, responsibilities, reason for leaving, interview impressions, ratings, consultant initials
3. Job Order (Client Information)
   position, date, firm, department, contact name, address, billing contact, type of business, phone, educational requirements, salary, starting date, experience requirements, new position (Y/N), duties, bonuses, travel requirements, car (allotment, allowance, requirements), career opportunities, information on each candidate considered or referred with date, time of interview and comments, order taker initials, fee for placement, counselor involved in ultimate placement, invoice #, date placement made, candidate hired, actual starting date
4. Occupation Descriptions
   common duties, career opportunities, educational requirements, salary range(s), experience requirements

The Project

In a team, role-play the following roles: Barbara McNamara, Grant Byers, and Karen Billingsgate (one of the consultants), a client, and an applicant. These role-plays will provide more information about the files and the processes involved. Use dBase III to **prototype** a system to assist in (1) referral decision-making and (2) reporting efficiency and effectiveness of the system.

Produce a high-level DFD, a structure chart, a series of HIPO charts, and other models as needed. Explode the DFDs and produce data dictionary entries for each data flow and data store. Given the existing system and the expressed data, organizational, and procedural needs, improve the models and propose a new logical design.

Working with this logical design, translate the design into physical designs for a sequence of prototypes that you work out with Barbara, Grant, and Karen. Use the DISPLAY STRUCTURE TO PRINT command to document your data designs. Your working system will consist of a set of .PRG files (programs), .DBF files (data files), and procedures for using and maintaining the system (documented in user's and operator's guides). Additional documentation for screens can be obtained by pressing Shift and PrtSC with the printer turned on. This will "dump" the screen to the printer without (unfortunately) the reverse image blank space for data entry. You can fill this in later.

CASE 3: BIGTOWN CONVENTION CENTER BOOKING MANAGEMENT

Background

The Bigtown Convention Center (BCC) was built over fifteen years ago by the City of Bigtown to attract national and international conventions. The Center's Board of Directors consists of the Mayor, the Commissioner of Finance and Administration, two aldermen, and six citizens appointed by City Council. Operating under a budget of $2 million, it is estimated that the Center directly generated over $8 million from visitors last year. The Center has aggreed to allow the Ritz hotel to be the exclusive caterer for functions held at the Center, which is located immediately adjacent to the Ritz.

Bob Sykes is the General Manager of the Center, and Marta Thomlison is the Assistant General Manager (AGM). Jeremy Cross is the Controller and Gabriel Kaplan is the Sales Director. There are eighty other people employed by the Center.

All information is recorded manually on paper forms, and bookings are coordinated through a large Booking Manual that maps out dates and times by the quarter hour by room for the entire year. This manual system proved efficient for the first ten years, but the burden on the AGM to maintain these files has increased to the point that she spends hours each day merely looking up information. So far she has made few major errors, but she is aware of the lack of controls in the booking system and the possibility for error. For example, clients can book up to six years in advance. If Marta looks in the wrong year's entries and makes a commitment, the possibility for double-booking is enormous. In addition, sales people have noted that it takes a while — often several hours — for Marta to get back to them concerning a proposed booking.

Statement of Scope and Objectives

Problem: The manual system is tedious, error-prone, and slow. A lot of paper is generated and has to be filed. Changes are difficult to record and their volume is growing.

Objectives: A faster, more controlled, less error-prone system that produces less paper and that can be queried rapidly and accurately regardless of the number of bookings.

Criteria: No increase in clerical staff; system must be open to audit at any time; training time must be minimal; use restricted to AGM, Controller, Sales Director, and Facilities Service Manager. Double-booking *must* be avoided.

Scope: Booking subsystem only (see below); however, it must interface well with facilities management, sales, and accounting subsystems.

Booking Procedure

According to Marta, the booking procedure is crucial to understanding how the center functions. Here is the essence of her comments about the booking procedure:

Clients are either contacted by sales representatives or they phone in directly and are assigned a sales representative based on their geographical location or function type. Sales representatives who work with clients on functions are called *function coordinators*. Over the phone, the function coordinator checks the proposed booking dates with the AGM for availability of function rooms of the appropriate size and type and for resources (tables, AV, lighting, and sound equipment). The AGM checks the booking manual and negotiates dates, times, and rooms with the function coordinator, who prepares a function sheet for each room detailing AV equipment, meals, room requirements, length in hours, seating arrangements, head table requirements, and so forth. If the client is new, client information is also prepared for a credit check. Space is then blocked out on an interim basis and an entry is created in the booking manual for the client. Following a 48-hour client credit check and the receipt of a 10 percent deposit, the reservation is confirmed. Five weeks before the event, the function sheet file is pulled and the function coordinator contacts the client to make final arrangements. Cancellations less than one month prior to the event are charged the full rate. Before that, the cancellation charge is computed by the formula:

$$Penalty = Charges * (25 - W) / 25$$

for $5 <= W <= 25$ where W is the number of calendar weeks to the date for the function. Food can be cancelled or changed at no charge up to three days prior to the event. After that, the penalty is 100% of the charges.

Each Thursday, function sheets for the next two weeks' events are finalized and a summaries are typed. At this time, the AGM checks the booking information against the function sheets for errors, inconsistencies, and logic. Since she looks at over 150 function sheets each week, there is a lot of work to be done. When the sheets are finally verified, Marta prepares a summary list and sends a copy to each of the departments involved (food, utensils, seating, lighting, sound, bar, custodial, three departments at the Ritz, accounting, sales management, facilities service management, signage, security, warehousing, cleanup, transportation, and archives). Forty-eight hours before the function, the function coordinators call the client to guarantee the number of people who will attend the function. A daily event list is prepared for the following day from the information in the guarantees. Daily event lists are distributed to the departments. They must match, and since they often do not, the AGM must reconcile them.

High-Level Analysis

1. Initial Booking: in response to client needs as conveyed by sales representative. Assign function coordinator and initiate credit check. Prepare credit check form and confirmation letter.
2. Interim Booking: after credit check is complete and deposit received
3. Prepare function sheet.
4. Prepare weekly function summary.
5. Prepare daily event list.
6. Process cancellations and changes.

Current Data Forms

1. Booking Manual
   by date, time, and room — indicates function holding the space up to six years in advance.

2. Client Data
   name, address, business, contact name, phone number, previous functions at BCC by date and attendance, comments, results of prior credit checks

3. Room List
   name and number of room, maximum capacity, possible seating arrangements, stage availability (Y/N), variable lighting (Y/N), color scheme, number of windows, dance floor (Y/N)

4. Meal List
   code and cost for meals per person or flat rate. Minimum guarantees required.

5. Resources Available
   AV [over-head projectors, slide projectors, 16 mm, video (playback and record), amplifiers, screens, tape recorders], lecterns, spotlighting, display stands, registration tables, charges for resources (where applicable), alternative sources for resource if needed, repair or service arrangements

6. Function Sheet
   date of event, organization, contact person, alternate contact, telephone, function coordinator, date entered, status (initial, interim, confirmed, guaranteed), type of contact (via sales representative, direct phone), type of organization (corporation, association, local/regional/national/international, public), length of time in days and hours, detailed outline of each event for the function including date, room, table arrangement, number of attendees, food and drink revenue estimation, security requirements, door arrangements for greeting attendees

7. Weekly Function Summary
   by date, time, and room: type of function, table arrangements, attendees, door arrangements for greeting attendees, security requirements, AV, and other resource requirements

8. Daily Event List
   same format as Weekly Function Summary

The Project

Complete the logical design for the Bigtown Convention Center's booking system, preparing specifications and a logical design model package, a user's guide, and an operator's manual. Then, using dBase III, design the file structures and programs needed to book and cancel events, and prepare the required reports. The preparation of credit check forms, confirmation letters, and bills will *not* be part of this system.

Because the educational version of dBase III is limited to 32 records per file, it is not feasible to create a booking manual of six years' duration. In fact, you could not handle even one days' booking at a real convention center. Instead, concentrate on booking up to 32 events per week (seven days) in mornings, afternoons, and evenings (21 slots for each room) for three rooms, ten different meals, ten resources, and ten separate clients.

While this severely limits the files, it does not trivialize the problem, which is easy to generalize from this example. If you have access to an unrestricted database manager, these limitations can be removed. It is not hard to imagine information for this project, since most of us have attended functions, meetings, or conventions. In order to prepare function summaries and event lists, you should use these commands to print the lines:

SET PRINT ON   Directs ? output to printer

$? a + b + c + x$ Display a set of values

GLOSSARY

Abstraction: Reducing and arranging a large amount of data in a systematic fashion.

Active listening: A variety of techniques designed to increase an interviewer's ability to understand and interpret the respondent. These techniques include: **encouraging**, **paraphrasing**, **reflecting** and **summarizing**.

Activity scheduling: Determining the order of activities to be performed during a project.

AND-THEN: In structured programming, two or more activities performed in sequence.

Application generation: User creation of an application (with an application generator) consisting of a small number of specific parameters that describe the desired output.

Approval meeting: A meeting at which approval for subsequent IRMaint steps is solicited by the IRU.

Backup: A copy of data used in the event the active data is destroyed or lost.

Base system: The system being modeled through a simulation.

Black box: A system described only in terms of its inputs and outputs.

Black-box design: A design technique that treats subsystems as black boxes.

Bottleneck: A point in the flow of work where the volume or variety of tasks exceeds the system's capacity to perform them.

Bottom-up: Analysis, design, or specification of a system in terms of its operational components.

Boundary effects: Problems stemming from observation of the wrong aspect of a system, too narrow a definition of a system, or ignorance of the aspects of the system boundary. See also **observation**, **observation effects**, and **system boundary**.

Brainstorming: A group technique used to uncover a large number of ideas regarding a problem or system.

Buck slip: A log attached to any form whose progress through processing stations is being monitored. At each station, the arrival and departure times and the type of processing performed are noted on the buck slip. See also **forms tracing**.

CPBX: See **Computer-Based Private Branch Exchange**.

CPM: See **Critical Path Method**.

CPM Diagram: A chart illustrating how events are scheduled relative to one another in a network of events. See also **precedence chart**.

Career path: A sequence of jobs, vocations, or positions held by an individual. In the IRU, there seems to be two tiers: one leading to managerial responsibilities, and the other to increasing breadth and depth of technical responsibility.

Cell: A subsystem or group of elements with a specialized function within a system. Cells include **defender**, **ingestor**, **locomotor-manipulator**, **memory**, **repair**, **sense**, and **control cells**.

Charge-back policy: An economic policy that charges IRU users for computing, consulting, and operating services.

Cohesion: The extent to which a specific logical function actually performs one and only one task.

Collection point: A single person, activity, or data item or a combination thereof that is observed, queried, or sampled during data collection. See also **observation**.

Component-based design: A design technique that focuses on input, output, data, boundary, or processes.

Computer-Based Private Branch Exchange (CPBX or CBX): Computer-controlled telecommunications switching equipment which is located at a customer's site. See also **office automation**.

Concern agenda: A list of IRU client concerns that stimulated the request for services.

Concurrency: The property of a system that allows more than one process to occur at any moment in time.

Conformance: The correspondence between a model and the base system it models. See also **model**.

Contractor: An individual outside a firm who contracts to build either part or all of a system for a firm.

Control loop: A cycle of activities in cybernetic systems that senses environmental conditions and selects appropriate actions to change the environment to a more desirable state. See also **system architecture**.

Conviviality: A design criterion that dictates that systems be enjoyable or comfortable to use.

Coupling: The degree to which elements within a system mutually dictate or limit one another's goals.

Crisis management: Planning for and countering the effects of crises.

Critical incident technique: A variation of an interview which focuses on critical events rather than day-to-day operations. See also **interview**.

Critical Path Method (CPM): A technique for determining events whose slippage would delay the entire project.

Critical variable: A variable sensed by a cybernetic system that determines whether or not countering action is necessary. See also **system architecture**.

DFD: See **Data flow diagram**.

DSS: See **Decision support system**.

DSSG: See **Decision support system**.

Data analysis: The act of determining overall trends (aggregational) or case examples from a large volume of data.

Database administrator: The individual responsible for defining an organization's use of data.

Database management: A collection of data processing techniques that makes data available to programmers and users regardless of its actual physical location. See also **Information center**.

Data capture: A subsystem that regulates the manual entry of information into the system through prompts, menus, forms, and data checks.

Data checks: Tests performed on input to ensure its validity. These include **access**, **authority**, **batch**, **format**, **presence**, **range**, **reasonableness**, and **size checks**.

Data dictionary: A list of data elements in a logical design. The data dictionary defines each data flow, data store, external entity, and process. See also **HIPO chart**.

Data flow: The movement of data between processes, data stores, and external entities.

Data flow diagram (DFD): A model of data relationships among processes that disregards time and volume. See also **data flow**, **data store**, **process**, **external entity**, and **morph**.

Data reduction: Part of data abstraction aimed at building models. Data can be reduced through cataloging, categorizing, statistical characterization, or case study exemplification.

Data store: A collection of data at rest. Equivalent to a file.

Data structure: A group of data elements that appear together in a data flow diagram.

Decision support system (DSS): A system that assists managers and users form conclusions based on data. A Decision Support System Generator (DSSG) is an information system that assists in the creation of a Specific Decision Support System (SDSS). See also **model base** and **office automation**.

Decision table: A diagram that illustrates a single condition. For each condition in the condition stub, a single action from the action stub is selected through a connection table.

Decision tree: A diagram that illustrates single conditions through branched networks beginning at a root and working through to terminal conditions, specifying actions to be taken.

Defense mode: See **mode**.

Delphi: An interactive data collection technique used to gather data from a panel. The mean results for each question are then fed back to the panel for further comment.

Design: A set of actions that results in specifications for the construction of a new system or improvement of an existing subsystem. There are two types of design. **Logical design** results in the specification of functions and data elements; **physical design** produces operational specifications for the system, its hardware and software components, data layouts, and operating procedures.

Design review: A meeting at which current logical or physical system design is reviewed by the IRU and its management.

Design walkthrough: A technical meeting consisting of a subsystem-by-subsystem discussion of logical or physical design and testing to weed out errors.

Destination verification: An output check to verify that the sink to which data is directed is the appropriate destination.

Detailed investigation: The phase of systems analysis in which details of the planned or existing system are collected through investigation and detailed models are constructed. See also **preliminary investigation** and **report analysis**.

Dialogue manager: The part of a decision support system that controls user interaction with the data and model bases. See also **model base**.

Diary: A data collection technique in which an individual observes and periodically records data about his or her activities. See also **participant-observer**.

Direct access: An intrinsic file organization that grants access to a record through its own unique address on the storage device. See also **sequential file**.

Discount rate: A factor that decreases (discounts) the value of money received or spent at a later date.

Distributed processing: Information processing that is geographically dispersed among several machines and sites.

DO-UNTIL: In structured programming a set of activities performed until a certain condition exists.

DO-WHILE: A version of DO-UNTIL; in structured programming, a set of activities performed as long as a certain condition exists.

Document analysis: An investigation technique used to examine the content and function of information resource documents already in use.

Document management: The activities involved in acquiring, cataloguing, circulating, and controlling the flow of documents.

Dummy event: An event shown in a CPM diagram to synchronize the onset of two or more other events.

E-Mail: See **Electronic mail**.

Editability: Whether or not and at what level the information contained in a medium can be altered.

Electronic Mail (E-Mail): Computer storage of messages for later retrieval by the intended recipients. See also **office automation**.

Element: The basic building block of a system.

Encouraging: Reinforcing responses during an interview through words or phrases such as "Uh-huh," "I understand," or "Tell me more about that."

End-User Computing [also called **End-user programming** and **End-user software development**]:

The creation of software or "near" programs by the same individuals who will ultimately use it. See also **information center** and **user driven**.

End-user programming: See **End-user computing**.

End-user software development: See **End-user computing**.

Enhancement: An IRMaint activity which increases the functionality or usefulness of an information resource.

Entity-relationship diagram: A chart that shows how specific system elements (entities) are related to one another in the following form: "A affects [has/does/contains, etc.] B..."

Environment: That within which a system "resides" and shares transactions. Environments can be **placid** (predictable), **disturbed** (statistically predictable), or **turbulent** (unpredictable or undefined).

Equilibrium: A condition of equal exchange of resources holding between either a system and its environment or between two systems.

Ergonomics: The science of designing systems for human use.

Error generator: A morph within an information system that is a source of errors.

Evaluation apprehension: The tendency of a respondent being questioned or observed to "act" in order to be liked by the observer.

Executable specifications: Specifications that are interpreted by computer software and build systems directly.

Expenditure tracking: Charting expenses over time graphically, rather than budgeting project targets.

Experimenter expectancy: The tendency of a respondent being questioned or observed to act a manner consistent with the observer's preconceptions.

Expert system: An information system that collects knowledge from experts and processes it against others' requests for advice. See also **logical modeling**.

External entity: A system in the same environment as the information system being investigated. See also **Data flow diagram**.

Feasibility study: An investigation effort that attempts to discover which of several implementation alternatives is the most feasible in terms of cost, scheduling, or technology.

Feedback: Information about past actions that tends to influence current choices of action. Positive feedback tends to reinforce actions that resulted in specific past outcomes. Negative feedback tends to discourage those actions or reinforce others to avoid specific past outcomes. See also **System architecture**.

File organization: The way data is organized on a medium. **Intrinsic** organization refers to the actual layout of the data; **extrinsic**, to the way data is accessed through software or procedures. See also **sequential file**.

Filtering: In a questionnaire, asking questions to guide individuals into reading or avoiding specific questions.

Flowchart: A model that depicts the flow of information, materials, or control.

Forms distribution chart: A model depicting the flow of a specific form through processing stations, offices, or individuals.

Forms management: A close relative of systems analysis devoted to maintenance a set of paper (or increasingly computerized) forms.

Forms tracing: An investigation technique in which forms are tagged and traced through processing by logging each destination on a **buck slip**, which is later examined by the systems analyst.

Frame of reference: The particular set of views, opinions, theories, and ideas that an observer brings to the observation situation.

Frequency distribution: A chart depicting the distribution of a set of data across a range.

Function point analysis: A mathematical technique for estimating the time it will take to develop specific items of software.

Functional analysis: A technique for decomposing an activity into a set of constituent activities. See also **structural-functional design**.

GST: See **General Systems Theory**.

Gantt chart: A chart depicting the group activities broken down into components over time.

General systems theory (GST): A complex field of study concentrating on system behavior.

Goal: The component of the definition of a system that identifies the state the system will strive to achieve through interaction with its environment.

Goal analysis: The decomposition of a system's high-level goals into a set of lower-level goals; at each level, the achievement of a set of goals implies previous achievement of the immediately higher-level one. See also **structural-functional design**.

Grid Management: See **matrix management**.

Growth mode: See **mode**.

Hierarchy chart: Also called **structure chart**.

HIPO: See **Hierarchy-Input-Process-Output chart**.

Halo effect: An effect on the observer caused by perception of a single event that colors perception of all subsequent events, either positively or negatively. See also **observer effects**.

Handover meeting: A meeting held formally to approve a phase of system development, terminate it, and move on to the next phase. Also called **sign-off meetings**.

Hierarchy-Input-Process-Output (HIPO) chart: A graphical representation of where a process falls within the subsystem structure of an information system, the inputs to and outputs from the process, and a systematic description of what the process does. The HIPO chart may be considered the data dictionary entry for a process.

Human resource management: The management of people in a system.

IC: See **Information center**.

IR: See **Information resource**.

IRM: See **Information resource management**.

IRMaint: See **Information resource maintenance**.

IRU: See **Information resource unit**.

IF-THEN-ELSE: In structured programming, a tested (IF) condition; if it is true, then one action is performed (THEN...); otherwise, the other is (ELSE...).

Implementation: The construction of a system; for computer systems, this mean writing and testing software, acquiring and testing hardware, writing user procedures, and training users.

Implementation independence: The property of a design or specification that no particular way of building the system or subsystem is implied by the design or assumed in its specification.

In-house development: Implementation of an information system by a firm's own staff. See also **turn-key** and **contractor**.

In-situ observation: Collecting work data on site as it occurs. See also **observation scheme** and **task analysis**.

Index: A table that indicates where specific records in a file can be located based on one or more data elements from each record. See also **key**.

Indexed file: A file whose records are accessed through an index.

Indexed sequential file: A file whose index points to areas of the file within which records can be sequentially accessed.

Information center (IC): An extension of the IRU that provides tools for non-professionals to create their own software or access corporate databases through terminals or microcomputers. See also **database manager** and **end-user computing**.

Information laws: [1] **Law of Conservation of Information**: information must have a source outside the information system; [2] **Law of Utilization of Information**: information must eventually be used to supply a sink; and [3] **Law of Logical Data Flow**: a process is not complete until all the inputs have been processed to produce all the outputs.

Information overload: Results from processing too much information in too quickly in an effort to make a decision.

Information resource (IR): The set of procedures, users, software, hardware, data, and human resources that work together to provide necessary information to a firm.

Information resource maintenance (IRMaint): Activities of the IRU aimed at improving the functioning of the IR.

Information resource management (IRM): Activities of the IRU aimed at managing the IR to achieve corporate goals.

Information resource unit (IRU): Those individuals in charge of carrying out IRMaint, including systems analysts, programmers, system operators, librarians, and administrators, among others.

Information resource unit relationships: The reporting relationship of the IRU to the rest of the firm. There are three classes of relationships: **collateral** (the IRU is a separate division or department with its own goals); (**dependent** the IRU is a subdepartment of a primary user group); and **service** (the IRU serves the needs of all functional units in the firm).

Information system: A system whose elements, relationships, and goals involve information.

Instrument: A procedure for collecting data, including a questionnaire form, a diary, or an interview schedule.

Intelligence: The ability to create information.

Interpretation: The set of activities of a systems analyst are aimed at creating specifications based on observational data, information about technology and management and client requirements. See also **observation**, **representation**, and **specifications**.

Interview: A technique for gathering system experience data through a structured conversation with one or more individuals. See also **Critical incident technique**, **Interview schedule**, **Interview structure**, and **Question construction**.

Interview schedule: A series of questions to be asked in an interview. See also **interview**.

Interview structure: The order in which questions are presented in an interview. There are three: **funnel** (questions go from the general to the specific); **fan** (questions go from the specific to the general); and **flip-flop** (questions begin general, become specific, and then general again). See also **interview**.

Intrusion: The effects an observer has on the system being observed.

Investigation: The collection of data about a system. Investigation data in the form of models is used in **representation**. See also **model** and **representation**.

Key: The set of data elements within a record used to identify the record uniquely for indexed retrieval.

Knowledge-based design: Design based solely on knowledge of the ultimate user of the system and a resulting model of that user's behavior and needs. See also **user-driven**.

LAN: See **Local area network**.

LCI: Law of Conservation of Information. See **Information laws**.

LLDF: Law of Logical Data Flow. See **Information laws**.

LUI: Law of Utilization of Information. See **Information laws**.

Linked-list file: A file in which each record contains the direct address of related records.

Local area network (LAN): A network of processors and data storage devices linked through telecommunications and confined to a small "local" area, generally under a mile in radius, from a central point.

Logical design: 1) Specification of the functions that a system or system element is to carry out, or 2) the activities of a systems analyst that lead to producing such specifications. See **model-based design** and **physical design**.

Logical modeling: Construction of computer-based models that relate facts in logical (true-false) terms rather than mathematical terms, as in most spreadsheets.

Logical relationship diagram: A chart that specifies necessary relations among system elements, such as a data flow diagram.

MIS: See **Management information system**.

Maintainability: The quality of a system that relates to the ability of the IRU to keep it running to meet user goals.

Maintenance mode: See **mode**.

Management information system (MIS): An information system intended to assist management in improving decision making by means of regular reports on specific aspects of the firm. A management information system tends to report variances and assists in management by exception.

Manpower loading: A planning technique that assists in assigning costs to the labor aspects of projects.

Materials flow diagram: A graphical representation of the flow of materials through a production system.

Matrix management: A management technique that divides administrative responsibilities from day-to-day technical ones. Also called **grid management**.

Medium: The physical substance on which information is recorded and transported. See also **reproducibility** and **shelf life**.

Milestone chart: A chart that represents specific reporting or meeting deadlines across time for a project.

Mode: A set of activities that contribute to system functioning. **Maintenance** activities keep the system functioning the way it has in the past; **growth** activities add system elements from the environment; **defense** activities prevent disorganization of the system from forces within the environment.

Model: A simple representation of a system with corresponding representations of some elements and relationships. See also **base** and **investigation**.

Model base: A file of models used in decision support systems. See also **Decision Support System** and **dialogue manager**.

Model-based design: A logical design technique that subjects system models to a variety of tests and measurements in an attempt to uncover weaknesses. See also **Logical design**.

Morph: A common configuration of elements in a DFD that can be combined with others to form more complex configurations. See also **Data flow diagram**.

Nassi-Schneiderman chart: A graphical representation of a process that represents the three structures of structured programming through nested boxes. See also **structured programming**.

Navigation: Moving through a database from one document or record to another via a reference in the data.

Needs document: Part of the preliminary investigation report that lists user information, organizational goal, and procedural requirements. See also **preliminary investigation report**.

Normalization of database: Removal of redundancy from a database to simplify its design.

OA: See **Office automation**.

Observation: The systematic gathering of user experiences with a system. See also **interpretation** and **participant-observer**.

Observer effects: Observer influences on the system being observed. There are three types: **boundary**, **observer**, and **sampling** effects. These effects invalidate the observation.

Observation scheme: A procedure for observing a system. See also *in-situ* **observation**.

Office automation (OA): The introduction of computers and automated information processing into the office. See also **computer-based private branch exchange, decision support systems, electronic mail**, and **word processing**.

On-line system: Information resource hardware or software that is electronically connected to a central computer and is available for use at all times.

Operational model: A system model that depicts the procedural activities interacting with the environment.

Operator's guide: A manual that tells the user of an information resource how to operate it. Compare to **users' guide** and **technical manual**.

Opportunity cost: The figure representing the implied cost expense when compared to the expected gain had those funds been invested.

Out-of-pocket costs: A cost that involves the direct, tangible spending of funds.

P/A: See **programmer/analyst**.

PIR: See **preliminary investigation report**.

Package modification: Modifying an existing software information system to meet the specific needs of an organization. See also **prototyping**.

Parameterization: The creation of pseudocode sub-procedures whose activities vary based on one or more values called *parameters*.

Paraphrasing: The restating of a repondent's reply in the interviewer's own words.

Pareto's Law: An economic principle stemming from an unequal distribution of resources, usually stated in this form: $x\%$ of resources is controlled by $100 - x\%$ of the populace. Often the value for x is either 70 or 80.

Parsimony: A design principle that states that often the best design is the smallest or the most economical.

Participant-observer: A system observer who is also a system element. Participant-observers are characteristic of data collection through diaries. See also **diary** and **observation**.

Payback analysis: A decision-making technique that attempts to determine when a project's accumulated benefits will ultimately exceed its accumulated costs.

Payoff table: A decision-making tool that assists in comparing the benefits of a number of alternatives.

Performance appraisal: A management tool for determining how well an employee has performed during a period of time by means of a discussion with that employee.

Personal computing: The employment of a microcomputer by an individual to meet some of his or her information resource needs.

PERT chart: A diagram that represents the sequential dependence of activities on one another. See also **precedence chart**.

Physical design: Determining software and hardware specifications to be used by an operating information resource. Compare with **logical design**. See also **design**.

Precedence chart: A diagram that represents activities or events and the order in which they must be performed or should occur. See also **CPM diagram** and **PERT chart**.

Preliminary Investigation: A phase of IRMaint activities that outlines the boundary of the system under investigation, its higher-level goal structure, and the general nature of the problem or opportunity. See also **preliminary investigation report**, **problem statement**, and **structural-functional design**, and **detailed investigation**.

Preliminary Investigation Report (PIR): The report that results from the preliminary investigation and contains the **problem statement**, **scope and limitations statement**, **needs document**, and a recommendation for or against further investigation. See also **preliminary investigation**.

Primacy effect: An observer effect that causes the initial events noted by an observer to be more memorable.

Problem statement: A description of the problem or opportunity presented by the system being investigated. See also preliminary investigation report.

Process: In an information resource, an activity that transforms information over a period of time. A process has both input and output data flow. See also **data flow diagram** and **process logic**.

Process chart: A class of diagrams which represent how processes work. See also **process logic**.

Process logic: The manner in which an information process works to transform inputs to outputs. See also **process**.

Program flowchart: A process chart that represents the sequence of activities performed by a process over time. See also **structured programming**.

Programmer/Analyst (P/A): An individual who performs the work of a systems analyst and then codes the software that results from the logical and physical designs. See also **systems analyst**.

Project: A planned set of activities that leads to the development of a system.

Project brief: A short definition of a project produced at its start to document immediate responsibilities.

Project management: A set of activities necessary to guide a project to completion. Includes project **planning**, **monitoring**, and **control**.

Project status meeting: One of a set of regular meetings that stimulates the sharing of information about project progress with IRU staff and clients.

Project summary: A succinct description of project responsibilities, requirements, and goals.

Protocol analysis: A task analysis technique that combines features of interviewing and *in-situ* observation by having a worker discuss a task while performing it. See also **task analysis**.

Prototyping: A method of system development in which the user and the analyst cooperatively produce a series of mock-ups of the final product until the user is satisfied that the final product meets his or her needs. Compare with **end-user computing, package modification** and **project**.

Pseudocode: A program-like representation of process logic that expresses the structures of structured programming with English-like expressions. See also **structured programming**.

QC: See **quality circle**.

Quality circle (QC): A group technique that gathers teams of workers together to improve a product, a production system, or working conditions through a series of structured meetings.

Question construction: The style of one of six types of questions used in interviews and questionnaires: **criterion** questions collect information on specific points; **check** questions re-ask criterion questions to verify answers; **mirror** questions repeat answers to verify the wording of responses; **lead-in** questions introduce topics; **filter** questions direct questionnaire respondents to particular parts of the questionnaire; and **probe** questions obtain more detailed information, usually following criterion questions. See also **interview**.

Questionnaire: A printed form used to gather system experience data from individuals. Questionnaires are filled out and returned to the analyst. Compare with *in-situ* **observation** and **interview**.

ROI: See **Return on investment**.

Random file: A file composed of records whose locations depend on a formula based on some key element recorded in the record. Contrast with **indexed file**, **indexed sequential file**, **linked list file**, and **sequential file**.

Rapport: A feeling of cooperation and comfort that is important in interviewing.

Recency Effect: An observer effect that causes more recent events most memorable.

Records Management: Maintaining a set of usually non-computerized records or files.

Redirection: A management strategy for coping with crises involving a change in the overall plan.

Redundancy: A system element (process, data flow, or data store) that repeats all or part of another system element.

Reflecting: Telling a respondent the emotional impact of his or her previous response during the interview.

Report analysis: An analysis of the content of a report intended to contribute to information needs analysis during a detailed investigation. See also **detailed investigation**.

Report generator: A software system that guides the user to create a stored sequence of commands that, when later accessed and compared with a database, will result in a formatted report.

Report-based design: A design technique that begins with the required reports and works backward to discover the required inputs.

Representation: 1) An easily understood depiction of a system; 2) the activities of a systems analyst intended to construct representations. See also **interpretation**, **investigation**, and **model**.

Representative sampling: A non-systematically biased sampling technique that over the long term produces a sample truly representative of the sampling frame. See also **sample** and **sampling frame**.

Reproducibility: The quality of a medium that allows copies of the data it holds to be made. See also **medium**.

Request for services: A form filled out by a potential client of the IRU that describes his or her information problem or opportunity.

Respondent: Either the person being interviewed or the person filling out a questionnaire. See also **sampling**.

Return on investment (ROI): A single figure that represents the efficiency with which funds are invested in a product or project, equivalent to return divided by investment; ROI can also be "annualized" to be expressed as an annual rate of return on investment. See also **discount rate**.

Reusable software: Software that, after development for one system, can be directly employed in another.

SA: See **systems analyst**.

SDLC: See **system development life cycle**.

SDM: See **system development methodology**.

Sample: A subset of the sampling frame. See also **sampling**.

Sampling: The selecting a set of collection points from a sampling frame for investigation. Sampling can be *ad-hoc* (whatever happens to be available), **forced** (supplied by the client), **guided** or **progressive** (respondents recommend others to be included in the sample), **random** (selected randomly, without bias, by choosing every n-th collection point from the sampling frame), or **stratified random** (artificially increasing the odds that elements from an identifiable subset of the sampling frame will be selected). See also **representative sample**.

Sampling frame: The set of available or accessible collection points from which samples are selected.

Sampling effects: An influence on the validity of observation resulting from the selection of too small or unrepresentative a sample or by the overlooking of important events.

Scale type: The comparability of the data units collected during investigation. There are three common types: **nominal** scales contain only class names and are not numerically comparable; **ordinal** scales have data values and are comparable in terms of "more" or "less"; and **interval** scales, in which differences between adjacent scale items are equivalent across the scale.

Scenario-based design: Design that stems from a set of user activities; specifically refers to a user's creation of his or her own data capture screens.

Scope and limitations statement: A description of the extent of a project and the financial, organizational, or time limitations imposed on the it. See also **preliminary investigation report**.

Screen generator: A software tool that allows users and analysts to generate data capture screens quickly without programming. See also **scenario-based design**.

Semi-structured interview: An interview that has some flexibility of question order and content, but has a set of criterion questions to be asked. See also **interview structure**.

Sequential file: A file for which each record has one record immediately prior and one immediately following. Generally these records are physically adjacent to one another. See also **file organization**. Contrast with **direct access**.

Shelf life: The period of time that the medium holding recorded information can be trusted to maintain that information correctly. See also **medium**.

Sign-off meeting: Same as **handover meeting**.

Simulation: 1) in modeling, the creation of a model from a base system; 2) an **investigation technique** in which a user employs a simplified version of the system to be constructed to allow for the analyst to observe the user's activities. See also **task analysis**.

Sink: The information destination outside a system. Contrast with **source**. See also **external entity**.

Social accounting: Placing economic value on benefits and costs that do not normally carry a price tag.

Social facilitation: An increase in an individual's ability to work effectively due to the presence of others.

Software wear-in: An increase in the ability of users to employ software over time as they learn how and when to use it.

Source: An external entity that creates information for a system. Contrast with **sink**. See also **external entity**.

Specifications: See also **interpretation**.

State: A description of a system at a point in time.

Statistical characterization: Generating a single number or value that captures a trend or tendency in a set of data.

Structural-functional design: A design that uses goal analysis to create a structure chart of system goals; lowest-level goals are equated with system functions; the implementation of functions in the structure provided is guaranteed to meet the higher-level goals. See also **goal analysis, functional analysis**, and **preliminary investigation**.

Structure chart: A graphical representation of the goal or activity hierarchy structure of a system. Also called **hierarchy chart**.

Structured programming: 1) A theory that all programs written for a von Neuman processor can be reduced or equated to a nested structure of three basic types or structures, and 2) a technique of coding programs that uses only those three structures. See also **AND-THEN, DO-UNTIL, DO-WHILE, IF-THEN-ELSE, pseudocode, Nassi-Schneiderman chart**, and **program flowchart**.

Study boundary: The limits of data collection efforts during a detailed investigation. Compare with **system boundary**.

Subsystem: A subset of another system.

Summarizing: Providing a summary of an interview segment for the respondent.

Superfluity: The quality of being useless. If a subsystem is never used, it is *superfluous*.

System: A collection of elements whose activities are interrelated in an effort to meet a goal.

System architecture: The construction of a system able to cope with environmental events. There are three types: **simple**, (has no ability to select particular actions); **cybernetic**, (has the ability to select actions based on the results of specific environmental events); and **learning**, (has the ability to evaluate the process and content of selections to improve the selection process). See also **feedback**.

System audit: A systematic investigation of the effectiveness of an information resource in meeting its original goals.

System boundary: The set of elements of a system whose behavior depends at least partly on the activities of the environment. Metaphorically, the term is used to mean the *limits* of a system. Compare with **study boundary**.

System development life cycle (SDLC): A series of activities done when a new system is developed from scratch. Phases include **preliminary investigation, detailed investigation, physical design, implementation, installation**, and **audit**. See also **system development methodology**.

System development methodology (SDM): A technique for developing systems through the system development life cycle. SDMs may be either automated or semiautomated.

System flowchart: A graphical depiction of the flow of information between processes over time and the sequence of processing.

Systems analyst (SA): An IRU employee who investigates existing systems and defines work to be performed to maintain an information resource through logical and physical design. See also programmer/analyst. Compare with **user-analyst**.

Task analysis: The observation and subsequent modeling of the activities of individuals. See also *in-situ* **observation, protocol analysis**, and **simulation**.

Technical manual: A manual that informs programmers and analysts of how a system actually functions. Compare with **operators' guide** and **users' guide**.

Technology adaptation: The changing of existing technology to meet new needs.

Technology diffusion: The spread of the use of a particular set of tools or methods throughout an organization.

Technology dilution: The simplification of a complex technology formerly employed by specialists to make it useful to non-specialists.

Technology evolution: The slow change of technology over time.

Technology infusion: The spread of technology from outside the organization.

Technology invasion: The sudden forced adoption of a specific technology through forces beyond the organization's control.

Technology stampede: The unplanned adoption of technology simply because it is there or because of social pressures.

Technology trend: A set of responses to technology or changes in the use or employment of technology.

Toolsmith: An individual who creates coding tools for use in developing decision support systems.

Top-down: Analysis, design, or specification of a system beginning with the major goals of the system and working "down" through subgoals.

Transaction processing: The process of updating, adding, or deleting records to or from a file based on incoming information.

Transfer cost: A cost in developing an information resource resulting from the displacement of resources from one area of need to another.

Transparency: The quality of a system that makes its operation obvious to individuals who have used another system to accomplish the same goals in the past.

Turn-key: Construction and installation of a system by an outside organization. All the customer does is "turn the key" and the system is operable.

Uncertainty principle: The principle according to which it is difficult for systems analysts to have direct knowledge of a system's elements and relationships and for conscious elements within a system to have direct knowledge of the system's goals.

User-analyst: An individual employed by a group of IRU service users to help understand and translate their needs into terms the IRU can work with. Compare with **systems analyst**.

User-driven: Design or implementation of an information resource which is based on user needs or design or implementation actually performed by users. See also **end-user computing** and **knowledge-based design**.

User objects: Additional vocabulary added to pseudocode to simplify the specification of process logic and extend pseudocode to include common conditions or actions such as end-of-file or pauses.

Users' guide: A document that instructs users how to employ an information resource to meet specific needs. Contrast with **operators' guide** and **technical manual**.

Value-added approach: A technique for decision making that attempts to put a value on benefits by analyzing the specific value that design alternatives can provide.

von Neuman processor: A machine that is capable of performing one operation at a time and that operates on sequential commands. Most electronic computers are von Neuman processors. Spreadsheet calculators simulate non-von Neuman processors by seemingly performing several operations simultaneously in two, rather than one, dimensions. See also **structured programming**.

Warnier-Orr diagram: A two-dimensional graphical reprentation of process logic.

Wave: A repetition of the distribution of a questionnaire to all potential respondents used in questionnaire surveys and Delphi.

WP: See **word processing**.

Word Processing (WP): The application of automated information technology to the task of creating, storing, retrieving, editing, and distributing textual material. See also **office automation**.

BIBLIOGRAPHY

Alavi, Maryam. "End-User Computing: The MIS Managers' Perspective." *Information and Management* (9): 171-8, (1985).

Alavi, Maryam. "Some Thoughts on Quality Issues of End-User Developed Systems." Proceedings of the 21st Annual Computer Personnel Research Conference, Association for Computing Machinery, Minneapolis, May 1985.

Andrews, William. "Prototyping Information Systems." *Journal of Systems Management* 34(9): 16-18, (1983).

Arendt, Hannah. *The Human Condition*. Chicago: The University of Chicago Press, 1958.

Ashby, W. Ross. *An Introduction to Cybernetics*. London: University Paperbacks, 1965.

Awad, Elias. *Systems Analysis and Design, 2nd Ed*. Homewood, Ill.: Richard Irwin, 1985.

Bateson, Gregory. *Steps to an Ecology of Mind*. San Francisco: Chandler, 1972.

Behrens, Charles. "Measuring the Productivity of Computer Systems Development Activities with Function Points." *IEEE Transactions on Software Engineering* SE-9(6): 648-52, (1983).

Bellman, Richard. *Artificial Intelligence: Can Computers Think?* Boston: Boyd & Fraser, 1978.

Benbasat, Izak, Albert Dexter, Donald Drury, and Robert Goldstein. "A Critique of the Stage Hypothesis: Theory and Empirical Evidence." *Communications of the ACM* 27(5): 476-85, (1984).

Benjamin, Art, Tom Carey, and R. Mason. "ACT/1: A Tool for Information Systems Prototyping." Second Software Engineering Symposium: Workshop on Rapid Prototyping, ACM SIGSOFT, Columbia, Md., April 1982.

Berdie, Douglas, and John Anderson. *Questionnaires: Design and Use*. Metuchen, N.J.: The Scarecrow Press, 1974.

Berrisford, Thomas, and James Wetherbe. "Heuristic Redevelopment: A Redesign of Systems Design." *MIS Quarterly* 3(1): 11-19, (1979).

Blake, Robert, and J. Mouton. *Consultation*. Reading, Mass.: Addison-Wesley, 1976.

Blankenship, A. B. *Professional Telephone Surveys*. New York: McGraw-Hill, 1977.

Blanning, Robert. "Management Applications of Expert Systems." *Information and Management* (7): 311-6, (1984).

Bohm, C., and G. Jacobini. "Flow Diagrams, Turing Machines, and Languages with only Two Formation Rules." *Communications of the ACM* 9(5): 366-71, (1965).

Booz, Allen & Hamilton, Inc. "A Study of Managerial, Professional Productivity and Office Automation." New York: 1980.

Bostrom, Robert P. "Development of Computer-Based Information Systems: A Communication Perspective." *Computer Personnel* 9(4): 17-25, (1984).

Bottom, Joseph, Alan Bernard, and Kevin Anderson. "The Art of Modeling." *Datamation* 31(22): 140-50, (1985).

Brooks, Fred. *The Mythical Man-Month: Essays in Software Engineering*. Reading, Mass.: Addison-Wesley, 1975.

Burch, John, and Felix Strater. *Information Systems: Theory and Practice*. New York: John Wiley, 1974.

Burke, Richard. *Decision Making in Complex Times*. Hamilton, Ontario: Society of Management Accountants of Canada, 1984.

Canning, Richard. "'Programming' by End Users." *EDP Analyzer* 19(5), (1981).

Canning, Richard. "Supporting End User Programming." *EDP Analyzer* 19(6), (1981).

Carey, Tom, and R. E. A. Mason. "Information System Prototyping: Techniques, Tools, and Methodologies." *Infor* 21(3): 177-91, (1983).

Cheeseborough, P., and Gordon Davis. "Planning a Career Path in Information Systems." *Journal of Systems Management* 34(1): 6-13, (1983).

Cheyney, P., and N. Lyons. "Information Systems Skill Requirements: A Survey." *MIS Quarterly* 4(1): 35-43, (1983).

Colter, Mel. "A Comparative Examination of Systems Analysis Techniques." unpublished manuscript, College of Business and Administration, University of Colorado, n.d.

Couger, J. Daniel, and Robert Zawacki. *Motivating and Managing Computer Personnel*. New York: Wiley, 1980.

Coughlin, Clifford. "The Need for Good Procedures." *Journal of Systems Management* (6): 30-33, (1973).

Crossman, Trevor. "Taking the Measure of Programmer Productivity." *Datamation* 28(5): 144-7, (1982).

Czarnecki, Gerald. "The Case for Internal Consulting." *Journal of Systems Management* 32(1): 6-13, (1981).

Davis, William S. *Tools and Techniques for Structured Systems Analysis and Design.* Reading, Mass.: Addison-Wesley, 1983.

DeLone, William. "Firm Size and the Characteristics of Computer Use." *MIS Quarterly* 5(4): 65-77, (1981).

DeMarco, Tom. *Structured Analysis and System Specification.* Englewood Cliffs, N.J.: Prentice-Hall, 1979.

Dillman, Don. *Mail and Telephone Surveys.* New York: Wiley, 1978.

Drummond, Steve. "Measuring Applications Development Performance." *Datamation* 31(4): 102-11, (1985).

Ein-Dor, Philip, and Eli Segev. "Information Systems: Emergence of a New Organizational Function." *Information and Management* (5):279-86, (1982).

Emery, F. E., and E. L. Trist. *Towards a Social Ecology: Contextual Appreciations of the Future in the Present.* London: Plenum Press, 1973.

FitzGerald, Jerry. *Designing Controls into Computerized Systems.* Redwood City, Calif.: Jerry FitzGerald and Associates, 1981.

Franz, Charles, and Daniel Robey. "An Investigation of User-Led System Design: Rational and Political Perspectives." *Communications of the ACM* 27(12): 1202-9, (1984).

Gall, John. *Systemantics: How Systems Work and Especially How They Fail.* New York: Pocket Books, 1975.

Gane, Chris, and Trish Sarson. *Structured Systems Analysis: Tools and Techniques.* Englewood Cliffs, N.J.: Prentice-Hall, 1979.

Gerrity, Thomas, and John Rockart. "End User Computing: Are You a Leader or a Laggard?" *Sloan Management Review* 27(4): 25-34, (1986).

Gibson, C., and Robert Nolan. "Managing the Four Stages of EDP Growth." *Harvard Business Review* 52(1): 76-85, (1974).

Goldfarb, Stephen. "Writing Policies and Procedures Manuals." *Journal of Systems Management* 32(4): 10-11, (1981).

Gotlieb, Calvin. *The Economics of Computers*. Englewood Cliffs, N.J.: Prentice-Hall, 1985.

Guimaraes, Tor. "A Study of Application Program Development Techniques." *Communications of the ACM* 28(5): 494-9, (1985).

Hackman, J. R., and G. R. Oldham. "Motivation through the Design of Work: Test of a Theory." *Organizational Behavior and Human Performance* 16(2): 250-79, (1976).

Harmon, Paul, and David King. *Expert Systems*. New York: Wiley, 1985.

Heather, Pauline. *Questionnaires*. Sheffield, England: Sheffield University Press, 1984.

Henderson, John, and Michael Treacy. "Managing End-User Computing Competitive Advantage." *Sloan Management Review* 27(2): 3-14, (1986).

Hiltz, Starr Roxanne, and Murray Turoff. *The Network Nation: Human Communication Via Computers*. Reading, Mass.: Addison-Wesley, 1978.

Hirscheim, R. A. "Assessing Participative Systems Design: Some Conclusions from an Exploratory Study." *Information and Management* (6): 317-27, (1983).

Holtz, D. H. "A Nonprocedural Language for On-Line Applications." *Datamation* 25(4): 166-72, (1979).

Howes, Norman. "Project Management Systems." *Information and Management* (5):243-58, (1982).

Hurst, E. Gerald, David Ness, Thomas Gambino, and Thomas Johnson. "Growing DSS: A Flexible, Evolutionary Approach." In *Building Decision Support Systems*, edited by John Bennett. Reading, Mass.: Addison-Wesley, 1982.

International Business Machines Corporation. *Joint Application Design (JAD)*. Toronto: IBM Information Systems Services, 1986.

Isaacs, P. Brian. "Warnier-Orr Diagrams in Applying Structured Concepts." *Journal of Systems Management* 33(10): 28-33, (1982).

Ives, Blake, and Gerard Learmonth. "The Information System as a Competitive Weapon." *Communications of the ACM* 27(12): 1193-201, (1984).

Janz, Tom, and Paul Licker. "Transporting a Measure of Corporate Culture to the Information Services Area." *Journal of Systems Management*, 1987.

Jenkins, A. Milton, Justus D. Naumann, and James Wetherbe. "Empirical Investigation of Systems Development Practices and Results." *Information and Management* (7):73-82, (1984).

Jenkins, A. Milton. "Surveying the Software Generator Market." *Datamation* 31(17): 105-20, (1985).

Jones, P. F. "Four Principles of Man-Computer Dialogue." *Computer-Aided Design* 10(3): 197-202, (1978).

Kaiser, Kate, and Anath Srinivasan. "User-Analyst Differences: An Empirical Investigation of Attitudes Related to Systems Development." *Academy of Management Journal* 25(3): 630-46, (1982).

Keen, Peter G. W., and Michael S. Scott Morton. *Decision Support Systems*. Reading, Mass.: Addison-Wesley, 1978.

Kendall, K. E., and R. D. Losee. "Information System FOLKLORE: A New Technique for System Documentation." *Information and Management* 10(2): 103-12, (1986).

Kilmann, Ralph, and Ian Mitroff. "Problem Defining and the Consulting/Intervention Process." *California Management Review* 21(3): 26-33, (1979).

King, John, and Kenneth Kraemer. "Evolution and Organizational Information Systems: An Assessment of Nolan's Stage Model." *Communications of the ACM* 27(5): 466-75, (1984).

Kole, Michael. "Going Outside for MIS Implementation." *Information and Management* (6): 261-8, (1983).

Kraft, Philip. *Programmers and Managers: The Routinization of Computer Programming in the United States*. New York: Springer Verlag, 1977.

Kroenke, David M. *Business Data Processing*. Santa Cruz: Mitchell Publishing, 1981.

Langle, Gernot, Robert Leitheiser, and Justus Naumann. "A Survey of Applications Systems Prototyping in Industry." *Information and Management* (7): 273-84, (1984).

Lederer, Albert. "Information Requirements Analysis." *Journal of Systems Management* 32(12): 15-20, (1981).

Leeson, Marjorie. *Systems Analysis and Design, 2nd Ed*. Chicago: SRA, 1985.

Leitheiser, Robert, and James Wetherbe. "The Successful Information Center: What does it Take?" Proceedings from the 21st Annual Computer Personnel Research Conference, Association for Computing Machinery, Minneapolis, May 1985.

Licker, Paul. *The Art of Managing Software Development People*. New York: Wiley, 1985.

Licker, Paul. "DP/Company Interaction: Is Job Rotation the Answer?" *Journal of Information System Management*, 1985.

Licker, Paul, and Ron Thompson. "Consulting Systems: Briding the AI/DSS Gap." Proceedings from the Hawaii International Conference on Systems Science, Honolulu, Jan. 1986, Vol IA, pp. 471-8.

Linstone, Harold, and Murray Turoff (Eds). *The Delphi Method: Techniques and Applications*. Reading, Mass.: Addison-Wesley, 1975.

Lucas, Henry. "Organizational Power and the Information Services Department." *Communications of the ACM* 27(1): 58-65, (1984).

Martin, Daniel. *Three Mile Island: Prologue or Epilogue?* Cambridge, Mass.: Ballinger, 1980.

Martin, James. *Application Development without Programmers*. Englewood Cliffs, N.J.: Prentice-Hall, 1982.

Martin, Merle. "Management Reports." *Journal of Systems Management* 33(6): 32-39, (1982).

Mason, R. E. A., and Carey, T. "Prototyping Interactive Information Systems." *Communications of the ACM* 26(5): 347-54 (1983).

McFarlan, E. Warren. "Information Technology Changes the Way You Compete." *Harvard Business Review* 62(3): 98-103, (1984).

McLean, E. R. "End Users as Application Developers." *MIS Quarterly* 3(4): 37-46, (1979).

Milgram, Stanley. *Obedience to Authority*. New York: Harper & Row, 1974.

Miller, Marc. "Problem Avoidance in the User/Analyst Relationship, Part I" *Journal of Systems Management* 32(5): 14-18, (1981).

Miller, Marc. "Problem Avoidance in the User/Analyst Relationship, Part II" *Journal of Systems Management* 32(6): 34-39, (1981).

Molyneaux, Dorothy. *Effective Interviewing*. Boston: Allyn & Bacon, 1982.

Monaco, Ron. "Mechanized Filing: Cost Versus Expense." *ARMA Quarterly*, April 1984, pp. 10-12.

Moret, Bernard M. "Decision Trees and Diagrams." *Computing Surveys* 14(4): 593-623, (1982).

Morland, D. Verne. "Human Factors Guidelines for Terminal Interface Design." *Communications of the ACM* 26(7): 484-94, (1983).

Mosard, Gil. "Problem Definition: Tasks and Techniques." *Journal of Systems Management* 33(6): 16-21, (1983).

Munro, Malcolm. "Determining the Manager's Information Needs." *Journal of Systems Management* (6): 34-39, (1978).

Munro, Malcolm, and Sid Huff. "Information Technology Assessment and Adoption: Understanding the Information Centre Role." Proceedings from the 21st Annual Computer Personnel Research Conference, Association for Computing Machinery, Minneapolis, May 1985.

Newell, Gale. "Organizing a Successful Management Needs Analysis." *Journal of Systems Management* 32(6): 30-33, (1981).

Nolan, Richard L. "Managing the Computer Resource: A State Hypothesis." *Communications of the ACM* 16(7): 399-405, (1973).

Nolan, Richard L. "Controlling the Cost of Data Services." *Harvard Business Review* 55(4): 114-24, (1977).

Olson, Margrethe. "New Information Technology and Organizational Culture." *MIS Quarterly* 6 (Special Issue): 71-92, (1982).

Olson, Margrethe, and Norman Chervany. "The Relationship Between Organizational Characteristics and the Structure of the Information Services Function." *MIS Quarterly* 4(2): 57-68, (1980).

Peters, Thomas, and Robert Waterman. *In Search of Excellence: Lessons from America's Best-Run Companies*. New York: Harper & Row, 1982.

Porter, Michael, and Victor E. Millar. "How Information Gives You Competitive Advantage." *Harvard Business Review* 63(4): 179-85, (1985).

Powers, H., D. Adams, and H. Mills. *Computer Information Systems Development: Analysis and Design*. Cincinnati, Oh.: Southwestern Publishing, 1984.

Rivard, Suzanne, and Sid Huff. "An Empirical Study of Users as Application Developers." *Information and Management* (8): 89-102, (1985).

Roethlisberger, F. J., and William Dickson. *Management and the Worker*. Cambridge, Mass.: Harvard University Press, 1966.

Rubenstein, Ellis. "The Accident that Shouldn't Have Happened." *IEEE Spectrum*, (1979): 33-42.

Sackman, Harold. *Delphi Critique: Expert Opinion, Forecasting, and Group Process*. Lexington, Mass.: Lexington Books, 1984.

Scharer, Laura. "Systems Analyst Performance: Criteria and Priorities." *Journal of Systems Management* 33(2): 10-15, (1982).

Senn, James A. "A Management View of Systems Analysts: Failures and Shortcomings." *MIS Quarterly*, (1978): 25-42.

Shelly, Gary and Tom Cashman. *Learning to Use dBASE III: An Introduction*. Boston, Mass.: Boyd & Fraser, 1986.

Shore, Edwin. "Reshaping the IS Organization." *MIS Quarterly* 7(4): 11-17, (1983).

Simon, Herbert. *The New Science of Management Decision*. New York: Harper & Row, 1960.

Sprague, Ralph H. "A Framework for the Development of Decision-Support Systems." *MIS Quarterly* 4(4): 1-26, (1980).

Sprague, Ralph H. and E. Carlson. *Building Effective Decision-Support Systems*. Englewood Cliffs, N.J.: Prentice-Hall, 1982.

Stewart, Charles. *Interviewing, 4th Ed*. Dubuque, Iowa: W. C. Brown, 1985.

Tinker, Tony, and Charles Lindblom, eds. *Social Accounting for Corporations*. New York: Markus Wiener, 1984.

Van Duyn, J. *Documentation Manual*. Philadelphia: Auerbach, 1972.

Vassiliou, Yannis, ed. *Human Factors and Interactive Computer Systems*. Ablex: Norwood, N.J.: 1984.

Vitalari, Nicholas. "Knowledge as a Basis for Expertise in Systems Analysis: An Empirical Study." *MIS Quarterly* 9(3): 221-41, (1985).

Vogel, Douglas, and James Wetherbe. "Office Automation: End User Impact." Proceedings from the 21st Annual Computer Personnel Research Conference, Association for Computing Machinery, Minneapolis, May 1985.

Weinberg, Gerald. *The Psychology of Computer Programming*. New York: Van Nostrand Reinhold, 1971.

Wetherbe, James. *Systems Analysis and Design: Traditional, Structured, and Advanced Concepts and Techniques, 2nd Ed*. St. Paul, Minn.: West, 1984.

Whitehead, Alfred North, and Bertrand Russell. *Principia Mathematica*. Cambridge: The University Press, 1962.

Wiener, Norbert. *Cybernetics*. Cambridge, Mass.: MIT Press, 1948.

Willoughby, Ted. *Business Systems, 2nd Ed*. Cleveland: The Association for Systems Management, 1981.

Woodson, Wesley E. *Human Factors Design Handbook*. New York: McGraw-Hill, 1981.

Wooldridge, S. *Project Management in Data Processing*. New York: Petrocelli/Charter, 1976.

Wright, David. "Designing Terminals for the Human Factor." *Canadian Datasystems* 14(4): 37-41, (1982).

Yablonski, Lewis. *Robopaths*. Indianapolis: Bobbs-Merrill, 1972.

Yourdon, Edward. *Structured Walkthroughs*. Englewood Cliffs, N.J.: Prentice-Hall, 1979.

Yourdon, Edward. *Techniques of Program Structure and Design*. Englewood Cliffs, N.J.: Prentice-Hall, 1979.

Yourdon, Edward, and Larry L. Constantine. *Structured Design: Fundamentals of a Discipline of Program and System Design*. Englewood Cliffs, N.J.: Prentice-Hall, 1979.

Zaltman, Gerald, Robert Duncan, and Jonny Holbek. *Innovations and Organizations*. New York: Wiley, 1973.

INDEX